Human Development Report 2015

Work for Human Development

Published for the
United Nations
Development
Programme
(UNDP)

Empowered lives.
Resilient nations.

Human Development Report 2015 Team

Director and lead author
Selim Jahan

Deputy director
Eva Jespersen

Research and statistics
Shantanu Mukherjee (Team Leader). Milorad Kovacevic (Chief Statistician), Astra Bonini, Cecilia Calderon, Christelle Cazabat, Yu-Chie Hsu, Christina Lengfelder, Sasa Lucic, Tanni Mukhopadhyay, Shivani Nayyar, Thomas Roca, Heriberto Tapia, Katerina Teksoz and Simona Zampino

Outreach and production
Botagoz Abdreyeva, Eleonore Fournier-Tombs, Jon Hall, Admir Jahic, Jennifer Oldfield, Anna Ortubia and Michael Redante

Operations and administration
Sarantuya Mend (Operations Manager), Mamaye Gebretsadik, Fe Juarez Shanahan and May Wint Than

Foreword

Twenty five years ago the first Human Development Report in 1990 began with a simple notion: that development is about enlarging people's choices—focusing broadly on the richness of human lives rather than narrowly on the richness of economies. Work is a major foundation for both the richness of economies and the richness of human lives but has tended to be conceptualized in economic terms rather than in human development terms. The 2015 Human Development Report goes beyond that convention in directly linking work to the richness of human lives.

This Report starts with a fundamental question—how can work enhance human development? The Report takes a broad view of work, going beyond jobs and taking into account such activities as unpaid care work, voluntary work and creative work—all of which contribute to the richness of human lives.

The Report highlights impressive progress on human development over the past quarter century. Today people are living longer, more children are in school and more people have access to clean water and basic sanitation. Per capita income in the world has gone up, and poverty has gone down, resulting in a better standard of living for many people. The digital revolution has connected people across countries and societies. Work has contributed to this progress by building people's capabilities. Decent work has provided people with a sense of dignity and an opportunity to engage fully in society.

Considerable challenges remain, from persistent poverty and grinding inequalities to climate change and environmental sustainability in general, and to conflict and instability. These all create barriers to people fully engaging in decent work, and as a result huge amounts of human potential remain untapped. This is of particular concern for young people, women, people with disabilities and others who may be marginalized. The Report argues that if the potential of all people is harnessed through appropriate strategies and proper policies, human progress would be accelerated and human development deficits would be reduced.

The Report reminds us that there is no automatic link between work and human development. The quality of work is an important dimension of ensuring that work enhances human development. Issues such as discrimination and violence, however, prevent positive links between work and human development. Some work is very damaging to human development, such as child labour, forced labour and the labour of trafficked workers, all of which constitute serious violations of human rights. In many cases workers in hazardous conditions face serious risks of abuse, insecurity and loss of freedom and autonomy.

All these issues are becoming even more critical to address as the world of work, driven by globalization and technological revolution, is undergoing rapid changes. Globalization has generated gains for some and losses for others. The digital revolution has created new opportunities, but has also given rise to new challenges, such as irregular contracts and short-term work, which are asymmetrically distributed between highly skilled and unskilled workers.

The Report makes a strong case that women are disadvantaged in the world of work—in both paid and unpaid work. In the realm of paid work, they are engaged in the workforce less than men, they earn less, their work tends to be more vulnerable and they are underrepresented in senior management and decisionmaking positions. In terms of unpaid work, they bear a disproportionate share of the housework and care work.

The Report identifies sustainable work, which promotes human development while reducing and eliminating negative side effects and unintended consequences, as a major building block of sustainable development. Such work would expand opportunities for the present generation without shrinking those for future ones.

The Report argues that enhancing human development through work requires policies and strategies in three broad areas—creating work opportunities, ensuring workers' well-being and developing targeted actions. The first area focuses on national employment strategies and seizing opportunities in the changing world of work, while the second area covers such important issues as guaranteeing workers' rights and benefits, expanding social protection and addressing inequalities. Targeted actions should focus on sustainable work, addressing imbalances in paid and unpaid work and interventions for specific groups—for example, for youth and people with disabilities. Above all, there needs to be an agenda for action pursuing a New Social Contract, a Global Deal, and the Decent Work Agenda.

This year's Report is particularly timely, following shortly after the UN Sustainable Development Summit, where the new Sustainable Development Goals were adopted, including Goal 8's explicit emphasis on work: Promote sustained, inclusive and sustainable economic growth, full and productive employment, and decent work for all.

In this context there should be serious discussion about the challenges created by the ongoing changes in the world of work. Opportunities should be taken to strengthen the links between work and human development. During the past 25 years the human development concept, reports and indices have generated considerable debate, dialogue and discussions around the world on development challenges and policy issues. I expect this year's report to be no exception in its capacity to generate dialogue and debate around the concept of human development and strategies to advance it.

Helen Clark
Administrator
United Nations Development Programme

Acknowledgements

The 2015 Human Development Report is the product of the Human Development Report Office (HDRO) at the United Nations Development Programme (UNDP).

The findings, analysis and policy recommendations of the Report are those of HDRO alone and cannot be attributed to UNDP or to its Executive Board. The UN General Assembly has officially recognized the Human Development Report as "an independent intellectual exercise" that has become "an important tool for raising awareness about human development around the world."

The Report has benefited from a series of contributions by eminent people and organizations. Particular appreciation is due for the signed contributions by HE Mr. Benigno S. Aquino III (President of the Philippines), Leymah Gbowee (winner of the 2011 Nobel Peace Prize), HE Ms. Roza Otunbayeva (former president of Kyrgyzstan), Nohra Padilla (recipient of the 2013 Goldman Environmental Prize), Orhan Pamuk (winner of the 2005 Nobel Prize in Literature), Robert Reich (former United States Secretary of Labor), Kailash Satyarthi (winner of the 2014 Nobel Peace Prize) and HE Mr. Maithripala Sirisena (President of Sri Lanka).

Appreciations are also extended for contributions commissioned from the following authors: Antonio Andreoni, Marizio Atzeni, Fred Block, David Bloom, Jacques Charmes, Martha Chen, Diane Coyle, Christopher Cramer, Peter Evans, Nancy Folbre, Marina Gorbis, Kenneth Harttgen, Rolph Eric van der Hoeven, Rizwanul Islam, Patrick Kabanda, Claudio Montenegro, Nameera Nuzhat, Dani Rodrik, Jill Rubery, Malcolm Sawyer, Frances Stewart, Miguel Szekely, and Lanying Zhang.

Discussions with experts in many disciplines are equally necessary and valuable for the development of the Report, starting with an informal sounding round, followed by consultations with a designated advisory panel for the 2015 Report. The commitment of time, advice and reviews from the following individuals are highly valued: Amartya Sen, Sudhir Anand, Amy Armenia, Martha Chen, Mignon Duffy, Peter Evans, Nancy Folbre, Gary Gereffi, Enrico Giovannini, Marina Gorbis, James Heintz, Jens Lerche, José Antonio Ocampo, Samir Radwan, Akihiko Tanaka, Lester Salamon, Frances Stewart and Ruan Zongze.

The Statistical Advisory Panel plays a critical role in extending expert advice on methodologies and data choices related to the calculation of the Report's indices. Its members are Wasmalia Bivar, Martine Durand, Haishan Fu, Pascual Gerstenfeld, Ifeyinwa Isiekwe, Yemi Kale, Rafael Diez de Medina, Fiona Robertson and Michaela Saisana. The composite indices and other statistical resources in the Report also rely on the expertise of the leading international data providers in their specialized fields. To ensure accuracy and clarity, the statistical analysis has also benefitted from discussions of statistical issues with Gisela Robles Aguilar, Sabina Alkire, Jacques Charmes, Kenneth Harttgen, Claudio Montenegro and Yangyang Shen. I deeply appreciate their contributions.

The Report also benefitted from dialogues with representatives from national statistical offices to further refine and update source data used for the compilation of internationally-generated indicators.

The consultations convened during preparation of the Report relied on the generous support of many institutions and individuals who are too numerous to mention here. Events and consultations were held in Accra, Boston, Geneva and Singapore (participants and partners are listed at http://hdr.undp.org/en/2015-report/consultations).

A Report focusing on work would not have been possible without extensive consultations and generous investment of time by a wide range of colleagues at the International Labour Organization both in Geneva and New York. Colleagues at the United Nations Food and Agriculture Organization, United Nations Children's Fund, United Nations Volunteers and UN Women also offered much valued insights and commentary. Valuable financial contributions were received from the Governments of France and Germany.

Contributions, support and assistance from UNDP regional bureaus, regional service centres, global policy centres and country offices are acknowledged with gratitude. Special thanks are extended to those UNDP colleagues who constituted the Readers Group for the Report: Nathalie Bouche, Douglas Broderick, Pedro Conceição, George Ronald Gray, Sheila Marnie, Ayodele Odusola, Romulo Paes de Sousa, Thangavel Palanivel and Claudia Vinay. The political read of the Report was done by Ruby Sandhu-Rojon, Mourad Wahba and Kanni Wignaraja and their advice is thankfully acknowledged. Randi Davis, Mandeep Dhaliwal, Karen Dukess, Alberic Kacou, Patrick Keuleers, Brian Lutz, Abdoulaye Mar Dieye and Heather Simpson provided comments, suggestions and guidance on the Report. I am grateful to them.

HDRO interns Geneva Damayanti, Qiansheng Hou, Yiying Sana Riaz, Elizabeth Scheib and Elle Wang deserve recognition for their dedication and contribution.

The highly professional editing and production team at Communications Development Incorporated—led by Bruce Ross-Larson, with Joe Caponio, Christopher Trott and Elaine Wilson—and designers Gerry Quinn, Accurat Design and Phoenix Design Aid are of course of critical importance for producing a report that is both attractive and highly readable.

Most of all, I am grateful to UNDP Administrator Helen Clark for her leadership and vision and for her advice, guidance and support and to the entire HDRO team for their dedication and commitment in producing a report that strives to further the advancement of human development.

Selim Jahan

Selim Jahan
Director
Human Development Report Office

Contents

MAPS

TABLES

Overview

Work for human development

Infographic: Dimensions of human development

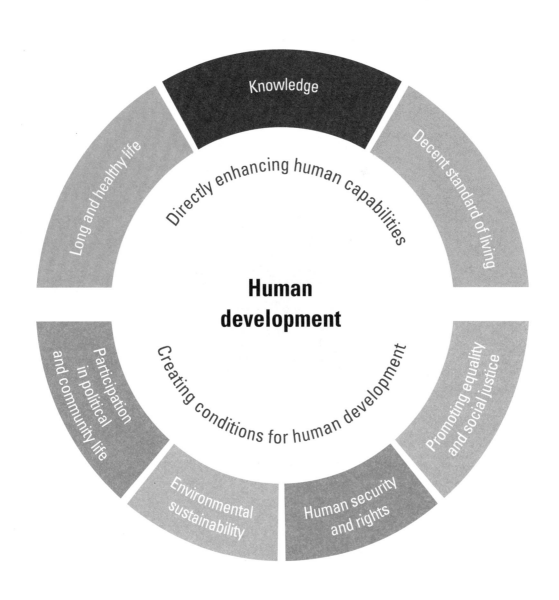

Overview
Work for human development

Human development is about enlarging human choices—focusing on the richness of human lives rather than simply the richness of economies (see infographic). Critical to this process is work, which engages people all over the world in different ways and takes up a major part of their lives. Of the world's 7.3 billion people, 3.2 billion are in jobs, and others engage in care work, creative work, voluntary work or other kinds of work or are preparing themselves as future workers. Some of this work contributes to human development, and some does not. Some work even damages human development (figure 1).

Work enables people to earn a livelihood and be economically secure. It is critical for equitable economic growth, poverty reduction and gender equality. It also allows people to fully participate in society while affording them a sense of dignity and worth. Work can contribute to the public good, and work that involves caring for others builds cohesion and bonds within families and communities.

Work also strengthens societies. Human beings working together not only increase material well-being, they also accumulate a wide body of knowledge that is the basis for cultures and civilizations. And when all this work is environmentally friendly, the benefits extend across generations. Ultimately, work unleashes human potential, human creativity and the human spirit.

This year's Human Development Report explores how work can enhance human development, given that the world of work is changing fast and that substantial human development challenges remain. The Report takes a broad view of work, including voluntary work and creative work, thus going beyond jobs. And it examines the link between work and human development, focusing on care work as well as paid work and discussing sustainable work.

The Report also makes the points that the link between work and human development is not automatic and that some work, such as forced labour, can damage human development by violating human rights, shattering human dignity and sacrificing freedom and autonomy. And without proper policies, work's unequal opportunities and rewards can be divisive, perpetuating inequities in society.

The Report concludes that work can enhance human development when policies expand productive, remunerative and satisfying work opportunities, enhance workers' skills and potential and ensure their rights, safety and well-being. The Report also pursues an action agenda based on a New Social Contract, a Global Deal and the Decent Work Agenda.

People are the real wealth of nations, and human development focuses on enlarging people's choices

Twenty-five years ago the first Human Development Report presented the concept of human development, a simple notion with far-reaching implications. For too long, the world had been preoccupied with material opulence, pushing people to the periphery. The human development framework, taking a people-centred approach, changed the lens for viewing development needs, bringing the lives of people to the forefront.

It emphasized that the true aim of development is not only to boost incomes, but also to maximize human choices—by enhancing human rights, freedoms, capabilities and opportunities and by enabling people to lead long, healthy and creative lives (box 1).

The human development concept is complemented with a measure—the Human Development Index (HDI)—that assesses human well-being from a broad perspective, going beyond income (box 2).

With this simple but powerful notion of people-centred development, nearly two

Human development focuses on the richness of human lives

FIGURE 1

Work engages people all over the world in different ways

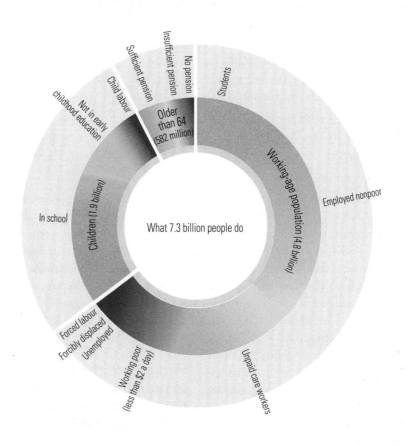

Source: Human Development Report Office.

Work unleashes human potential, human creativity and human spirit

BOX 1

Human development—a comprehensive approach

Human development is a process of enlarging people's choices—as they acquire more capabilities and enjoy more opportunities to use those capabilities. But human development is also the objective, so it is both a process and an outcome. Human development implies that people must influence the process that shapes their lives. In all this, economic growth is an important means to human development, but not the goal.

Human development is development of the people through building human capabilities, for the people by improving their lives and by the people through active participation in the processes that shape their lives. It is broader than other approaches, such as the human resource approach, the basic needs approach and the human welfare approach.

Source: Human Development Report Office.

BOX 2

Measuring human development

The Human Development Index (HDI) is a composite index focusing on three basic dimensions of human development: to lead a long and healthy life, measured by life expectancy at birth; the ability to acquire knowledge, measured by mean years of schooling and expected years of schooling; and the ability to achieve a decent standard of living, measured by gross national income per capita. The HDI has an upper limit of 1.0.

To measure human development more comprehensively, the Human Development Report also presents four other composite indices. The Inequality-adjusted HDI discounts the HDI according to the extent of inequality. The Gender Development Index compares female and male HDI values. The Gender Inequality Index highlights women's empowerment. And the Multidimensional Poverty Index measures nonincome dimensions of poverty.

Source: Human Development Report Office.

dozen global Human Development Reports and more than 700 national Human Development Reports have been produced over the past 25 years. They have contributed to the development discourse, assessed development results, spurred research and innovative thinking and recommended policy options.

Work, not just jobs, contributes to human progress and enhances human development

From a human development perspective, the notion of work is broader and deeper than that of jobs or employment alone. Jobs provide income and support human dignity, participation and economic security. But the jobs framework fails to capture many kinds of work that have important human development implications —as with care work, voluntary work and such creative work as writing or painting.

The link between work and human development is synergistic. Work enhances human development by providing incomes and livelihoods, by reducing poverty and by ensuring equitable growth. Human development—by enhancing health, knowledge, skills and awareness—increases human capital and broadens opportunities and choices (figure 2).

> The notion of work is broader and deeper than that of jobs

FIGURE 2

Work and human development are synergistic

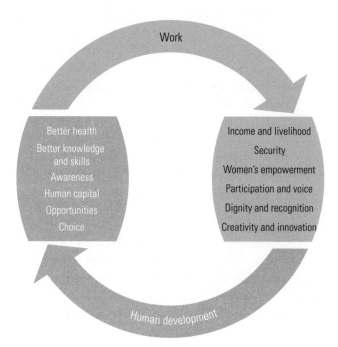

Source: Human Development Report Office.

Since 1990 the world has made major strides in human development. The global HDI value has increased by more than a quarter and that of the least developed countries by more than half. This progress has been fairly steady over time and across regions. The number of people living in low human development fell from 3 billion in 1990 to slightly more than 1 billion in 2014 (see table 8 in *Statistical annex*).

Today, people are living longer, more children are going to school and more people have access to clean water and basic sanitation. This progress goes hand in hand with rising incomes, producing the highest standards of living in human history. A digital revolution now connects people across societies and countries. Just as important, political developments are enabling more people than ever to live under democratic regimes. All are important facets of human development.

Between 1990 and 2015 income poverty in developing country regions fell by more than two-thirds. The number of extreme poor people worldwide fell from 1.9 billion to 836 million. The child mortality rate fell by more than half, and under-five deaths fell from 12.7 million to 6 million. More than 2.6 billion people gained access to an improved source of drinking water, and 2.1 billion gained access to improved sanitation facilities, even as the world's population rose from 5.3 billion to 7.3 billion.

Work in various forms by 7.3 billion people has contributed to this progress. Nearly a billion people who work in agriculture and more than 500 million family farms produce more than 80 percent of the world's food, improving nutrition and health. Worldwide, 80 million workers in health and education have enhanced human capabilities. More than a billion workers in services have contributed to human progress. In China and India 23 million jobs in clean energy are increasing environmental sustainability.

Work has a societal value that goes beyond the gains of individual workers. More than 450 million entrepreneurs are contributing to human innovation and creativity. Some 53 million paid domestic workers are addressing the care needs of people. Care work for children is preparing them for the future. Work that involves caring for older people or people with disabilities is helping them maintain their capabilities. Work by artists, musicians and writers is enriching human lives. More than 970 million people who engage in volunteer activity each year are helping families and communities, building social networks and contributing to social cohesion.

Yet human progress has been uneven, human deprivations are still widespread and much human potential remains unused

Human development has been uneven among regions, across countries and within countries. In 2014 Latin America and the Caribbean's HDI value was 0.748, compared with 0.686 in the Arab States. And the maternal mortality ratio was only 21 per 100,000 live births in Organisation for Economic Co-operation and Development countries, compared with 183 in South Asia (see table 5 in *Statistical annex*).

Globally women earn 24 percent less than men and hold only 25 percent of administrative and managerial positions in the business world—while 32 percent of businesses have no women in senior management positions. Women still hold only 22 percent of seats in single or lower houses of national parliament.

In Malaysia the richest 10 percent of the population had 32 percent of national income in 2012, the poorest 10 percent of the population had only 2 percent. In Moldova 69 percent of urban people have access to safe drinking water, compared with only 23 percent of rural people.

Added to the uneven human development achievements are widespread human deprivations. Worldwide 795 million people suffer from chronic hunger, 11 children under age 5 die every minute and 33 mothers die every hour. About 37 million people live with HIV and 11 million with tuberculosis.

More than 660 million people use an unimproved source of drinking water, 2.4 billion people use an unimproved sanitation facility and nearly a billion people resort to open defecation.

Worldwide 780 million adults and 103 million young people (ages 15–24) are illiterate. In developed countries 160 million people are functionally illiterate. Globally 250 million

The number of people living in low human development fell by nearly 2 billion children have not learned basic skills—even though 130 million of them have spent at least four years in school.

One critical human deprivation is not using, misusing or underusing the deep human potential of people for human development–enhancing work. In 2015, 204 million people were out of work, including 74 million young people—based on formal unemployment data. About 830 million people in the world are working poor—living on less than $2 a day—and more than 1.5 billion are in vulnerable employment, usually lacking decent working conditions and adequate voice and social security.

Unleashing this potential becomes even more important when considering the emerging human development challenges.

Take the rising inequalities in income, wealth and opportunity. Today around 80 percent of the world's people have only 6 percent of the world's wealth. The share of the richest 1 percent is likely to be more than 50 percent by 2016. In the world of work, wages lag behind productivity, and workers' shares in income have been falling.

Population growth, driven mostly by South Asia and increasingly by Sub-Saharan Africa, will have major implications for human development—for work opportunities, the care gap between care needs and care providers and the provision of social protection. Recent estimates indicate that there is a global shortage of 13.6 million care workers, causing extreme deficits in long-term care services for those over age 65. Greater longevity, ageing, the youth bulge and dependency ratios will all have impacts. In 2050 more than two-thirds of the world's population—or 6.2 billion people—are expected to live in urban areas, stressing the coping capacities of cities.

Human security is under threat from many sources. At the end of 2014, 60 million people had been displaced worldwide. Between 2000 and 2013 the cumulative death tolls from global and national violent extremism rose more than fivefold, from 3,361 to 17,958. Violence against women is one of the most brutal threats to human development. One in three women has been subject to physical or sexual violence.

Human development is undermined by multiple shocks, vulnerabilities and risks—by epidemics, by emerging health risks, by economic and financial crises and by food and energy insecurities. For example, noncommunicable (or chronic) diseases are now a global health risk, killing 38 million people each year, almost three-quarters of them (28 million) in low- and middle-income countries. Almost 30 percent (2.1 billion) of the world's people are obese, more than three-fifths of them in developing country regions.

Around the world communities are becoming more vulnerable to the effects of climate change, including the loss of biodiversity—the lifeline of many poor communities. Around 1.3 billion people live on fragile lands. Millions are affected by natural disasters.

Work can enhance human development, but some work damages it—the link between the two is not automatic

The link between work and human development is not automatic. It depends on the quality of work, the conditions of work, the societal value of work and so on. Whether people have a job is important, as are other issues. For example: Is work safe? Are people fulfilled and satisfied by their work? Are there prospects for advancement? Does employment support a flexible work–life balance? Are there equal opportunities for women and men?

The quality of work also includes whether a job provides dignity and a sense of pride and whether it facilitates participation and interaction. To strengthen the link with human development, work also has to enhance environmental sustainability. Work strengthens its link with human development when it goes beyond individual benefits to contribute to shared social objectives, such as poverty and inequality reduction, social cohesion, culture and civilization.

Conversely, the value of work is diminished and its link with human development becomes weaker when there is discrimination and violence at work. The most observable discrimination is along gender lines—in positions, pay and treatment. In the United States female financial specialists' salaries are only 66 percent of their male counterparts'. But discrimination also occurs along lines of race, ethnicity, disability

Worldwide, 11 children under age 5 die every minute, and 33 mothers die every hour

and sexual orientation. In Latin America the wage gap between indigenous ethnic groups and the rest of the population is estimated at 38 percent.

Workplace or occupational violence—in the form of threats and physical or verbal abuse—also weakens the work–human development link. In 2009 some 30 million EU workers experienced work-related violence, such as harassment, intimidation, bullying or physical violence—10 million in the workplace and 20 million outside it.

The link also weakens in conflict and post-conflict situations. Work under such conditions does not always have a definable content, and human development may entail simple survival.

Some work in some conditions damages human development. Many people are in work that restricts their life choices. Millions work in abusive and exploitative conditions that violate their basic human rights and destroy their dignity, such as child labourers, forced labourers and trafficked workers (figure 3). And millions of domestic, migrant, sex and hazardous-industry workers face various risks.

The world has around 168 million child labourers, almost 11 percent of the child population, some 100 million boys and 68 million girls. Around half are engaged in hazardous work.

In 2012 about 21 million people worldwide were in forced labour, trafficked for labour and sexual exploitation or held in slavery-like conditions—14 million were subject to labour exploitation and 4.5 million to sexual exploitation. Women and girls accounted for a larger share than men and boys. Forced labour is thought to generate around $150 billion a year in illegal profits.

After arms and drug trafficking, human trafficking is the most lucrative illicit business worldwide. Between 2007 and 2010 trafficked victims of 136 nationalities were detected in 118 countries, 55–60 percent of them women.

Trafficking of illegal migrants has recently surged. Networks of traffickers take money from desperate migrants who try to cross seas and land illegally into other countries. In 2014 some 3,500 people, maybe many more, lost their lives in the Mediterranean Sea when trafficking boats heading towards Europe, mainly from Libya, capsized or sank.

Paid domestic work is an important means of income for millions of workers, the majority

FIGURE 3

Corrosive and exploitative work shatters human development

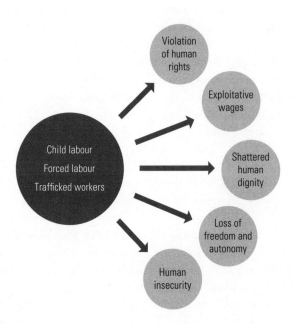

Source: Human Development Report Office.

of whom are women. With appropriate protections in place, domestic work can empower people and help lift their families out of poverty. But abuse is common in paid domestic work, particularly for female migrant workers. Sometimes if the legal framework is inadequate or unenforced, employers use threats and coercion to pay low or even no wages. They can force paid domestic workers to work long hours—up to 18 hours a day without days off. Working conditions are often poor, with little food and no access to medical care. Paid domestic workers may also be subject to physical or sexual abuse.

Mining is one of the most hazardous occupations in many countries. It accounts for only 1 percent of the global workforce (30 million workers) but is responsible for 8 percent of fatal accidents at work and for many injuries and disabling diseases, such as pneumoconiosis (black lung disease).

Globalization and the technological revolution are fast changing how we work and what we do

The context of work is changing, with implications for human development. Driving the transformation of work are globalization and technological revolutions, particularly the digital revolution. Globalization has fostered global interdependence, with major impacts on patterns of trade, investment, growth and job creation and destruction—as well as on networks for creative and volunteer work. We seem to be living through new and accelerated technological revolutions.

In the past 10 years global trade in goods and services almost doubled—reaching nearly $24 trillion in 2014, up from $13 trillion in 2005. The digital components of these flows have also been increasing.

The speed of adoption and penetration of digital technologies is mind-boggling. In the United States it took more than 50 years before half the population had a telephone. For cell phones it took only 10 years (figure 4). By the end of 2015 the planet will have more than 7 billion mobile subscriptions and more than 3 billion Internet users.

FIGURE 4

Speed of adoption of new technologies in the United States

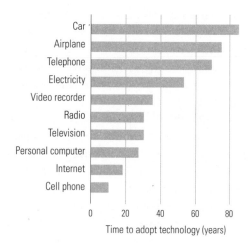

Note: Adoption refers to time for penetration of 50 percent of the population.
Source: Donay 2014.

Access to the digital revolution is uneven across regions, sexes, age groups and the urban–rural divide. In 2015, 81 percent of households in developed countries had Internet access, compared with only 34 percent in developing country regions and 7 percent in the least developed countries.

Globalization brings workers and businesses together in global networks through outsourcing and global value chains. Companies relocate or subcontract (or a bit of both) some functions or noncore activities to other countries where costs are lower. For example, Apple employs only 63,000 of the more than 750,000 people around the world who design, sell, manufacture and assemble its products.

Many economic activities are now integrated in global value chains that span countries, sometimes continents. This integration goes from raw materials and subcomponents to market access and after-sales services. Production is mainly of intermediate goods and services organized in fragmented and internationally dispersed production processes, coordinated by multinational companies and cutting across industries.

In recent years knowledge has become central to production. Even in manufacturing the value of finished goods comes increasingly from embodied knowledge. In 2012 trade in knowledge-intensive goods, services and finance—worth nearly $13 trillion—grew

The world has around 168 million child labourers and 21 million people in forced labour

1.3 times faster than trade in labour-intensive goods, to account for a larger proportion in total trade in goods and services.

The digital revolution has produced such new frontiers of work as the sharing economy (GrabTaxi), business process outsourcing (UpWork), crowdworking (Mechanical Turk) and flexible working. It has also revolutionized creative work and empowered small producers and artisans.

Technological advances have not only transformed work; they are also engines for new forms of creativity and innovation. Collaborative teams and visionaries have turned ideas into tangible goods and services. Innovations in computers and electronics were central to this growth: From 1990 to 2012 their share in all new patents more than doubled, from more than 25 percent to nearly 55 percent.

The digital revolution has also changed volunteering, which can now be done virtually (online or digitally). UN Volunteers' online volunteering system helped 10,887 volunteers (60 percent of them women) contribute their skills towards development work in 2014.

Some of the technologies with the highest potential to change work include cloud technology, 3D printing, advanced robotics, energy storage and the automation of knowledge work —which through intelligent software systems will transform the organization and productivity of knowledge work and enable millions to use intelligent digital assistants.

In the new world of work, workers need to be more flexible and adaptable—and be ready to retrain, relocate and renegotiate working conditions. They also need to dedicate more time to searching for new opportunities.

The people most linked to the new world of work are millennials—roughly the cohort born since 1980. This group has come of age at a time when digital technologies and advanced information and communication technologies penetrate all areas of life. They have also become adults at a time when flexibility, adaptability and unconventional work are increasingly common.

Many millennials are looking for work that goes beyond creating profits, hoping to solve environmental and social problems as part of their livelihoods.

Social entrepreneurs are also emerging as a new workforce. They are cause-driven people committed to addressing social problems, and they establish nonloss, nondividend companies (where all profits are reinvested back into the company) that aim to be financially self-sustainable and to maximize social benefits.

Globalizing work has generated gains for some and losses for others

With outsourcing, assembly jobs in developed countries began moving to export processing zones as developing countries adopted export-oriented industrialization. The impact on job creation in large developing countries such as China and Mexico, as well as smaller countries such as Costa Rica, the Dominican Republic and Sri Lanka, has been substantial and positive, often boosting local development, although the quality of the work and enforcement of labour standards have varied.

The global offshoring of service jobs started to pick up in the 1990s as advances in information and communications technology allowed many support services to be performed offsite. For example, between 2000 and 2010 the number of direct jobs in information and communications technology in India jumped from 284,000 to more than 2 million. Services are also growing in the Russian Federation, Latin America and Africa, in part matching companies' interests in diversifying into different time zones to enable 24-hour service. But outsourcing to developing countries has not benefitted all sectors and all workers.

While outsourcing in general seems beneficial to developing country regions, it has consequences for workers in developed countries. Estimates vary, and the long-term impacts are less clear than the short-term effects, but job losses are greater in manufacturing than services. Short-term job losses due to offshoring have been found to range from 0 in some countries to almost 55 percent of all job losses in Portugal.

Today, jobs that involve administrative support, business and financial operations, and computer and mathematical tasks are most likely to be outsourced. In Australia, Canada

We seem to be living through new and accelerated technological revolutions

and the United States 20–29 percent of all jobs have the potential to be offshored, though it is unlikely that all of them will be. Many jobs in this estimate are in medium- and high-skilled service professions that can be carried out at lower cost abroad as education levels rise and information and communications technology infrastructures improve.

So, while there may be immense benefits in access to new jobs in countries hosting offshore activities, individuals losing jobs may require training and new skills for a more competitive environment. To ease the adjustment, programmes are needed to help people find new work, enhance their skills and maintain access to a basic income. Training can also enhance the abilities of workers in developing countries to access the new jobs.

The integration of developing countries in global value chains has increased opportunities for paid work and prompted a shift in labour force participation for women (many find jobs in the garment industry). In 2013, 453 million workers (up from 296 million in 1995), including 190 million women, were involved in global value chains.

But such integration does not say much about the quality of work and whether workers have expanded their human capabilities. There are concerns about levels of labour protection and opportunities for skills upgrading.

The global value chain system generates winners and losers, within and across countries and industries. The footloose nature of global value chains can generate less job security and put even more pressure on governments and subcontractors to minimize costs. This in turn puts pressures on workers' wages and working conditions, particularly among the low skilled. Developing countries also face the risk of becoming locked into low value-added nodes of global value chains that limit work opportunities, skill development and technology exposure.

The transition to global value chains has introduced new complexities for workers in developed and developing countries alike. There are questions about how much workers gain by partaking in work contributing to global value chains versus work outside them. There is some evidence that productivity is higher in global value chain–oriented work but that wages are the same for workers inside and outside global value chains, raising questions about how the increases in productivity are shared between workers and capital.

Market pressures transmitted through global value chains tend to be absorbed by workers —whether in wages (driven down by global competition), in increased informalization and contractual insecurity (through multiple subcontracting chains) or in layoffs (during downturns). Multinational corporations increasingly rely on a disenfranchised workforce, using a mix of fixed-term employees, temporary workers, independent contractors, project-based workers and outsourced workers to provide production flexibility and manage costs. Participation in value chains provides some with secure, decent jobs and others with more precarious work (even in the same country and sector), in a type of "labour dualism."

Seizing the future in the digital revolution is not chance or fate— it is a matter of skill and foresight

The types of work that people do and the ways they do it are being transformed by new technologies. This change is not new, but it is reshaping the link between work and human development and the types of policies and institutions needed to foster positive outcomes for people.

The spread and penetration of digital technologies are changing the world of work everywhere, but the effects vary across countries. Some technological changes are cross-cutting, such as information and communications technologies and the spread of mobile phones and other handheld devices. Still, countries will continue to have divergent production and employment structures and different uses for digital technologies, largely reflecting the relative economic weights of agriculture, industry and services, as well as the resources invested in developing people's capabilities. Labour markets, the ratio of paid to unpaid work and the predominant types of workplaces in each country differ—so the impacts of digital technologies on work will vary, too.

The digital revolution may be associated with high-tech industries, but it is also

In recent years knowledge has become central to production

influencing a whole range of more informal activities from agriculture to street vending. Some may be directly related to mobile devices. In Ethiopia farmers use mobile phones to check coffee prices. In Saudi Arabia farmers use wireless technology to distribute scarce irrigated water for wheat cultivation. In some villages in Bangladesh, female entrepreneurs use their phones to provide paid services for neighbours.

Mobile phones now facilitate many aspects of work through a combination of voice calls, SMS and mobile applications. There are benefits for many other types of activity—formal and informal, paid and unpaid—from food vendors in Cairo to street cleaners in Senegal to care providers in London.

Internet and mobile phone access empowers people to harness their creativity and ingenuity. Much more is possible, particularly if inequalities in access between men and women and rural and urban areas are addressed. If Internet access in developing countries were the same as in developed countries, an estimated $2.2 trillion in GDP could be generated, with more than 140 million new jobs—44 million in Africa and 65 million in India. Long-term productivity in developing countries could be boosted by up to 25 percent.

The digital economy has enabled many women to access work that allows them to apply their creativity and potential. In 2013 about 1.3 billion women were using the Internet. Some have moved to e-trading as entrepreneurs, and some are employed through crowdworking or e-services. But this new world of work puts a high premium on workers with skills and qualifications in science and technology, workers less likely to be women.

Older workers also have new work options, as they continue to work either because they enjoy their work or because they cannot afford to retire. Most of the older and younger workers are in different labour markets (so there is no direct substitution) and the anxiety that young people will lose out because older people are encouraged to work may not be the case.

Still, there are risks and promises as yet unfulfilled. We may in fact be at an inflexion point, with both positive and negative impacts. The technological revolution presents skill-biased technical change: the idea that the net effect of new technologies reduces demand for less skilled workers while increasing demand for highly skilled ones. By definition, such change favours people with higher human capital, polarizing work opportunities.

At the top will be good jobs for those with high education and skills. For example, in the automobile industry those who benefit will be the engineers who design and test new vehicles. At the bottom there will still be low-skill, low-productivity, low-wage service occupations such as office cleaning. But the middle areas will see a steady hollowing out of many jobs in office cubicles and on factory floors. The biggest losers will thus be workers with less specialized, routine-work skills (figure 5).

Many cognitively complex jobs are beyond the abilities even of people with reasonable qualifications. Some industries could thus face skill shortages, so companies willing to pay high salaries for the best talent will look to the global market. And besides being polarized nationally, workforces are being stratified internationally, with low-skilled workers coming mainly from national markets and high-skilled workers from global markets.

Now is the time to be a worker with special skills and the right education, because these people can use technology to create and capture value. But there has never been a worse time to be a worker with only ordinary skills and abilities, because computers, robots and other digital technologies are acquiring these skills and abilities at an extraordinary speed.

An implied promise of the digital revolution was that it would increase labour productivity and thus would lead to higher pay. This does not seem to have happened on either front: productivity has not grown at the rates expected, and few of the gains have translated into higher wages. In many economies (for example, the Netherlands) the gap between productivity and wage growth has widened over the years, and the situation is even more serious as average wages mask the fact that as real pay for most workers stagnated, income for the highest earners soared.

The technological revolution has been accompanied by rising inequality. Workers are

FIGURE 5

The 20 jobs most and least likely to be replaced by automation

Least likely to be replaced	Most likely to be replaced
Recreational therapists	Telemarketers
First-line supervisors of mechanics, installers and repairers	Title examiners, abstractors and searchers
Emergency management directors	People working in sewers
Mental health and substance abuse social workers	Mathematical technicians
Audiologists	Insurance underwriters
Occupational therapists	Watch repairers
Orthotists and prosthetists	Cargo and freight agents
Health care social workers	Tax preparers
Oral and maxillofacial surgeons	Photographic process workers
First-line supervisors of fire fighting and prevention workers	New accounts clerks
Dietitians and nutritionists	Library technicians
Lodging managers	Data entry keyers
Choreographers	Timing device assemblers
Sales engineers	Insurance claims
Physicians and surgeons	Brokerage clerks
Instructional coordinators	Order clerks
Psychologists	Loan officers
First-line supervisors of police and detectives	Insurance appraisers
Dentists	Umpires, referees and sports officials
Elementary school teachers, except special education	Tellers

Note: The figure ranks occupations according to their probability of computerization (least likely to become automated in blue and most likely to become automated in red). Occupations correspond closely to the US Department of Labor Standard Occupational Classification.
Source: Frey and Osborne 2013.

getting a smaller share of total income. Even people with better education and training who can work more productively may not receive commensurate rewards in income, stability or social recognition.

The declining share of workers' income can be seen as part of the slowdown in the growth of average real wages: As the income shares of high-skilled labour (and of capital) have been going up, the share of other labour has been going down.

The sharp increase in work compensation to top salary earners has benefited a minority, whether the top 10 percent, 1 percent or even 0.1 percent. The global elite, the world's richest 1 percent, had an average wealth of $2.7 million per adult in 2014.

Are workers, employers and policymakers prepared to respond to the challenges of the emerging world of work? In such a world, specific technical knowledge quickly becomes obsolete, and the policies and rules of yesterday might not serve the challenges of today or tomorrow.

Imbalances leave women at a disadvantage in the realm of work—whether paid or unpaid

In the two worlds of work—unpaid care work and paid work—there continue to be pronounced imbalances across genders, reflecting local values, social traditions and historical gender roles. Care work includes housework, such as preparing meals for the family, cleaning the house and gathering water and fuel, as well as work caring for children, older people and family members who are sick—over both the short and long term. Across most countries in all regions, women work more than men. Women are estimated to contribute 52 percent of global work, men 48 percent.

But even if women carry more than half the burden, they are disadvantaged in both realms of work—paid as well as unpaid work—in patterns that reinforce each other.

In 2015 the global labour force participation rate was 50 percent for women but 77 percent for men. Worldwide in 2015, 72 percent of

Women are estimated to contribute 52 percent of global work, men 48 percent

working-age (ages 15 and older) men were employed, compared with only 47 percent of women. Female participation in the labour force and employment rates are affected heavily by economic, social and cultural issues and care work distributions in the home.

Of the 59 percent of work that is paid, mostly outside the home, men's share is nearly twice that of women—38 percent versus 21 percent. The picture is reversed for unpaid work, mostly within the home and encompassing a range of care responsibilities: of the 41 percent of work that is unpaid, women perform three times more than men—31 percent versus 10 percent.

Hence the imbalance—men dominate the world of paid work, women that of unpaid work. Unpaid work in the home is indispensable to the functioning of society and human well-being: yet when it falls primarily to women, it limits their choices and opportunities for other activities that could be more fulfilling to them.

Even when women are in paid work, they face disadvantages and discrimination. The evidence of the glass ceiling is just one of them. Women are underrepresented in senior business management globally: They hold only 22 percent of senior leadership positions, and 32 percent of businesses do not have any female senior managers, with regional variations (figure 6). Occupational segregation has been pervasive over time and across levels of economic prosperity—in both advanced and developing countries men are over-represented in crafts, trades, plant and machine operations, and managerial and legislative occupations; and women in mid-skill occupations such as clerks, service workers and shop and sales workers.

Even when doing similar work, women can earn less—with the wage gaps generally greatest for the highest paid professionals. Globally, women earn 24 percent less than men. In Latin America top female managers earn on average only 53 percent of top male managers' salaries. Across most regions women are also more likely to be in "vulnerable employment"—working for themselves or others in informal contexts where earnings are fragile and protections and social security are minimal or absent.

Women bear an unequal share of care work

Women worldwide undertake most of the unpaid care work, which includes mainly housework (such as preparing meals, fetching firewood, collecting water and cleaning) and care work (such as caring for children, the sick and older people) in the home and community.

Due to their disproportionate share of care work, women have less time than men for other activities, including paid work and education. This includes less discretionary free time. In a sample of 62 countries, men devoted on average

> Men dominate the world of paid work, women that of unpaid work

FIGURE 6

Women's representation in senior management in business, by region, 2015

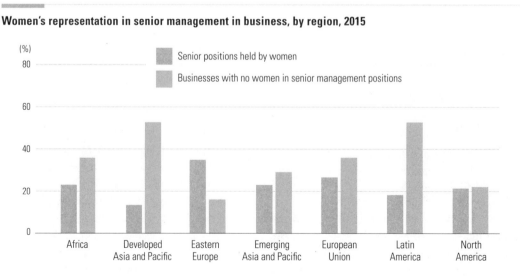

Source: Grant Thornton 2015.

4.5 hours a day to social life and leisure, and women 3.9 hours. In low human development countries men spend nearly 30 percent more time on social life and leisure than women. In very high human development countries the difference is 12 percent.

Women are also disproportionately involved in paid care work. The demand for paid domestic workers has risen. Globally an estimated 53 million people ages 15 and older are in paid domestic work. Of these, 83 percent are women —some, migrant workers. And so a global care work chain has emerged where migrant domestic workers undertake housework and provide care to children and others in households abroad. But they often leave their own children and parents behind in their homeland, creating a care gap often filled by grandparents, other relatives or hired local helpers.

Despite the possible abuse in domestic work— low wages, poor working conditions, no access to medical care, and physical or sexual abuse— many workers feel obliged to remain with abusive employers because they need the work.

Despite the importance for human development, care work often goes unrecognized. This is partly because, being unpaid, it is not reflected in economic indicators such as GDP. But valuing unpaid care work would highlight women's contributions in households and communities and draw attention to their material conditions and well-being, with a possible implication for policymaking. Among all countries attempting to measure the value of unpaid care work, estimates range from 20 percent to 60 percent of GDP. In India unpaid care is estimated at 39 percent of GDP, in South Africa 15 percent.

When women have no choice but to give priority to unpaid work and stay out of the labour force, they make large sacrifices, perhaps missing the chance to expand their capabilities in the workplace. They also lose opportunities for economic independence.

Addressing imbalances in unpaid and paid work benefits both current and future generations

Imbalances in the division of work between women and men have to be changed. Yes, many

societies are experiencing a generational shift, particularly in educated middle-class households, towards greater sharing of care work between men and women. Yet much remains to be done, and action needs to happen quickly to address deep gender inequalities. Longstanding patterns of inequalities can reinforce each other, trapping women and girls across generations in realms of limited choices and opportunities. Steps are needed along four policy axes—reducing and sharing the load of unpaid care work, expanding opportunities for women in paid work, improving outcomes in paid work and changing norms.

Time spent in unpaid care work needs to be reduced overall and shared more equally. Universal access to clean water, modern energy services for household needs, quality public services (including those related to health and care), workplace arrangements that accommodate flexible schedules without penalizing professional advancement and a shift in mindsets about gender-specific roles and responsibilities can all contribute to reducing the load of care work for families and women in particular.

Legislation and targeted policies can increase women's access to paid work. Access to quality higher education in all fields and proactive recruitment efforts can reduce barriers, particularly in fields where women are either underrepresented or where wage gaps persist.

Policies can also remove barriers to women's advancement in the workplace. Measures such as those related to workplace harassment and equal pay, mandatory parental leave, equitable opportunities to expand knowledge and expertise and measures to eliminate the attrition of human capital and expertise can help improve women's outcomes at work.

Paid parental leave is crucial. More equal and encouraged parental leave can help ensure high rates of female labour force participation, wage gap reductions and better work–life balance for women and men. Many countries now offer parental leave to be split between mothers and fathers.

Social norms also need to evolve to reflect the equal potentials of women and men. Promoting women to visible positions of seniority, responsibility and decisionmaking in both public and private spheres and

Globally, women earn 24 percent less than men

encouraging the engagement of men in traditionally female-dominated professions can help shift deep-seated views.

Sustainable work is a major building block for sustainable development

Sustainable work promotes human development while reducing and eliminating negative side effects and unintended consequences. It is critical not only for sustaining the planet, but also for ensuring work for future generations (figure 7).

For such work to become more common, three parallel changes are needed:
- Termination (some work will end or be reduced).
- Transformation (some work will be preserved through investment in adaptable new technologies and retraining or skill upgrading).
- Creation (some work will be new).

Some occupations can be expected to loom larger—railway technicians, for instance, as countries invest in mass transit systems. Terminated workers may predominate in sectors that draw heavily on natural resources or emit greenhouse gases or other pollutants. About 50 million people are employed globally in such sectors (7 million in coal mining, for example).

In many occupations, how output is produced also needs to change, as in ship breaking, by implementing and enforcing standards.

New areas of work include solar photovoltaic technologies, an important part of many countries' renewable energy strategies. Their potential for human development differs radically depending on whether they replace grid-based electricity generated by conventional means, as in many developed countries, or expand off-grid energy access, as in many developing countries. Renewable energy could become a key vehicle towards achieving Sustainable Development Goal 7.1 to ensure universal

> Sustainable work promotes human development

FIGURE 7

The matrix of sustainable work

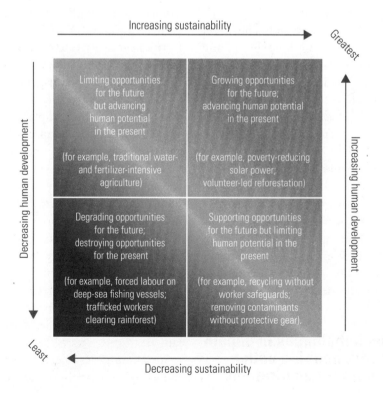

Source: Human Development Report Office.

access to affordable, reliable and modern energy services by 2030 (table 1).

The Sustainable Development Goals have key implications for sustainable work

The Sustainable Development Goal with the most direct implications for sustainable work is goal 8 (promote sustained, inclusive and sustainable economic growth, full and productive employment, and decent work for all), and its associated targets, which spell out some of the implications for sustainable work. Target 8.7 is to take immediate and effective measures to eradicate forced labour, end modern slavery and human trafficking and secure the prohibition and elimination of the worst forms of child labour, including the recruitment and use of child soldiers and by 2025 to end child labour in all its forms.

Target 8.8—to protect labour rights and promote safe and secure working environments for all workers, including migrant workers, in particular women migrants, and those in precarious employment—aims to strengthen the human development outcomes of workers, avoiding a race to the bottom. Target 8.9—to devise and implement policies to promote sustainable tourism that creates jobs and promotes local culture and products by 2030—advocates for a particular kind of (sustainable) work.

Target 3.a—to strengthen the implementation of the World Health Organization Framework Convention on Tobacco Control in all countries, as appropriate—seeks to reduce work associated with tobacco production and distribution while improving the health of workers. Target 9.4—to upgrade by 2030 infrastructure and retrofit industries to make them sustainable, with increased resource-use efficiency and greater adoption of clean and environmentally sound technologies and industrial

> Renewable energy could become a key vehicle towards achieving sustainable development

TABLE 1

Sustainable Development Goals

Goal 1	End poverty in all its forms everywhere
Goal 2	End hunger, achieve food security and improved nutrition and promote sustainable agriculture
Goal 3	Ensure healthy lives and promote well-being for all at all ages
Goal 4	Ensure inclusive and equitable quality education and promote lifelong learning opportunities for all
Goal 5	Achieve gender equality and empower all women and girls
Goal 6	Ensure availability and sustainable management of water and sanitation for all
Goal 7	Ensure access to affordable, reliable, sustainable and modern energy for all
Goal 8	Promote sustained, inclusive and sustainable economic growth, full and productive employment and decent work for all
Goal 9	Build resilient infrastructure, promote inclusive and sustainable industrialization and foster innovation
Goal 10	Reduce inequality within and among countries
Goal 11	Make cities and human settlements inclusive, safe, resilient and sustainable
Goal 12	Ensure sustainable consumption and production patterns
Goal 13	Take urgent action to combat climate change and its impacts[a]
Goal 14	Conserve and sustainably use the oceans, seas and marine resources for sustainable development
Goal 15	Protect, restore and promote sustainable use of terrestrial ecosystems, sustainably manage forests, combat desertification, and halt and reverse land degradation and halt biodiversity loss
Goal 16	Promote peaceful and inclusive societies for sustainable development, provide access to justice for all and build effective, accountable and inclusive institutions at all levels
Goal 17	Strengthen the means of implementation and revitalize the global partnership for sustainable development

a. Acknowledging that the United Nations Framework Convention on Climate Change is the primary international, intergovernmental forum for negotiating the global response to climate change.
Source: UN 2015b.

processes, with all countries taking action in accordance with their respective capabilities—implies a specific direction to upgrading skills and possibly to new areas of work.

A large number of the Sustainable Development Goal targets intend to focus on work that has negative implications for human development. Target 8.7, if reached, would improve the lives of 168 million child labourers and 21 million in forced labour. Target 5.2 would assist 4.4 million sexually exploited women, and target 3.a would affect an estimated 100 million workers in tobacco. Active policies and programmes will be needed to support the people formerly engaged in these kinds of work.

Other goals and targets involve transforming current modes of work and introducing new approaches. Goal 2—ending hunger and achieving food security and improved nutrition and promoting sustainable agriculture—has the potential to transform how the very large number of people engaged in agriculture carry out their activities.

Work in some primary industries—such as farming, fisheries and forestry—engages more than a billion people worldwide, including most of those living on less than $1.25 a day. The sector is responsible for a large proportion of greenhouse gas emissions, is associated with unsustainable patterns of water and soil use, is linked to deforestation and loss of biodiversity and is especially susceptible to the disruptions of climate change.

Transforming the way farmers grow and process crops is thus crucial. Technologies and farming methods that can make a difference exist but need to be adopted faster. For example, about a third of total food production is lost or wasted, with cereals accounting for the largest share. Broad efforts are needed to expand demonstrable, immediate gains—and to create new products for industrial or artisanal manufacture.

Much of the work tied to a move to environmental sustainability (target 9.4) will involve infrastructure and construction. Energy projects (goal 7) can drive long- and short-term jobs, directly and indirectly, when they enable other industries to grow and flourish. In 2014 renewable energy (excluding large hydropower, which had roughly 1.5 million direct jobs) employed an estimated 7.7 million people directly and indirectly. In renewable energy the field of solar photovoltaics is the largest employer worldwide, with 2.5 million jobs.

By strengthening health and education outcomes, especially among children, the Sustainable Development Goals can set the basis for people to acquire the skills to move to occupations that involve sustainable work.

Enhancing human development through work needs concrete policies and an agenda for action

Policy options for enhancing human development through work have to be built around three broad dimensions: creating more work opportunities to expand work choices, ensuring workers' well-being to reinforce a positive link between work and human development and targeted actions to address the challenges of specific groups and contexts. An agenda for action to build momentum for change is also needed, pursuing a three-pillar approach—a New Social Contract, a Global Deal and the Decent Work Agenda (figure 8).

FIGURE 8

Policy options for enhancing human development through work

Strategies for creating work opportunities	Strategies for ensuring workers' well-being
Seizing opportunities of the changing world of work	Guaranteeing workers' rights and benefits
Formulating national employment strategies to address crisis in work	Extending social protection
	Addressing inequalities
Strategies for targeted action	**An agenda for action**
Moving towards sustainable work	Decent Work Agenda
Balancing domestic and outside work	Global Deal
Undertaking group-specific initiatives	New Social Contract

Source: Human Development Report Office.

Creating work opportunities requires well formulated employment plans as well as strategies to seize opportunities in the changing world of work

Work for human development is about more than just jobs, but human development is also about expanding people's choices and making sure that opportunities are available. This includes ensuring that adequate and quality paid work opportunities are available and accessible for those who need and want paid work. National employment strategies are needed for addressing the complex challenges in work in many countries. About 27 developing countries have adopted national employment strategies, another 18 are doing so and 5 are revisiting their policies to better respond to new employment challenges. Major policy instruments of a national employment strategy might include:

- *Setting an employment target.* More than a dozen countries have employment targets (including Honduras and Indonesia). Central banks may pursue dual targeting—moving beyond a focus primarily on inflation control to emphasize employment targets. They may also consider specific monetary policy instruments (such as credit allocation mechanisms) for creating more work opportunities, as in Chile, Colombia, India, Malaysia and Singapore.

- *Formulating an employment-led growth strategy.* Employment can no longer be considered to be simply a derivative of economic growth. Some policy interventions would entail strengthening links between small and medium-sized enterprises in need of capital and large capital-intensive firms to boost employment, upgrading workers' skills over the lifecycle, focusing investments and inputs on sectors where poor people work (such as agriculture), removing barriers critical to employment-led growth (such as removing biases towards small and medium-sized enterprises in access to credit), implementing solid legal and regulatory frameworks and addressing the distribution of capital and labour in public spending to emphasize technologies that create jobs .

- *Moving to financial inclusion.* An inclusive financial system is essential for structural transformation and work creation. In developing

Employment can no longer be considered to be simply a derivative of economic growth

countries the lack of access to finance is a major hindrance to enterprise operation and growth, particularly for women. Policy options might encompass extending banking services to disadvantaged and marginalized groups (as in Ecuador), steering credit towards unserved, remote areas and targeted sectors (as in Argentina, Malaysia and the Republic of Korea) and lowering interest rates and providing credit guarantees and subsidized credit to small and medium-sized enterprises and export-oriented sectors.

- *Building a supportive macroeconomic framework.* Some policy instruments to reduce volatility and create secure jobs include keeping the real exchange rate stable and competitive, managing capital accounts prudently, restructuring budgets towards job-creating sectors, building fiscal space for public spending, promoting an enabling business environment, ensuring high-quality infrastructure and adopting a regulatory framework that encourages competition, enhances efficiency and ensures transparency and accountability for business.

National employment strategies are needed for addressing the complex challenges in work in many countries

Seizing opportunities in the changing world of work requires policy actions to help people thrive in the new work environment. Individuals can flourish if they are equipped with skills, knowledge and competencies to harness new technologies and capitalize on emerging opportunities. Some of the policy actions here would require:

- *Heading off a race to the bottom.* Given the realized and potential benefits that globalization brings to work, a race to the bottom —ever lower wages and worsening working conditions—is not the only outcome. Global attention to ensuring decent wages, maintaining workers' safety and protecting their rights can pre-empt such a race and make business more sustainable in the long run, as can fair trade, because work conditions are becoming increasingly critical in consumers' minds.

- *Providing workers with new skills and education.* Higher and specific skills will be needed for science and engineering jobs and for many other jobs, as will be an aptitude for creativity, problem solving and lifelong learning.

Strategies for ensuring workers' well-being must focus on rights, benefits, social protection and inequalities

Guaranteeing the rights and benefits of workers is at the heart of strengthening the positive link between work and human development and weakening the negative connections.

Policies could include:

- *Setting legislation and regulation.* These should be on collective bargaining, unemployment insurance, the minimum wage, protection of workers' rights and worker safety. Steps to ratify and implement the eight International Labour Organization conventions on work and to put in place legal frameworks for enforcement are also needed (figure 9).

- *Ensuring that people with disabilities can work.* Measures can induce employers to provide an appropriate working environment. States can make efforts to change norms and perceptions, enhance the capabilities of people with disabilities, ensure

workplace accessibility and access to appropriate technology and adopt affirmative action policies.

- *Making workers' rights and safety a cross-border issue.* Measures may include regulatory frameworks that extend to migrants, subregional remittance clearinghouses and more support to migrant source countries. These frameworks may constitute regional or subregional public goods.

- *Promoting collective action and trade unionism.* Given globalization, the technological revolution and changes in labour markets, support is needed for emerging forms of collective action (such as the Self-Employed Women's Association in India), innovative organizations for flexible workers (such as the Freelancers Union in the United States) and collective bargaining, including peaceful protests and demonstrations.

FIGURE 9

Number of countries having ratified International Labour Organization conventions, 1990 and 2014

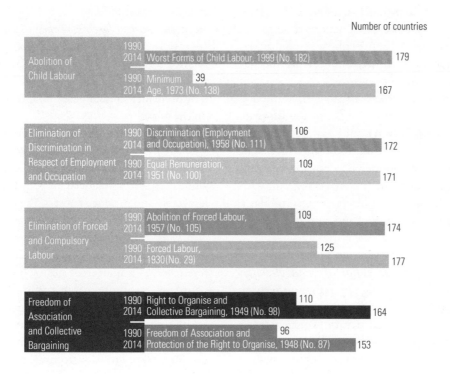

Source: Human Development Report Office calculations based on ILO (2014c).

Only 27 percent of the world's population is covered by comprehensive social protection, which means that the security and choices of many workers is severely limited. Action to extend social protection should focus on:

- *Pursuing well designed, targeted and run programmes.* A basic and modest set of social security guarantees can be provided for all citizens through social transfers in cash and kind. Resources can be mobilized through, for example, progressive taxes, restructured expenditures and wider contributory schemes.
- *Combining social protection with appropriate work strategies.* Programmes would provide work to poor people while serving as a social safety net.
- *Assuring a living income.* This would be a basic minimum income for all, independent of the job market, through cash transfers. Such a policy would help make unpaid work a more feasible and secure option.
- *Tailoring successful social protection programmes to local contexts.* Programmes for cash transfers or conditional cash transfers have provided a source of social protection, particularly in Latin America (such as Bolsa Família in Brazil and Oportunidades, now called Prospera, in Mexico) and could be replicated in other parts of the world.
- *Undertaking direct employment guarantee programmes.* Countries have also pursued employment guarantees. The best known is the National Rural Employment Guarantee Scheme in India.
- *Targeting interventions for older people.* Older people's scope for choice in work is limited by access to pensions. Policy choices include expanding noncontributory basic social pensions systems and exploring fully funded contributory pension systems (as in Chile, for example).

Because workers are getting a smaller share of total income and inequalities in opportunities are still substantial, policy options should focus on:

- *Formulating and implementing pro-poor growth strategies.* This would entail creating work in sectors where most poor people

Guaranteeing rights and benefits of workers is at the heart of strengthening the positive link between work and human development

work, improving poor households' access to such basic social services as health, education, safe water and sanitation and providing access to such productive resources as inputs, credit and finance. These actions can also free up time spent in unpaid care work. Subsidies, targeted expenditures and pricing mechanisms are other options.

- *Providing complementary support.* Marketing facilities, investments in physical infrastructure (particularly in rural areas), expansion of extension services and labour-intensive technologies are conducive to equalizing work opportunities. The private sector can, with the right incentives, be encouraged to play a major role in building and running physical infrastructures.

- *Democratizing education, particularly at the tertiary level, nationally and globally.* Countries place a high premium on tertiary education, but access is unequal and can perpetuate inequalities in work, as seen within countries (most workers with a tertiary education come from higher income families) and between countries (countries with greater increases in tertiary education are industrial, with already high attainment in this segment).

- *Pursuing profit sharing and employee ownership.* Profit sharing with labour and giving employees shares in enterprises may help cut income inequality.

- *Adopting and enforcing proper distributive policies.* These could include progressive taxes on income and wealth, regulations to reduce rent extraction, stricter regulation (particularly of finance) and targeted public spending on the poor.

- *Regulating the financial sector to reduce the regressive effects of cycles.* Promoting investments in the real economy can generate secure jobs, while increases in financial investment can be less stable and produce fewer jobs.

- *Removing asymmetries between the mobility of labour and of capital.* Labour mobility does not match capital's given intrinsic differences. As a matter of policy, industrial countries promote capital mobility but discourage that of labour. Nonetheless, regulating capital movements can reduce macroeconomic instability and middle-income traps in developing countries, preventing capital from moving overseas when wages become too high. Migration policies can at a minimum reduce the risks of migration.

Targeted actions are needed for balancing care and paid work, making work sustainable, addressing youth unemployment, encouraging creative and voluntary work and providing work in conflict and post-conflict situations

Addressing imbalances in paid and unpaid work opportunities between women and men may benefit from the following policy measures:

- *Expanding and strengthening gender-sensitive policies for female wage employment.* Programmes should address skills development through education, particularly in math and science, training that matches market demands and access to continuing professional development.

- *Actions to increase representation of women in senior decisionmaking positions.* Representation can be enhanced in public and private sectors through policies on human resources, selection and recruitment,

and incentives for retention. The criteria for moving men and women into senior positions should be identical. Mentoring and coaching can empower women in the workplace, for example, by using successful senior female managers as role models.

- *Specific interventions.* Legislative measures are needed to reduce inequalities between women and men in harassment in the workplace, discrimination in hiring, access to finance and access to technology.

- *Focusing on maternal and paternal parental leave.* Rather than pursuing a totally gender-neutral approach, if a bonus is granted to parents who share parental leave more equally, fathers may be induced to make more use of paternal leave.

- *Enlarging care options, including day-care centres, after-school programmes, senior citizens' homes and long-term care facilities.* Employers can also offer child-care onsite. Another alternative is to subsidize care work through vouchers and tickets.
- *Encouraging flexible working arrangements, including telecommuting.* There should be sufficient incentives to return to work after giving birth. These may encompass reservation of jobs for women on maternity leave for up to a year. Women could also be offered benefits and stimulus (for example, salary increases) to return to work. Telecommuting and flexible hours can also allow women and men to address imbalances in paid and unpaid work.
- *Valuing care work.* Efforts would help raise policy awareness about the value care work brings to society and could encourage different options for rewarding such work.
- *Gathering better data on paid and unpaid work.* National statistical systems, using more female investigators and appropriate samples and questionnaires, should gather better data on the distribution of paid and unpaid work.

Targeted measures for sustainable work may focus on terminating, transforming and creating work to advance human development and environmental sustainability. Policy measures may focus on:

- *Adopting different technologies and encouraging new investments.* This would require departing from business as usual, pursuing technology transfer and quickly moving to more sustainable work.
- *Incentivizing individual action and guarding against inequality.* This requires recognizing and incentivizing the positive externalities in people's work—for example, using a social wage, which goes beyond a private wage to reward workers when their work is of value to society (for example, conservation of forests).
- *Managing trade-offs:* For example, supporting workers who lose their jobs due to an end of activities in their sector or industry (for example, mining), implementing standards (as in the ship-breaking industry), addressing intergenerational inequality and managing and facilitating change.

Additionally, a mechanism is needed to translate the desired global outcomes into country actions (box 3).

Policy options mentioned earlier, particularly for education and skills building, are especially relevant to addressing youth unemployment. But given the severity of this challenge and its multidimensional (economic, social and political) impacts, it also requires targeted interventions. Exciting work opportunities for young people should be created so that they can unbridle their creativity, innovation and

Targeted measures for sustainable work may focus on terminating, transforming and creating work

---

BOX 3

Possible measures at the country level for moving towards sustainable work

- Identify appropriate technologies and investment options, including leapfrogging opportunities.
- Set up regulatory and macroeconomic frameworks to facilitate adoption of sustainable policies.
- Ensure that the population has the appropriate skills base—combining technical and high-quality skills with core abilities for learning, employability and communicating.
- Retrain and upgrade the skills of large numbers of workers in informal sectors, such as agriculture. While some workers may be reached through the market, others will need the help of the public sector, nongovernmental organizations and others.

These programmes can be a means to support women and other traditionally disadvantaged groups.

- Manage the adverse impacts of the transition by offering diversified packages of support and levelling the playing field to break the transmission of intergenerational inequality.
- Continue to build the skill base of the population. This will require a lifecycle approach that recognizes the cumulative nature of interventions that lead to learning. Large investments in the number and quality of health and education workers will be necessary, underscoring the continuing role of the public sector in transforming skills.

Source: Atkinson 2015.

entrepreneurship in the new world of work. Methods for doing so include:

- *Providing policy support to the sectors and entities creating new lines of work.* Such initiatives are ongoing, and new opportunities are being discovered every day, but they need policy support.
- *Investing in skills development, creativity and problem solving.* Special support should be extended to young women and men in apprenticeships, trade and vocational training, and on-the-job learning.
- *Providing supportive government policies to help young entrepreneurs.* Areas include advisory services for establishing businesses and initiatives and better instruments and channels for financing. Recently, crowdsourcing has emerged as a means of generating funds for small initiatives.
- *Making tertiary learning more widely available through the Internet.* Massive open online courses are linking world-renowned academic institutions and students around the world.
- *Using cash transfer programmes to provide employment for local young people and poor people.* In India and Uganda these programmes have provided resources for funding job searches and for supporting high-quality training and skills development. They have also increased access to other sources of credit for entrepreneurship.

Creative work requires an enabling work environment, including financial support, and opportunities to collaborate and cross-fertilize ideas. Some key requirements for creativity and innovation to thrive are:

- *Innovating inclusively.* Here, new goods and services are developed for or by those living on the lowest incomes or by women, extending creative opportunities to groups that may be underrepresented.
- *Assuring democratic creativity.* Workplaces and online platforms can be organized in ways that encourage innovation at all levels.

Exciting work opportunities for young people should be created

- *Funding experimentation and risk.* This entails solving intractable social and environmental problems that may require foundations and public institutions to take funding risks on less proven approaches.
- *Innovating for the public good.* Creativity and innovation can advance many objectives. Policies that direct innovation towards the greater social good, including volunteer work, can enhance human development.

Voluntary work can be encouraged by tax rebates, subsidies and public grants to voluntary organizations. Public support to create and protect space for voluntary work can bring social benefits, particularly during emergencies like conflicts and natural disasters.

In conflict and post-conflict situations it is important to focus on productive jobs that empower people, build agency, provide access to voice, offer social status and increase respect, cohesion, trust and people's willingness to participate in civil society. Some policy options are:

- *Supporting work in the health system.* In many conflict-afflicted countries the health system has collapsed, and support for emergency health services is critical for workers and the wounded.
- *Getting basic social services up and running.* This has social and political benefits. Communities, nongovernmental organizations and public–private partnerships can be the drivers.
- *Initiating public works programmes.* Even emergency temporary jobs, cash for work and the like can provide much needed livelihoods and contribute to the building of critical physical and social infrastructures.
- *Formulating and implementing targeted community-based programmes.* Such programmes can yield multiple benefits, including stability. Economic activities can be jumpstarted by reconnecting people, reconstructing networks and helping restore the social fabric.

Beyond the policy options, a broader agenda for action is needed

- *Developing a New Social Contract.* In the new world of work participants are less likely to

have long-term ties to a single employer or to be a member of a trade union than their forebears. This world of work does not fit the traditional arrangements for protection. How does society mobilize funds to cover a widening population that is not always in work, reach those working outside the formal sector, accommodate new labour market entrants (especially migrants) and cover those unable to work? There may be a need for a New Social Contract in such circumstances involving dialogue on a much larger scale than took place during the 20th century. Denmark is making strides providing security alongside reskilling and skills upgrading in an increasingly flexible job market (box 4).

- *Pursuing a Global Deal.* In an era of global production, national policies and social contracts may not work outside of global

commitments. Further, true globalization rests on the idea of sharing—we should share the responsibility for a "global working life."

A Global Deal would require mobilizing all partners—workers, businesses and governments—around the world, respecting workers' rights in practice and being prepared to negotiate agreements at all levels. This will not require new institutions, merely reoriented attention in the world's strong international forums.

A Global Deal can guide governments in implementing policies to meet the needs of their citizens. Without global agreements, national policies may respond to labour demands at home without accounting for externalities. This implies that a global–national compact is also necessary. International conventions such as the International Labour Organization Convention Concerning Decent Work for Domestic Workers, which entered into force in September 2013, was a groundbreaking agreement that stands to establish global standards for the rights of domestic workers worldwide. This kind of agreement offers guiding principles to signatories but leaves space for national governments to implement policies within national contexts to meet commitments. Motivated by global actions, national policies create real change in local communities.

- *Implementing the Decent Work Agenda.* The Decent Work Agenda has four pillars (box 5). The agenda and the human development framework are mutually reinforcing. Decent work enhances human development

Implementing the Decent Work Agenda will help work enhance human development

BOX 4

Flexicurity in Denmark

The Danish labour market has a lot of what is often called "flexicurity": coexistence of flexibility, in the form of low adjustment costs for employers and employees, and security, a by-product of Denmark's well developed social safety net, ensuring high coverage and replacement rates.

The principal aim of flexicurity is to promote employment security over job security, meaning protection focuses on workers rather than their jobs. Consequently, employers benefit from all the advantages of a flexible labour force while employees can take comfort in a robust social safety net applied with active labour market policies.

Source: World Bank 2015b.

BOX 5

The four pillars of the Decent Work Agenda

- *Employment creation and enterprise development.* This requires acknowledging that a principal route out of poverty is jobs and that the economy needs to generate opportunities for investment, entrepreneurship, job creation and sustainable livelihoods.
- *Standards and rights at work.* People need representation opportunities to participate, to voice their views in order to obtain rights and to earn respect. The International Labour Organization's normative work is key for compliance and measuring progress.

- *Social protection.* Basic social protection, such as health care and retirement security, is a foundation for participating productively in society and the economy.
- *Governance and social dialogue.* Social dialogue among governments, workers and employers can resolve important economic and social issues, encourage good governance, establish sound labour relations and boost economic and social progress.

Source: ILO 2008b.

through each of its pillars. Employment creation and enterprise development provide income and livelihoods to people, are crucial instruments for equity, are a means for participation and facilitate self-esteem and dignity. Workers' rights support human development by ensuring human rights, human freedom and labour standards. Social protection contributes to human development by ensuring safety nets, protecting people from risks and vulnerabilities and providing care work. And social dialogue helps human development through broad-based participation, empowerment and social cohesion.

Conversely, human development contributes to the four pillars. Expanding capabilities through human development enhances opportunities for employment and entrepreneurship. The participation aspect of human development helps enrich social dialogue. Human development also emphasizes the promotion of human rights, which boosts workers' rights and enhances human security. Given all these interlinks, implementing the Decent Work Agenda will help work enhance human development.

The world has changed dramatically, but the concept of human development remains as relevant as ever—if not more so

The world today is very different from the world in 1990, when the notion of human development and its measures to assess human well-being were launched. Since then, the development canvas has changed, global growth centres have shifted, important demographic transitions have materialized and a fresh wave of development challenges has emerged.

The global economy is changing. The influence of emerging economies is rising. Developed economies' share of global GDP (based on purchasing power parity dollars) fell from 54 percent in 2004 to 43 percent in 2014. Politically, the desire for freedom and voice has swept different parts of the world. The digital revolution has changed the ways we think and operate. Inequalities have gone up. Human security has become more fragile. And climate change is affecting more human lives.

So, is the notion of human development still relevant for development discourse and as a measure of human well-being? Yes—even more so in today's world.

Even with all the economic and technological advancements at the world's disposal, people do not have equitable benefits from progress, human capabilities and opportunities do not always flourish, human security is at stake, human rights and freedoms are not always protected, gender inequalities remain a challenge, and future generations' choices do not get the attention they deserve. So the notion of human development—enlarging choices, emphasizing a long, healthy and creative life and highlighting the need for expanding capabilities and creating opportunities—assumes a new importance as a development framework, with people at the centre of development.

Similarly, as a measure of human well-being, the human development framework still provides perhaps the broadest perspective of human progress, while contributing to policymaking.

Yet after a quarter of a century, the time has come to revisit both aspects—the notion and the measures.

The notion and measures of human development should be reviewed to make them more relevant for today's challenges and tomorrow's world

The conceptual angle of human development requires a fresh look for dealing with emerging challenges in a fast-changing world, especially in dialogue with the new 2030 Agenda for Sustainable Development and the Sustainable Development Goals.

The basic focus should be kept intact. But issues such as individual and collective choices, their probable trade-offs in conflicting situations, hierarchies among such choices and the balance between choices of present and future generations need to be looked after. Similarly, issues of human development in relation to shocks and vulnerabilities and the relationships among human development, human rights and human security will have to be revisited.

The 2030 Agenda for Sustainable Development and the Sustainable Development Goals require fresh assessment tools for monitoring progress—measuring environmental sustainability and integrating it with overall measures for human well-being are priorities.

Three other challenges stand out. First, measures and indicators have to be identified that allow policy impacts to be more quickly captured. Second, measures are often inadequate for assessing human well-being at times of shocks and crises and should thus be revamped to fill this need. Third, "quick guidance" policy measures should be explored.

All these efforts require robust, consistent and credible data. Taking that into account and also considering a much more ambitious international agenda, the High Level Panel on the Post-2015 Agenda convened by the UN Secretary-General in 2014 called for a data revolution. It emphasized the need to monitor progress. Three issues need highlighting:

- First, huge amounts of real-time data can provide better information on, say, the attendance of students at school. Sensors, satellites and other tools produce real-time data on people's activities. These can be harnessed to inform policymaking.

- Second, big data holds the promise of producing statistics almost instantaneously and allowing disaggregation to levels of detail hitherto undreamt of outside population censuses. Such data are expanding the understanding of causation in an increasingly complex world and enabling rapid responses in some humanitarian situations. But the data have risks—they could do harm where privacy and anonymity are not respected. Still, many researchers are identifying how this large volume of information—generated both incidentally and deliberately as billions of people go about their daily lives—can support sustainability and provide usable insights for improving lives.

- Third, it is possible to combine traditional and new methods of data collection for censuses, ranging from administrative registers to mobile devices, geospatial information systems and the Internet. Many countries have already done this.

In this changed and changing world, with a new development agenda and new development goals, the need for revisiting the notion and measures of human development is vital. Next year's Human Development Report, the 25th in the series, will be devoted to it.

In this changed and changing world the need for revising the notion and measures of human development is vital

Chapter **1**

Work and human development—analytical links

Infographic: Links between work and human development

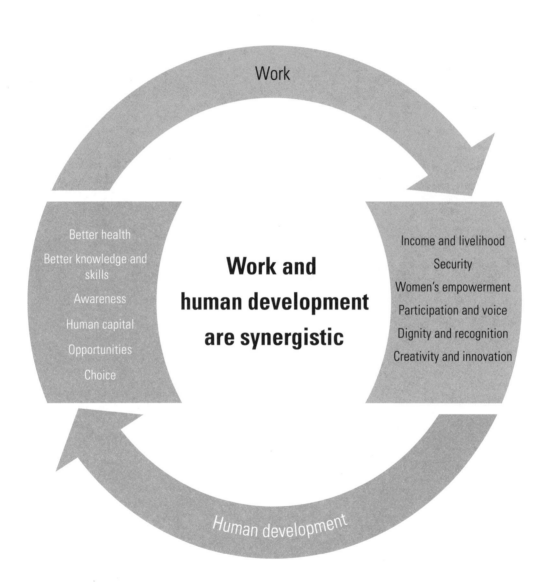

Work

**Work and
human development
are synergistic**

Better health
Better knowledge and
skills
Awareness
Human capital
Opportunities
Choice

Income and livelihood
Security
Women's empowerment
Participation and voice
Dignity and recognition
Creativity and innovation

Human development

1.

Work and human development— analytical links

The true aim of development is not only to boost incomes, but also to maximize human choices—by enhancing human rights, freedoms, capabilities and opportunities and by enabling people to lead long, healthy and creative lives. Critical to this process is work, central to human existence. Human beings prepare for work as children, engage in work as adults and expect to retire from work in later life. Through the human lifecycle, quality of life is thus closely bound to the quality of work.

From an economic perspective, work enables people to earn a livelihood and achieve economic security. But from a human development perspective, it also allows people to enhance their capabilities by providing them with acquired skills and knowledge. Income from work helps workers achieve a better standard of living and allows them to have better access to health and education—critical ingredients to enhancing capabilities. Work also provides people with expanded opportunities and choices in their economic and social lives. It allows workers to participate fully in society while affording them a sense of dignity and worth. And work that involves caring for others builds social cohesion and strengthens bonds within families and communities. To contribute to human development, work needs to be productive, worthwhile and meaningful —and to unleash human potential, creativity and spirit.

Work also strengthens societies. By working together, human beings not only increase material well-being, they also accumulate a wide body of knowledge that serves as the basis for cultures and civilizations. In addition, work should make societies fairer by providing opportunities for poor people to strive for a better living. When work is environmentally friendly, the benefits extend through generations. Work thus enhances human development. Needless to say, by improving human capabilities, opportunities and choices, human development also contributes to work. In short, work and human development are synergistic and mutually reinforcing (see info-graphic at the beginning of the chapter).

However, the link between work and human development is not automatic. Work can be tiring, boring, repetitive and dangerous. And just as the right kind of work enhances human development, so the wrong kind can be deeply destructive. Around the world millions of people, many of them still children, are forced into exploitative labour. Some are trapped in bonded labour that robs them of their rights and dignity. Some work in hazardous conditions deprived of labour rights and social protection, spending their days in drudgery that stifles their potential (see table A1.1 at the end of the chapter). And while work should create fairer societies, it can also be divisive if vast differences in opportunities and rewards perpetuate divisions and inequalities.

Work is broader than jobs

The notion of work is broader and deeper than that of jobs or employment (box 1.1). Jobs provide income and support human dignity, participation and economic security. But the jobs framework is restrictive. It fails to capture many kinds of work that are more flexible and open-ended, including care work, voluntary work and creative expression, such as writing or painting—all of which are important for human development. Embracing these other dimensions requires a broader notion of work (figure 1.1).

Seen in this way, work is very diverse. It can be paid or unpaid, formal or informal, and carried out within households or outside (see table A1.2 at the end of the chapter). And it can take place in very different circumstances, pleasant or unpleasant, offering a wide range of rights and opportunities, all reflecting different contexts and levels of development.

The notion of work is broader and deeper than that of jobs or employment

BOX 1.1

What is work?

For this report, work is any activity that not only leads to the production and consumption of goods or services, but also goes beyond production for economic value. Work thus includes activities that may result in broader human well-being, both for the present and for the future.

Work involves four sets of people: workers themselves; other entities such as employers who provide complementary inputs; consumers of the goods or services produced; and the rest of the world, which encompasses other people, society and the natural environment as well as future generations and the workers' future selves. Work has both monetary and nonmonetary returns, tangible and intangible, with expressed and unintended consequences.

Consider someone who is cooking. If cooking for himself, the cost as a producer is the opportunity cost of his time, which could have been spent doing something else; the return to him as a consumer is the nutrition from the meal. No other actors are involved, though there could be an impact on the environment. Now consider an individual cooking for his family. In this case the list of consumers grows. The tangible returns now include family nutrition, combined with the intangible

feeling of satisfaction along with their approbation—or feeling of frustration at being expected to perform the task or to forgo other activities such as earning money, pursuing education or participating in public life. Work, from a human development perspective, is about the degree of freedom individuals have in making choices about the work they do.

However, if a domestic helper does the cooking, the situation changes. In this case there is a financial return to the worker, with the family not only acting as consumers, but also providing the complementary inputs. This activity may be differently remunerated depending on whether it is performed by a man or woman, a citizen or an immigrant. An individual who is coerced into performing the activity or who receives lower returns than those available to another is being exploited.

This approach to work can also be applied to creative workers and volunteers. For example, in a restaurant a chef may pursue creativity in addition to income and experience professional satisfaction, self-esteem and dignity. Similarly, a volunteer at a community kitchen is not working for money but may be working for altruistic satisfaction.

Source: Human Development Report Office.

The quality of work is as important as the quantity.

Care work (discussed in chapter 4) is fundamentally important to human development. Care work includes housework, such as preparing meals for the family, cleaning the house and gathering water and fuel, as well as work caring for groups who cannot care for themselves—children, older people and family members who are sick—over both the short and long term.

Voluntary work has been defined as unpaid noncompulsory work—that is, the discretionary free time individuals give without pay to activities performed either through an organization or directly for others outside their own household. Voluntary work is by definition an expression of free choice, intrinsically enhancing human development from the individual perspective.

Creative expression contributes new ideas to society in works of art and cultural products, functional creations, scientific interventions and technological innovations. Creative work is desirable to many individuals because it

affords them the opportunity to be innovative and self-expressive while striving to earn a living. Although creative work can be improvisational and derivative, it involves originality and uniqueness.

Work in different stages of the lifecycle

Periods in a lifecycle are demographically, physically and culturally dependent and change over time. There are overlaps between the different stages, and the stages may vary among individuals. Work needs to be seen in the context of a human lifecycle and how it changes in different stages of that cycle. In many cases it is not a question of choice; decisions triggered by cultural expectations or absence of appropriate support force people to be in different work situations. Thus early marriage for girls or cuts in state-provided child care may result in pulling girls out of school—with lifelong effects on

Creative expression contributes new ideas

FIGURE 1.1

Work encompasses more than just jobs

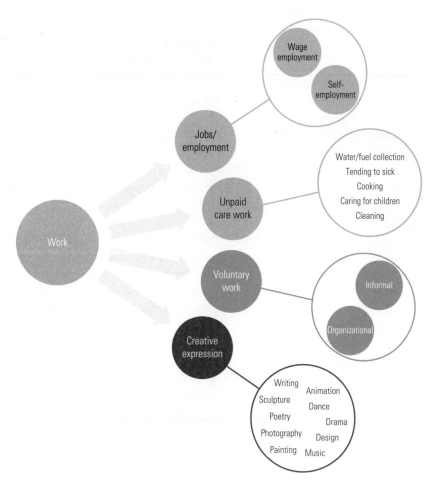

> Work needs to be
> seen in the context
> of a human cycle

Source: Human Development Report Office.

their ability to enter the labour market, earn a decent income and become economically empowered. Work is a source of, as well as a response to, risk and vulnerability, which people experience differently over the course of their lives.

At the prime working age, a worker's priority may be maximizing the economic returns of work over other things—not only to provide for present needs, but also to save for future requirements. But under other conditions people (including older workers and many young workers) may wish to move to areas of work where the economic returns are not very high but where they feel happier and content in their work. At the same time, young people may be constrained in their work options. With youth unemployment rising, options

beyond low-wage, unsecure livelihoods may be scarce for many. There are similar risks and vulnerabilities for older workers, particularly in developing countries, where paid work remains a necessity in old age but work choices may be constrained.

People are living longer and healthier lives in most countries, which expands the productive potential of older workers. But there are trade-offs for individuals who may wish to have more control of their time, for discretionary leisure or for different types of paid and unpaid work activities. The right age to make the transition is probably not the same for every worker, which complicates policy decisions about increasing the retirement age.

Strong forces on the work choices of older people come from national pension policies and programmes. From a global perspective, workers from countries with low pension coverage tend to stay in the labour market after age 65, while workers from countries with high pension coverage tend to retire earlier (figure 1.2). This has clear impacts on how people make work decisions and has deep human development implications. People with low pension coverage or low pensions—a typical condition in most developing countries—are forced to remain in paid employment longer. In contrast, people with access to pensions with substantial benefits—typical in developed countries—tend to retire earlier or move towards work that may not maximize economic returns but may offer other rewards.

In Sri Lanka most workers in the formal sector retire in their 60s, and a relatively small fraction are employed part- or full-time. But casual workers and self-employed workers tend to keep their full-time jobs for many more years. Formality is thus a strong determinant of the length of the working lifetime.[1]

In a world where life expectancy is increasing and technology enables an active place in society for far longer, many older people are seeking active engagement in work—sometimes paid, sometime voluntary. Many countries are responding with initiatives to keep older people engaged in work without blocking the opportunities of others—particularly younger people (see chapters 3 and 6).

Work enhances human development

Human development is a process of enlarging people's choices, so it is closely bound with work. The positive relationship between work and human development goes both ways (figure 1.3). Human development is affected by work through many channels, all of which can be mutually reinforcing:

- *Income and livelihood.* People work primarily to achieve a decent standard of living. In market-based economies they generally do so through wage or self-employment. In more traditional and subsistence economies they sustain their livelihoods through specific cycles of activities. Work can also be a major factor in ensuring that economic growth is equitable and poverty-reducing.
- *Security.* Through work people can build a secure basis for their lives, enabling them to make long-term decisions and establish priorities and choices. They can also sustain stable households, particularly if they use

FIGURE 1.2

Workers from countries with low pension coverage tend to stay in the labour market after age 65, while workers from countries with high pension coverage tend to retire earlier

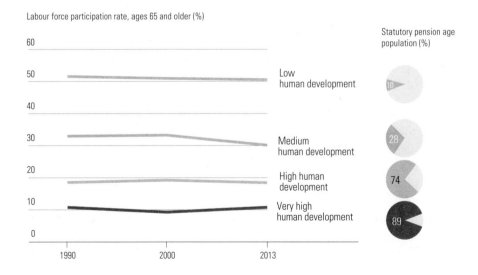

Source: Human Development Report Office calculations based on ILO (2015d, 2015e).

FIGURE 1.3

Work and human development are synergistic

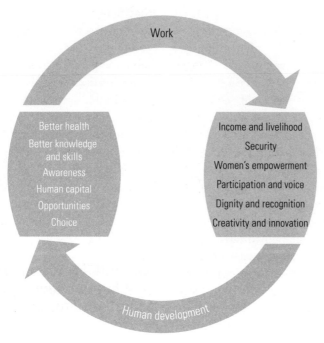

Work

Income and livelihood
Security
Women's empowerment
Participation and voice
Dignity and recognition
Creativity and innovation

Better health
Better knowledge
and skills
Awareness
Human capital
Opportunities
Choice

Human development

Source: Human Development Report Office.

their income prudently on food and nutrition for their family, on education and health for their children, or for savings.

- *Women's empowerment.* Women who earn income from work often achieve greater economic autonomy and decisionmaking power within families, workplaces and communities. They also gain confidence, security and flexibility.
- *Participation and voice.* By interacting with others through work, people learn to participate in collective decisionmaking and gain a voice. Workers also encounter new ideas and information and interact with people of different backgrounds and can engage more extensively in civic issues.
- *Dignity and recognition.* Good work is recognized by co-workers, peers and others and provides a sense of accomplishment, self-respect and social identity. People have historically defined and named themselves by their occupation: Miller in English or Hurudza (master farmer) in Shona.
- *Creativity and innovation.* Work unleashes human creativity and has generated countless innovations that have revolutionized many aspects of human life—as in health,

education, communications and environmental sustainability.

Workers can also benefit from higher human development, again operating through multiple channels:

- *Health.* Healthier workers have longer and more productive working lives and can explore more options at home and abroad.
- *Knowledge and skills.* Better educated and trained workers can do more diverse work —and to a higher standard—and be more creative and innovative.
- *Awareness.* Workers who can participate more fully in their communities will be able to negotiate at work for better conditions and higher labour standards, which in turn will make industries more efficient and competitive.

In the work–human development nexus volunteerism, as well as creativity and innovation, warrants particular attention.

Volunteerism, by its nature, reflects agency and the capability to choose. Volunteers benefit from their work, either because they value altruism or through the personal enrichment they gain from community involvement. Volunteer work also tends to have substantial

Volunteerism, by its
nature, reflects agency

social value. Participation in volunteer activities can enable people to contribute to their communities and the public good in ways that markets or public institutions may not.

Volunteers can be great innovators, forging the way towards new ways of working and organizing workers, paid and unpaid. Wangari Maathai, who won the Nobel Peace Prize in 2004, mobilized grassroots movements to promote sustainable development, democracy, women's rights and peace. Her legacy has been influential in preparing the Post-2015 Development Agenda.

Volunteer organizations can create bridges between political, geographical and cultural realities and can coordinate international efforts and solidarity to pursue humanitarian causes. Among others, the International Committee of the Red Cross, devoted to protecting human life and health, has been awarded the Nobel Peace Prize three times: in 1917, 1944 and 1963. Similarly, Médecins Sans Frontières, which won the prize in 1999, mobilizes doctors and nurses to address health emergencies around the globe.[2] UN Volunteers encourages the integration of volunteerism into development and peace processes in developing countries, partnering with governments and other local and international bodies. In all these ways volunteerism promotes cross-cutting approaches for human development.

Creative work is equally important for expanding people's capabilities and enhancing their opportunities. There are instrumental contributions, which result in direct economic gain, and noninstrumental contributions, which enhance knowledge and expand social cohesion. These contributions are not static; they interact with each other to enlarge human choices and empower people.[3]

Creative work is not only self-satisfying to the worker but can also extend happiness, pleasure, satisfaction and well-being to others, making it an important public good. Artwork from ancient civilizations continues to function as a foundation for new knowledge. Works of past musicians continue to inspire new music. Creative work can build bridges of social welfare and cohesion. And this welfare can also extend across international borders, linked by creativity in cultural tourism, for example.

Given the centrality of creativity and innovation in the world of work, attention is shifting to types of workplaces and working conditions that may be more conducive to innovation than others. Recognizing that agency is a key factor in worker engagement and creativity, some companies have set aside time for workers to be creative. Since as early as 1948, 3M has had a 15 percent rule that allows employees to dedicate almost a full day each week to their own projects. One outcome of this approach was the invention of the Post-it.[4] More recently, Google, Facebook, LinkedIn and Apple implemented different versions of a 20 percent rule for engineers to explore areas of their own interest to spark creative thinking.[5]

The link between work and human development is not automatic

The foregoing links between work and human development are not automatic. They are subject to various conditions, ranging from the quality of work for individuals to the societal value of work. The link can vary depending on the scale of opportunities for workers' voice and participation and whether discrimination or even violence is common at work.

Quality of work

Whether people have jobs is important, but so are the nature and conditions of their work. For example: Is work safe? Are people fulfilled and satisfied by their work? Does their work provide a secure livelihood? Are there opportunities for training, social dialogue and progression? Does employment support a flexible work–life balance? These are standard questions for what makes a job a good job. From a human development perspective, the quality of work also depends on whether a job provides dignity and a sense of pride and whether it facilitates participation and interaction. Does it present a platform for voice, and does it respect workers' rights and human rights more widely?

Of course, what constitutes the quality of work can vary subjectively by country, personal circumstances and frame of reference—a reality that complicates quality of work measurement at the global level. Even if a universal definition of work quality were agreed on, comparable cross-national datasets would probably be limited. Despite these challenges, several initiatives are in place to put forward regionally—and in some cases, globally—comparable measures of job quality, including composite indicators that capture multiple dimensions of work quality (box 1.2).

While work is generally beneficial for people, the quality of work can be affected by doing too much work. A culture of overwork is increasingly common, facilitated by all kinds of mobile devices that enable constant access to work. The pressure of a round-the-clock work culture is particularly acute in highly skilled, highly paid professional service jobs such as law, finance, consulting and accounting.[6] The overwork culture can lock gender inequality in place, because work–family balance is made more difficult for women, who bear a disproportionate share of care work.

Among Organisation for Economic Co-operation and Development countries, the highest prevalence of overwork—defined as a work week of 50 hours or more—is in Turkey (close to 41 percent of the formal workforce), followed by Mexico (nearly 29 percent).[7] Pressure for overwork is also high in some Asian countries. In Japan the term karoshi means death from overwork.

Working very long hours can cause death due to stroke, heart attack, cerebral haemorrhage or other sudden causes.[8] Even from an economic perspective, overwork is generally counterproductive, as it undermines labour productivity. Shorter and flexible working hours may be beneficial from both a human development perspective and an economic perspective.

Workers' satisfaction and happiness

Workers' satisfaction and happiness are critical elements in ensuring a stronger link between work and human development. But the work–satisfaction–happiness nexus is not simple and straightforward.

Yes, being out of work reduces happiness. For example, in the United Kingdom the relationship between unemployment and lower subjective well-being is well established across a range of measures, including lower life satisfaction,

> Workers' satisfaction and happiness are critical elements in ensuring a stronger link between work and human development

BOX 1.2

Measures of quality of work

At the regional level the European Council proposed a portfolio of 18 statistical indicators in 2011 to measure job quality, and the European Trade Union Institute developed the Quality of Jobs Index in 2008.[1] Eurofound has established the European Working Conditions Survey.[2] A dashboard of indicators has been proposed to measure progress towards the International Labour Organization decent work agenda.[3] Each of these endeavours has been challenged by intense debates among workers, employers and governments on conceptualizing employment quality and agreeing on objectives. There are questions of different priorities for workers and employers, as well as whether to focus on individual workers, the regulatory environment or the nature of jobs.

Existing measures draw attention to the need for policies to enhance the quality of employment, particularly when trends in employment quality are depicted, as in the case of the Eurofound reports. Noting that simple indicators that summarize only a few variables are typically most successful in conveying information to policymakers, some entities (such as the United Nations Economic Commission for Europe) have proposed a global Quality of Employment Indicator, based on dimensions including income, degree of formality, participation in social security systems, duration of contracts and availability of training.[4] While composite measures like this inevitably leave out some dimensions of work quality, they can convey the urgency of employment-quality issues to policymakers and be used to advocate for more systematic data collection on elements of work quality. At the same time, additional steps can be taken—as has been done by the Canadian Imperial Bank of Commerce through the employment quality index it has constructed—to identify, assess and diagnose the causes of poor job quality, which can vary by national, local and even workplace context.[5]

Notes
1. ETUI 2015. 2. Eurofound 2013. 3. ILO 2012a. 4. UNECE Expert Group on Measuring Quality of Employment 2012. 5. Tal 2015.
Source: Human Development Report Office.

lower happiness and increased anxiety.[9] In the United States the nonincome impacts of unemployment have been estimated to be five times greater than the income impact.[10] The unhappiness of becoming unemployed is not something that people adjust to quickly, if at all. In Germany life satisfaction declines sharply on losing a job, and there is little evidence of any improvement among men even after three or more years of unemployment.[11] So unemployment is miserable, and work is satisfying. But is it merely having a job or some other aspect of work that matters in terms of human happiness (box 1.3)?

Voice and participation

Unions, political parties, women's groups and other collective entities have long provided workers with a means for voice and participation—critical for value and quality of work. Through these platforms, workers have formed shared values and collective interests and pursued them. This has resulted in real gains, even in the face of powerful opposition, as for example, in the 1980s in Brazil when workers gained better rights and wages and in South Africa when movements helped legalize Black trade unions.[12] Through such processes, workers have substantially strengthened the link between work and human development, including protections against health hazards

at work, higher compensation and social insurance.

Over recent decades the potential for strong worker negotiating power has steadily atrophied—because of globalization, which pits workers in different countries against each other; because of technological innovations, which have atomized the workforce; and because of new working patterns, which include more short-term work that blurs the lines between formal and informal work. It is now harder for workers to unite for mass action, to some extent a desired management outcome. Some employers have deliberately shaped innovation so as to weaken labour power, using new technologies to monitor workers more closely and reduce their scope for controlling their work environment.

Today, traditional forms of worker organizations like trade unions are weak and represent fewer members than in the past. Looking at union members as a percentage of employees, variations across countries are considerable. While such membership is quite strong in Argentina, Kenya and South Africa (more than 30 percent), it is very low in Niger and Uganda (less than 5 percent).[13] But even this measure may overestimate the real influence of trade unions in the workforce, as a high proportion of workers in many countries are self-employed or not in the formal economy and may not be captured in the measures of union participation.

> Today, traditional forms of worker organizations like trade unions are weak

BOX 1.3

What brings happiness—having a job or something beyond it?

What do we know about the ways in which work—beyond having a job—promotes happiness? Some types of work appear to be associated with happiness more than others, and the income gained through work is not the only thing that matters. One study of the relationship between different occupations and life satisfaction found a wide variation in satisfaction that did not directly correlate with income, with members of the clergy the most satisfied, followed closely by chief executives and senior officials and agricultural managers. The least satisfied were those working as bar managers, in low-skill construction jobs or as debt collectors.[1]

Much remains to be learned about the causes of happiness at work, although some factors are increasingly well accepted. For example, the relationship with one's manager is important. Indeed, trust in the workplace generally seems to be an important factor. In Canada a one-third standard-deviation increase in trust in management did as much for happiness as an income increase of 31 percent. Job fit—having an opportunity at work to do what one does best—is also important for subjective well-being. Better job fit has been associated with higher life evaluations and better daily experiences in all seven regions of the world.[2]

Notes
1. Easton 2014. 2. Helliwell and Huang 2011b.
Source: Human Development Report Office.

The weakness or underdevelopment of worker organizations can be detrimental to human development, impeding the functioning of labour and social institutions as a whole.[14]

There have been international efforts to support workers in addressing some of these challenges. For example, the International Trade Union Confederation and the International Domestic Workers Federation are working on uniting global workers.

Sustainable work

To strengthen the link with human development, work has to be sustainable. Sustainable work is defined as work that promotes human development while reducing or eliminating negative externalities that can be experienced over different geographic and time scales. It is not only critical for sustaining the planet, but also for ensuring work for future generations.

Sustainable work is not just about paid work (as argued in chapter 5); it also encompasses the often impactful efforts of caregivers, volunteers, artists, advocates and others, which have positive impacts on human development. Furthermore, sustainable work concentrates on activities that can achieve the dual mutual goal of high sustainability and high human development.

To forge a stronger link with human development, sustainable work has to follow three pathways, as detailed in chapter 5: the termination and transformation of some existing forms of work, and the creation of new forms of work.

Societal value of work

Work strengthens the work–human development link when it goes beyond individual benefits and contributes to shared social objectives. Work that boosts income and lowers poverty reduces inequality and can relieve social tension, while money earned from work can be taxed to generate resources for human development. Work that involves caring for older people or people with disabilities helps maintain their capabilities and strengthens social cohesion, while child care builds capacities for future generations. Workers also establish social and economic ties and build trust in others and in institutions, which also builds social cohesion.

Creative work is valuable to a society by contributing to its tradition, culture and heritage. Similarly, volunteering provides a means to build social relationships and networks outside paid employment and can contribute to the social good. One survey in Australia found that for 83 percent of volunteers, their contributions had increased their sense of belonging to their communities.[15]

Work thus has a value for society as a whole. Often individual and societal values of work converge, as environment-friendly work benefits not only workers, but also other people, including future generations. But sometimes they diverge—for example, the work of poachers generates income for individuals but is detrimental to society.

Discrimination and violence at work

Sometimes, the value of work is diminished and its links with human development become weaker because of discrimination and violence at work. The most observable discrimination is along gender lines—in positions, pay and treatment (chapter 4). But discrimination also occurs on grounds of ethnicity, race, disability and sexual orientation.

In the United States racial discrimination is the most frequently reported form of discrimination at work, accounting for 35 percent of charges presented to the United States Equal Employment Opportunity Commission.[16] In the United Kingdom 22 percent of poll respondents had witnessed racism in the workplace, and 34 percent of Black people and 29 percent of Asian people reported first-hand experience of racial or religious discrimination at work.[17] In Hungary 64 percent of migrant respondents to a survey reported discrimination when looking for a job, and wages paid to Roma are lower than the Hungarian minimum wage.[18]

The consequences of discrimination in developed countries (where employment is predominantly formal) are lower labour force participation, higher unemployment and a considerable wage gap between marginalized groups and the rest of the population. Two layers of discrimination influence these gaps in outcome: uneven access to a good education and prejudice when accessing jobs and promotions in the workplace.

Work has a value for society as a whole

In developing countries discrimination is often against indigenous ethnic groups.[19] Labour outcomes of disadvantaged ethnic groups reflect different forms of inequality, restricted opportunities and discrimination throughout the lifecycle. Unlike in developed countries, labour participation tends to be higher for these disadvantaged groups because of a high reliance on self-employment and informality, which in turn results in high vulnerability and limited social protection. For example, in Latin America the wage gap between indigenous ethnic groups and the rest of the population is estimated at 38 percent, with more than 10 percent typically remaining unexplained.[20]

Work discrimination against people with disabilities is also common. More than 1 billion people—one person in seven—have some form of disability.[21] Most people with disabilities cannot make full use of their capabilities. Discrimination starts as early as in school. In developing countries students with special needs rarely have access to rehabilitation education or training, and later they are less likely to find work. In both low- and high-income countries, their employment rates are typically lower than those of workers without disabilities (table 1.1). Access to labour markets is even more difficult for women with disabilities. Thus in low-income countries the employment rate for women with disabilities is about one-third that of men with disabilities, and in high-income countries, it is a little more than a half.

Also facing discrimination in the workplace are lesbian, gay, bisexual, transgendered, queer and intersex people. Only a few countries have laws to protect them, and those laws are often poorly enforced. Research in Argentina, Hungary, South Africa and Thailand finds that many face discrimination and harassment in the workplace and that the discrimination often begins in school. In Thailand discrimination occurs at all stages in the employment process, including education and training, access to jobs, advancement in opportunities, social security and partner benefits.[22] Participation of transgendered workers in the workforce is often limited to a few professions such as beauticians and entertainers.

Workplace or occupational violence—in the form of threats and physical or verbal abuse—is also a concern for many workers. The numbers are startling. In 2009, 572,000 occurrences of nonfatal workplace violence—including various forms of assault—were reported in the United States, and 521 individuals were victims of homicide in the workplace.[23] Some 80 percent of survey respondents in South Africa reported that they had suffered from hostility at work (bullying in all its forms) at some point.[24] In 2009 some 30 million EU workers experienced work-related violence (such as harassment, intimidation, bullying or physical violence)—10 million in the workplace and 20 million outside the workplace. The violence takes a direct toll on physical and emotional health, leading to greater absenteeism.[25]

Work in conflict and post-conflict situations

Some 1.5 billion people, or one in four people on the planet, live in fragile or conflict-afflicted countries.[26] Conflict situations range from high criminal violence to civil wars to other forms of internal conflict. Work in these situations and the work–human development link take on different dimensions in such exceptional states of emergency. With increasing degrees of violence, human priorities change in the sense that survival and physical integrity of the self and family members become higher priorities than economic needs.

TABLE 1.1

Employment rate of people with disabilities, low- and high-income countries, most recent year available (%)

	Low-income countries		High-income countries	
	Not disabled	Disabled	Not disabled	Disabled
Male	71	59	54	36
Female	32	20	28	20

Source: WHO and the World Bank 2011.

The links between conflict and work are mutually reinforcing. Work can help with peace building, and unemployment, when overlapping with other social discontent, can be destabilizing. When people are unemployed, they may be more susceptible to participation in violent activities. In extreme cases and in unstable environments this may boost recruitment by insurgents, thus increasing the risk of civil war. Unemployment and the discontent and grievances it stirs—as well as reduced opportunity cost of violence—may be considerable.

Violent conflicts have many complex effects on work. Conflicts not only lead to suffering and death; they also destroy livelihoods. As a result of violence and looting, people lose property, land and businesses. They also find it difficult to work when they are displaced, especially to neighbouring regions or countries where natural and other resources are already scarce. This not only leads to immediate poverty, but also reduces the prospects for recovery—particularly if the conflict is exacerbated, for example, by the proliferation of small arms, cattle raiding and militia activities. When violent conflict severs major infrastructural arteries, destroys power plants and fuel facilities and starves industries of foreign exchange for imported inputs, work in manufacturing, manufacturing-related services, tourism and some agriculture can decline.

Violent conflict can and often does unleash mechanisms of "primitive accumulation": the use of extra-economic or noneconomic coercion to wrest assets from their owners or occupiers (for example, through forced displacement and wartime land accumulation) and to force people to join wage labour markets for their survival.

The redistribution of work opportunities and of workers during conflict is not restricted to national boundaries. Regional conflict complexes, as they have been called, often generate complicated labour markets and frequently abusive labour relations and working conditions across borders, helping create displacement economies. Many refugee women during the Mozambican war ended up working as illegal—and cheap—workers on globally competitive agro-business farms in South Africa during the late 1980s and early 1990s. Ugandan rural labour markets contain many people with Congolese or Rwandan conflict-related origins, and they may experience discrimination in hiring practices and treatment.

The effect of conflicts on work varies across groups. For example, young people may have fewer opportunities for productive work, leaving them more vulnerable to recruitment by gangs, rebels or terrorist groups. Even after conflicts end, young people are the ex-combatants most likely to be out of work.

Conflicts can also add to the workload for women. If men are absent due to fighting, women face even greater paid and unpaid work responsibilities at home. But women may also be actively involved in conflicts. In Sudan women and girls were on the front lines of the two north–south civil wars as both combatants and peace activists.[27]

Women are crucial in the work of peace making and conflict prevention, including peace talks, conflict mediation and all aspects of post-conflict resolution. This is widely recognized in the UN Security Council Resolution 1325 on Women, Peace and Security, adopted in October 2000. The resolution also acknowledges the dramatic impact violent conflicts have on women and urges governments to increase the participation of women in peace and security efforts.[28] Actions to implement the resolution have already been planned and implemented by some 25 countries, from Liberia to Norway, from Nepal to the Philippines.[29]

Even so, few women have been present in peace negotiations, amid 39 active conflicts over the past 10 years, and only a small proportion of peace treaties contains direct references to women (16 percent of 585 peace treaties).[30] And a study of 31 major peace processes between 1992 and 2011 revealed that only 9 percent of the peace negotiators were women.[31] Women's participation in the peace process is not just a matter of morality or equality, but one of efficiency. If women are not part of the peace process globally, half the world's potential for building peace is lost. Instrumental in resolving conflict in Liberia, 2011 Nobel Peace Prize winner Leymah Gbowee underscores the significant involvement of women in peace building in a special contribution (signed box).

Violent conflicts
have many complex
effects on work

Building peace, restoring hope: women's role in nation building

Just like women's invisible work in the household, women's work in community building and conflict resolution, which has significant human development implications, often goes unacknowledged.

Women's peace-making efforts are not a strange phenomenon; they have been happening since the outbreak of violence and wars at the end of the cold war. From Bosnia and Herzegovina to Liberia, Burundi to Sierra Leone, Congo to Uganda women have been mobilizing at the local level to effect change, including during and after civil wars. In fact, I can unequivocally state that women's active engagement in ending wars and all types of violence is critical. The only sad fact is that policymakers at the global and national levels are too slow to embrace this reality.

My own experience began with the civil war that ravaged Liberia for almost two decades. There I saw how war can strip a whole society of its humanity, its deference for life and the respect for the rights and dignity of persons. How war can tear the fabric of society, break down the norms that keep in check the terror that comes from man's ego. There I lived the rule by man, and all its diabolical manifestations.

It was in this vacuum of order, of rights, of civility that the women of Liberia stood up to reclaim the soul of our country, of our society. Thousands of women across Liberia, under the banner of the Liberia Mass Action for Peace, risked their safety to protest against the war, against the despicable acts perpetrated in the name of war.

But this story is not unique to Liberia. It reverberates in most war-affected societies. About a year ago, when I had the opportunity to travel to three provinces in the Democratic Republic of the Congo, I saw that the story of women mobilizing for peace was the same. I met the same category of women that were part of the Liberian peace movement—the context is different, but the characters are the same; the virtues, qualities and principles they bring to their work are the same. The women of these communities are also stepping up and responding to their sisters' need for counselling and humanitarian assistance and their community's need for peace. Women in Nigeria are also persistently protesting for the 200 Chibok girls that are still missing. The stories of women's struggle for peace, for human rights and for justice are consistent.

When the wars end, women's role in recovery and peace building diminishes. Patriarchy steps in and forgets about the efforts of women. In no time patriarchy lays the foundation for the next war, for the next institutionalized violations of the rights of women, children and the vulnerable. This has remained the tragic tale. This is the story that women must begin to tackle—to insist on a place in the rebuilding of the polity, the economy and the state that is created after wars. In many communities post-conflict reconstruction fails to take women's unique needs and concerns into consideration, leaving women who had already borne the brunt of these conflicts to further impoverishment.

To address this situation, in many communities women are also mobilizing and forming cooperatives for farming and other enterprises as a means of providing for them and their wards through their work. This is fundamental to sustaining human rights, consolidating gains for peace.

The hope for this new agenda in women's search for human rights and peace is strong. This is evident from the growing number of women who are assuming leadership roles in public offices; more women's voices are heard against the violations against them and their children thanks to new platforms made possible through technology.

I strongly believe that it is time for the world and global bodies to recognize the critical role that women can and are playing in conflict resolution and community building. While recognition is important, I believe engaging women's expertise is also crucial to finding solutions to global conflicts. If we must reverse the tidal waves of conflict and insecurity globally, I believe we urgently need all hands on deck to accomplish the task.

Leymah Gbowee
Peace activist and 2011 Nobel Peace Prize winner

Work that damages human development

Most people work for their livelihoods. They also work to give their life greater value and meaning, but not all work achieves this. Many people are in work that restricts their life choices. Millions work under abusive and exploitative conditions that violate their basic human rights and destroy their dignity, such as child labourers, forced labourers and trafficked workers (figure 1.4). Such corrosive and exploitative activities shatter human development. And millions of domestic, migrant, sex and hazardous-industry workers make their living in ways that are dangerous, also eroding their well-being.

Child labour

In most societies children help their families through work in ways that are neither harmful nor exploitative. But child labour deprives children of their childhood, their potential and their dignity—and can be mentally, physically, socially or morally harmful. It can also interfere with education, forcing children to combine school with excessively long and heavy work or to leave school early or even to not attend. In its most pernicious forms children are enslaved, separated from their families, exposed to serious hazards and illnesses or left to fend for themselves on the streets of large cities, often at a very early age.

The world has around 168 million child labourers, almost 11 percent of the child population, some 100 million boys and 68 million girls. Around half are engaged in hazardous work. About 23 percent of child labour in developing countries occurs in low-income countries, with the largest numbers in Asia and the Pacific. The highest prevalence is in Sub-Saharan Africa, at one child in five.[32]

But there has been substantial progress, partly reflecting efforts to ensure that children attend school. Over 2000–2012 the number of child labourers fell by 78 million, or almost a third. Progress was greatest among younger children and children in hazardous work. The drop among girls was 40 percent, compared with 25 percent among boys.[33]

The 2014 Nobel Peace Laureate Kailash Satyarthi shares views and perspectives from his lifelong work on child labour in a special contribution (signed box).

Forced labour

Forced labour is one of the most detrimental types of work for human development, destroying people's choices and freedoms. It includes work or services exacted from any person under the menace of any penalty and for which that person has not offered himself or herself voluntarily. Exceptions are work required by compulsory military service, normal civic obligations or a conviction in a court of law, so long as the work or service is under the supervision and control of a public authority. Human trafficking broadly refers to recruiting and transporting people for the purpose of exploitation.[34] This typically involves entrapping workers and moving them to another country for sex work, domestic service or exploitative work in agricultural sweatshops.

FIGURE 1.4

Corrosive and exploitative work shatters human development

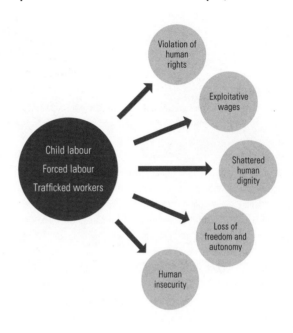

Source: Human Development Report Office.

In 2012 about 21 million people worldwide were in forced labour, trafficked for labour and sexual exploitation or held in slavery-like conditions; 14 million (68 percent) of them were subject to labour exploitation and 4.5 million (22 percent) to sexual exploitation. Private individuals and enterprises exploited 90 percent of these people.[35] Women and girls accounted for a larger share than men and boys, and adults accounted for a larger share than children (figure 1.5). Forced labour is thought to have generated around $150 billion a year in illegal profits since 2006.[36] The amount of profit is highest in Asia and the Pacific and lowest in the Middle East (figure 1.6). However, the profit per victim is highest in developed economies ($34,800) and lowest in Africa ($3,900).[37]

The most extreme forced labour is slavery, still prevalent in some parts of the world. A more common working relationship close to slavery is debt bondage. In some parts of the world in times of scarcity, poor peasants and indigenous people may accept wage advances or small loans that steadily accumulate until they are impossible to pay off, bringing them into bondage with their landlord or employer.

FIGURE 1.5

Women and girls account for a larger share of forced labour than men and boys, 2012

(millions)

Men and boys	Women and girls
9.5	11.4

Children under 18	Adults
5.5	15.4

Source: ILO 2014e.

Working towards the elimination of child labour

How has the world of child labour changed since you first got involved in the field? What are some of the positive changes and some of the less positive ones?

When I started in 1981, there was complete ignorance about this serious problem. People easily ignored it with a common perception that they are poor children, and they are working, nothing new in it. The notion of the rights of children was not institutionalized, not even conceptualized yet. The Convention on the Rights of the Child was adopted in 1989 and rolled out in 1990.

Along with my colleagues, I raised the issue that they are not simply poor children and compelled to work, it is denial of their freedom in childhood, denial of their dignity and their development for the future—including education and health.

The first reaction was denial, from powerful sections including business and governments. They did not want to touch this issue. It was not well researched or well documented; there were no laws except for old British laws. I found that the situation was the same in the rest of South Asia as I started to work in Nepal and Pakistan. Eventually I realized that the gross ignorance, neglect and denial was a global problem.

And then we moved to a situation of opposition, we approached the judiciary, raised the issue in the media and also challenged business practices of those who were using child slaves and trafficked children. Then there was sharp opposition and violent reactions—two colleagues were killed, and I was attacked several times.

Finally, ordinary people, citizens started to understand the problem, that this was more than poverty and something that cannot be simplified and ignored. There was a change in perception in all corners of society—also the judiciary, which reacted quite positively and proactively in India and other parts of the world as well. Then the trade unions also started getting involved.

The perceptional change led to changed practices as well. In the past 15–17 years, child labour is down from 260 million to 168 million. But it remains a serious problem. There is still a lack of global political will, which is also reflected at the national level. National laws have been made, and there is the International Labour Organization convention that was unanimously adopted by the international community in 1999. Since then it has been ratified by 179 countries. Yet, concrete change at the country level is still wanting, and a serious amount of political will is needed.

How do you see this political will developing? The path that you and your colleagues have taken forward in changing perceptions, do you see that continuing to be the mode of engagements with national governments and partners?

Definitely. One of the biggest things that happened during the past 20–30 years was the emergence of a strong civil society. It has emerged from the conventional charity work and evolved into critical policy partners and found more space—for example, in UN discourses. I have been part of this, involved deeply in two of the largest movements—Global Campaign for Education and Global March against Child Labour. As far as more political will is concerned, a more vibrant and dynamic civil society, a more engaged and strong civil society can act as watchdogs and be critical policy partners for governments.

The social mobilization is equally important, where the media has a crucial role.

We also have to address the corporate sector directly, making them more responsible, making them more accountable for the children. That will also, in turn, help in creating much more political will. Corporations have played a role in building political momentum against global warming and ecological change, resulting in the growth of political will, and the same can be done for child labour in particular, but also child rights in general.

You have touched on something that is very central to our discussions of how the world of work is changing—there seems to be a power shift between the various actors. You mentioned that there are good examples where the corporate side has taken on issues such as global warming and ecological changes. Are there similar examples in the field of child labour where corporate endorsement has come in and supported the cause of child rights?

There has definitely been a significant paradigm shift, though we need more concrete efforts and outcomes from them. For example, we led a campaign in the mid- and late 1980s in the carpet industry in India, Nepal and Pakistan. That led to the first ever social certification monitoring and labelling mechanism in South Asia, known as GoodWeave (earlier known as Rugmark). This was the first very concrete engagement of the corporate world along with local producers, civil society organizations, nongovernmental organizations, some UN agencies and more importantly, consumer organizations. This combination has led to remarkable results in the South Asian carpet industry, with the number of child workers in the industry down from 1 million to 200,000.

I was also personally engaged in a campaign against trafficking and slavery in the cocoa industry, in Ghana and Cote d'Ivoire. We are also trying to engage the corporate world in the garment industry.

Corporate engagement is vital. Ideally, we want to be more positive and constructive, rather than ban or boycott any products or industries. Corporate consciousness and the whole notion of corporate responsibility have to trickle down from the boardroom level to the workshop level, through the supply chain. We are working on it.

That's very encouraging for us to hear. Now, you have touched on what the ordinary citizen or a common person can do in order to eradicate this issue. Would you like to develop a little bit further what the person on the streets or someone who is a student can do within their own sphere to help remove this problem?

Child labour is not an isolated problem and cannot be solved in isolation. It is very directly connected with overall human development and with human rights. If the rights of human beings, families and communities are not protected, children are not protected. Similarly, without sustainable and inclusive development, child labour cannot be eradicated in isolation.

I have been emphasizing for over 30 years that poverty, child labour and illiteracy form a kind of "vicious triangle." They are the cause and consequence of each other, and one cannot be solved without solving the others. While 168 million children are in full-time jobs, millions of adults are jobless. Many of these adults in developing countries are the very parents of these child labourers. Children are preferred because they are the cheapest —or even free, if bonded—labour. They are physically and mentally docile, they do not form unions or go to court; so no challenge for the employer and thus perpetuating this vicious cycle.

Child labour is also interconnected with health, education and poverty eradication issues. I'm working to pose the agenda of the complete eradication of child labour, slavery and trafficking, as part of the Sustainable Development Goals.

This is very important as we live in a world where children are not safe. On the one hand, a lot of growth, a lot of development, a lot of legal protections are being talked about. On the other hand, the world is becoming more and more difficult and unsafe, frightening for the children, for reasons that include terrorism and intolerance. We are not able to assure a safer future for our children, especially in developing countries. The violence is increasing.

My strong belief is that violence and human development cannot coexist. We have to put an end to violence to ensure human development. I am very confident that until and unless we make a safer world for our children, we cannot protect human rights. Protection of human rights, inclusive development for society, as well as getting rid of violence are interconnected issues, and we have to begin with our children.

People should raise their voices against all kinds of violence against children. Denial of education is also a violation of children's rights beside child labour, child trafficking and the like. Ordinary people can get engaged in many ways. One is to help in creating, spreading and deepening the knowledge among their own peers and friends and relatives. Things are happening around the world against children. Let us spread the knowledge and create more consciousness and concern. Second, as responsible citizens, one can definitely ask local politicians, parliament members and the government to give more attention to the cause of children in both developing and developed countries, with a very strong notion that no problem in the world is an isolated problem and no problem on the planet could be solved in isolation. The solutions are interconnected, as are the problems. Problems can begin in one part of the world, and other parts would suffer.

We have to think globally and after 9/11, and after knowing more about global warming, one must know that we have to be united as global citizens in solving the problem of child labour and illiteracy of children. As consumers, one can demand that we buy only those products that are free of child labour—from toys to sporting goods to aprons to shoes. This will help in

generating pressure on the industries to ensure that they are much more responsible and much more accountable.

As citizens, as consumers, as someone who is really connected with the social media, we can use that to raise voices on behalf of children. Young people, especially those in universities and colleges, can gather information about child labour, and as responsible young people they can just push some of the buttons on their smartphones and computers and can help in raising awareness of the problem of child labour and violence against children.

With the Sustainable Development Goals in the offing, what would be your message for the international community, particularly for advancing the rights of all children?

Even before the Millennium Development Goals, were rolled out, I was very concerned about their place in the international agenda and organized a number of meetings in the corridors of the UN. We tried to send a clear message that we cannot attain many of the goals without ensuring the abolition of child labour and violence against children—whether education for all, poverty reduction through reducing unemployment, ecological issues and health issues (85 million children are in hazardous employment). We have seen some progress but also failures among these goals.

Now there is a campaign with like-minded organizations and trade unions asking for:

- *Strong language against child labour in the Sustainable Development Goals, which has been introduced and should be retained.*
- *Strong language for education, which has also been included.*
- *Explicit language against child slavery and children in forced labour, which does not exist right now.* The number of child slaves did not go down, in spite of the number of child labourers having gone down, as I mentioned, from 260 million to 168 million. According to the International Labour Organization, the number of children in forced labour or, what we call, child slaves is still stagnant at 5.5 million. My sense is that there are more.

The international community must realize that if we cannot protect our children, we cannot protect our development. We have to end violence against children.

When it comes to implementation, I have also been urging UN agencies to work in much closer cooperation on children's issues, because sometimes we see a compartmentalized approach and compartmentalized reactions, segmented by the respective agency mandates. One child gets fragmented into different agencies. We have to be more strongly coordinated and be more proactive, rather than just reacting to things that happen against children.

How will your Nobel Peace Prize impact the most marginalized children in terms of protection from violence, inclusion in education and improving their health and human development?

My spontaneous reaction to the prize on the day the announcement was made was that this is the first ever, the biggest ever recognition for the most marginalized children on the planet. This is a prize for them; it just went through me. That is why, personally, I feel much more morally responsible now. This is a recognition that "peace" cannot be restored without the safety of children and the protection of children. The prize has connected peace and child rights at the highest levels of morality, peace and society.

Within three hours of the prize, the issue got more attention than in past three decades! Now, we are able to interact and influence the political debate, a little bit, at the highest level. I have been meeting a number of prime ministers and presidents and have had very good meeting with the UN Secretary-General and also with other UN agencies. The whole discourse, the whole debate has gone to the highest level, and it cannot be put aside now. It is at a stage where the government and intergovernmental agencies have to prioritize children because every single minute matters, every single child matters, every single childhood matters.

Kailash Satyarthi
Activist and 2014 Nobel Peace Prize Laureate

FIGURE 1.6

Annual profits from forced labour have been highest in Asia and the Pacific since 2006

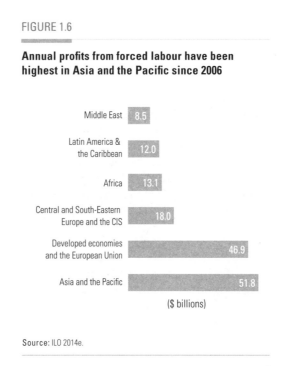

($ billions)

Source: ILO 2014e.

A more recent variant of debt bondage has arisen through international migration. To pay for travel and secure work abroad, aspirant migrants may have to borrow large amounts from agents. The agents and employers can then manipulate this credit to entrap workers. Other people may be forced to work as a result of incarceration or physical or sexual violence. Thus bonded work often happens when workers are given wage advances and are then pushed deeper and deeper into debt they cannot pay off.

Trafficked workers

After arms and drug trafficking, human trafficking is the most lucrative illicit business worldwide. The majority of victims are women, but men are also affected. For example, men from Bolivia have been trafficked with their families to Argentina to work in garment factories. Deprived of their passports, they have been locked in factories and forced to work up to 17 hours a day.[38]

Trafficking occurs on a large scale, but its extent is difficult to assess. It can be tricky to judge whether migration is voluntary or forced and difficult to extract data specifically on trafficking from data on other forms of illegal migration and exploitation. And because the activity is illegal, victims are unwilling to report abuse for fear of being deported. There is also a grey area between trafficking and smuggling.

Between 2007 and 2010 trafficked victims of 136 nationalities were detected in 118 countries. Some 55–60 percent of the victims were women.[39] Most were trafficked for sexual exploitation or for working as forced labour. More than 60 percent of victims in Europe and Central Asia and more than 50 percent in the Americas were trafficked for sexual exploitation, nearly half in Africa and the Middle East and more than 45 percent in South Asia, East Asia and the Pacific were trafficked for forced labour (figure 1.7).

Workers at risk

While all workers may find themselves in abusive situations, certain groups are particularly vulnerable, including illegal migrant workers, those in domestic service, sex workers and those in hazardous industries.

FIGURE 1.7

Over 2007–2010 a substantial number of trafficked victims were trafficked for sexual exploitation

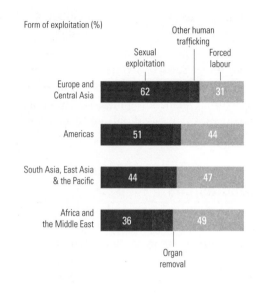

Source: UNODC 2012.

Migrant workers—illegal and legal

Among workers at risk of exploitation are illegal migrants from other countries. They often become the prey of human traffickers and thus go through a risky and sometimes life-threatening process to reach their destination. Trafficking of illegal migrants has recently surged. Networks of traffickers take money from aspiring migrants who try to cross illegally into other countries. In 2014 some 3,500 people, maybe many more, lost their lives in the Mediterranean Sea when trafficking boats heading towards Europe, mainly from Libya, capsized or sank.[40]

In many instances, legal migrant workers—particularly low-skilled and low-paid ones—are also subjected to rights violations, unsafe conditions, indignity and even abuse. Some work long hours for low wages with little time off. They can be trapped in their workplace if their employer seizes their passport or other documents.[41]

Domestic workers

Paid domestic work is an important means of income for millions of workers, the majority of whom are women. With appropriate protections in place, domestic work can empower people and help lift families out of poverty. But exploitation is common in paid domestic work, particularly for female migrant workers in developing countries. In people's homes exploitation often remains outside the scope of labour law. Employers use threats and coercion to pay low or even no wages. They can force domestic workers to work long hours—up to 18 hours a day without days off—limiting their movements and potential for social interaction. Working conditions are often poor, with little food and no access to medical care. Domestic workers may also be subject to physical or sexual abuse.[42] Even so, many domestic workers feel obliged to remain with abusive employers because they need to work to support their own families.

Sex workers

According to the UN General Assembly definition, sex workers include consenting female, male and transgender adults and young people ages 18 and older who receive money or goods in exchange for sexual services, either regularly or occasionally.[43] Most countries have laws prohibiting buying, selling or facilitating sexual services as well as soliciting sex, operating brothels and living off the earnings of sex work. Recommendation 200 on HIV and the World of Work of the International Labour Organization covers sex workers and recommends that governments recognize sex work as an occupation, so that it can be regulated to protect workers and clients.[44] Sex work involves exploitation, abuse, violence and insecurity, all of which damage human development in terms of agency and choice.

Legal restrictions on sex work and related activities displace sex workers to unsafe environments, heightening their risk of violence and sexually transmitted infections, including HIV. Recent modelling by The Lancet suggests that decriminalizing sex work could avert 33–46 percent of HIV infections in the next decade across generalized and concentrated epidemics.[45] Evidence from India suggests that activities led by sex worker collectives have helped identify 80 percent of trafficked minors and women in West Bengal, showing

Trafficking of illegal migrants has recently surged

that sex workers can be allies in assisting such victims.[46]

Workers in hazardous industries

Mining is one of the most hazardous occupations in many countries. Although it accounts for only 1 percent of the global workforce (30 million workers), it is responsible for 8 percent of fatal accidents at work and for many injuries and disabling diseases such as pneumoconiosis (black lung disease).[47] Official statistics probably underestimate the number of cases, and many miners work in informal mines, where accident rates are often far higher.

Construction is another hazardous industry. In developing countries construction workers face multiple health risks, such as exposure to dust, and are three to six times more likely than other workers to die from accidents at work. This is due in part to a lack of safety measures or lax enforcement. In many developing countries construction workers wear no protection. Similarly, in many instances the booming construction sector relies heavily on rural migrant workers who earn low wages, an intense pace of work and long and irregular hours.[48]

Many dangers are also evident in factories, and risks seem to have intensified in recent years due to increased global movements and aggravated by the global financial crisis. Faced with intense competitive pressures, firms are cutting costs in a race to the bottom. The most unscrupulous bypass safety standards to keep costs low, which increases the danger to factory workers—consider, for example, the 2013 Rana Plaza tragedy in Bangladesh. The government has since made progress in supervising the readymade garment and knitwear industry to guarantee labour rights and reinforce safety measures, but much remains to be done.

Older workers also face greater risk. They need more recovery time between shifts, especially when performing physically or psychologically demanding jobs. For example, in Europe most fatal work-related accidents happen in the 55–65 age group.[49]

Conclusions

Work is critical to human existence and human development. Work has contributed considerably to impressive human development achievements over the years, but a large amount of human potential remains unused—because people are out of work, engaged in vulnerable work or working but still in poverty. Many younger people are out of work, and women face lower wages and opportunities for paid work while bearing a disproportionate burden of unpaid care work. The world is deprived of their contributions, creativity and innovation. Creating opportunities to put the work potentials of all people to use could further accelerate human progress, help overcome the remaining human development deficits and address emerging human development challenges. These issues are addressed in the following chapter.

Work with exploitation, risks and insecurities

		Work that is a risk to human development					Occupational injuries		Security from employment		
		Child labour	Domestic workers		Working poor at PPP$2 a day	Low pay rate	Nonfatal	Fatal	Unemployment benefits recipients	Mandatory paid maternity leave	Old age pension recipients
			(% of total employment)						(% of unemployed ages 15–64)		(% of statutory pension age population)
		(% ages 5–14)	Female	Male	(% of total employment)	(% of total employment)	(thousands)	(cases)		(days)	
HDI rank		2005–2013[a]	2000–2010[a]	2000–2010[a]	2003–2012[a]	2001–2011[a]	2005–2013[a]	2005–2013[a]	2005–2013[a]	2014	2004–2012[a,b]
VERY HIGH HUMAN DEVELOPMENT											
1	Norway	..	0.1	0.1	..	..	15.2	48	61.8	..	100.0
2	Australia	..	0.1	0.1 c	..	16.1	100.1	212	52.7	..	83.0
3	Switzerland	..	2.8	0.4	..	12.2	93.8	192	61.9	98	100.0
4	Denmark	..	0.3	0.1 c	..	13.4	41.7	40	77.2	126	100.0
5	Netherlands	..	0.1	0.1 c	..	..	831.8	49	61.9	112	100.0
6	Germany	..	1.1	0.1	..	20.5	1,007.2	664	88.0	98	100.0
6	Ireland	..	1.0	0.1	..	..	..	40	21.6	182	90.5
8	United States	..	0.9	0.1	..	25.1	1,191.1	4,383	26.5	..	92.5
9	Canada	..	0.9	0.1 c	..	20.3	..	..	40.5	105	97.7
9	New Zealand	..	0.2	0.1 c	..	12.6	..	48	32.9	98	98.0
11	Singapore	..	..	..	..	..	11.8	59	0.0	112	0.0
12	Hong Kong, China (SAR)	..	..	..	..	..	37.8	188	16.9	70	72.9
13	Liechtenstein	..	..	..	..	..	..	..	..	..	..
14	Sweden	..	..	..	..	..	30.5	33	28.0	..	100.0
14	United Kingdom	..	0.6	0.3	..	20.6	79.9	148	62.6	273	99.5
16	Iceland	..	..	..	..	16.7	1.6	6	28.6	90	100.0
17	Korea (Republic of)	..	1.5	0.1 c	..	22.2	..	1,292	45.5	90	77.6
18	Israel	..	3.5	0.3	..	20.3	66.3	62	29.4	98	73.6
19	Luxembourg	..	..	..	..	19.1	8.5	22	43.8	..	90.0
20	Japan	..	0.1	0.1 c	..	14.4	..	1,030	19.6	98	80.3
21	Belgium	..	1.9	0.1	..	12.7	71.0	72	80.2	105	84.6
22	France	..	4.1	0.7	..	..	658.8	529	56.2	112	100.0
23	Austria	..	0.5	0.1 c	..	16.5	59.5	91	90.5	112	100.0
24	Finland	..	0.3	0.3	..	5.7	..	26	59.1	147	100.0
25	Slovenia	..	0.1	0.1 c	..	..	15.2	18	30.8	105	95.1
26	Spain	..	8.4	0.6	..	16.2	402.7	232	46.9	112	68.2
27	Italy	..	4.0	0.4	..	9.5	402.9	621	55.8	150	81.1
28	Czech Republic	..	0.1	0.1 c	..	17.1	42.9	105	21.2	196	100.0
29	Greece	..	4.8	0.1	..	13.3	15.2 d	107 d	43.1	119	77.4
30	Estonia	..	..	..	..	..	4.2	20	27.6	140	98.0
31	Brunei Darussalam	..	28.8	1.9	..	..	..	..	0.0	..	81.7
32	Cyprus	..	9.7	0.1	..	..	1.5	9	78.7	..	85.2
32	Qatar	..	38.9	2.8	..	..	0.1	..	0.0	..	7.9
34	Andorra	..	..	..	..	..	..	..	11.1	..	..
35	Slovakia	..	0.4	0.1 c	..	20.0	8.5	53	11.2	238	100.0
36	Poland	..	0.1	0.1 c	..	24.3	..	348	16.8	182	96.5
37	Lithuania	..	0.1	0.1	..	..	3.1	60	21.5	126	100.0
37	Malta	..	0.2	0.1 c	..	..	3.1	6	..	..	60.5
39	Saudi Arabia	..	47.1	3.9	..	..	..	..	0.0	70	..
40	Argentina	4.4	18.3	0.3	4.0	25.6	441.1	562	4.9	90	90.7
41	United Arab Emirates	..	42.4	6.0	..	..	..	..	0.0	45	..
42	Chile	6.6 e	14.3	2.0	3.8	18.5	215.0	322	29.9	126	74.5
43	Portugal	3.4 e,f	7.2	0.1	..	10.3	173.6	276	42.1	..	100.0
44	Hungary	..	0.1	0.1 c	..	21.0	17.0	62	31.4	168	91.4
45	Bahrain	4.6 f	42.2	5.8	..	..	1.0	23	7.9	..	40.1
46	Latvia	..	..	..	..	31.5	1.6	29	19.5	112	100.0
47	Croatia	..	0.6	0.1	..	..	15.4	27	20.0	208	57.6
48	Kuwait	..	53.3	11.3	..	..	..	..	0.0	70	27.3
49	Montenegro	9.9	0.1	0.1 c	1.8	..	..	..	..	45	52.3
HIGH HUMAN DEVELOPMENT											
50	Belarus	1.4	..	..	2.7	..	..	141	46.1	126	93.6
50	Russian Federation	..	0.1	0.1 c	3.4	..	..	1,699	20.6	140	100.0
52	Oman	..	59.3	2.8	..	..	..	..	0.0	42	24.7
52	Romania	0.9 f	0.5	0.2	..	..	3.4	223	..	126	98.0
52	Uruguay	7.9 e	18.5	1.4	3.9	27.7	22.9	51	27.9	84	68.2
55	Bahamas	..	6.4	3.0	..	..	..	..	25.7	..	84.2
56	Kazakhstan	2.2	0.4	0.1	3.7	..	2.6	341	0.5	126	95.9
57	Barbados	..	..	..	..	..	0.8	0	..	..	68.3

Work with exploitation, risks and insecurities (continued)

		Work that is a risk to human development					Occupational injuries		Security from employment		
		Child labour	Domestic workers		Working poor at PPP$2 a day	Low pay rate	Nonfatal	Fatal	Unemployment benefits recipients	Mandatory paid maternity leave	Old age pension recipients
			(% of total employment)						(% of unemployed ages 15–64)		(% of statutory pension age population)
		(% ages 5–14)	Female	Male	(% of total employment)		(thousands)	(cases)		(days)	
HDI rank		2005–2013[a]	2000–2010[a]	2000–2010[a]	2003–2012[a]	2001–2011[a]	2005–2013[a]	2005–2013[a]	2005–2013[a]	2014	2004–2012[a,b]
58	Antigua and Barbuda	..	6.3	1.4	..	..	..	..	0.0	..	69.7
59	Bulgaria	..	..	..	..	..	2.2	81	25.6	410	96.9
60	Palau	..	..	..	..	..	..	..	..	..	48.0
60	Panama	5.6[e]	13.8	1.0	8.5	11.6	0.0	24	0.0	98	37.3
62	Malaysia	..	5.9	0.4	8.4	..	41.5	274	0.0	60	19.8
63	Mauritius	..	8.1	0.6	5.7	..	1.1	3	1.2	84	100.0
64	Seychelles	..	..	..	..	..	0.1	1[d]	5.0	..	100.0
64	Trinidad and Tobago	0.7	..	..	..	..	0.8	9	0.0	..	98.7
66	Serbia	4.4	0.5	0.1	1.8	..	..	..	..	135	46.1
67	Cuba	..	..	..	5.9	..	4.9	88	0.0	..	..
67	Lebanon	1.9	..	..	..	..	..	..	0.0	49	0.0
69	Costa Rica	4.1	17.3	1.1	4.3	21.4	134.8	95	0.0	120	55.8
69	Iran (Islamic Republic of)	11.4[e]	0.4	0.1[c]	7.4	..	..	..	..	90	26.4
71	Venezuela (Bolivarian Republic of)	7.7[f]	14.4	0.9	8.4	12.1	3.0[g]	31[g]	..	182	59.4
72	Turkey	5.9[e]	2.1	0.4	2.5	..	2.2	745	7.7	112	88.1
73	Sri Lanka	..	2.5	0.6	20.4	..	1.5	141	0.0	84	17.1
74	Mexico	6.3[e]	10.3	0.6	2.8	17.4	542.4	1,314	0.0	84	25.2
75	Brazil	8.3[e]	17.0	0.9	3.4	21.5	636.1	2,938	8.0	120	86.3
76	Georgia	18.4	1.2	0.2	20.6	..	..	..	0.0	126	89.8
77	Saint Kitts and Nevis	..	..	..	..	..	..	..	0.0	..	44.7
78	Azerbaijan	6.5[e]	2.0	1.5	5.0	40.0	0.1	66	2.5	126	81.7
79	Grenada	..	..	..	..	..	..	..	0.0	..	34.0
80	Jordan	1.6[e]	2.2	0.2	4.0	..	15.3	87	0.0	70	42.2
81	The former Yugoslav Republic of Macedonia	12.5	0.3	0.1	4.1	..	..	..	..	270	52.2
81	Ukraine	2.4	..	..	1.8	..	15.6	474	20.9	126	95.0
83	Algeria	4.7[e]	1.2	0.3	..	..	..	912	8.8[d]	98	63.6
84	Peru	33.5[e]	6.7	0.3	9.5	26.3	19.0	178	0.0	90	33.2
85	Albania	5.1[e]	..	..	3.9	..	..	..	6.9	365	77.0
85	Armenia	3.9	0.6	0.2	12.2	25.8	0.0	12	15.8	140	80.0
85	Bosnia and Herzegovina	5.3	..	..	2.0	..	..	..	2.0	365	29.6
88	Ecuador	8.6[e]	..	..	10.3	17.2	13.7	..	6.7[h]	84	53.0
89	Saint Lucia	3.9	7.0	0.9	..	..	..	..	0.0	..	26.5
90	China	..	..	..	20.4	21.9	3.8[g]	14,924[g]	14.0	98	74.4
90	Fiji	..	..	..	19.4	..	..	..	0.0	84	10.6
90	Mongolia	10.4	1.1	1.2	..	..	..	..	10.0	120	100.0
93	Thailand	8.3	1.2	0.1	5.6	..	..	619	28.5	45	81.7
94	Dominica	..	..	..	..	..	..	..	0.0	..	38.5
94	Libya	..	..	..	..	..	..	..	0.0	..	43.3
96	Tunisia	2.1	..	..	4.5	..	43.2[i]	155[i]	..	30	68.8
97	Colombia	9.7[e]	13.0	0.6	10.2	20.5	..	..	0.0	98	23.0
97	Saint Vincent and the Grenadines	..	..	..	..	..	0.2	1	0.0	..	76.6
99	Jamaica	3.3	10.2	1.5	6.7	..	..	..	0.0	56	55.5
100	Tonga	..	2.4	1.3	..	..	..	..	0.0	..	1.0
101	Belize	5.8	12.8	2.3	..	..	1.8	..	0.0	..	64.6
101	Dominican Republic	12.9	14.4	0.8	7.3	..	..	313	0.0	84	11.1
103	Suriname	4.1	..	..	..	..	..	..	0.0	..	..
104	Maldives	..	..	..	12.6	..	..	..	0.0	..	99.7
105	Samoa	..	3.1	0.8	..	..	..	..	0.0	..	49.5
MEDIUM HUMAN DEVELOPMENT											
106	Botswana	9.0[e]	7.0	2.6	23.9	..	1.2	24	0.0	84	100.0
107	Moldova (Republic of)	16.3	0.6	0.1	3.2	23.6	..	36	11.4	126	72.8
108	Egypt	9.3[e]	0.3	0.3	9.6	..	26.9[d]	208	..	90	32.7
109	Turkmenistan	..	..	..	..	..	..	..	..	..	..
110	Gabon	13.4	..	..	15.2	..	1.1[i]	20[i]	0.0	98	38.8
110	Indonesia	6.9[e]	4.4	0.9	38.1	29.0	..	..	0.0	90	8.1
112	Paraguay	27.6[e]	..	..	8.1	..	..	..	0.0	63	22.2
113	Palestine, State of	5.7	0.2	0.1	3.0	..	0.7	20	..	70	8.0
114	Uzbekistan	..	..	..	..	..	..	..	0.0	126	98.1
115	Philippines	11.1[e]	11.5	1.4	36.8	14.5	..	161	0.0	60	28.5

		Work that is a risk to human development					Occupational injuries		Security from employment		
		Child labour	Domestic workers		Working poor at PPP$2 a day	Low pay rate	Nonfatal	Fatal	Unemployment benefits recipients	Mandatory paid maternity leave	Old age pension recipients
			(% of total employment)						(% of unemployed ages 15–64)		(% of statutory pension age population)
		(% ages 5–14)	Female	Male	(% of total employment)		(thousands)	(cases)		(days)	
HDI rank		2005–2013[a]	2000–2010[a]	2000–2010[a]	2003–2012[a]	2001–2011[a]	2005–2013[a]	2005–2013[a]	2005–2013[a]	2014	2004–2012[a,b]
116	El Salvador	8.5[e]	10.2	0.7	9.9	..	..	96	0.0	84	18.1
116	South Africa	..	15.5	3.5	13.0	32.4	9.4	185	13.5	120	92.6
116	Viet Nam	6.9	0.9	0.1	13.8	..	..	..	8.4	180	34.5
119	Bolivia (Plurinational State of)	26.4[e]	7.4	0.2	11.6	..	..	..	0.0	84	100.0
120	Kyrgyzstan	3.6	0.7	0.7	14.5	..	0.2	29	0.9	126	100.0
121	Iraq	4.7	0.2	0.1	10.6	..	..	..	0.0	..	56.0
122	Cabo Verde	6.4[e]	..	..	28.6	..	..	..	0.0	..	55.7
123	Micronesia (Federated States of)	..	..	..	..	..	..	..	..	..	..
124	Guyana	16.4	7.1	0.8	..	..	2.1[h]	1[h]	0.0	..	100.0
125	Nicaragua	14.9[f]	12.1	1.7	14.2	..	25.8	42	0.0	84	23.7
126	Morocco	8.3	..	..	13.3	..	..	..	0.0	98	39.8
126	Namibia	..	19.4	4.2	30.9	..	0.6[i]	10[i]	0.0	84	98.4
128	Guatemala	25.8[e]	8.8	0.3	18.8	..	..	..	0.0	84	14.1
129	Tajikistan	10.0	0.1	0.2	19.1	..	..	..	9.2	140	80.2
130	India	11.8	2.2	0.5	55.5	..	6.0	2,140	0.0	84	24.1
131	Honduras	14.0[e]	..	..	18.8	33.4	2.1	..	0.0	84	8.4
132	Bhutan	2.9	..	..	14.1	..	..	..	0.0	..	3.2
133	Timor-Leste	4.2[f]	..	..	66.9	..	..	..	..	..	100.0
134	Syrian Arab Republic	4.0	..	..	11.8	..	9.7	612	0.0	120	16.7
134	Vanuatu	..	..	..	..	..	..	..	0.0	..	3.5
136	Congo	18.4	..	..	50.7	..	..	..	0.0	105	22.1
137	Kiribati	..	..	..	..	..	..	..	0.0	..	..
138	Equatorial Guinea	27.8[f]	..	..	19.3	..	..	..	0.0	..	..
139	Zambia	40.6[e]	..	..	84.7	..	..	..	0.0	84	7.7
140	Ghana	33.9	0.3	0.4	44.3	..	..	..	0.0	84	7.6
141	Lao People's Democratic Republic	10.1[e]	..	..	65.0	..	..	..	0.0	90	5.6
142	Bangladesh	12.8	2.3	0.2	76.4	..	..	..	0.0	112	39.5
143	Cambodia	18.3[e]	0.8[f]	0.9[f]	40.2	..	..	..	0.0	90	5.0
143	Sao Tome and Principe	7.5	3.3	0.1	..	..	..	..	0.0	..	41.8
LOW HUMAN DEVELOPMENT											
145	Kenya	25.9[f]	0.6	0.7	..	..	..	..	0.0	90	7.9
145	Nepal	33.9[e]	0.3	0.3	49.5	..	..	..	0.0	52	62.5
147	Pakistan	..	1.2	0.3	45.8	..	0.1[g]	110[g]	0.0	84	2.3
148	Myanmar	..	..	..	66.9	..	0.2	32	0.0	..	..
149	Angola	23.5[f]	..	..	56.4	..	..	..	0.0	90	14.5
150	Swaziland	7.3	..	..	36.0	..	..	..	0.0	..	86.0
151	Tanzania (United Republic of)	21.1[e]	1.2	0.3	73.9	..	..	..	0.0	84	3.2
152	Nigeria	24.7	0.6	0.4	76.6	..	0.1[i]	5	0.0	84	..
153	Cameroon	41.7	..	..	52.9	..	..	..	0.0	98	12.5
154	Madagascar	22.9[e]	..	..	93.0	..	..	..	0.0	98	4.6
155	Zimbabwe	..	3.6	1.2	84.6	..	4.6	91	0.0	98	6.2
156	Mauritania	14.6	..	..	40.8	..	..	..	0.0	98	9.3[f]
156	Solomon Islands	..	5.6[f]	0.2[f]	52.1	..	..	..	0.0	..	13.1
158	Papua New Guinea	..	..	..	..	..	11.8[d]	180[d]	0.0	..	0.9
159	Comoros	22.0	..	..	70.7	..	..	..	0.0	..	..
160	Yemen	22.7	2.5	0.4	25.1	..	..	..	0.0	70	8.5
161	Lesotho	22.9[f]	8.1[f]	1.3[f]	63.9	..	..	..	0.0	84	100.0
162	Togo	28.3	..	..	70.5	..	0.3[i]	10[i]	0.0	98	10.9
163	Haiti	24.4	..	..	..	..	..	..	0.0	42	1.0
163	Rwanda	28.5	..	..	83.7	..	1.0[h]	406[h]	0.0	84	4.7
163	Uganda	16.3[e]	1.8	0.6	57.7	..	..	..	0.0	60	6.6
166	Benin	15.3	..	..	72.3	..	0.7[i]	4[i]	0.0	98	9.7
167	Sudan	..	..	..	35.0	..	..	..	..	56	4.6
168	Djibouti	7.7	41.6[f]	1.9[f]	..	..	..	..	0.0	..	12.0[f]
169	South Sudan	..	..	..	..	..	..	..	..	..	..
170	Senegal	14.5	6.7	1.4	58.4	..	..	..	0.0	98	23.5
171	Afghanistan	10.3	..	..	88.1	..	..	..	0.0	..	10.7
172	Côte d'Ivoire	26.4	..	..	59.0	..	..	..	0.0	98	7.7
173	Malawi	25.7	..	..	88.4	..	..	..	0.0	56	4.1

Work with exploitation, risks and insecurities (continued)

HDI rank	Work that is a risk to human development					Occupational injuries		Security from employment		
	Child labour	Domestic workers		Working poor at PPP$2 a day	Low pay rate	Nonfatal	Fatal	Unemployment benefits recipients	Mandatory paid maternity leave	Old age pension recipients
		(% of total employment)						(% of unemployed ages 15–64)		(% of statutory pension age population)
	(% ages 5–14)	Female	Male	(% of total employment)		(thousands)	(cases)		(days)	
	2005–2013ᵃ	2000–2010ᵉ	2000–2010ᵉ	2003–2012ᵃ	2001–2011ᵉ	2005–2013ᵃ	2005–2013ᵃ	2005–2013ᵃ	2014	2004–2012ᵃ,ᵇ
174 Ethiopia	27.4	1.5	0.1	73.8	..	..	..	0.0	90	9.0
175 Gambia	19.2	..	..	56.0	..	..	..	0.0	..	10.8
176 Congo (Democratic Republic of the)	15.0	..	..	90.4	..	..	..	0.0	98	15.0
177 Liberia	20.8	0.6	1.0	94.0	..	..	..	0.0	90	..
178 Guinea-Bissau	38.0	..	..	..	..	..	..	0.0	..	6.2
179 Mali	21.4	8.7	1.4	78.1	..	..	..	0.0	98	5.7
180 Mozambique	22.2	..	..	85.8	..	..	..	0.0	60	17.3
181 Sierra Leone	26.0	0.5	0.4	82.8	..	..	..	0.0	84	0.9
182 Guinea	28.3	0.4ᶠ	0.3ᶠ	73.8	..	..	..	0.0	98	8.8
183 Burkina Faso	39.2	0.6	0.3	70.1	..	3.4ʰ	8ʰ	0.0	98	3.2
184 Burundi	26.3	..	..	94.8	..	..	..	0.0	84	4.0
185 Chad	26.1	..	..	62.2	..	..	..	0.0	98	1.6
186 Eritrea	..	..	..	77.4	..	..	..	0.0	..	..
187 Central African Republic	28.5	..	..	83.9	..	0.1	9	0.0	..	..
188 Niger	30.5	..	..	72.2	..	..	..	0.0	98	6.1
OTHER COUNTRIES OR TERRITORIES										
Korea (Democratic People's Rep. of)	..	..	..	59.6	..	..	..	0.0	..	..
Marshall Islands	..	..	..	..	..	..	..	0.0	..	64.2
Monaco	..	..	..	..	..	..	..	..	..	..
Nauru	..	..	..	..	..	..	..	0.0	..	56.5
San Marino	..	..	..	..	..	0.6	0	..	..	..
Somalia	49.0	..	..	86.8	..	..	..	0.0	..	..
Tuvalu	..	..	..	..	..	..	..	0.0	..	19.5
Human development groups										
Very high human development	..	4.3	0.4	..	..	..	..	43.4	123	89.4
High human development	8.3	..	..	14.7	..	..	..	6.0	125	73.9
Medium human development	11.6	..	..	46.9	..	..	..	1.7	98	27.7
Low human development	23.8	..	..	67.5	..	..	..	0.0	85	9.8
Developing countries	14.5	..	..	33.8	..	2,273.9ᵏ	29,292ᵏ	2.5	99	51.0
Regions										
Arab States	10.5	..	..	17.3	..	..	..	1.7	70	35.7
East Asia and the Pacific	..	..	..	23.8	..	..	..	1.6	..	65.3
Europe and Central Asia	5.4	..	..	3.8	..	..	..	6.2	161	86.1
Latin America and the Caribbean	10.8	14.2	0.8	5.6	..	..	..	5.3	92	60.8
South Asia	12.3	2.2	0.5	54.9	..	..	..	0.0	84	23.9
Sub-Saharan Africa	24.7	..	..	70.5	..	..	..	2.3	90	21.9
Least developed countries	21.7	..	..	71.7	..	..	..	0.0	..	19.7
Small island developing states	..	..	..	..	..	..	..	0.9	..	28.0
Organisation for Economic Co-operation and Development	..	3.0	0.3	..	17.1	6,098.6ᵏ	13,210ᵏ	38.7	131	87.1
World	**14.5**	..	..	**26.4ᵀ**	..	**7,691.9ᵏ**	**42,493ᵏ**	**12.2**	**109**	**65.0**

NOTES

a Data refer to the most recent year available during the period specified.

b Because statutory pension ages differ by country, cross-country comparisons should be made with caution.

c Less than 0.1

d Refers to 2003.

e Differs from standard definition or refers to only part of the country.

f Refers to a year earlier than that specified.

g Refers to 2002.

h Refers to 2000.

i Refers to 2004.

j Refers to 2001.

k Unweighted sum of the reported cases.

T From original data source.

DEFINITIONS

Child labour: Percentage of children ages 5–11 who, during the reference week, did at least one hour of economic activity or at least 28 hours of household chores, or children ages 12–14 who, during the reference week, did at least 14 hours of economic activity or at least 28 hours of household chores.

Domestic workers: Percentage of the employed population that performs work in or for a household or households.

Working poor at PPP$2 a day: Employed people who live on less than $2 (in purchasing power parity terms) a day, expressed as a percentage of the total employed population ages 15 and older.

Low pay rate: Percentage of employees whose hourly earnings at all jobs were less than two-thirds of the median hourly earning.

Nonfatal occupational injuries: Number of occupational accidents that do not leading to death but that entail a loss of working time.

Fatal occupational injuries: Number of occupational accidents that lead to death within one year.

Recipients of unemployment benefits: Percentage of unemployed people ages 15–64 receiving unemployment benefits (periodic and means-tested benefits).

Mandatory paid maternity leave: Number of days of paid time off work that a female employee is entitled to in order to take care of a newborn child.

Old age pension recipients: People above the statutory pensionable age receiving an old age pension (contributory, noncontributory or both), expressed as a percentage of the eligible population.

MAIN DATA SOURCES

Column 1: UNICEF (2015).

Columns 2 and 3: ILO (2013).

Column 4: ILO (2015a).

Column 5: ILO (2012).

Columns 6 and 7: ILO (2015b).

Columns 8 and 10: ILO (2015c).

Column 9: World Bank (2015c).

TABLE A1.2

Different types of employment

In 1993 the 15th International Conference of Labour Statisticians adopted the most recent revision to the classification of employment status.

Paid employment

Paid employment covers jobs where the incumbents hold explicit (written or oral) or implicit employment contracts, which give them a basic remuneration that does not depend directly on the revenue of the unit for which they work (this unit can be a corporation, a nonprofit institution, a government unit or a household).

Wage and salaried employees are employees who hold paid employment jobs.

Self-employment

Self-employment covers jobs where the remuneration depends directly on the profits (or the potential of profits) derived from the goods and services produced (where own consumption is considered to be part of the profits). The incumbents make the operational decisions affecting the enterprise or delegate such decisions while retaining responsibility for the welfare of the enterprise.

Own-account workers are workers who, working on their own account or with one or more business partners, hold the type of jobs defined as self-employment.

Contributing (unpaid) family workers are workers who hold a self-employment job in a market-oriented establishment operated by a related person living in the same household who cannot be regarded as a business partner.

Self-employment comprises a vast array of types of work with varying degrees of insecurity, vulnerability and remuneration. The self-employed range from highly skilled "freelance" professionals to unskilled street vendors to small business owners.

Within the category of self-employed, the International Labour Organization has identified own-account workers and contributing family workers as vulnerable workers.

Nonstandard employment

Nonstandard employment covers work arrangements that fall outside the realm of the standard employment relationship, understood as work that is full-time, indefinite and part of a subordinate but bilateral employment relationship. Nonstandard forms of employment include workers in their formal or informal employment arrangements, as long as their contractual status covers one of the four categories included in the International Labour Organization definition. More specifically:

Temporary employment covers workers that are engaged for a specific period of time, including under fixed-term or project- or task-based contracts, as well as seasonal or casual work, including day labourers.

Contractual arrangements involving multiple parties, including temporary agency work, covers the situation in which worker is deployed and paid by a private employment agency and service provider, but the work is performed by the user firm.

In *part-time employment* the normal hours of work are fewer than those of comparable full-time workers.

Ambiguous employment is when the respective rights and obligations of parties concerned are not clear.

Self-employment may include workers who have access to some form of social protection and unemployment insurance through national programmes and workers whose survival depends solely on their ability to sell their goods and services. Vulnerable workers—a number of whom are women and children—are more likely than other self-employed people to lack contractual arrangements, and they often lack economic independence and are vulnerable to household power relations and economic fluctuations. They frequently lack unemployment insurance, social security and health coverage. As self-employment and own-account work become increasingly common around the world, better working conditions and social protection for them have emerged as major concerns.

Source: ILO 2015h.

Chapter **2**

Human development and work: progress and challenges

Infographic: Work engages the majority of people in the world

What **7.3 billion** people do:

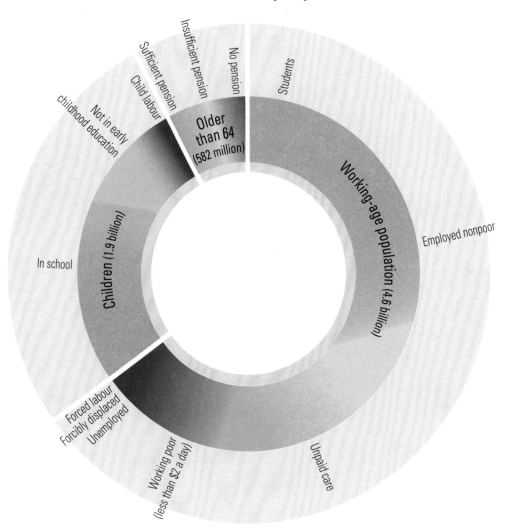

Insufficient pension

Sufficient pension

Child labour

Not in early childhood education

No pension

Students

Older than 64 (582 million)

In school

Children (1.9 billion)

Working-age population (4.6 billion)

Employed nonpoor

Forced labour
Forcibly displaced
Unemployed

Working poor (less than $2 a day)

Unpaid care

and what it means for **human development**:

Enabling human potential

Thwarting human potential

Working-age population	Children	Older than 64
Students	In school	Sufficient pension
Employed nonpoor		
Unpaid care workers		
Working poor (less than $2 a day)	Not in early childhood education	Insufficient pension
Unemployed		
Forcibly displaced		
Forced labour	Child labour	No pension

2.

Human development and work: progress and challenges

Over the past two decades, the world has made major strides in human development. Today, people are living longer, more children are going to school and more people have access to clean water and basic sanitation. This progress goes hand in hand with increasing incomes, leading to the highest standards of living in human history. A digital revolution connects people across societies and countries. Just as important, political developments enable more people than ever to live under democratic regimes. All these are important facets of human development.

The activities of 7.3 billion people have contributed to this human progress in various forms (see infographic at the beginning of the chapter). Work has helped people achieve a better standard of living—not only in terms of income, but also in terms of health and education, major ingredients for enhancing capabilities. Work has also provided people with security, contributed to women's empowerment and enhanced participation in different spheres of human lives.

However, there are marked differences in progress among regions, across countries and within countries. Furthermore, serious human deprivations remain, and the world faces persistent inequality, along with climate change and social, economic and environmental vulnerabilities that could reverse human advances and constrain wider choices for everyone.

At the same time, considerable human potential remains unused. Millions of people are either out of work or working but still in poverty. Even though young people account for around half the global population (and an even larger share in some regions because of a youth bulge), millions of young people are unemployed, depriving the world of their dynamism, creativity and innovation. Similarly, the employment potential of women remains substantially unused in various parts of the world.

Overcoming the existing human deprivations and addressing the emerging human development challenges will require optimal use of the world's human potential. Such use would also accelerate the achievements in human development to date and impart human progress with greater resilience.

The scale of human progress and contributions of work

Progress is evident for many measures of human development—such as those for health, education, income, security and participation —and for such composite indicators as the Human Development Index (HDI).

Trends on the Human Development Index

In almost 25 years the world HDI value has increased by more than 20 percent and that of the least developed countries by more than 40 percent. Every region of the world has seen HDI gains. Over time and across all developing regions, progress has been fairly steady, though at a slower pace during the last 15 years, with most having moved up through the human development classifications (figure 2.1).

These changes are also reflected in the number of countries in each human development classification. Between 1990 and 2014 for 156 countries with comparable data covering 98 percent of the world's population in 2014, the number of countries in the very high human development classification rose from 12 to 46, as the population in that group climbed from 0.5 billion to 1.2 billion. Over the same period the number of countries in the low human development classification fell from 62 to 43, as the population in that group fell from 3.2 billion to 1.2 billion (figure 2.2).

Progress on the HDI has been considerable at the country level. For example, Ethiopia increased its HDI value by more than half; Rwanda by nearly half; five countries, including Angola and Zambia, by more than a third;

Work has helped people achieve a better level of human development

FIGURE 2.1

Progress on the Human Development Index since 1990 has been fairly steady over time and across all developing regions

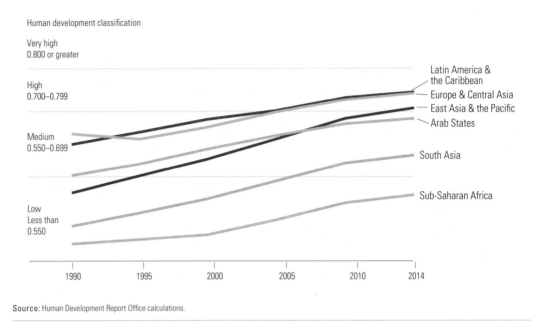

Source: Human Development Report Office calculations.

FIGURE 2.2

Between 1990 and 2014 the number of people living in countries in higher human development classifications rose, while the number of people living in countries in the low human development classification fell

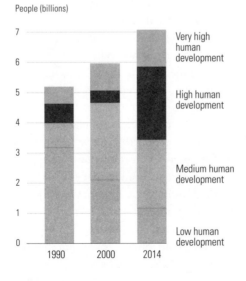

Note: Data are from a panel of 156 countries.
Source: Human Development Report Office calculations.

> There is no automatic link between work and human development

and 23 countries, including Bangladesh, the Democratic Republic of the Congo and Nepal, by more than a fifth. It is also encouraging that the fastest progress was among low human development countries.

But it also reminds us that there is no automatic link between income and human development. Income enters into the HDI but is just one of its four indicators. Economic growth does not automatically translate into higher human development. Equatorial Guinea and Chile have similar gross national incomes per capita (in purchasing power parity terms) but different HDI values; by contrast, Gabon and Indonesia have different incomes but similar HDI values (figure 2.3).

Improvements in overall HDI value were also paced by progress in all HDI component indicators, as well as in many of the non-HDI dimensions of human development—income poverty and hunger, health, education, gender equality, access to basic social services, environmental sustainability and participation (see table A2.1 at the end of the chapter).

Work has contributed to human development achievements

The various activities of 7.3 billion people have contributed to human progress. Nearly a billion people who work in agriculture and more than 500 million family farms, which produce more than 80 percent of the world's food, have contributed to better health and nutrition of the global population.[1] Worldwide, the work of 80 million workers in health and education has helped enhance human capabilities.[2] More than a billion workers in services have contributed to human progress.[3] In China and India 23 million jobs were created in clean energy, allowing workers to contribute to environmental sustainability.[4]

Work undertaken for the care of others has contributed not only to accelerating and protecting human development for present generations (for example, through care for older people and people with disabilities), but also to creating human capabilities (for example, through care work for children). Care work can be paid or unpaid and includes household activities such as cooking, cleaning and collecting water and fuelwood as well as tending to children, older people and the sick. When care work is paid, it is often referred to as domestic work. The International Labour Organization estimates that there are at least 53 million paid adult domestic workers worldwide, 83 percent of them women, and the number may be rising. Between 1995 and 2010 the share of domestic workers in total employment rose from nearly 6 percent to 8 percent in Latin America and the Caribbean; it also rose in the Middle East and in Asia and the Pacific.[5] This work is critical for present and future human development.

Work has a societal value that goes beyond the gains of individual workers. More than 450 million entrepreneurs have contributed to human innovation and creativity.[6] Art contributes to social cohesion and cultural identity while at times generating income. The global market for artisan handcrafts alone was worth an estimated $30 billion in 2011. Handicraft production and sales account for a substantial share of GDP in some countries. In Tunisia 300,000 craft workers account for 3.8 percent of GDP. In Thailand the number of craft workers is estimated at 2 million. And in Colombia craft production generates an annual income of roughly $400 million, including some $40 million in exports.[7] In developing countries the export of visual art in 2011 was around $9 billion and that of publishing $8 billion.[8]

Innovations driven by creative expression are taking place in many other areas of work, including health, energy and finance, with the potential to improve lives. When people are innovative and creative in their work, they can advance human progress by making leaps in health care, education and other areas rather than limiting advances to small increments. For example, immunotherapy tries to enhance the immunity of healthy cells, rather than focusing on cancer cells, so that a body's enhanced immunity can isolate the cancer cells and destroy them. This has been hailed as a new pillar in the war against cancer, particularly for lung cancer, which kills 354,000 Europeans and 158,000 Americans every year.[9] In energy the glass for buildings and cars has been designed to make electricity from sunlight. Solar Impulse is an experiment in solar-powered flight. Automated wealth managers can offer sound financial advice for a small fraction of the cost of a real-life adviser.

More than 970 million people—62 percent of them in low- and lower middle-income countries, 12 percent in upper middle-income countries and 26 percent in high-income countries—who engage in volunteer activity each year have helped families and communities, improving social cohesion and contributing to the public good. Around two-thirds of volunteer work is performed in informal settings or through community mechanisms, such as watching a neighbour's children, house sitting and the like. The rest is channelled through organizations, typically nonprofits. The economic size of the volunteer economy is estimated as equivalent to 2.4 percent of global GDP.[10] Older people in many countries also spend a considerable amount of time in voluntary work. In 2011, 33 percent of people ages 55–75 in the Netherlands and 30 percent of people ages 65–74 in the United Kingdom were involved in volunteering.[11]

The work of overseas workers and their remittances have helped advance human development in both source and destination countries. Between 1990 and 2013 the number of international migrants worldwide rose more than 92 million, to 247 million, with most of the growth between 2000 and 2010. It is expected to surpass 250 million in 2015. Of the 143 million in developed countries, 40 percent

FIGURE 2.3

There is no automatic link between income and human development, 2014

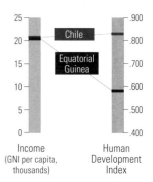

Similar income, different HDI value

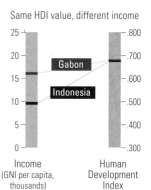

Same HDI value, different income

Source: Human Development Report Office calculations.

Shrinking contrasts in
human development
achievement remain
across and within
regions and countries

were born in a developed country. Around half of international migrants are women—the proportion is slightly higher (around 52 percent) and rising in developed countries and lower (around 43 percent) and falling in developing countries.[12] These people have contributed to the economic growth and income generation of destination countries and brought skills, knowledge and creativity to these economies.

Remittances have been a major source of foreign exchange reserves for many developing countries, with considerable macroeconomic implications. But at the microeconomic level remittances have been the lifeline of many households in terms of income, as well as in terms of resources for better health and education. Officially recorded global remittances totalled $583 billion, or more than four times the global official development assistance, in 2014 and are projected to reach $586 billion in 2015. Officially recorded remittances to developing countries are expected to increase from $436 billion in 2014 to $440 billion in 2015.[13] The top recipients in 2014 were India ($70 billion, or 4 percent of GDP), China ($64 billion, less than 1 percent), the Philippines ($28 billion, 10 percent) and Mexico ($25 billion, 2 percent). Remittance flows are even more important in some Eastern European and Commonwealth of Independent States countries: Remittances in Tajikistan were equivalent to 49 percent of GDP in 2013.[14] Migrants also send remittances through unofficial channels or carry them when they return to their home country. Figure 2.4 reveals their importance for such human development outcomes as poverty reduction in Kyrgyzstan.

Serious human deprivations, but considerable human potential not in use

Despite overall gains in human development, not everyone has benefited. And serious human deprivations remain in various areas of human lives. Furthermore, a huge amount of human potential is not in use because of widespread unemployment, particularly among young people, and because many people are working but remain impoverished, with limited work choices. The paid work potential of women

around the world is also not at the optimum level, with gender-based discrepancies in wages, job opportunities and career advancement.

Uneven human development achievements

Striking contrasts in human development achievements remain across and within regions and countries. By 2014 all regions except Sub-Saharan Africa had reached the medium human development classification. Within the low human development classification the HDI value ranges from 0.348 for Niger to 0.548 for Nepal.

National HDI values can mask large differences within countries, but countries may disaggregate the HDI to unmask the differences and use the results for proactive policies to provide more support to less developed areas. Ethiopia has done so through its 2015 National Human Development Report (figure 2.5).

Progress is also uneven across various human development indicators: Between 1990 and 2015 while the global prevalence of extreme poverty was reduced from 47 percent to

FIGURE 2.4

Income poverty in Kyrgyzstan would be much higher without remittances

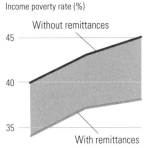

Income poverty rate (%)

Note: Data refer only to official flows.
Source: Data from a household budget survey by the Kyrgyzstan National Statistical Committee.

FIGURE 2.5

Disaggregated Human Development Index values can unmask national averages: Ethiopia

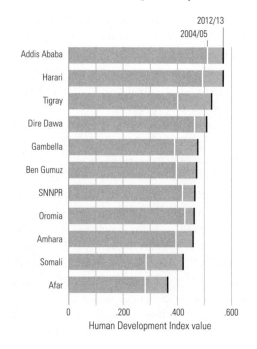

SNNPR is Southern Nations, Nationalities, and Peoples' Region.
Source: UNDP 2015a.

14 percent (a 70 percent reduction), in Sub-Saharan Africa it was reduced from 57 percent to only 41 percent (a 28 percent reduction). Between 2014 and 2016 less than 5 percent of people in Latin America are estimated to be malnourished, compared with 16 percent in South Asia and 23 percent in Sub-Saharan Africa. In terms of stunting, children from the poorest quintile of the population are more than twice as likely to be stunted as those from the richest quintile.[15]

In 2015 the child mortality rate was 11 deaths per 1,000 live births for East Asia and 86 for Sub-Saharan Africa. The highest maternal mortality ratio is in Sub-Saharan Africa (510 deaths per 100,000 live births), followed by South Asia (190).[16] In 2014 women accounted for more than half the world's people living with HIV. And more than 70 percent of the world's people living with HIV were in Sub-Saharan Africa. Nearly 1.4 million newly infected people were also in the region.[17] Of the 57 million out-of-school children at the primary level, 33 million are in Sub-Saharan Africa. In developing countries children in the poorest households are four times as likely to be out of school as those in the richest households.[18] In East Asia and the Pacific, Europe and Central Asia, and Latin America and the Caribbean the gross secondary enrolment ratio has surpassed 80 percent, but in Sub-Saharan Africa it is less than 50 percent. Access to safe drinking water and basic sanitation is uneven between rural and urban areas around the world. Rural access has improved, but most progress has been in urban areas. For example, 96 percent of urban populations have access to an improved source of drinking water, compared with 84 percent of rural populations.[19]

In all regions the HDI value is lower for women than for men (table 2.1)—women fare better in only in 14 countries, including Barbados, Estonia, Poland and Uruguay.

Even though the digital revolution has contributed substantially to human development and revolutionized the world of work, as chapter 3 shows, access to the digital revolution remains uneven, constraining the effects it could have on human lives (box 2.1).

Within countries differences can be considerable across income, age and ethnic groups and between rural and urban locations. In Malaysia in 2012 the richest 10 percent of population had 32 percent of national income, while the poorest 10 percent had only 2 percent.[20] In Nepal about 10 percent of Hill Brahmans (the upper caste) lived in income poverty in 2014,

> In all regions the Human Development Index value is lower for women than for men

TABLE 2.1

Gender Development Index values by region, 2014

	Human Development Index (HDI) value		Gender Development Index value (female HDI value/ male HDI value)
	Female	**Male**	
Arab States	0.611	0.719	0.849
East Asia and the Pacific	0.692	0.730	0.948
Europe and Central Asia	0.719	0.760	0.945
Latin America and the Caribbean	0.736	0.754	0.976
South Asia	0.525	0.655	0.801
Sub-Saharan Africa	0.480	0.550	0.872

Source: Human Development Report Office calculations.

BOX 2.1

Uneven access to the digital revolution

- Developed and developing countries: In 2015, 81 percent of households in developed countries had Internet access, compared with 34 percent in developing countries and 7 percent in the least developed countries.
- Urban and rural areas: In 2015, 89 percent of the world's urban population had 3G mobile broadband coverage, compared with 29 percent of its rural population.
- Women and men: In 2013, 1.3 billion women (37 percent) and 1.5 billion men (41 percent) used the Internet.

- Young and old: In 2013 people ages 24 and younger accounted for 42.4 percent of the world's population but 45 percent of Internet users.
- Website content production: This is dominated by developed countries, which in 2013 accounted for 80 percent of all new domain-name registrations. Registrations from Africa were less than 1 percent.

Source: ITU 2013, 2015.

compared with 44 percent of Hill Dalits (lower caste).[21] In 2012, 43 percent of adults in South Africa were below the national poverty line, compared with 57 percent of children.[22] The prevalence of child poverty was also highest among the poorest households (figure 2.6). In Moldova rural–urban disparities in access to basic social services are sharp (figure 2.7). In the Czech Republic the male unemployment rate is 33 percent among Roma and 5 percent among non-Roma.[23]

Human development deficits

Apart from uneven achievements in human development, deficits remain in many dimensions, impairing capabilities, opportunities and thus human well-being.

Human Development Index shortfall

The HDI is a composite index with an upper limit of 1.0. In 2014 the average global value was 0.711, so there was a shortfall of 0.289. One way of considering HDI progress is to consider how some regions have cut their shortfall. In East Asia and the Pacific the shortfall was 0.484 in 1990 and 0.290 in 2014, a narrowing of 0.194 (40 percent) and the steepest reduction among developing country regions (figure 2.8). In Sub-Saharan Africa the HDI shortfall was whittled down by only about 20 percent over the same period. One major challenge for human development in the coming years is to reduce the HDI shortfall.

FIGURE 2.6

Children in South Africa are not only disproportionately poor, but also more concentrated in the poorest households, 2012

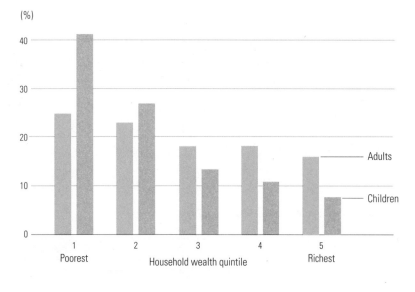

Source: Hall 2015.

FIGURE 2.7

Access to basic social services in Moldova is unequal by rural–urban locale, 2014

(% of population)

Absolute poverty rate

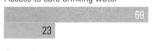
5 Urban
19 Rural

Access to safe drinking water

69
23

Access to sewerage

50
1

Source: UNDP 2014c.

FIGURE 2.8

East Asia and the Pacific had the greatest reduction in Human Development Index shortfall among developing country regions over 1990–2014

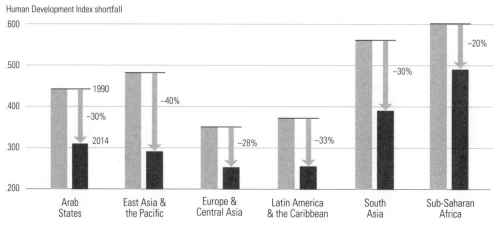

Source: Human Development Report Office calculations.

Multidimensional poverty

The conventional measure of poverty considers only income: People in extreme poverty live on less than $1.25 a day. But people can also be deprived of schooling, be undernourished or lack access to safe drinking water. This broader concept of poverty is reflected in the Multidimensional Poverty Index, a weighted average of 10 indicators. A person is considered to be in multidimensional poverty if she or he is deprived in at least a third of these indicators, with each indicator having a defined deprivation level.

The Human Development Report has been reporting Multidimensional Poverty Index estimates since 2010. This year the Multidimensional Poverty Index is estimated for 101 countries (see table 6 in *Statistical annex*). The estimates suggest that about 1.5 billion people live in multidimensional poverty. Table 2.2 lists the five countries with the largest populations in multidimensional poverty; however, the countries with the highest proportions of their population in severe poverty (deprived in more than half the dimensions) are Niger, South Sudan, Chad, Ethiopia, Burkina Faso and Somalia, at more than 60 percent, and Guinea-Bissau and Mali, at more than half.

The Multidimensional Poverty Index usually represents a national average, which can mask disparities and large areas of deprivation within countries. In China an estimated 72 million people live in multidimensional poverty, but its prevalence varies greatly across provinces, with rural areas having a higher prevalence than urban areas (figure 2.9).

In addition to considering indices of human development and multidimensional poverty, it is useful to consider the absolute number of people who are deprived (figure 2.10).

Huge human potential not in use

Despite impressive achievements in many areas, huge human potential remains unused. In 2015, 204 million people worldwide—including 74 million young people (ages 15–24)—were unemployed.[24] About 830 million workers in developing country regions live on less than $2 a day.[25] Half of workers and their families in developing countries live on less than $4 a day.[26] The work potential for these people is not fully used or rewarded.

Two groups whose work potential is not fully used are women (particularly in the context of paid work) and younger people. Women are less likely than men to seek or find paid work and are less likely to have secure employment or senior positions when they are employed. Worldwide, the labour force participation rate (a measure of those active in the labour market) in 2015 was around 50 percent among women, compared with 76 percent among men.[27] In the same year 47 percent of working-age (ages 15 and older) women were employed, compared with 72 percent of men. Nearly 50 percent of employed women are in vulnerable employment, compared with 44 percent of men.[28] Worldwide, women hold only 18 percent of ministerial

Despite impressive achievements in many ways, huge human potential remains unused

TABLE 2.2

Countries with the most people in multidimensional poverty

Country	Year	Population in multidimensional poverty	
		(millions)	(%)
Ethiopia	2011	78.9	88.2
Nigeria	2013	88.4	50.9
Bangladesh	2011	75.6	49.5
Pakistan	2012/2013	83.0	45.6
China	2012	71.9	5.2

Source: Human Development Report Office calculations using data from Demographic and Health Surveys, Multiple Indicator Cluster Surveys and national household surveys.

FIGURE 2.9

Rural areas of China have a higher prevalence of multidimensional poverty than urban areas, 2012

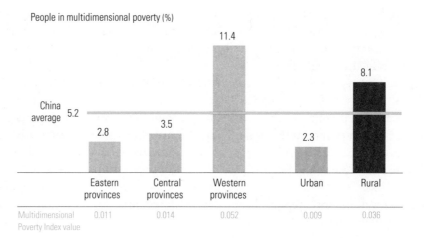

People in multidimensional poverty (%)

	Eastern provinces	Central provinces	Western provinces	Urban	Rural
Multidimensional Poverty Index value	0.011	0.014	0.052	0.009	0.036

China average 5.2

Source: Human Development Report Office calculations based on data from the 2012 China Family Panel Studies.

FIGURE 2.10

Extent of human deprivations in the world

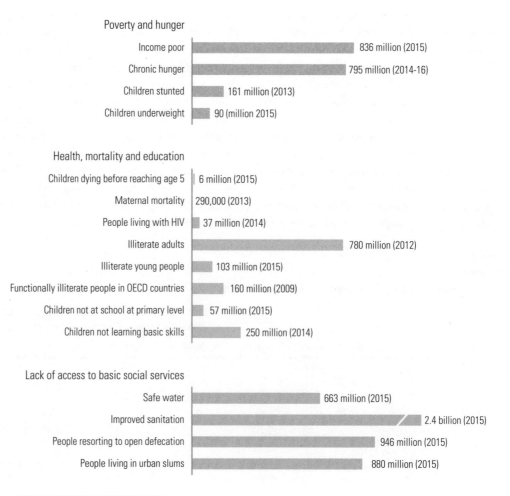

Poverty and hunger

- Income poor — 836 million (2015)
- Chronic hunger — 795 million (2014-16)
- Children stunted — 161 million (2013)
- Children underweight — 90 (million 2015)

Health, mortality and education

- Children dying before reaching age 5 — 6 million (2015)
- Maternal mortality — 290,000 (2013)
- People living with HIV — 37 million (2014)
- Illiterate adults — 780 million (2012)
- Illiterate young people — 103 million (2015)
- Functionally illiterate people in OECD countries — 160 million (2009)
- Children not at school at primary level — 57 million (2015)
- Children not learning basic skills — 250 million (2014)

Lack of access to basic social services

- Safe water — 663 million (2015)
- Improved sanitation — 2.4 billion (2015)
- People resorting to open defecation — 946 million (2015)
- People living in urban slums — 880 million (2015)

Source: UN 2015b; UNAIDS 2015; UNESCO 2013a, 2014.

More than 1.5 billion workers worldwide are in vulnerable employment

positions, 22 percent of parliamentary seats and 25 percent of administrative and managerial positions.[29] Some 32 percent of businesses do not have any female senior managers.[30] See chapters 4 and 6 for more details on these issues.

Workers are also staying unemployed longer. Even in the United States, where the recovery seems to have been stronger, at the end of May 2015 nearly 30 percent of jobseekers were unemployed for more than six months.[31] This is damaging for workers and for the economy, as people out of work for a long period lose their skills and struggle to find jobs at a similar skill level. Longer term unemployment (12 months or more) pushes people towards informal sector, low-level, low-productivity jobs. It also makes them "discouraged workers"—that is, people who would like to work but who do not actively look for work for various reasons.

Global employment growth has slowed to around 1.4 percent a year since 2011, or far below the average annual growth rate of 1.7 percent during the pre-crisis years (2000–2007).[32] Some 61 million fewer people worldwide had jobs in 2014 than would have been the case if the pre-crisis trend had continued. And on current trends, the number of global jobseekers could reach 215 million by 2018, up from 206 million in 2014.[33]

Vulnerable work

Worldwide, only half of workers were in wage or salaried employment in 2014, with wide variations across regions. For countries with data covering 84 percent of global employment, 26 percent of employed people have a permanent contract, 13 percent a temporary one and 61 percent work without any contract. Part-time work is widespread. For 86 countries with data covering 65 percent of global employment, more than 17 percent of employed people work part time. Women are more likely to work part time than men (24 percent versus 12 percent).[34]

More than 1.5 billion workers worldwide are in vulnerable employment. Vulnerable employment accounts for almost 80 percent of work in low human development countries.[35] Most underemployed and underpaid workers work informally. In most developing country regions informal work accounts for more than half of nonagricultural work—the highest

share is in South Asia (82 percent), followed by Sub-Saharan Africa (66 percent), East and South-East Asia (65 percent) and Latin America (51 percent). Some 700 million informal workers live in extreme poverty. In many countries informal work accounts for a considerable proportion of national nonagricultural output—for example, 46 percent in India in 2008. The proportion is also very high in some West African countries—more than 50 percent in Benin, Niger and Togo.[36]

Youth unemployment

Today more than half the world's population is under age 30.[37] These young people are likely to be healthier and better educated than their parents and can take advantage of modern communications technologies and media that enable them to engage more fully in global society. So they have higher work expectations —but many of them cannot find work.

In 2015, 74 million young people (ages 15-24) were unemployed.[38] The youth-to-adult unemployment ratio is at a historical peak and is particularly high in the Arab States as well as in parts of Southern Europe and Latin America and the Caribbean.[39] For example, the youth unemployment rate in 2014 was 3.4 times the adult unemployment rate in Italy, nearly 3 times higher in Croatia and nearly 2.5 times higher in the Czech Republic, Portugal and Slovakia. In absolute terms the youth unemployment rate in these countries was also high—53 percent in Spain, 46 percent in Croatia, 35 percent in Portugal and 30 percent in Slovakia.[40]

The youth unemployment rate is highest in the Arab States, where in many countries of the region insufficient numbers of jobs are being created for the increasingly educated workforce (figure 2.11). For example, between 1995 and 2006 Egypt produced 5 million college graduates but generated only 1.8 million jobs in skill-intensive service sectors (excluding construction and wholesale and retail trade).[41]

Youth unemployment is not confined to countries with a young population. In countries with an ageing population, such as Greece and Spain, more than half of economically active young people are unemployed, a situation worsened by persistently high school dropout rates. Early school leavers account for 28 percent of the

Today more than half the world's population is under age 30

FIGURE 2.11

The youth unemployment rate is highest in the Arab States, 2008–2014

> In 2015, 74 million young people (ages 15–24) were unemployed

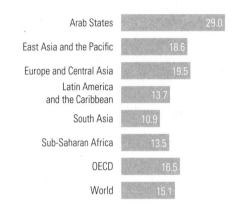

Arab States	29.0
East Asia and the Pacific	18.6
Europe and Central Asia	19.5
Latin America and the Caribbean	13.7
South Asia	10.9
Sub-Saharan Africa	13.5
OECD	16.5
World	15.1

Source: Human Development Report Office calculations based on ILO (2015e).

population ages 18–24 in Spain, compared with 14 percent across the European Union.[42] Young unskilled workers have a much lower chance of securing full-time work in the formal economy.

Youth unemployment, exacerbated by the 2008 financial crisis and the global recession, seems set to continue during the slow post-crisis growth. Between 2012 and 2020 almost 1.1 billion young jobseekers are expected to enter the job market, many of them in South Asia and Sub-Saharan Africa.[43] There is also mismatch between skills demanded and skills available. A mismatch in skills is associated with a high proportion of young people not in education, employment or training. One outcome of this phenomenon is high youth unemployment, as in Spain, where the youth unemployment rate was more than 50 percent in early 2015.[44] Developing countries face many of the same issues but have to employ even more young people—absorbing not just the highly skilled and technologically savvy, but also the growing bulge of young people with fewer skills, many of whom will need to work in agriculture and basic services.

For many young workers the only route to employment in the formal economy is through temporary work. They are thus more likely to be laid off, because many countries, particularly in Europe, prioritize job protection for older permanent workers. Since the start of the crisis, numerous countries have seen a steep rise in

the proportion of young people who are neither employed nor in education or training.[45] Indeed, youth workforce participation is decreasing everywhere. One consequence of low participation and high youth unemployment is limited contributions—or the late start of contributions—for retirement benefits. Another is that rising youth unemployment, particularly of the long-term sort, if left unaddressed, could leave economies with a lost generation of workers (box 2.2).

As described in chapter 1, in 2012 some 21 million people worldwide were in forced labour, 14 million of them in labour exploitation and 4.5 million in forced sexual exploitation (including forced prostitution and forced pornography).[46] And there were about 168 million child labourers.[47] This human potential is wasted.

Further human development challenges

There are human development challenges beyond uneven human progress, existing human

BOX 2.2

Impact of long-term youth unemployment

Young people unemployed for long periods will not only suffer financially, but also start to lose their skills, reducing their future work potential. And they lose self-esteem, with impacts on their well-being and that of their family members. Youth unemployment can also reduce a country's economic potential by underusing people's capabilities. In addition, it undermines social cohesion, with the potential for increases in crime, violence and social unrest, affecting entire communities.

One study of 69 countries found that being without a job translates into negative views about the effectiveness of democracy, particularly among the long-term unemployed.[1] In North Africa postponing the transition from youth to adulthood has fuelled resentment and unrest. In Somalia, where youth unemployment stands at 67 percent, feelings of being socially excluded and financially marginalized make young people more vulnerable to recruitment by extremist groups.[2]

Note
1. Altindag and Mocan 2010. 2. UNDP 2012c.
Source: Human Development Report Office.

deprivations and lost human potentials. Some are already looming large (such as rising inequality and climate change), some have been changing rapidly in their extent and nature (such as human insecurity, shocks and vulnerabilities) and some have full implications that are yet to be captured (such as population growth, growing urbanization and global epidemics).

Rising inequality

In recent years rising incomes around the world have been accompanied by rising inequality in income, wealth and opportunities. Inequality can be measured by the Gini coefficient, which is expressed as a value between 0 (everyone has the same) and 1 (one person has everything).

From 1990 to 2010 on average and taking into account population size, income inequality increased 11 percent in developing countries. A substantial majority of households in developing countries—more than 75 percent of the population—live in a society where income is more unequally distributed than in the 1990s. Income inequality is also a serious issue in developed countries. Between the 1990s and 2010 household income inequality increased 9 percent in high-income countries.[48] Although income inequality across households has risen in many countries, some estimates show that it has fallen for the world as a whole as the average incomes of developing and developed regions have been converging.[49]

Inequality can be measured by the Inequality-adjusted HDI. Each component is discounted according to the extent of inequality in that component. The Inequality-adjusted HDI value is 33 percent lower than the HDI value in

Sub-Saharan Africa and more than 25 percent lower in South Asia and the Arab States (table 2.3). In four countries—the Central African Republic, Comoros, Namibia and Sierra Leone—the Inequality-adjusted HDI value is more than 40 percent lower. In 35 other countries it is 30–40 percent lower.

There have been sharp increases at the top of the income distribution. Between 1976 and 2011 the share of total annual income received by the richest 1 percent of the population in the United States rose from 9 percent to 20 percent. Similarly, between 1980 and 2007 the share received by the richest 1 percent in the United Kingdom rose from 7 percent to 13 percent.[50] Some developing countries have similar patterns. In Colombia the richest 1 percent of earners receives 20 percent of national income, and the situation is similar in South Africa.[51] The increasing income inequalities in economies have affected labour more and more, a point made by former United States Labor Secretary Professor Robert Reich in a special contribution (signed box).

With regard to global wealth, inequality is substantial. In fact, a small elite takes a large share of global wealth. The richest 1 percent held 48 percent of global wealth in 2014, a share projected to be more than 50 percent in 2016.[52] Around 80 percent of the world's people have just 6 percent of global wealth (figure 2.12). Indeed, just 80 individuals together have as much wealth as the world's poorest 3.5 billion people. Such inequality has become a serious problem—both for economic efficiency and for social stability.[53]

Inequalities extend beyond income. Widespread inequality in opportunities persists,

FIGURE 2.12

Around 80 percent of the world's people have just 6 percent of global wealth, 2014

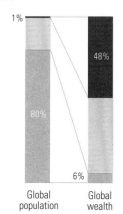

Source: Oxfam 2015.

Just 80 individuals together have as much wealth as the world's poorest 3.5 billion people

TABLE 2.3

Human Development Index and Inequality-adjusted Human Development Index values for selected regions, 2014

	Human Development Index (HDI) value	Inequality-adjusted HDI value	Loss due to inequality (%)
Sub-Saharan Africa	0.518	0.345	33.3
South Asia	0.607	0.433	28.7
Arab States	0.686	0.512	25.4
Latin America and the Caribbean	0.748	0.570	23.7

Source: Human Development Report Office calculations.

Inequality and labour markets

The argument voiced in the late 19th century over alleged "freedom of contract" was that any deal between employees and workers was perfectly fine if both sides voluntarily agreed to it.

So workers were worth no more than the wages they could command in the labour market. And if they toiled 12 hour days in sweatshops for lack of any better alternative, it was not a problem. They had "voluntarily" agreed to those conditions.

It was a time of great wealth for a few and squalor for many. And of corruption, as the lackeys of robber barons deposited sacks of cash on the desks of pliant legislators.

Eventually, after decades of labour strife and political tumult, the 20th century brought an understanding that capitalism requires minimum standards of decency and fairness—workplace safety, a minimum wage, maximum hours (and time-and-a-half for overtime) and a ban on child labour.

We also learned that capitalism needs a fair balance of power between big corporations and workers.

We achieved that through antitrust laws that reduced the capacity of giant corporations to impose their will and through labour laws that allowed workers to organize and bargain collectively.

By the 1950s, when 35 percent of private sector workers in the United States belonged to a labour union, they were able to negotiate higher wages and better working conditions than employers would otherwise have voluntarily provided.

But now America seems to be heading back to the 19th century.

Corporations are shifting full-time work onto temps, freelancers and contract workers who fall outside the labour protections established decades ago.

Meanwhile, the nation's biggest corporations and Wall Street banks are larger and more potent than ever.

And labour union membership has shrunk to fewer than 7 percent of private sector workers.

So it is not surprising we are once again hearing that workers are worth no more than what they can get in the market.

But as we should have learned a century ago, labour markets do not exist in nature. They are created by human beings. The real question is how they are organized and for whose benefit.

In the late 19th century they were organized for the benefit of a few at the top.

But by the middle of the 20th century they were organized for the vast majority.

During the 30 years after the end of World War II, as the economy doubled in size, so did the wages of most Americans—accompanied by improved hours and working conditions.

Yet since around 1980, even though the economy has doubled once again (the Great Recession notwithstanding), the wages of most Americans have stagnated. And their benefits and working conditions have deteriorated.

This is not because most Americans are worth less. In fact, worker productivity is higher than ever.

It is because big corporations, Wall Street and some enormously wealthy individuals have gained political power to organize the market in ways that have enhanced their wealth while leaving most Americans behind.

That includes trade agreements protecting the intellectual property of large corporations and Wall Street's financial assets but not American jobs and wages.

Bailouts of big Wall Street banks and their executives and shareholders when they cannot pay what they owe, but not of homeowners who cannot meet their mortgage payments.

Bankruptcy protection for big corporations, allowing them to shed their debts, including labour contracts. But no bankruptcy protection for college graduates overburdened with student debts.

Antitrust leniency towards a vast swathe of American industry—including Big Cable (Comcast, AT&T, Time-Warner), Big Tech (Amazon, Google), Big Pharma, the largest Wall Street banks and giant retailers (Walmart).

But it also includes less tolerance towards labour unions—as workers trying to form unions are fired with impunity and more states adopt so-called "right-to-work" laws that undermine unions.

We seem to be heading full speed back to the late 19th century. What will reverse it this time?

Robert Reich
Former United States Secretary of Labor

particularly for education. In the late 2000s in South Asia children in the wealthiest quintile were twice as likely as those in the poorest to complete primary school.[54]

There are also serious inequalities in health. In Latin America and the Caribbean and in East Asia children in the poorest asset quintile are about three times more likely than those in the richest to die before age 5. Health disparities are also wide between urban and rural areas: In Latin America and the Caribbean child mortality is a third higher in rural areas than in urban areas.[55]

Population growth and structure

Between 2015 and 2050 the world's population is projected to rise from 7.3 billion to 9.6 billion. Most of this growth will be in developing countries—from 6.0 billion to 8.2 billion. A substantial part of this is attributable to high fertility in 15 countries, mostly least developed countries, in Sub-Saharan Africa. By 2050, 85 percent of the world's people are projected to live in developing country regions.[56] Population growth has major implications for human development and has a direct impact on the lives of women and girls—including for the labour force and jobs, care work and social protection, which are considered in chapters 4 and 6. Ageing is also a major issue.

One dimension of population growth is the expansion of a sizeable global middle class, defined as households with daily expenditure of $10–$100 per capita (in purchasing power parity terms).[57] That global middle class is expected to be 4.9 billion—nearly 57 percent of the global population—in 2030, with 3.2 billion in Asia and the Pacific, mostly in China and India. Consumption by the middle class in the 10 countries with the largest middle-class population (Brazil, China, France, Germany, India, Indonesia, Japan, Mexico, the Russian Federation and the United States) is forecast to be $38 trillion in 2030.[58] These shifts will have major implications for the consumption patterns and living standards of a large share of the global population and as a result for environmental sustainability and other aspects of human development.

Thanks to technological advances in medicine, nutrition and sanitation infrastructures, people in most countries are living longer, with advances often improving the quantity and quality of basic social services, particularly in poor countries. Over the first decade of the 21st century, global life expectancy increased by more than three years.[59] Greater longevity is a welcome sign of human development, but it raises new issues for public policies on work,

health care and social security, as well as retirement ages and pensions—issues that are taken up in chapters 3 and 6.

As a result of greater longevity and declining fertility, older people now make up an increasing proportion of national populations. This ageing is reflected in the old-age dependency ratio: the ratio of the number of people ages 65 and older to the number of people of working age (ages 15–64). In developing country regions the ratio was about 10 percent in 2015. It is expected to rise to 47 percent by 2050.[60] Although ageing is a global phenomenon, countries are at different stages depending on their levels of development and their stage in the demographic transition (figure 2.13). In low human development countries the main driver of a high dependency ratio is a young population (children under age 15). In very high human development countries the demographic pressures of old and young populations are similar, but the older population will predominate fairly soon.

Growing urbanization

The world is experiencing an unprecedented transition from rural to urban living. In 1950 a third of the world's population lived in cities, in 2000, nearly half were city dwellers, and in 2050 more than two-thirds of the world's population is projected to be living in urban areas (figure 2.14). This rapid increase will take place mainly in developing countries. Africa and Asia —both still less urbanized than other regions— will have the fastest urban growth rates. Africa's urban population is projected to jump from 40 percent today to 56 percent by 2050, and Asia's from 48 percent to 64 percent.[61]

One factor driving rapid urbanization in developing countries is rural–urban migration motivated by better employment opportunities in cities ("pull factors"). The largest rural–urban migration is in China, where about 275 million people, or more than a third of the labour force, are migrant workers from the countryside.[62] In many developing countries migration from rural areas to cities is driven at least partly by natural disasters (such as floods) and increasing land degradation and desertification that make agriculture difficult ("push factors"). For example, many people are migrating from

FIGURE 2.13

In low human development countries the main driver of a high dependency ratio is a young population, 2014

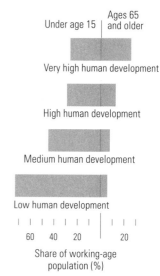

Source: Human Development Report Office calculations based on UNDESA (2013b).

FIGURE 2.14

In 1950 a third of the world's population lived in cities, in 2000 nearly half were city dwellers and by 2050 more than two-thirds will be

1950 Urban population—1.2 billion

Urban share of the world population (%)

2000 Urban population—2.6 billion

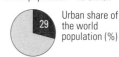

2050 Urban population—6.2 billion

Source: WEF 2015.

drought-stricken areas in northeast Brazil to favelas in Rio de Janeiro.[63]

Urbanization has the potential to improve the economic well-being of societies. More than half the world's people live in cities, but they generate more than 80 percent of global GDP.[64] In most countries the health of city dwellers has improved through access to better education and health care, better living conditions and targeted public health interventions. Cities hold promises for improving human well-being.[65]

Yet urbanization also presents many human development challenges. It puts pressure on cities' infrastructures—such as housing, electricity, drinking water and sanitation—and may adversely affect residents' quality of life. Nearly 40 percent of the world's urban expansion may be in slums, exacerbating economic disparities and unsanitary conditions. Almost 700 million urban slum dwellers lack adequate sanitation, which, along with the paucity of safe drinking water, raises the risk of communicable diseases such as cholera and diarrhoea, particularly among children.[66] And urbanization creates income and opportunity disparities, not only between rural and urban areas but also among socioeconomic groups in cities, boosting social tensions.[67]

Human security

Human development is under threat from many sources. In 2014 the world witnessed widespread conflicts, violence and human rights violations that resulted in the massive displacement of people, within and outside their countries. At the end of 2014 almost 60 million people had been displaced worldwide, the highest level recorded since the Second World War. If these people were a nation, they would make up the 24th largest country in the world. Multiple conflicts took place in the Democratic Republic of the Congo, Iraq, Nigeria, Pakistan, South Sudan, the Syrian Arab Republic and Ukraine.[68]

About a third of the 60 million were refugees or asylum seekers, and more than 38 million were displaced within the borders of their own countries. Based on available evidence, children accounted for half the global refugee population, the highest proportion in 10 years.

The increase has been driven principally by the growing number of Afghan, Somali and Syrian refugee children.[69]

Between 2000 and 2013 the worldwide cumulative death toll from violent extremism —acts intended to provoke a state of terror in the general public, a group of people or particular people—rose from 3,361 to 17,958.[70] The Islamic State and associated armed groups have carried out brutal attacks on civilians in Iraq, Syria, Turkey and other countries, including killings, rape and sexual slavery, torture, forced religious conversion and the conscription of children. Boko Haram has kidnapped and attacked civilians in Benin, Chad, Cameroon, Nigeria and Niger, and Al Shabaab militants have made similar attacks in Somalia and Kenya. Originating in Uganda but operating in the Central African Republic, the Democratic Republic of the Congo and South Sudan, the Lord's Resistance Army, a militant cult group, has carried out human atrocities.

People living with the threat of violence from violent extremism and conflict not only lose their freedom, but also have fewer opportunities to expand their capabilities. With basic infrastructures—physical and social —destroyed, they have fewer opportunities to earn a livelihood and less access to health services and schooling. Parents who fear for the physical and sexual safety of their children are likely to keep them out of school.[71] Taken together, these conditions increase care responsibilities within the home.

Violent extremism and conflict also erode collective human development. Intimidation and violence can demoralize communities and destroy social structures and political institutions, undermining cohesion and weakening states. And violence targeted at particular groups can magnify existing societal cleavages. Extremist groups regularly attack or threaten women and sexual, ethnic and religious minorities, often using rape as an expression of power and brutality.[72] These threats are global, and their effects cut across national borders—not just because of the flows of refugees, but also because of uncertainty about when and where a terrorist attack might occur.

Violent extremist activities and large-scale conflicts make it difficult for governments to provide essential social services such as basic health care and education. An estimated 30 million children are out of school in conflict-affected countries.[73] Undereducated youth with few economic opportunities are then susceptible to recruitment by extremists, increasingly over the Internet. For example, in Somalia youth engagement with Al Shabaab is driven by high unemployment, inadequate education and weak political participation.[74]

Women's empowerment

Women's empowerment requires female autonomy in all areas of life—financial, economic, political, social and cultural, in and outside the home. Measures of gender parity are improving, but gender parity may not lead to real empowerment in countries where overall development is low.

Improvements in gender equality—the socially constructed relations between men and women—have also occurred. Women are better represented in political decisionmaking, and in the workplace glass ceilings are at least cracking, if not breaking. Stereotyped gender roles are changing in many societies, as is sharing of responsibilities within households.

Even so, women are actively disempowered by mutually reinforcing processes in every walk of life. They are held back by biases in social beliefs, norms and cultural values. They face discrimination in economic, political and social structures as well as policies, institutions and strategies. Too often they are constrained by real and perceived physical insecurity.

Violence against women, including domestic violence, is evident in all societies, among all socioeconomic groups and at all levels of education. According to a 2013 global review, one in three women (35 percent) has experienced physical or sexual intimate partner violence or non–intimate partner sexual violence.[75] But in some countries that share rises to 70 percent.[76] More often than not, the violence goes unreported to the police. Interviews with 42,000 women across the 28 EU member states revealed that only 14 percent of women had reported their most serious incident of intimate partner violence, and merely 13 percent had reported their most serious incident of non–intimate partner violence.[77]

Such violence impacts women's empowerment. The full effects on human development are difficult to measure, but there have been efforts to quantify some of the costs in financial terms. In Australia the cost of domestic and nondomestic violence against women and children has been estimated at A$14 billion a year. In Viet Nam domestic violence against women costs nearly 1.4 percent of GDP in lost earnings and out-of-pocket spending to treat health-related violence outcomes, leading to an overall output loss of 1.8 percent of GDP.[78]

Shocks, vulnerabilities and risks

For many societies, and particularly for poor communities, human development is undermined by multiple shocks, vulnerabilities and risks: economic and financial crises, rising food prices and food insecurities, energy price hikes, emerging health risks and epidemics. Any of these can slow, reverse or completely derail human development, as documented by the financial and economic crisis of 2008–2009. In Latin America an additional 3.2 million people live in poverty over what was expected without the crisis.[79] Globally there were at least 61 million fewer jobs in 2014 than expected.[80]

Emerging health risks

Noncommunicable (or chronic) diseases have become a global health risk. They kill 38 million people a year, almost three-fourths of them (28 million) in low- and middle-income countries. Cardiovascular disease accounts for the most deaths due to noncommunicable diseases (about 18 million), followed by cancer (8.2 million), respiratory disease (4 million) and diabetes (1.5 million).[81]

Tobacco accounts for around 6 million deaths a year (including deaths due to secondhand smoke), a figure projected to increase to 8 million by 2030.[82] Two million women and children—four a minute—die prematurely due to illness caused by indoor air pollution, primarily from smoke produced while cooking.[83]

Obesity, though preventable and reversible, accounts for a growing share of noncommunicable diseases. Over the past 20 years being overweight or obese has emerged as a worldwide scourge on health, initially in developed

Violent extremism and conflict also erode collective human development

countries but now in developing countries, particularly in urban settings (figure 2.15). Almost 30 percent of the world's population, or 2.1 billion people, are overweight or obese, 62 percent of them in developing countries, with wide variations by region. The number of overweight children is projected to double by 2030.[84]

Noncommunicable diseases are not only a matter of lifestyle, but also strongly related to poverty and deprivation. Poor people often have to consume low-priced, low-nutrition, highly processed foods. So their intake is made up largely of calories without micronutrients. The spread of noncommunicable diseases is inherently linked to social influences.

Epidemics

The year 2014 saw a serious outbreak of the Ebola virus. As of late May 2015 almost 27,000 reported, confirmed, probable and suspected cases and 11,000 deaths had been accounted for in Guinea, Liberia and Sierra Leone. More than 5,000 children were infected, and 16,000 children lost one or both parents or their primary caregiver.[85]

Closed schools, threats to past health gains and economic declines are among the many impacts of the Ebola outbreak. An estimated 5 million children were deprived of education in Guinea, Liberia and Sierra Leone, where schools were closed for months.[86] In Sierra Leone the number of children under age 5 receiving treatment for malaria declined 39 percent between May and September 2014. Liberia saw the proportion of women giving birth with the assistance of a skilled health care provider tumble from 52 percent in 2013 to only 37 percent between May and August 2014.[87]

Guinea, Liberia and Sierra Leone felt a total fiscal impact from Ebola of over $500 million in 2014, nearly 5 percent of their combined GDP. In 2015 lost output was estimated to be more than 12 percent of GDP.[88] In addition, the countries have suffered from reduced agricultural production, possible food insecurity, lower wages and pauses in investment plans by external partners.

Climate change

Around the world communities are steadily becoming more vulnerable to the adverse effects of climate change, including loss of biodiversity. Most exposed are those who live in arid zones, on slopes, in areas with poor soils and in forest ecosystems. Estimates show that in 2002, these fragile lands supported around 1.3 billion people (double the number 50 years ago) and have been under increasing pressure.[89] Global emissions of carbon dioxide, which has an impact on climate change, have increased 50 percent since 1990.[90]

Climate change exacerbates all these vulnerabilities, fettering the choices of present and

Noncommunicable (or chronic) diseases have become a global health risk

FIGURE 2.15

Undernourishment and obesity rates vary by region, most recent year available

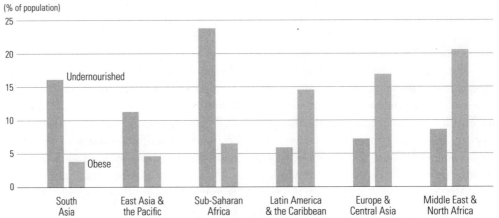

Source: World Bank 2015a.

future generations. Impacts are particularly severe on developing countries and their poorest people, who often live in the most fragile ecological areas and depend directly on the natural environment for their lives and livelihoods. Climate change poses an existential threat to small island developing states.

Water availability will be increasingly affected by climate change, which in Africa could expose 250 million people to greater water stress. In some countries drought could halve the yields from rainfed agriculture by 2020.[91] Across Sub-Saharan Africa and South and East Asia drought and rainfall variations could lead to large productivity losses in cultivated food staples.

The climate changes of recent decades have probably already affected some health outcomes. These changes were held responsible for 2.4 percent of worldwide diarrhoea cases and 6 percent of malaria cases in a few middle-income countries in 2000.[92] Small changes against a "noisy" background of other causal factors are difficult to confirm. But once identified, cases of causal attribution are strengthened by similar observations in different population settings.

The first detectable changes in human health may be alterations in the geographical range (latitude and altitude) and seasonality of certain infectious diseases, including vectorborne infections such as malaria and dengue fever, and foodborne infections such as salmonellosis, which peak in the warmer months. Warmer average temperatures combined with increased climatic variability could alter the pattern of exposure to thermal extremes, with resultant health impacts in both summer and winter.

By contrast, the public health consequences of the disturbance of natural and managed food-producing ecosystems, rising sea levels and population displacements due to physical hazards, land loss, economic disruption and civil strife may not become evident for several decades.

Millions of people are also affected by natural disasters. Between 2003 and 2013 there were on average 388 natural disasters a year, affecting 216 million people and killing 106,654. But in 2013 natural disasters were at their lowest level in 16 years, with 330 disasters that affected 97 million people and killed 21,610.

The total damage was also less—down from the 2003–2013 annual average of $157 billion to $119 billion in 2013 dollars. By number of deaths, of the 10 most affected countries in 2013, five were low- or lower middle-income.[93] Since 2008 an estimated one person every second has been displaced by a disaster, with 19.3 million people forced to flee their homes in 2014 alone.[94]

Human development— looking forward

The record on human development has been mixed—quite uneven across regions, among countries and between groups. While progress has been impressive on some fronts, considerable deprivations remain, all presenting tremendous challenges to human development. Table A2.1 at the end of the chapter presents a balance sheet of human progress and deprivations.

These dynamics must be seen in the context of a changed world. The world today is very different from the world in 1990, when the notion of human development and its measures to assess human well-being were launched. The development canvas has changed, global growth centres have shifted, important demographic transitions have materialized and a fresh wave of development challenges has emerged.

The notion of human development needs to be contextualized to make it a more relevant analytical foundation for dealing with emerging challenges in a changing world. Issues of individual and collective capabilities and choices, their probable trade-offs in various situations, the possible hierarchies among such choices, the intergenerational (for present as well as future generations) choices and capabilities (which reflect the notion of sustainability) and human development in situations of shocks and vulnerabilities will have to be revisited to make the current framework more robust and relevant for the future.

Thinking on some of these issues has already begun. New notions of human well-being (such as subjective well-being and happiness) have been proposed, and alternative options for measuring it have been constructed (box 2.3). Even for human development measures, several innovative approaches have been experimented

Climate change fetters the choices of present and future generations with in some national human development reports (for example, using child mortality as the indicator for a long and healthy life in calculating the HDI for Madhya Pradesh, India).

The above considerations imply that assessing human well-being in a changed world has to go beyond what was developed 25 years ago. The need for exploring new measures and tools for monitoring and assessing human well-being has become more pronounced as the world has been discussing the Post-2015 Development Agenda and the Sustainable Development Goals, successors to the Millennium Development Goals. The agenda and the goals are expected to provide a big push in overcoming the remaining human development gaps and challenges in coming years, but fresher, innovative and relevant assessment tools are needed for monitoring. Measuring environmental sustainability and integrating it with overall measures for human well-being are priorities.

Three other challenges need to be addressed. First, measures and indicators have to be identified so that the impacts of policy measures can be quickly captured. Second, available measures are often inadequate for assessing human well-being at times of shocks and vulnerabilities; measures and indicators that can forecast and capture the impacts of shocks and vulnerabilities should be considered. Third, measures that are agile enough to provide quick policy guidance have to be explored.

All these measures will require robust, consistent and credible data. Taking that into account and also considering a much

> The notion of human development needs to be contextualized for dealing with emerging challenges in a changing world

more ambitious international agenda like the Post-2015 Development Agenda and the Sustainable Development Goals, the High Level Panel on the Post-2015 Agenda convened by the UN Secretary-General in 2014 called for a data revolution. It emphasized the need to leave no one behind in the context of the Post-2015 Development Agenda and to monitor the progress in achieving the agenda. Three issues need highlighting:

- First, substantial quantities of real-time data can provide better information on, say, spikes in prices of food staples that will disproportionately impact the near-poor. Sensors, satellites and other tools produce real-time data on activities and vulnerabilities—for example, on deforestation, urbanization, flooding and poverty indexing via images of tin roofs in Africa.

- Second, "big data" analysis holds the promise of producing highly relevant information almost instantaneously to levels hitherto undreamt of outside population censuses.[95] For example, anonymized mobile phone data can show flows of crowds and be used in urban planning. Trends in millions of searches on the Internet and users' posts on social media offer insights into people's opinions and priorities. Big data has the potential to be harnessed for diverse uses, including measurement of important indicators, capturing attitudinal trends, monitoring health epidemics and providing real-time feedback on the efficacy of policymaking.[96]

Each day people send 294 billion emails, upload 100,000 gigabytes[97] of data to

BOX 2.3

Alternative measures of human well-being

To inform the public and broaden debate over the past 15 years or so, several national governments have released comprehensive reports on national well-being. Some of the earliest initiatives came from Australia, Ireland and the United Kingdom. Rather than using composite indicators, as with the HDI, they have generally involved a collection of individual indicators.

Other countries support subjective measures of well-being or happiness. For example, Bhutan has a Gross National Happiness Index. The United Kingdom

has considered talking not just about GDP but about GWB—general well-being. This would involve generating a single number that summarizes people's feelings about many aspects of well-being. That might avoid the need to assign weights to components, but it raises questions of data reliability and makes cross-country comparisons difficult.

All these approaches of well-being are being used by a range of actors from government, academia, civil society and beyond.

Source: Human Development Report Office.

Facebook, generate 230 million tweets and send and receive 1.3 billion gigabytes of data over mobile phones. According to some estimates, should current trends continue, there will be 100 billion times more data in 2035 than there is today.[98]

Such data are expanding understanding of causation in an increasingly complex world and enabling rapid responses in some humanitarian situations. But such data have risks—the same data could be used to do harm where privacy and anonymity are not fully respected. Notwithstanding the evident issues related to privacy and security, many researchers are engaged in identifying how this large volume of information—generated both incidentally and deliberately as billions of people go about their daily lives—can support sustainability and provide usable insights for improving lives.

- Third, many countries have combined traditional and new methods of data collection for censuses, ranging from administrative registers to mobile devices, geospatial information systems and the Internet. Birth registration coverage has benefited from new technologies: In Albania, Kosovo, Pakistan and other countries geo-mapping technologies have helped collect and visualize birth registration data.

In this changed and changing world, with a new development agenda and new development goals, human progress will depend critically on work by humankind. And making full use of available human potential will be an important ingredient in overcoming existing human development deficits and addressing emerging human development challenges. And that has to be done in a changing world of work —the focus of the next chapter.

Many countries have combined traditional and new methods of data collection

Balance sheet of human development

Progress	Deprivations

Health

- The global child mortality rate in 2015 is less than half the 1990 rate, having dropped from 90 deaths per 1,000 live births to 43—or from about 12.7 million to 6 million in absolute terms.

- Between 1990 and 2013 the global maternal mortality ratio dropped 45 percent, from 380 deaths per 100,000 live births to 210.

- Between 1995 and 2013 tuberculosis prevention, diagnosis and treatment interventions saved 37 million lives worldwide. Between 2000 and 2015 malaria interventions saved 6.2 million lives. And since 2000, global responses to HIV have averted 30 million new infections.

- In 2015 almost 6 million children worldwide died before reaching age 5. Almost 3 million of those deaths occurred during the first 28 days of life, the neonatal period.

- In 2013 nearly 290,000 women worldwide died from causes related to pregnancy and childbirth.

- In 2015 about 214 million cases of malaria occurred around the world, and the disease killed about 472,000. In 2013, 11 million people were living with tuberculosis. And in 2014 an estimated 37 million people were living with HIV.

Education

- The global youth literacy rate (ages 15–24) increased from 83 percent in 1990 to 91 percent in 2015. The adult literacy rate (ages 15 and older) increased from 76 percent to 86 percent.

- Between 1990 and 2015 the number of children enrolled in primary education increased in all regions—and more than doubled in Sub-Saharan Africa.

- Worldwide 780 million adults in 2012 were illiterate, as were 103 million young people in 2015.

- There are 57 million out-of-school children of primary school age around the world. One in six adolescents (ages 14–16) does not complete primary school.

Women's empowerment

- In 42 countries women hold more than 30 percent of seats in at least one chamber of national parliament.

- Globally women earn 24 percent less than men and nearly 50 percent of the employed women are in vulnerable employment.

Access to basic social services

- Access to an improved drinking water source has become a reality for 2.6 billion people since 1990.

- Between 1990 and 2012 almost 2.1 billion additional people worldwide gained access to an improved sanitation facility.

- Some 663 million people worldwide draw water from an unimproved source.

- In 2015, 2.4 billion people worldwide did not use an improved sanitation facility, and 946 million people resorted to open defecation.

Income and poverty

- Between 1990 and 2015 the number of people living below $1.25 a day in developing country regions fell from 1.9 billion to 836 million.

- The proportion of undernourished people—individuals unable to obtain enough food regularly to conduct an active and healthy life—decreased in developing regions from 23.3 percent in 1990–1992 to 12.9 percent in 2014–2016. The prevalence of stunting among children under age 5 fell from 40 percent in 1990 to 25 percent globally in 2013.

- The world's gross national income per capita has gone up from PPP $8,510 in 1990 to PPP $13,551 in 2013.

- In 2013 about 370 million people in developing country regions were working but living on less than $1.25 a day.

- In 2014–2016 nearly 795 million people worldwide, about 780 million of whom were in developing countries, suffered from chronic hunger. One in seven children worldwide—an estimated 90 million children under age 5—is underweight, and one in four children was stunted in 2015.

- In 2014 the richest 1 percent of the world's people owned 48 percent of global wealth, a share that could rise to more than 50 percent in 2016.

Participation

- By the end of 2015 there will be 7.1 billion mobile subscriptions and 3.2 billion Internet users in the world. By the end of 2014 there were more than a billion active users of Facebook per month and more than 300 million active users of Twitter per month.

- Of the world's 7.3 billion people in 2015, nearly 4 billion people, the majority of whom live in developing country regions, do not have access to the Internet. The unconnected are typically the world's poorest and most disadvantaged populations.

Environmental sustainability

- Between 2000 and 2010 the global net loss in forest fell from an average of 8.3 million hectares a year to an average of 5.2 million hectares a year thanks to afforestation and the natural expansion of forests.

- The world has also almost eliminated ozone-depleting substances, consumption of which fell 98 percent between 1986 and 2013.

- About 1.3 billion people in the world live on ecologically fragile land.

- Global emissions of carbon dioxide increased 50 percent between 1990 and 2013.

- Water scarcity affects more than 40 percent of people around the world.

Source: Human Development Report Office.

Chapter **3**

The changing world of work

Infographic: Globalization and digital revolution transforming work

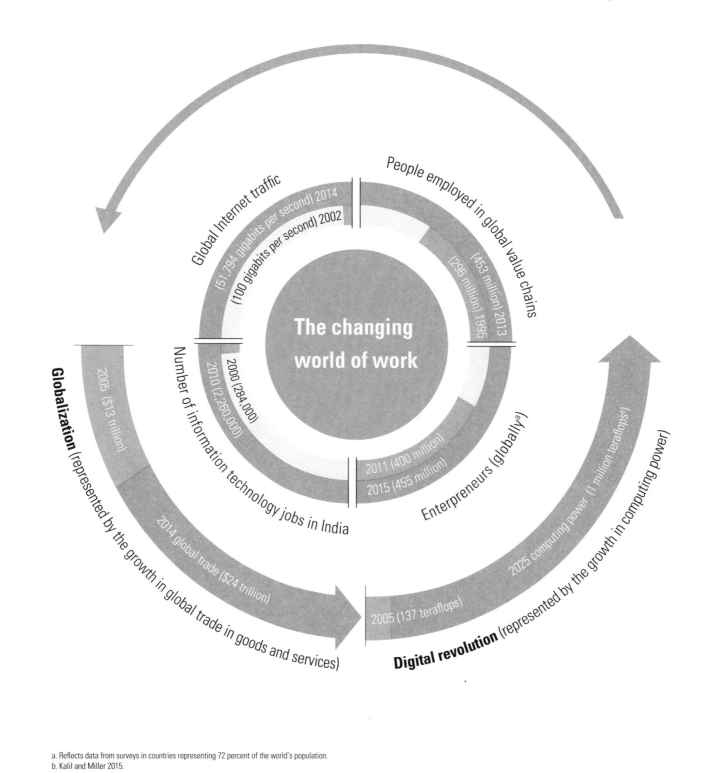

The changing world of work

Global Internet traffic
(51,794 gigabits per second) 2014
(100 gigabits per second) 2002

People employed in global value chains
(453 million) 2013
(296 million) 1995

Number of information technology jobs in India
2010 (2,260,000)
2000 (284,000)

Globalization (represented by the growth in global trade in goods and services)
2005 ($13 trillion)
2014 global trade ($24 trillion)

Enterpreneurs (globally[a])
2011 (400 million)
2015 (455 million)

Digital revolution (represented by the growth in computing power)
2005 (137 teraflops)
2025 computing power (1 million teraflops[b])

a. Reflects data from surveys in countries representing 72 percent of the world's population.
b. Kalil and Miller 2015.

3.

The changing world of work

Throughout history the nature of work has evolved. Changes in social, economic and political structures have changed the when and where of work, the what of goods and services produced and the how of organizing work. The effects on human development have been wide-ranging and complex. Some of the most dramatic changes started in the 18th century with the industrial revolution in Europe (box 3.1).

Today, work transformations are driven by globalization and technological revolutions, particularly the digital revolution—the shift from mechanical to digital technology. Workers and employers around the globe are increasingly linked by complex webs of trade and migration, while financial assets cross borders instantly. Over time companies have restructured and dispersed their production systems. Taking advantage of lower wages and other costs, and in some cases operating closer to emerging markets, they have divided their activities into multiple units and distributed them around the globe, locating them in countries offering not just lower wages, but also the necessary skills and infrastructure.

A transformative force during the past 30 years, globalization has fostered worldwide interdependence, with major impacts on patterns of trade, investment, growth, and job creation and destruction, as well as on networks of creative and volunteer work. We now seem to be living through a new and accelerated technological revolution—or even multiple revolutions at the same time.

These technological revolutions are changing wages and productivity in labour markets and workplaces through new ways of contracting and subcontracting, new conditions of work and new business and organizational models. They are influencing the distribution of labour demand across sectors, with implications for the processes of structural transformation. They also influence the quantity and quality of jobs in some sectors and enterprises as well as the distribution of incomes and wealth at all levels. And they create new opportunities for creativity and innovation and in some ways bring more unpaid work into the public sphere. This pace of change will not slacken—the next 20 years will see a continuing revolution in work and workplaces, marked by complexity, uncertainty and volatility.

Against this backdrop a critical question is: What do all these patterns mean for human development? And more important: Are workers, employers and policymakers prepared to respond to the challenge of the emerging world of work? In such a world, specific technical knowledge quickly becomes obsolete, and the policies and rules of yesterday might not serve the challenges of today or tomorrow. National and international businesses, education providers and policymakers are compelled to keep up with these accelerated changes, to make sense of the trends and to translate them into sound strategies and rules that can deliver more work—and work of higher quality—for present and future generations. Workers face new types of competition and challenges to traditional forms of collective organization.

Traditional paths to development seem less viable today. So seizing the future is not a game of chance or fate—it is a matter of skill, foresight and understanding.

Structural transformation of work

During the past century development in advanced economies followed a path that was more or less linear. Activity moved from agriculture to industry to services. The sectoral distribution of economic activity is reflected both in value added as a proportion of GDP and in employment by sector. Today some trends raise questions about whether such linearity will persist for other workers and economies.

First, despite a persistent drop in the contribution of the agricultural sector to the economy, many workers remain in agriculture,

> Traditional paths to development seem less viable today

BOX 3.1

An encapsulated history of work

At one point in human civilization there were farmers and animal husbandmen. Life was, per Hobbes, "nasty, brutish and short." Taxes and other requirements imposed by chiefs, landlords or the state were onerous. Many people were serfs or slaves, devoid of autonomy and dignity. Save for the lucky few, poverty and injustices were the norm.

Then came the industrial revolution. Men and women flocked from the countryside to towns to satisfy factories' growing demand for labour. The new technologies in transport, cotton textiles, and iron and steel delivered steadily rising labour productivity. But for decades few of these benefits trickled to the workers who worked long hours in stifling conditions, were jammed into packed housing and saw little rise in earnings.

Eventually capitalism transformed itself, and its gains began to be shared more widely. This was in part because wages naturally began to rise as the surplus of workers from the countryside dried up. But equally important, workers organized themselves and began to claim their rights. It was not just their grievances that gave their demands urgency. The conditions of modern industrial production also made it more difficult for the elites to pursue their usual tactics of divide and rule. Factory work, concentrated in major cities, facilitated coordination among labourers, mass mobilization and militant activism.

Fearing revolution, the industrialists compromised. Political rights and the right to vote were extended to the working class. And democracy in turn tamed capitalism. Conditions in the workplace improved as state-mandated or negotiated arrangements reduced working hours and increased safety, along with vacations, family health care and other benefits. Public investment in education and training made workers more productive and freer to exercise choice. Labour's share of the enterprise surplus rose. Factory jobs provided blue-collar work that enabled a middle-class existence, with all its lifestyle opportunities.

Technological progress fostered industrial capitalism but eventually undermined it. Labour productivity rose much faster in manufacturing industries than in the rest of the economy. This meant that the same or higher quantity of steel, cars or electronics could be produced with far fewer workers. Manufacturing's share of total employment began to decline steadily in all the advanced industrial countries after the Second World War. Workers moved to service industries—education, health, entertainment and public administration. Thus was born the post-industrial economy.

Work became more pleasant for some but not all. For those with the skills, capital and savvy to prosper in the post-industrial age, services offered inordinate opportunities. Bankers, consultants and engineers earned much higher wages. Equally important, office work allowed a degree of freedom and personal autonomy that factory work had never provided. Hours may have been long—longer perhaps than in factory work—but service professionals enjoyed much greater control over their daily lives and workplace decisions. Teachers, nurses and waiters were paid not nearly as well, but they too were released from the humdrum mechanical drudgery of the shop floor.

However, the post-industrial economy opened a new chasm between those with good jobs in services—stable, high-paying and rewarding—and those with bad jobs—fleeting, low-paying and unsatisfying. Two things determined the mix between the two types of jobs and the extent of inequality that the post-industrial transition produced. First, the greater the education and skill level of the workforce, the higher the wages in general. Second, the greater the institutionalization of labour markets in services (not just manufacturing), the higher the quality of service jobs in general. So inequality, exclusion and duality became more marked in countries where skills were poorly distributed, and many services approximated the textbook ideal of spot markets.

Source: Rodrik 2015b.

particularly in South Asia and Sub-Saharan Africa. In other words, although the importance of agriculture to economies· may be dropping—in 2010 it accounted for only 3.1 percent of GDP—the importance of agriculture to workers, albeit declining, remains high—in 2010, 33.1 percent of the world's labour force worked in agriculture (figure 3.1).[1] Second, job growth and economic activity in industry have been sluggish in recent years.

Global value added in industry declined from 32.8 percent of GDP in 1995 to 26.9 percent in 2010,[2] and employment rose only 1.2 percentage points.[3] Third, the service sector is growing rapidly, and many people are finding work there, but the jobs are not all or even primarily in high-tech advanced services.

Each of these trends has implications for how policymakers and individuals prepare for the future of work and for how efforts should be

FIGURE 3.1

Although the importance of agriculture to economies may be dropping, the importance of agriculture to workers, albeit declining, remains high

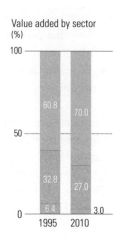

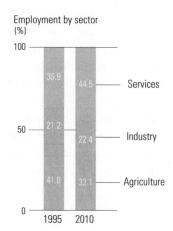

Source: World Bank 2015f; ILO 2015e.

focused to bolster positive links between work and human development.

Agriculture

Though now with a smaller proportion of national output, agriculture is still an important source of work, with 1.34 billion people worldwide working in or seeking work in agriculture.[4] Most of this work is on family farms. Some 70–80 percent of the world's agricultural land is managed by more than 500 million family farms whose workers, mostly family, produce more than 80 percent of the world's food.[5] Around 43 percent of the agricultural labour force in developing countries are women, and in parts of Africa and Asia women account for more than 50 percent of farmers.[6]

Family farms range from plots of less than 1 hectare (72 percent of family farms) to more than 50 hectares (1 percent of family farms).[7] The largest are often highly mechanized and use improved seeds and fertilizers as well as agricultural extension services. By contrast, small and medium-sized farms in developing countries often have limited access to resources and lower productivity. Many workers on family farms supplement their income with off-farm work.

Wages and productivity on these farms are typically low, working conditions can be unsafe and hours are unpredictable. Work is often seasonal: Harvesting and planting demand long work hours of all family members, including children—around 60 percent of child labourers work in agriculture.[8] During some parts of the year, there may be little work or income. Given the high number of people who depend on agriculture for their livelihoods and the vulnerable conditions many of these workers face, efforts to improve productivity and working conditions in agriculture could have considerable positive impacts on human development. (Chapters 5 and 6 discuss these points in detail.)

Industry

In developed and developing economies, industry—particularly manufacturing—remains an important source of work, accounting for 23.2 percent of global employment.[9] But since 1990, manufacturing as a share of total employment has declined in many countries, even in strong exporting countries.[10] Between 2000 and 2010 employment in manufacturing fell 8 percent in Germany and 11 percent in the Republic of Korea.[11]

This is partly because manufacturing is becoming more capital intensive. The use of robots is on the rise. Every year an additional 200,000 industrial robots come into use. The automotive industry, a key export industry for

Manufacturing is becoming more capital intensive

many countries, is a leading purchaser of industrial robots.[12]

Manufacturing is also more skill intensive, and digital technology producers are demanding a different set of skills. One study reports that for a selection of corporations, 10 million jobs in manufacturing cannot be filled because people with the right skills are unavailable.[13]

At the same time, the line between manufacturing and service work has become increasingly blurred. Manufacturing firms are now integrated production systems with such service activities as research, sales and customer support. In the United States 30–55 percent of manufacturing jobs are now service activities.[14]

So, in the future, industry is unlikely to absorb large numbers of rural workers. In the past manufacturing provided work for millions who migrated to cities—as in China. But today these jobs are harder to find. With pressures on manufacturing jobs from globalization and labour-saving technologies, many countries, particularly in Sub-Saharan Africa, face "premature deindustrialization," a phenomenon where opportunities in industry shrink sooner and at much lower levels of income than for early industrializers.[15] The implications are enormous for creating decent jobs for growing populations and in expanding the opportunities that people have to maintain a dignified life.

Services

Since 2002 the leading employment sector globally has been services, which in 2013 accounted for 46 percent of all jobs.[16] The rise of service work stems from the growth of knowledge-intensive work such as financial and business services and technology-intensive sectors, as well as low-skill work such as food services, care work and construction—areas crucial to human well-being but where workers are among the most vulnerable.[17] The global exchange of goods and services and the logistics in trade have created many new jobs. Employment in trade and distribution services has absorbed most workers in the service sector since 1960, with the highest increases between 1990 and 2010.[18]

In many respects the service sector masks a vast array of tasks, skill requirements and work conditions that influence human development in different ways. Highly skilled, highly remunerated knowledge workers apply and produce new technologies that can enhance human progress, as discussed below. Care workers provide essential services that enhance human welfare, but the conditions of their work vary (see chapters 1 and 4). In addition, service work includes innovative, creative and cultural services (see chapter 1).

The growing dominance of work in services calls for policy attention to the sector, which is likely to continue employing most people. The aims should be to ensure that essential services for human development and technological progress are sustained, to enable workers providing these services to gain the requisite skills and to protect service workers against inadequate wages and exploitative work conditions.

Technological revolution

Today the types of work that people do and the ways they carry out tasks are being transformed by new technologies. This is not new, but it is reshaping the links between work and human development and the types of policies and institutions needed to foster positive outcomes for people's well-being.

History has witnessed two industrial revolutions, each associated with a general-purpose technology—a small group of technological innovations so powerful that they interrupt and accelerate the normal march of progress. The first was driven by steam, the second by electricity. The third revolution is unfolding now—driven by computers and networks. If directed towards expanding opportunities and building capabilities, the current technological revolution stands to vastly improve people's lives. But transformations that are rapid and widespread—that reshape the fabric of societies and revolutionize work—can be inherently disruptive in the short run, however positive in the long. Some workers will be in a position to adjust faster and see more positive outcomes than others.

Some of the technologies with the highest potential to change the nature of work include:[19]

- *Mobile Internet* affects the lives of 3.2 billion Internet users. There are more than

7.1 billion mobile subscriptions, which are changing the way people work, innovate, interact and do business.

- *Automation of knowledge work,* through intelligent software systems, is transforming the organization and productivity of knowledge work and could enable millions to use intelligent digital assistants.

- *Cloud technology* has the potential to improve access to online information technology services for businesses and governments at low costs—and to enable new online products and services for billions of consumers and millions of businesses.

- *3D printing* is reshaping work because it can produce anything from industrial prototypes to human tissue. One of the largest global networks of 3D printers operates in 110 countries, with 9,000 machines that it rents out by the hour.[20] It enables on-demand production and has the potential to affect the jobs of 320 million manufacturing workers in the world today—12 percent of the global workforce. Disabled workers have also benefited. The world's first 3D lab for printing prosthetics is in South Sudan: Project Daniel was started in 2013 to make prosthetic limbs for Daniel Omar, a young man who lost both arms in a bomb explosion. Today the printers can produce only a narrow range of goods, but the future could be very different. These machines could permanently disrupt the previous model of long runs of identical goods in factories, opening opportunities for individuals and smaller companies to participate in decentralized production. As the capacity of these machines advances, some expect that human workers, particularly those who are less skilled, may suffer in tighter job markets.[21]

- *Advanced robotics* are taking the automation of manufacturing to new levels. People have long anticipated that technology using artificial intelligence would displace the need for human thought. At higher thought levels, this has proved difficult. But for less complex tasks, greater knowledge and skills are being embedded in individual items of technology. Previously "dumb" equipment from vacuum cleaners to weighing scales, fitted with cameras, sensors and processors, is becoming more responsive to human interaction. And pieces of equipment once confined to military applications, such as drones, are now appearing in civilian life. Many of these objects are being connected to each other, allowing them to intercommunicate through the "Internet of things."

Automation is taking place at a fast rate (sales of industrial robots are estimated to have grown 170 percent between 2009 and 2011, despite the financial crisis),[22] and the total number of robots is expected to reach 1.5 million in 2015.[23] Robots have helped make the workplace safer, as some take on jobs that are too dangerous, such as locating radiation sources. Robots are also crucial in efforts to revive or preserve manufacturing where labour is scarce or expensive. Robots that carry out routine tasks may cost a company less than labour—for example, in some German plants robots doing routine work costs about €5 an hour over their lifetime (including maintenance and energy costs), whereas the cost of a German worker (including wages, pensions and health care) runs to about €40 an hour.[24] Furthermore, robots will soon be able to carry out more complex tasks as artificial intelligence takes hold in factories. The potential implications for jobs are obvious. All these phenomena have been referred to in recent times as the fourth industrial revolution (box 3.2).[25]

- *Energy storage* will allow maximum use of solar and wind energy, potentially bringing access to affordable electricity to the 1.2 billion people who do not have it. In due time, energy storage could also make electric cars more affordable and transform electricity grids—providing new jobs.

In addition, driverless vehicles, advanced materials, advanced oil and gas exploration and recovery technologies, big data, biotechnologies and renewable energy technologies all feature in the technological revolution.

A few characteristics of the current technological revolution present unique challenges and opportunities for human development, as achieved through work. First, the speed of technological change and penetration is mind boggling, marked by impressively quick adoption of new technologies (figure 3.2). Just two years after Apple shipped the first iPad, it sold 67 million units. It took 24 years to sell that

The current technological revolution presents unique challenges and opportunities for human development

BOX 3.2

The fourth industrial revolution

During the fourth industrial revolution, not only individual machines, but also entire factories become smart and automated, making the production process more precise and the products more customized. Digital technologies will allow products to control their own assembly as they communicate specific production requirements (colour, size, material) and steps to machines that can in turn communicate with each another to control the speed and flow of assembly lines.

Factories in Germany are leading the way in experiments with smart factories as part of the country's Industrie 4.0 vision. The state has allotted over €200 million for research in academia, business and government to support the digitalization of traditional industry.[1] Siemens has already built a digitized pilot factory in Amberg to produce electronics to be used in other smart factories. Production at the factory is largely automated—people handle only 25 percent of the work, while machines and computers manage 75 percent of the value chain. Human hands touch the product only at the beginning of the process when a circuit board is placed on the assembly line. The plant was established

in 1989, and although the size of the plant and the number of employees (1,200) has not changed, output has increased eightfold, and production quality is an unprecedented 99.9988 percent.[2]

Smart factories have the potential to use real-time data to provide high- and consistent-quality customized products at competitive costs. But what are the implications for jobs? Many argue that humans will remain relevant, but the value added in manufacturing will come from the programming and servicing of machines and computers rather than manual labour. In this scenario skills and knowledge may become as or even more important than wages, and the centres of manufacturing may shift towards countries with an educated labour force and plenty of capital to invest in smart factories and sophisticated machines.

The impacts of the fourth industrial revolution may not be limited to developed countries; in fact, the effects of robotic sewing machines are expected to be much wider, including on jobs in developing countries that traditionally count on low-cost, low-skilled labour.[3]

Notes
1. Germany Trade & Investment 2014. 2. Siemens AG 2015. 3. The Economist 2015b.
Source: Human Development Report Office.

FIGURE 3.2

Adoption of new technologies in the United States has been impressively quick

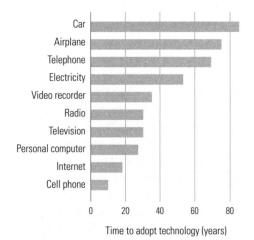

Time to adopt technology (years)

Note: Adoption refers to the amount of time it takes a technology to penetrate 50 percent of the US population.
Source: Donay 2014.

many Macs, five years to sell that many iPods and more than three years to sell that many iPhones.[26] The pace of technology penetration is illustrated in figure 3.3 for Internet use and mobile phones. Not only is the change substantial, adoption is widespread and holds promise to benefit people around the world.

This is because many technologies are in some sense universal machines that have applications in almost all sectors, industries and tasks, sweeping across all areas of production: manufacturing, services and agriculture. So agriculture can no longer be considered low-tech. Workers in all sectors will be challenged to be more educated, more flexible and more technologically savvy.

But some argue that recent progress in computing and automation could be less transformative for human development than such past innovations as electrification, cars and perhaps even indoor plumbing.[27] Previous advances that enabled people to communicate and travel quickly over long distances connected the world and may end up being more important to

FIGURE 3.3

The change in technology penetration around the world between 1995 and 2015 was substantial and holds promise to benefit people around the world

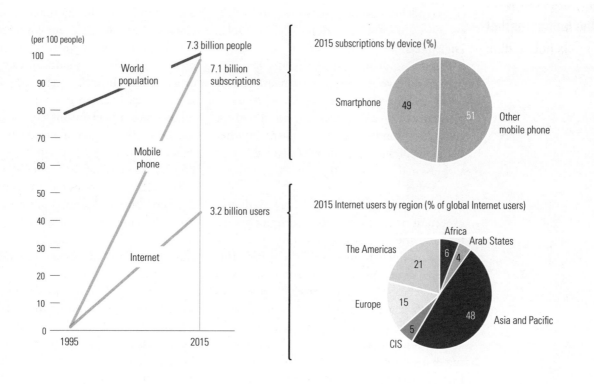

Note: Regions are International Telecommunication Union categories.
Source: Human Development Report Office calculations based on World Bank (2015f) and ITU (2015).

society's advancement than anything to come in the 21st century.[28] But this also depends on future innovations and how they are applied. Others note that the pace of technological development has stalled in fields that have tended to push the boundaries of knowledge and expand human progress, including energy, pharmaceuticals, space exploration and nanotechnology.[29]

All emerging technologies are expected to have massive economic and social impacts. Technologies that combine with and reinforce globalization are expected to transform how people live and work, create new opportunities and business models, drive growth and change the geography of comparative advantages for countries. These forces may also have negative impacts on work and workers.

Either way, the technological revolution is unlikely to be business as usual. There are revolutionary elements in this new wave, especially in a globalized world of production and work.

Will something new affect the future of work? And if so, what is it?

Globalizing work

Technology is affecting the nature of work by introducing new ways of communicating, new products and new demands for skills. New technologies are also reinforcing and deepening previous trends in economic globalization, bringing workers and businesses into a global network through outsourcing and global value chains. These processes are reshaping work and testing national and international policies.

Until recently, workers with specific skills were associated with particular activities in different sectors. They competed with other workers for jobs on a national scale. They gained skills through their work in particular sectors and industries, and for the most part the pace of change in workplace organization

Globalization is reshaping work and testing national and international policies

and products was slow enough for them to adjust.

In many areas of work the labour market is now global. Multinational corporations have access to labour around the world, and workers must compete on a global scale for jobs. Digital technologies heighten the competition by removing geographical barriers between workers and work demands—in many cases it is not even necessary for a company to move physically or for a worker to migrate. The work connection can be made through the Internet or mobile phones. That there is a global labour surplus makes competition among workers even fiercer.

Consumer demands have also evolved with expectations for low-priced consumer goods, for fresh and new products and for digital access to products from around the world. This has increased competition for companies to provide cheap, innovative products that cater to rapidly changing trends, all the more so as digital technologies allow companies immediate and constant access to information on consumer habits and interests. A flexible approach to production and cost cutting, including labour costs, has been the producer response. Low labour costs and flexible commitments to workers allow companies to quickly and efficiently respond to shifts in consumer needs and in the location of demand.

For workers these trends are aligning to create a world of work where creativity, skills, ingenuity and flexibility are critical. But even for those who are well positioned to compete in the emerging work system, security is lacking. Around 26 percent of workers worldwide have permanent contracts, around 13 percent have temporary or fixed-term contracts and 61 percent are working without a contract.[30] With just 30 percent of the world's labour force covered by unemployment protection, a world of work that values flexibility may be a challenge to the stability of worker's lives.[31]

Outsourcing

One way companies have responded to increasing market competition and cost pressure is to relocate some functions to countries with lower wages or to subcontract some noncore activities to companies in countries where costs

In many areas of work the labour market is now global

are lower (or to mix the two approaches). For example, Apple directly employs only 63,000 of the more than 750,000 people around the world who design, sell, manufacture and assemble its products.[32] In the hotel industry many employees are not direct employees of the hotel brand but temporary workers of other companies that do work in laundry, catering, cleaning or gardening. In other sectors the goods and services supply chain has been sliced up and subcontracted to many different employers.[33]

One of the biggest consequences of globalization from a work perspective is business process outsourcing, which has decentralized business services away from corporate headquarters. Business process outsourcing works through companies—for example, in India it works through major information technology enterprises. Such phenomena generate gains for some and losses for others.

Assembly jobs in developed countries began moving to export processing zones as developing countries adopted export-oriented industrialization. The impact on job creation in large developing countries such as China and Mexico, as well as smaller countries such as Costa Rica, the Dominican Republic and Sri Lanka, was considerable and positive, particularly for women, who often worked in garment factories.[34] In many cases outsourcing has provided a boost to local development, although the quality of the work and enforcement of labour standards have varied (box 3.3).

Global offshoring—outsourcing production or service provision abroad—of service jobs started to pick up in the 1990s as advances in information and communications technology allowed many support services to be performed offsite. With the burst of the dotcom bubble in 2001–2002, technology companies further explored cost-cutting measures involving relocating noncore activities to lower wage countries with high technical skills, particularly India. Between 2000 and 2010 the number of direct jobs in information and communications technology in India jumped from 284,000 to 2.26 million.[35] In recent years India has maintained a strong position as an offshoring destination for service jobs, but offshoring destinations are becoming increasingly diverse.[36] Services are growing in the Russian Federation, Africa and Latin America, in part matching

BOX 3.3

Bosnia and Herzegovina—local development through outsourcing

As Poland and Slovakia become epicentres of the services industry in Europe and emerge as leaders of the digital revolution in Central and Eastern Europe, Bosnia and Herzegovina is developing into an outsourcing target for large industrial and automotive corporations. In the never-ending search for skilled and low-cost labour, the production of automotive parts is steadily moving from centres in Germany to the Czech Republic and Slovakia and farther east towards Bosnia and Herzegovina.

Outsourcing in Bosnia and Herzegovina is still relatively new and limited, but progress is visible through a decrease in local unemployment and an increase in entrepreneurship and local start-ups. For example,

opportunities for people in the municipalities of Gorazde and Zepce have been greatly enhanced by an influx of investment from Western European automotive corporations. Local economic transformation bolsters local human development. A notable result in Gorazde is the unemployment rate among women, which is much lower than in the country as a whole. Successful women-owned businesses employ hundreds of workers, more than 40 percent of them women (above the national average of 34 percent), many in top managerial positions. These patterns show promise, reminiscent of Slovakia in the mid-1990s, when relatively small outsourcing investments came in, paving the way for bigger inflows later on.

Sources: Human Development Report Office.

companies' interests in diversifying into different time zones to enable 24-hour service.[37]

In developed countries offshoring has been viewed as a cause of job losses, raising fears that all such jobs will move away. In manufacturing the process started in the 1960s and 1970s as production began shifting towards national, regional and global production networks, the most modern and pervasive form of industrial production. However, estimates of the consequences of offshoring for workers in developed countries vary, and the long-term impacts are less clear than the short-term ones. Job losses are greater in manufacturing than in services. A 1 percent increase in imported intermediate manufactured goods reduces manufacturing employment in the importing country an estimated 0.15 percent, and a 1 percent increase in imported intermediate services reduces services employment 0.08 percent.[38] Offshoring-related job losses vary across countries, too, with short-term losses found to range from zero in some countries to 0.7 percent of all job losses in the Netherlands to almost 55 percent in Portugal.[39]

Today jobs that involve administrative support, business and financial operations, and computer and mathematical tasks are most likely to be outsourced. In Australia, Canada and the United States 20–29 percent of jobs have the potential to be offshored, though it is unlikely that all of them will be.[40] Many of those jobs are medium- and high-skill service

professions that can be carried out at lower cost abroad as education levels rise and communications infrastructures improve in developing countries.

So, while there may be immense benefits in access to new jobs in countries hosting offshore activities, individuals losing jobs, particularly in developed countries, may require training and new skills for a more competitive environment. To ease the adjustment for workers whose livelihoods are threatened by trade-related job displacement, programmes are needed to help people find new work, enhance their skills and maintain access to a basic income. Likewise, training can enhance the abilities of workers in developing countries to access jobs created through offshoring.

Global value chains

Many economic activities are integrated into global production networks and global value chains that span multiple countries and sometimes continents. This integration goes from provision of raw materials and subcomponents to market access and after-sales services. Thus production is performed within complex and dynamic economic networks made up of inter- and intra-firm relationships and global networks, where the relationships are "many-to-many" rather than "one-to-the-next."[41]

With the transition to global value chains, production is no longer about one company

One of the biggest consequences of globalization is business process outsourcing

in one industry offshoring a particular activity to a single destination. It is about intermediate goods and services organized in fragmented and internationally dispersed production processes that are coordinated by multinational companies and cut across industries. The production of any given component or service is produced by a network of affiliates, contractual partners and arm's-length suppliers that are often in developing countries and operating under a headquarters that is often in a developed country. There has been a shift from sector-based globalization to the globalization of stages and tasks of production.[42] The coordination required to make this form of production work has been facilitated by the digital revolution and advances in information and communication technology.

The number of people employed in these global value chains is high and rising: In 40 countries with data, an estimated one in five people was working in global value chains in 2013—or 453 million people (up from 296 million in 1995), including 190 million women.[43]

The integration of developing countries into global value chains has increased opportunities for paid work and prompted a shift in labour force participation for women (many of whom find jobs in the garment industry). Investments benefit young people who can learn new skills that they can use throughout their working lives. But jobs are needed for women of all ages and skill levels. There are concerns that many factories tend to employ only young women with low skill levels, with fewer opportunities for older female workers or those with high skill levels. There are also concerns about levels of labour protection.

The benefits to workers and economies associated with participation in global value chains are not preordained. Integration in global value chains affects dimensions of human development differently and often in contradictory ways. Such integration does not say much about the quality of work in globally integrated factories and whether workers have—or have not—expanded their human capabilities.

Moreover, the global value chain system generates winners and losers, within and across countries and industries. The footloose nature of global value chains can generate less job security and put pressure on governments and subcontractors to minimize costs. This in turn puts pressure on workers' wages and working conditions, particularly among the low-skilled. Developing countries also face the risk of becoming locked into low-value added nodes of global value chains that limit work opportunities, skills development and technology exposure.

The transition towards global value chains has introduced new complexities for workers in developed and developing countries alike. There are questions about how much workers gain by partaking in work that contributes to global value chains versus work that falls outside them. There is some evidence that productivity is higher in global value chain–oriented work but that wages are the same for workers inside and outside global value chains,[44] raising questions about how the increases in productivity are shared between workers and capital.

Market pressures that are transmitted through global value chains tend to be absorbed by workers—whether as wages driven down by global competition, increased informalization and contractual insecurity brought about by multiple subcontracting chains (creating exposure to price volatility that originates beyond local markets and national boundaries) or in the form of layoffs during downturns. In a competitive global economy multinationals increasingly rely on a disenfranchised workforce, using a mix of fixed-term employees, temporary workers, independent contractors, project-based workers and outsourced workers to provide production flexibility and manage costs.[45] Participation in value chains provides some with secure, decent jobs and others with more precarious work (even in the same country and sector). Temporary workers often work alongside those with long-term positions, creating a type of labour dualism.

Despite the challenges, policy attention coordinated at the national, regional and global levels can help people flourish in a global work environment shaped by global value chains. But this will take far more than business-as-usual policies or small policy steps. Chapter 6 provides some examples of the types of policies that can help workers and countries thrive when work becomes more flexible and cuts across national borders.

The global value chain system generates winners and losers

Work in the digital age

The digital revolution deserves attention in its own right because of the changes it is creating in the world of work and because of the way it is accelerating globalization. In recent years the digital revolution has accelerated the global production of goods and services, particularly digital trade (figure 3.4). In 2014 global trade in goods reached $18.9 trillion and trade in services $4.9 trillion.[46]

The knowledge-intensive portion of global flows increasingly dominates—and is growing faster than—capital- and labour-intensive flows. Today knowledge-intensive flows account for half of global flows, and shares are increasing: Knowledge-intensive goods flows are growing at 1.3 times the rate of labour-intensive goods flows.[47] As a result, the digital components of goods and services flows have also increased (figure 3.5). Indeed, many goods today, as demonstrated by the "app economy," are entirely virtual. Much of the data pass through the Internet, often on smartphones. Today there are more than 7 billion mobile subscriptions, 2.3 billion people on smart phones and about 3.2 billion people connected to the Internet.[48]

The spread and penetration of digital technologies are changing the world of work everywhere, but the effects vary across countries according to their own social and development contexts. Some technological changes are cross-cutting, such as information and communication technologies and the spread of mobile phones and other handheld devices. Still, countries will continue to have divergent production and employment structures and different uses for digital technologies, largely reflecting the relative economic weights of agriculture, industry and services, as well as the resources invested in developing people's capabilities. Labour markets, the ratio of paid to unpaid work and the predominant types of workplaces in each country differ—so the impacts of digital technologies on work will vary accordingly.

The digital revolution may be associated with high-tech industries, but it is also influencing a whole range of more informal activities from agriculture to street vending. Some may be directly related to mobile devices. In Ethiopia farmers use mobile phones to check coffee prices.[49] In Saudi Arabia farmers use wireless technologies to carefully distribute scarce irrigated water for wheat cultivation.[50] In some villages in Bangladesh female entrepreneurs use their phones to provide paid services for neighbours. Many people sell phone cards or sell and repair mobile phones across developing countries.

> The digital revolution is accelerating globalization and changing the world of work

FIGURE 3.4

The digital revolution has accelerated the global production of goods and services, particularly digital trade

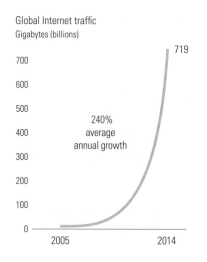

Global Internet traffic
Gigabytes (billions)

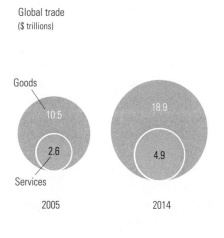

Global trade
($ trillions)

Source: Human Development Report Office calculations based on data from UNCTAD (2015) and Cisco (2015).

FIGURE 3.5

The digital component of global flows has increased—selected examples

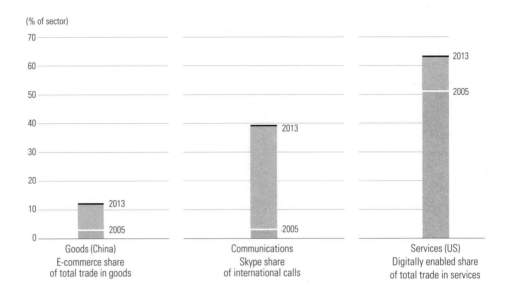

(% of sector)

Goods (China)
E-commerce share
of total trade in goods

Communications
Skype share
of international calls

Services (US)
Digitally enabled share
of total trade in services

Note: Digital component refers to flows of data and communication. For example, cross-border exchanges from books to design files would represent a digital component of flows.
Source: McKinsey Global Institute 2014.

Mobile phones empower workers with information

Mobile phones now facilitate many aspects of work through a combination of voice calls, SMS and mobile applications. Some uses of mobile phones in agriculture are shown in figure 3.6. But there are also benefits for many other types of activities, formal and informal, paid and unpaid, from food vendors in Cairo to street cleaners in Senegal to care providers in London. Mobile phone–based economic activity is likely to keep expanding rapidly. In Sub-Saharan Africa unique mobile subscriptions are predicted to rise from 311 million in 2013 to 504 million in 2020 (figure 3.7).[51]

Mobile phones and mobile Internet service offer many new opportunities and advantages to workers and to economies more generally:

- *Access to dynamic price information.* In India farmers and fishers who track weather conditions and compare wholesale prices through mobile phones increased their profits 8 percent, and better access to information resulted in a 4 percent drop in prices for consumers.[52] Similarly, in Niger the use of mobile phones reduced differences in grain prices across markets within the country by 10 percent.[53] Mobile phones empower workers with information.

- *Productivity gains.* In countries as diverse as Malaysia, Mexico and Morocco, small and medium-sized enterprises with Internet access averaged an 11 percent productivity gain by reducing transaction costs and barriers to market entry.[54]

- *Job creation.* The Internet and mobile technologies create new jobs directly through demand for labour from new technology-based enterprises and indirectly through demand from the wider ecosystem of companies created to support technology-based enterprises. Indirect jobs include network installation, maintenance providers and providers of skill-based services such as advertising and accounting.

- *Supply chain management.* Small businesses can keep track of supplies and deliveries and increase efficiency. This can help with everything from reducing food waste to increasing access to jobs in global value chains.

- *Better services.* Mobile phones are extending the reach of agricultural extension services. In India, Kenya and Uganda farmers can call or text hotlines to ask for technical agricultural services.[55] One application developed in Kenya is iCow, which helps cattle farmers maximize breeding potential by tracking their animals' fertility cycles.

- *Labour market services.* Mobile services can match employees with vacancies. In South

FIGURE 3.6

Opportunities for mobile applications for agriculture and rural development

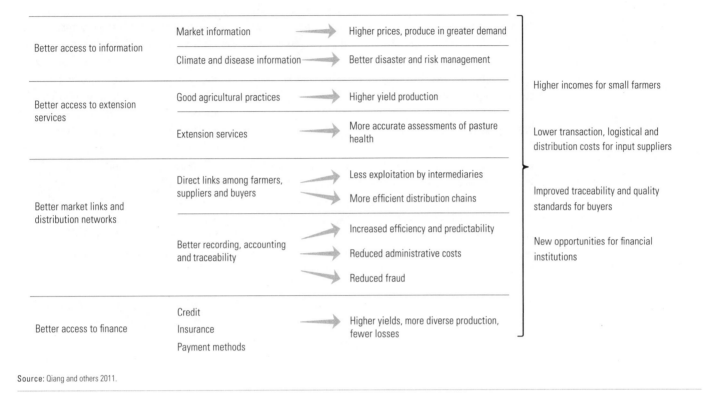

Source: Qiang and others 2011.

Africa the extension of mobile phone coverage is associated with a 15 percent increase in employment, mostly for women.[56] Many job-matching companies allow jobseekers to have real-time information on vacancies, while helping employers extend recruitment systems to entry-level and low-skill jobs.[57] Voice messages are particularly useful for recruiting jobseekers who have difficulty reading and writing.

- *Mobile banking.* Mobile phones can transfer funds and make payments. So garment workers or fruit vendors working in urban areas can quickly make transactions and send money back to rural households. Some of the most advanced services for mobile money have been developed in South Asia, as with bKash in Bangladesh, and in Sub-Saharan Africa, as with M-Pesa in Kenya.

- *Access to finance.* Small businesses can use online services to access finance from interested parties around the world. CARE International has a system that allows potential investors to browse the profiles and small business ideas of people in 10 countries, including Bosnia and Herzegovina. Individuals can invest as little as $25.[58]

These are among the myriad advantages of having Internet and mobile phone access for workers and economies. Access empowers people to harness their creativity and ingenuity through work for productivity- and human development–enhancing purposes. Much more is possible, particularly if efforts are made to ensure equitable access to the Internet and mobile phones, especially for women and people in rural areas. One study estimates that if Internet access in developing countries were the same as in developed countries, an estimated $2.2 trillion in GDP and more than 140 million new jobs—44 million of them in Africa and 65 million of them in India—could be generated. Long-term productivity in developing countries could be boosted up to 25 percent.[59]

New frontiers for work

The digital revolution, reshaping economic output and employment, has the potential to

FIGURE 3.7

Mobile subscriptions and connections in Sub-Saharan Africa are predicted to rise substantially between 2013 and 2020

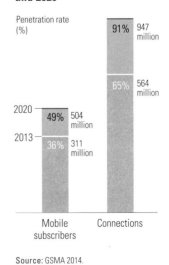

Source: GSMA 2014.

empower millions with new ways of working and new jobs. But activities are also moving beyond employment and jobs in the traditional sense, and the nature of work as an individual market-based activity is bound to shift. Artificial intelligence will most certainly disrupt business as usual. And in some cases new forms of collaboration, sharing and innovation will shift work towards a more social experience.

New producers

A distinctive feature of the digital economy is the prospect of zero marginal cost, where digitized knowledge in data and applications, once generated, can be reproduced endlessly at almost no extra cost. Low- or no-cost reproduction expands access to the fruits of work but may create few additional jobs. Twitter had 302 million monthly active users as of March 2015, who created or transmitted information and news through 500 million Tweets per day,[60] but only 3,900 employees, half of them engineers.[61]

A second major feature of the digital economy that affects whether work translates into jobs and employment is that some of the goods and services people consume are produced by consumers themselves—they have become "prosumers." The most direct example is probably Wikipedia, which has more than 73,000 active voluntary contributors.[62] The free online encyclopaedia competes directly with paid information services such as Encyclopaedia Britannica, which in 2012 ceased publication of its print edition after 244 years.[63]

Similar movements in work are occurring in the physical world. New technologies, some of them mentioned earlier in the chapter, enable radically new and generally more decentralized forms of production and consumption. One example is energy. Households have long been able to generate their own electricity using wind turbines, but now with smart grids they may be able to sell their surplus power to the grid, gaining financial returns from previously unpaid work.

Personalized services and goods

Technology has also been transforming markets as many personal services move online.

Customers can now use the Internet to buy groceries, order from restaurants, make hotel and airline bookings and hire help for house-cleaning or childcare. Online services require less commitment than employing service providers full-time and allow customers to use services occasionally. The online system can also provide temporary work opportunities for those who seek extra paid work or women who want a flexible schedule. Online task service companies allow people to pay providers to run errands such as shopping or queuing for theatre tickets. Online translation services offer clients the option to employ translators case by case, some of whom are students looking for flexible work.

The digital revolution has also revolutionized creative work and empowered small producers and artisans. It is possible, through sites such as eBay and Etsy, for artisans to find buyers looking for specific or niche products. Authors and artists can self-publish and share their creations around the world, whether as e-books, music downloads or video clips. And smartphones have created a new mass market for small-scale software designers targeting specific needs. Online stores sell individual apps that enable users to do everything from monitor their health to learn languages to play games. The resulting app economy has developed at break-neck speed. One study claims that in 2013 the app economy provided some form of work for 750,000 people in the United States alone.[64]

New business services

With the extension of the Internet to households it has become possible for individuals to provide business process services from their homes. This often involves specialized white-collar skills such as computer programming, copywriting and back-office legal tasks. These tasks, generally for firms in developed countries, are carried out by people in such developing countries as Bangladesh, India and the Philippines.

Much of this business is mediated by companies that coordinate freelancers with small and medium-sized firms that require business services. Coordination companies collect a commission from the freelancer, but often charge no fee to those offering the jobs. Most of this

Technology has also been transforming markets

work has been done in urban areas, but there are now efforts to extend the opportunities to more deprived areas through "impact sourcing"—a form of socially responsible outsourcing that seeks to create jobs for disadvantaged groups.[65] This would create potential for more rural employment.

The sharing economy

Another emerging trend with the potential to reshape work is the sharing economy. It is now possible to match demand and supply between individual producers and consumers. Alternatives to taxis allow people to use their own cars to provide ride services, blurring the distinction between professional drivers and those who have a spare seat in their private car. Technology is also allowing traditional taxi drivers to work more efficiently finding customers via online services such as Uber and GrabTaxi, which operates in several countries in South-East Asia. The same principle is being used by auto rickshaw drivers in India via mGaadi. Other companies allow people to rent out accommodation in their private homes (such as Airbnb).

These arrangements enable people to make better use of capital assets such as cars or homes. But they can also replace more traditional jobs if they compete with conventional hotels and transport services, such as taxi drivers and hotel staff, who are generally low skilled and poorly paid. There are also new challenges to regulating services, ensuring consistent quality and protecting consumers. In some ways the professionalization of work is reversing.

Start-ups

Technology has made it easier to start a business, an attractive option for young people, some of whom are leaving fairly prestigious jobs to do just that. When individuals have identified a good idea in the course of their work and want to pursue it on their own, they have more tools at their disposal to support their entrepreneurial efforts. Indeed, a recent estimate indicates that in countries with 73 percent of the world's population, there are 455 million entrepreneurs today, up from 400 million in 2011.[66]

These (often young) people see entrepreneurship as a viable alternative to traditional jobs and as a means to pursue their dreams. Start-ups are taking root in both developed and developing countries. Asia is embracing them quite rapidly. Young people see a lot of opportunities, bolstered by financial technology and big data.[67] Yet start-ups face challenges. Access to capital is one, ideas that are sustainable another. In developing countries weaker legal institutions pose a problem. And long-term viability is their largest challenge (box 3.4).

Crowdworking

In addition to working as individual contractors online, employees are also finding work through more casual channels as crowdworkers. This work generally involves "human intelligence tasks," and pay and work conditions are less than ideal. Major players in the market include Clickworker, Cloud Work, Casting Words and Amazon's Mechanical Turk. Amazon's Mechanical Turk is the largest market, with a global pool of half a million workers.[68] As of July 2015, there were more than 325,000 human intelligence tasks available for workers.[69]

Service requesters are free to withhold payment if they are dissatisfied with the work that Mechanical Turk provides. They can also give a bad rating: A Turker (a worker with Mechanical Turk) who receives several bad ratings is barred from similar tasks. There are attempts to improve the quality of work and the quality of services, so that the two reinforce one another (box 3.5).

The knowledge economy

In recent years knowledge has become central to production. Even in manufacturing, the value of finished goods increasingly derives from embodied knowledge. For example, the price of a top-end smartphone is driven less by the cost of components and assembly and more by the high charge for sophisticated design and engineering.[70] In 2012 research and development, combined with skilled labour, made up a large share of the value in nearly half the total trade in knowledge-intensive goods, services and finance—worth nearly $13 trillion.[71] And this proportion is growing steadily

Another emerging trend with the potential to reshape work is the sharing economy

BOX 3.4

Challenges for start-ups in the Arab States

A 2013 survey of the challenges facing start-ups in the Arab States that covered more than 700 entrepreneurs, nearly half in technology (including software development and services, e-commerce and online services, gaming, and telecoms and mobile services) yielded numerous findings.

Most entrepreneurs were male who started their companies in their late 20s or early 30s, held university degrees, had studied or worked abroad and partnered with co-founders. The average age of surveyed entrepreneurs was 32.5, and most companies were less than five years old. Over 75 percent of the companies had male founders; only 23 percent of entrepreneurs were female. Nearly all entrepreneurs surveyed had at least a bachelor's degree. In the next one or two years, 70 percent planned to open new offices, either in new countries or in countries where they already operated.

Many hoped to expand into the United Arab Emirates (39 percent) and Saudi Arabia (38 percent). In terms of financing, nearly all entrepreneurs had used their personal savings or support from family or friends, while 24 percent benefited from angel investment—financial backing from an affluent individual. A notable share reported not receiving support from commercial banks.

A quarter of the entrepreneurs indicated that obtaining investment was a challenge. A larger pool of capital, comprising different types and sources of funding, could improve access to finance. Many companies pointed to challenges in finding partners to help expansion abroad, as well as general costs and legal hurdles. Female representation was low in all start-ups. This lack of inclusion suggests that enabling diversity in education and gender could help expand the pool of innovation.

Source: Wyne 2014.

> In recent years knowledge has become central for production

BOX 3.5

Initiatives to improve crowdworking

CloudFactory engages around 3,000 crowdworkers in Kenya and Nepal. The company aims to improve its service by treating its contractors well. Rather than offer an open marketplace, it handpicks, trains and supervises its workers. CloudFactory workers need not spend much time searching for work and generally earn between $1 and $3 an hour, fairly high for crowdwork.

MobileWorks, a company with similar principles, launched its LeadGenius platform in 2010 and now has several hundred full-time workers in 50 countries. The company targets disadvantaged and marginalized groups, from military veterans to refugees. Unlike most crowdwork companies, it pays by the hour. By removing the incentive to complete assignments hastily, it aims to ensure high-quality work. MobileWorks crowdworkers can expect up to 40 hours of work a week. Pay is almost always above the national minimum wage.

Source: Pooler 2014.

while the proportion is falling for products and services that are intensive in labour, capital or resources rather than knowledge. This is in part because of differences in wages across countries participating in global value chains and the distribution of global value chain tasks between high- and low-wage countries.

Agrarian economies still exist, although in declining numbers. And industrial activities persist, although they are being transformed and replaced by computer-based technologies and workplaces. The reality is that the growth of knowledge societies and innovations in business models are bringing enormous transformations to work environments and in the skills demanded for many jobs. Work environments are technology-rich, and a whole new set of occupations has emerged based on the production, analysis, distribution and consumption of information.[72]

The challenge of the skills mismatch in the knowledge economy is due to the pace of technological innovation and to the rapid growth in demand for new and higher skills, which is are not forthcoming. In a survey of businesses in eight European countries, 27 percent of potential employers reported that they could not fill vacancies because applicants lacked the necessary skills.[73] Some 45 percent of employers in Greece and 47 percent in Italy also reported that their businesses were hampered by a lack of entry-level skills.[74]

Today jobs that are growing in developed countries demand complex interaction skills that require deep knowledge, judgement and experience—more than routine transaction or production skills. For example, in the European Union 16 million jobs employing highly qualified people are expected to be added between 2010 and 2020, while jobs employing people with few or no formal qualifications are expected to decline by around 12 million.[75] Many of the jobs being created are in nontradeable services such as health care, education and public services, which are also areas that are fundamentally human development enhancing.

In developing countries, transitioning into higher value-added services and manufacturing is increasing demand for workers with at least a secondary education and some vocational training, as well as for highly qualified professionals and technicians.[76] In an extremely dynamic global market where products and processes change rapidly, a basic education in literacy and numeracy as well as fast and efficient continued learning are critical. Having a core set of capabilities promotes success in many aspects of life. Cognitive and noncognitive skills (such as conscientiousness, self-regulation, motivation and far-sightedness) interact dynamically to shape the evolution of subsequent capabilities. Interventions in early childhood have the greatest impact towards promoting these essential skills and reflect the investments in human capital made by parents and children.[77] Enriching the early environments of disadvantaged children can improve child outcomes and can positively affect cognitive and noncognitive skills. Increasingly important are the quality of education, ability to learn and solve problems, and e-literacy.[78] Most developing countries have attained near universal primary education, but secondary education and high-quality education are essential foundations for an employable workforce for the future.

The knowledge intensity of work allows firms to reduce staff in some areas. Firms can now cut the number of workers engaged in routine tasks, both manual and cognitive, that are fairly easy to program and automate. A worker who welds metal sheets can often be displaced by a robot, which can do the job faster, more cheaply and more precisely. Similarly, a bookkeeper who performs cognitive tasks such as performing calculations can generally be replaced by a computer programme.

In the past many successful economies moved from low-margin, labour-intensive goods to electronic assembly and then on to higher manufacturing, design and management. As described earlier, countries coming late to the development process face what has been referred to as "premature deindustrialization" or even "nonindustrialization."[79] They can no longer look to manufacturing to absorb their jobless millions, since much of that work can now be automated.

Flexible working

The digital revolution allows for more flexible forms of work, enabling people to fit their livelihoods and work activities around their lives. Many people whose tasks are computer based can theoretically work anywhere, in cafés, trains and especially their homes. Indeed, some employers pay for home offices. Even using their own laptops, workers can connect with their office systems to access emails and files and communicate with colleagues through teleconferencing. In 2014, 14 percent of UK workers spent at least half their working hours at home.[80] Similarly, in the United States in 2012, 64 million people had jobs compatible with working at home at least 50 percent of the time.[81]

Flexibility at work is believed to make workers more productive and less likely to switch jobs (according to research on policies for a flexible work–life balance).[82] Recent research from telecommunications firms showed that flexible work schedules and telework had a positive effect on performance through knowledge sharing, cross-functional cooperation and interorganizational involvement.[83]

But workplace flexibility is usually an option only for those in formal employment. Workers in temporary short-term work arrangements, often paid by the hour, can attend to family and related needs only by directly losing wages (although freelancers with higher levels of skills and earnings are an exception). Workers in developing countries, many in or on the brink of poverty, can ill afford any flexibility.

Decentralized workplaces and telework have pros and cons. Such work arrangements save commuting time and offer flexible schedules and greater freedom. But they reduce the

There is a skill mismatch in the knowledge economy

Technological
advances are
engines for new
forms of creativity
and innovation

opportunities for teamwork and social interaction and make it harder for employers to manage workers.

These new ways of working are also reshaping offices. In a traditional workspace, the allocated desks might be unoccupied at least half the time. A popular alternative is hot-desking—making, say, seven desks available for every 10 staff—freeing space for other activities. When knowledge workers come to their workplace, they generally do so to brainstorm, solve problems and generate ideas—activities that can best be done in spaces that are cosier, with ample cafés, comfortable seating and quiet spots.

Some employers try to make offices so attractive that workers do not feel the need to leave. With kitchens, restaurants, gyms, game rooms and nap pods, employees can work and rest more flexibly and efficiently—and interact virtually with partners and clients across different time zones. Some law firms in the United Kingdom have sleeping pods, where late-working lawyers can rest or occasionally spend the night.

But even when away from the office, escaping work is difficult. Constant connectivity over mobile devices has blurred the distinction between work and leisure and narrowed the privacy boundaries between public and private life. Many workers are still connected to their office while on vacation, so work never stops. For freelancers and flexible workers, making a distinction between work and other areas of life may be particularly challenging because it is never clear when the next job will arrive. Such arrangements may appear to foster a better work–life balance, but they may in fact be unconducive to a meaningful and satisfying life.

Creative innovations

Technological advances have not only transformed work—they are engines for new forms of creativity and innovation. The digital revolution, and the information and communications technology systems that support it, emerged from creative and innovative work. Collaborative teams and visionaries turned ideas into tangible goods and services. Digital technologies are fast and cheap enough to foster all kinds of new innovations from auto-generated books to driverless cars to flexible factory robots.[84] This is reflected in the growth of patents granted over the past decades. Between 1970 and 2012 the number of patents (United States and other countries) granted by the United States Patent and Trademark Office increased almost fivefold (figure 3.8).[85] Innovations in computer and electronics fields were central to this growth: From 1990 to 2012 their share in all new patents more than doubled, from 25.6 percent to 54.6 percent.[86] Figure 3.9 shows the patents granted in the top 12 granting countries in 2013.[87]

The digitization of paintings, ceramic works, sculptures and prints is also expanding access to works of art. Many major museums have sweeping digitization projects: Amsterdam's Rijksmuseum has digitized 95 percent of its paintings and ship models, 60 percent of its sculptures and 50 percent of its ceramics. The Smithsonian Institution in Washington, DC, has already captured 2.2 million objects out of a collection of 138 million, developing new automated digitization methods such as conveyor-belt scanners.[88]

Creativity at work enhances productivity and increases innovation, but also contributes to the satisfaction and well-being that people have with their work. Almost 80 percent of workers acknowledge some level of creativity (even if very modest) in their work, according to the

FIGURE 3.8

Between 1970 and 2014 the number of patents granted by the United States Patent and Trademark Office increased almost fivefold

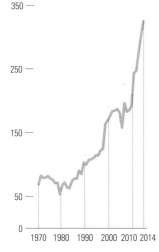

(thousands)

Source: Human Development Report Office calculations based on data from USPTO (2015).

FIGURE 3.9

Japan led the way in total number of patents granted in 2013

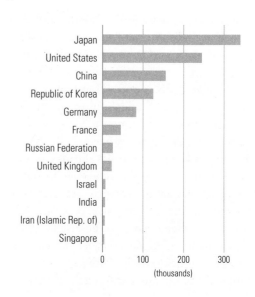

Source: WIPO 2015.

World Values Survey, but the proportion varies considerably and is linked in particular to level of education. Only 10 percent of workers with less than a primary education consider their work very creative, compared with 34 percent of those with a college degree.[89]

In many cases marginal adjustments of processes can have big cumulative effects on innovation in the workplace. Take Toyota. Innovation is an element of every worker's job, not a task assigned to selected managers and leaders. Small ideas accumulate and, by some figures, Toyota implements 1 million new ideas a year, most from shop-floor workers, who suggest, for example, ways to make parts easier to reach. In a work culture where continuous improvement is expected, Japanese companies are estimated to receive 100 times more suggestions from their workers than US companies.[90]

Taking the notion of creative labour beyond its traditional boundaries, 2006 Nobel Prize winner in literature Orhan Pamuk makes a special contribution on creative work (signed box).

Virtual volunteerism

Forms of work enabled by the digital revolution extend beyond paid employment. There are new opportunities for volunteering, social activism, political engagement and artistic expression. Individuals can work together virtually across borders and time zones to pool their resources and ingenuity. Online platforms for interaction, often less hierarchical than traditional organizations, can allow recognition of more diverse perspectives.

The digital revolution has changed the nature of volunteering, opening opportunities for those who may not have previously volunteered. There is now virtual (online or digital) volunteering—using the Internet, offsite from the organization or the groups of people assisted.

These new forms of volunteering can be particularly enabling for those with disabilities that may limit mobility or those who may have only a few minutes to an hour a day for volunteering. UN Volunteers's online volunteering system helped 10,887 volunteers (60 percent of them women) contribute their skills to development work in 2014.[91] Virtual volunteers mentor small business owners, write reports and proposals and teach online classes. "Micro-volunteers"

help crowdsource through the Internet, SMS mobile technology and smartphone apps.

After Typhoon Haiyan hit the Philippines in early November 2013, about 700 digital volunteers worked with the International Red Cross to identify the areas and extent of destruction, and the kinds of help and support needed through OpenStreetMap, which can provide information to guide actions.[92] In the aftermath of the Gorkha earthquake in Nepal in April 2015, global networks of volunteers contributed to disaster maps by identifying damage locations and pleas for help, drawing information from social media posts.[93] And Ushahidi, an open-source software company, supports information collection, mapping and data visualization linked originally to tracking conflicts in Kenya.[94] Efforts like these help fill information gaps for responders on the ground during disasters and other crises.

The modern workforce

In the new world of work, workers need to be more flexible and adaptable—and prepared to retrain, relocate and renegotiate work conditions. They also need to dedicate increasing amounts of time to searching for new opportunities. In addition to constantly thinking about their current work, they have to think about their next job.

The millennials

The people most linked to the new world of work are young adults sometimes referred to as "millennials"—roughly the cohort born since 1980. This group has come of age at a time when digital technologies and advanced information and communications technology penetrate all areas of life, and many of them have had access to these technologies since childhood, particularly in developed countries. They have also become adults at a time when flexibility, adaptability and unconventional work are increasingly common.

A 2013 survey of 7,800 millennials from 26 countries in North America, Western Europe, Latin America and East Asia found that millennials anticipated that their working lives would be flexible and diverse. About 70 percent

The digital revolution has changed the nature of volunteering

Creative work

There is a charming anecdote about the German mathematician Carl Friedrich Gauss. According to the version our mathematics teacher in Istanbul told us, a class of high school students in Germany was misbehaving (much like we were), and to punish them, their teacher told them to add up all of the numbers from one to one hundred. One of the students—Gauss himself—suddenly realized that the sum of the first and last number, the second and second-to-last, and so on, was always 101. Noting, too, that there were 50 such pairs within the first hundred numbers, it took him two minutes to work out the total (5,050) and come up with his famous formula to find it—saving him hours, possibly even days of calculations.

To me, this story is not just about mathematics, but about creativity and "creative labour" of all kinds—artistic and literary, too. Examining this anecdote can help us to better conceptualize the notion of "creative work" and to uncover and discuss its links to human development.

To most of us, the connection between creativity and human history is so clear that it is not even up for debate. Already in the 1960s, history textbooks in Turkey were teaching us that even though political and military triumphs are the most important measures of a civilization's worth, we should also consider its artistic, literary and creative achievements.

Though in 1960s Turkey, not all teachers would have appreciated Gauss's classroom creativity, for the point of the task assigned to his class was not to encourage them to be creative or to play around with numbers, but to punish them by forcing them to spend a given length of time on grinding out the solution. Perhaps this is the very definition of creativity, and its fundamental complication: that it is always unexpected and surprising.

To get to the heart of the matter, we might take a bold step and define creativity as a force that reduces the amount of effort required to achieve something, transforming those rules and traditions that would otherwise make the process more laborious. Our common sense tells us that creativity works against rules and regulations, traditions, bureaucracies and habits. Creative individuals may be involved in literary or artistic pursuits; they may be working in an advertising agency or on a factory assembly line; but none of this will change the most obvious feature of the nature of creativity. In this context, "creative labour" becomes an inherently oxymoronic concept.

At the same time there is more to creativity than just something that reduces the expenditure of time and effort; it also changes the intent behind that effort. We know, for instance, that the art of photography grew and spread in tandem with the rise of impressionist painting. This is because the spread of photography rendered obsolete any form of art whose purpose was still to mimic nature. Despite resistance from state-sponsored salons and galleries, bureaucrats, traditionalists and classicists, the impressionists' creativity rapidly transformed the very purpose of art. Artists stopped painting the world as it was and started looking for ways to present it as it appeared to the artists'—to the human—eye.

Another complication is that any line we might try to draw between creative work and uncreative, ordinary work soon begins to look arbitrary and unfair. There are many who would argue that the work done at an advertising firm is less creative than the work of a poet, but is this correct? Is it fair to say that an artist is more creative than a car designer or a teacher? The translator of a book is as creative as its author; translators too have the right to exercise creativity in their work and to be accepted for this aspect of their humanity. Creative labour allows us to express our individuality and singularity, and the right to follow this very human urge is as fundamental as the right to freedom of expression and the acceptance of our differences. My moral perspective tells me that all kinds of labour, all types of work, should be creative, or should be intended as such.

The notion of creative work may be problematic and difficult to conceptualize, but this should not deter us from treating creativity as a key measure of human development and labour. Gauss's teacher praised him for discovering a formula for the total of all the numbers from 1 to 100 rather than dutifully adding them up one by one; that is the kind of teacher we would all want, for we all wish our jobs would let us be as creative as an artist or a mathematician. Our respect and admiration for highly creative individuals hints at our desire to release our inner creativity and be innovative in our work, no matter what that work may be.

The enduring appeal of the story of Gauss's mathematical creativity does not simply lie in the usefulness of the formula he discovered. We value Gauss's creativity for its own sake. More than just the benefits of human creativity, ours is a reverence towards humanity itself, its ingenuity and its powers of imagination, capabilities and potential. On an intellectual if not a legal level, this understanding of creativity brings to mind the concepts of freedom of thought and freedom of expression. So when Gauss's teacher greets his student's discovery with enthusiastic approval and—instead of punishing him for disrupting the class—relays the story to others until it becomes the stuff of legend, we are delighted. It is a joy similar to what we feel when we know that our identity, our traditions, our personal stories and our choices are being treated with respect.

When we assess working conditions and environments, we need to measure and explore the extent to which there are mechanisms in place for human creativity to be taken into consideration, appreciated and harnessed. Where we work, are our discoveries and our new ideas respected and applied? Or do we simply replicate existing models and patterns that have been successfully implemented elsewhere? When we come up with an original solution at work, is it accepted, and are there mechanisms allowing us to voice it? Or are we expected to stick to pre-established norms and methods? If we work in a place where we are routinely expected to be creative, is it actual creativity that we are being asked for, or simply to produce hurried imitations of old blueprints? Does our work encourage us to be unique and original, or is there an understanding that we will be more successful if we mimic and replicate previously successful models? Does our workplace view creative individuals as troublemakers, or are they respected as Gauss was? I think the answers to these questions can be quantified and measured.

Half a century ago, colonial and post-colonial societies dramatically worried about imitation, authenticity and originality. Today, we value our creativity as an essential component of human development and wonder how we might begin to measure creative labour.

Orhan Pamuk
Novelist, screenwriter, academic and recipient of the 2006 Nobel Prize in Literature
Translated by Ekin Oklap

---

expected to be self-employed at some point.[95] Many millennials are also looking for work beyond creating profits, hoping to help solve environmental and social problems as part of their livelihoods. The same survey found that 63 percent had given money to charity, 43 percent were members of, or had volunteered for, community organizations and 52 percent had signed petitions. This generation seems particularly keen to take a communal view of work.

Social entrepreneurs

Social businesses are emerging as new areas of work. These are cause-driven entities designed to address a social problem—nonloss, nondividend companies (where all profits are reinvested back into the company) that aim to be financially self-sustainable, with a primary goal of maximizing social benefits as opposed to maximizing profits (though they are desirable).[96]

Inspired by a particular cause and by the desire to give something back to society, a number of successful commercial entrepreneurs in different parts of the world are transitioning from for-profit ventures to engaging in social change. A survey of 763 commercial entrepreneurs in India who experienced a transition from commercial to social entrepreneurship between 2003 and 2013 and a quantitative analysis of a final sample of 493 entrepreneurs indicate that 21 percent of the successful entrepreneurs shifted to social change efforts.[97] Most are skilled organization builders, independently wealthy, often from outside the establishment, some from the diaspora.

Opportunities for women

Globalization, digital technologies and new ways of working are also ushering in new notions of men's and women's work. The digital economy has enabled many women to access work that lets them apply their creativity and potential. In 2013 about 1.3 billion women were using the Internet.[98] Some have moved to e-trading as entrepreneurs or are employed through crowdworking or e-services. Moreover, women are now more likely to be found in senior jobs (see chapter 4), but if the glass ceiling —the invisible barrier to women's advancement in the workplace—seems to be cracking, it has not yet broken. Still, even poor female entrepreneurs in developing countries can now use mobile phones to gain access to market information and sources of finance. They can also do so flexibly from home.

One of the earliest examples emerged in 1997 in Bangladesh, when female entrepreneurs set up village pay phones. Using mobile phones leased from Grameen Bank, they could sell services to other villagers.[99] In Andhra Pradesh, India, women run e-Seva centres that provide a wide range of online services.[100] In addition to Internet browsing and access to online auctions, customers can use these centres to pay bills, obtain land and birth certificates, file complaints and grievances, and gain access to tele-medicine and tele-agriculture.

This new world of work puts a high premium on workers with skills and qualifications in science and technology, workers historically less likely to be women. Women are vastly underrepresented in these subjects at the secondary

> Social businesses are emerging as new areas of work

and tertiary education levels and in the overall technical workforce.[101] Countries looking to spur innovation will thus need to boost female participation in technology-oriented education and jobs. One way is through online education services such as EdX, a nonprofit online education group backed by Harvard University and the Massachusetts Institute of Technology. EdX is working with the Saudi Ministry of Labour, for example, to develop online classes for young people and women.[102]

New horizons for older workers

By 2050 an estimated 2.1 billion people will be over age 60.[103] Older workers can be a vital force in the economy, particularly in areas where the size of the labour force is expected to decline as the population ages (as in Europe and Japan). Many are continuing to work beyond normal retirement age, just as many societies face high youth unemployment. In the United Kingdom the employment rate among people ages 50–64 rose from 57 percent in 1995 to 69 percent in 2015, and the rate among people ages 65 or older rose from 5.2 percent to 10.4 percent.[104]

Older people continue to work because they enjoy their jobs or because they cannot afford to retire. Curiosity and interest do not necessarily diminish with age, and those who continue to work can improve their well-being by maintaining social connections that prevent isolation and loneliness. Work gives older people a purpose and a social function. They can also mentor younger workers, passing on their insights gained through years of experience.

For the most part, older and younger workers are in different labour-market segments, so there is no direct substitution. The anxiety that young people will lose opportunities for work when older people are encouraged to work has been termed "the lump of labour fallacy."[105] Yet retirement can have a cascade effect, opening space for younger workers.

Promises as yet unfulfilled

The new world of work is creating fresh opportunities. So far, however, many of the promises of higher productivity and better jobs have yet to be met, and some of the downsides are already evident. The implications of the new world of work for human development are promising, but those promises are, for now, not wholly fulfilled.

Technological revolution— job gains or losses?

Economists have historically rejected the argument that an increase in labour productivity reduces employment in the long run. The argument would be sound if there were a finite amount of work, but the view is that new technology creates new demands for labour. Indeed, in the two centuries since the industrial revolution began, workforces have continued to grow, and productivity and living standards have risen dramatically.

Some fear job losses through automation. Indeed, many jobs are already disappearing or are vulnerable (figure 3.10). Broad swathes of middle management risk being eliminated. Rows of desks could become empty, not because workers are unfit for their purpose, but because that purpose no longer exists. Some estimates indicate that by 2025 almost 50 percent of today's occupations could become redundant.[106] New jobs will require creativity, intelligence, social skills and the ability to exploit artificial intelligence.

But others argue that computers are very far from being able to use creativity, intuition, persuasion and imaginative problem solving, and they may never get there. The view is that certain occupations are insulated from any displacement effect of computerization. Further, the complementarity between machines and people is crucial: Most work activities require a mixture of tasks that must be accomplished jointly, some to be completed by a computer and some by a human.[107]

Either way, technology will leave many people behind, and some human skills will be more valuable than ever. We may be at an inflexion point, with both positive and negative impacts. The technological revolution represents "skill-biased technical change": the idea that the net effect of new technologies is to decrease demand for less skilled workers while increasing demand for skilled labour. By definition, such change favours people with more human capital, polarizing work opportunities.

The implications of the new world of work on human development are promising—but these promises are not yet wholly fulfilled

FIGURE 3.10

The 20 jobs most and least likely to be replaced by automation

Least likely to be replaced	Most likely to be replaced
Recreational therapists	Telemarketers
First-line supervisors of mechanics, installers and repairers	Title examiners, abstractors and searchers
Emergency management directors	People working in sewers
Mental health and substance abuse social workers	Mathematical technicians
Audiologists	Insurance underwriters
Occupational therapists	Watch repairers
Orthotists and prosthetists	Cargo and freight agents
Healthcare social workers	Tax preparers
Oral and maxillofacial surgeons	Photographic process workers
First-line supervisors of fire fighting and prevention workers	New accounts clerks
Dietitians and nutritionists	Library technicians
Lodging managers	Data entry keyers
Choreographers	Timing device assemblers
Sales engineers	Insurance claims
Physicians and surgeons	Brokerage clerks
Instructional coordinators	Order clerks
Physchologists	Loan officers
First-line supervisors of police and detectives	Insurance appraisers
Dentists	Umpires, referees, and sports officials
Elementary school teachers, except special education	Tellers

Note: Occupations are ranked by their probability of computerization (least likely to become automated in blue and most likely to become automated in red). Occupations correspond closely to the US Department of Labor Standard Occupational Classification.
Source: Frey and Osborne 2013.

At the top will be good jobs for those with the necessary education and skills. For example, in the automobile industry the engineers who design and test new vehicles will benefit. At the bottom there will still be low-skill, low-productivity, low-wage service occupations such as office cleaning. But the middle areas will see a steady hollowing-out of many jobs in office cubicles and factory floors. The biggest losers will thus be workers with less-specialized skills. Many cognitively complex jobs are beyond the abilities even of people with reasonable qualifications. Some industries could therefore face skill shortages, so companies willing to pay high salaries for the best talent will look to a global market. And besides being polarized nationally, workforces are being stratified internationally, with low-skilled workers coming mainly from national markets and high-skilled workers from global markets.

There has never been a better time to be a worker with special skills and the right education, because these people can use technology to create and capture value. But there has never been a worse time to be a worker with only ordinary skills and abilities, because computers, robots and other digital technologies are acquiring those skills and abilities at an extraordinary rate. The role of policy in equalizing the life chances of people to have decent work has never been more important.

Productivity and wages— not what was expected

An implied promise of the digital revolution was that it would increase labour productivity and thus lead to higher pay. This does not seem to have happened. Productivity has not grown at the rates expected, and the gains have not translated into higher wages for the most part. Take the United States and the Netherlands as examples (figure 3.11).[108] In the United States increases in productivity and wages began to diverge around 1973, and the gap has since been widening, with productivity gains in manufacturing of almost 75 percent through 2013 and wage increases of less than 10 percent. The

Technology will leave many people behind, and some human skills will be more valuable than others·

FIGURE 3.11

In the Netherlands and the United States productivity gains have not translated into higher wages for the most part

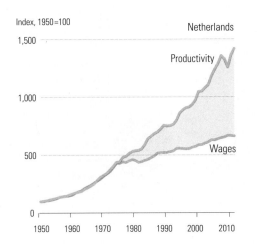

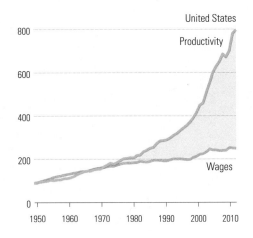

Source: Human Development Report Office calculations based on data from BLS (2012).

widening gap between productivity and wages since the middle of the 1970s has also been observed in the Netherlands. In some cases wages have remained flat. Between 2007 and 2013 real wages in Japan, Italy and the United Kingdom went down.[109] But these averages mask the fact that as real pay for most workers stagnated, incomes for the highest earners soared.

Although productivity has grown, the growth rate has not had the exceptional boost expected with the advent of the digital revolution (figure 3.12). This phenomenon has been termed the Solow Paradox. Several explanations have been provided for this paradox—the digital revolution has been less fundamental, thus resulting in smaller benefits, which have been further offset by demographic change and rising inequality; there is a longer lag involved; or the productivity boost due to the digital revolution will not surface in manufacturing but in services, where economies of scale can be dramatic.

Rising inequality in income shares —no question on this one

The technological revolution has been accompanied by rising inequality. Even people with better education and training who can work more productively may not receive commensurate rewards in income, stability or social recognition. Workers overall are getting

> The technological revolution has been accompanied by rising inequality

FIGURE 3.12

The growth rate of productivity has not had the exceptional boost expected with the advent of the digital revolution

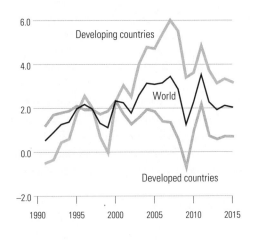

Source: The Conference Board 2015.

a smaller share of total corporate income based on analysis from 27 developed countries and 28 developing countries (figure 3.13). This result is confirmed by another study, which found that in developed countries the share of corporate income going to wages declined roughly 8 points between 1980 and 2015.[110] Developing countries have seen a sharp decline since 1990.[111] This decline may be seen as part of the slowdown in growth of average real wages, as the income shares of high-skill labour (and of capital) have been going up, while the share of medium- and low-skill labour has been going down (figure 3.14).

The sharp increase in work compensation to top salary earners has benefited a minority, whether the top 10 percent, 1 percent or even 0.1 percent (figure 3.15). Over time, those at the top in advanced economies have enjoyed a larger share of the corporate income distribution.

These figures have raised economic and ethical questions, related to the productivity of work and the "value of work," in particular after some of those highly paid workers were behind the financial collapse in 2008. What has changed so dramatically in the past 50 years that can justify a jump in the relative compensation of chief executive officers? Do they generate such high value to their respective companies compared with typical workers?

The sharp increase in work compensation to top salary earners has benefited a minority

FIGURE 3.13

Workers overall are getting a smaller share of total corporate income based on analysis from 27 developed countries and 28 developing countries

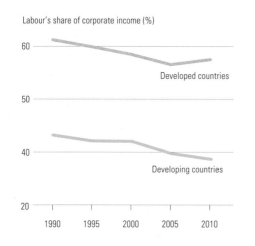

Note: Labour's share of corporate income equals the compensation of employees in the corporate sector divided by corporate gross value added.
Source: Karabarbounis and Neiman 2014.

FIGURE 3.14

The income shares of high-skill labour have been going up, while the share of medium- and low-skill labour has been going down

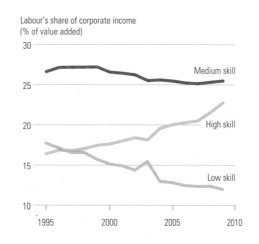

Note: Simple average of 40 countries.
Source: Human Development Report Office calculations based on WIOD 2014.

FIGURE 3.15

The sharp increase of work compensation to top salary earners has benefited a minority, cumulative change since 1980

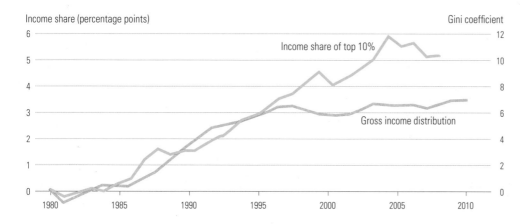

Note: The Gini coefficient equals 0 when all income is equally shared within a country and 100 when one person has all the income.
Source: Jaumotte and Buitron 2015.

Higher rewards for higher skilled workers are contributing to a disproportionate increase in income and wealth for workers at the top of the distribution. The world's richest 1 percent saw their share of global wealth increase from 44 percent in 2009 to 48 percent in 2014, set to reach more than 50 percent in 2016. Members of this global elite had an average wealth of $2.7 million per adult in 2014.[112]

For example, in the United States the ratio of chief executive officer compensation to worker compensation (including stock options) has risen steeply: from 20 to 1 in 1965 to 30 to 1 in 1978 to 383 to 1 in 2000 to 296 to 1 in 2013.[113] These figures raise fundamental economic and ethical questions about the value of different forms of work.

Myriad forces are behind the decline of labour income, including financialization, globalization, technological change, institutional issues (such as waning union membership) and retrenchment of the welfare state.[114] Technology is part of these processes, but its effect is not easily identified.

The prominent role of financial actors is tied to the expansion of credit and highly mobile flows. The greater mobility of capital has provided more investing options to capitalists (increasing their space for business), boosted their bargaining power relative to workers (as financial commitments receive legal priority over labour commitments) and expanded their outside options. Based on the Latin American experience, the macroeconomic volatility led by financial flows has had a negative effect on real wages, employment and equality.[115]

By changing the scale and the way goods and services are interchanged and produced, globalization—as with global value chains—has produced inequalities in income shares. There is evidence that in these chains, the contributions (measured in value added) of capital and of high-skill labour have increased consistently.[116] These trends have implications for income distribution.

Implications for human development

As the preceding discussion suggests, the changing world of work, propelled by globalization and the digital revolution, has considerable human development implications—some positive and some negative. This world has contributed to new capabilities of and opportunities for people, creativity and innovation. In many instances globalization has created new work opportunities for people, particularly women (though opportunities for women in some cases have not reached older women), but it has also resulted in job losses. Global value chains have helped younger people learn new skills that they can use throughout their working lives, but in many cases the networks have not created the promised work opportunities for them.

Participation in value chains has provided some with secure and decent jobs but others with precarious jobs. Three out of four people worldwide do not work with a full-time permanent contract; three out of five workers in wage or salaried employment are in part-time or temporary work. Global value chains are thus associated with economic insecurity. By creating winners and losers, globalization of work has had asymmetrical human development impacts across countries, within countries and among workers.

Similarly, digitization of work has given rise to enormous opportunities for some but has created risks for others. It has made work much more flexible and opened new frontiers for creativity. But not everyone has benefited. Those with the greatest skills and aptitudes have been able to take advantage of the opportunities, while those with more routine skills have seen their jobs extinguished. Digitization has touched the work of traditional sectors (such as agriculture), but not to the extent that it has the modern sector of finance. Knowledge workers apply and produce new technologies that can enhance human progress, but the fruits of the technologies have not been shared by all.

Personalized goods and services benefit people by providing targeted services, which save time and have facilitated improved quality of life. The sharing economy addresses some environmental concerns and contributes to community bonding. Flexible work arrangements allow people to spend more time with families. But at the same time, constant connectivity over mobile devices blurs the distinction between work and leisure, between connectedness through machines and direct human interactions and between public and private space.

Finally, both globalization and digitization have created inequalities in sharing the fruits of work. The shares of high-skilled workers and of capital are going up, while those of other labour have been going down. The reward to top earners and their share in income are hard to rationalize by considering their work or productivity. Such inequalities have substantial adverse impacts on human development opportunities and outcomes.

Conclusions

The changing world of work will take decades to fully play out and will lead to sharp changes in the path of history and human development. The twists and turns will not always be easy to navigate.

The market alone is unlikely to guide digital technologies and systems of global connectivity in the direction of higher human development. Public policies and actions, national and global, are needed now to take better advantage of these opportunities. And inclusive institutions are needed to provide incentives and opportunities for innovation and economic activity for a broad cross-section of society.

Ultimately, the challenge of this changing world of work is to ensure that a globalizing, revolutionizing world ensures equitable opportunities and enhanced human development for all—women and men, present and future generations. In that context, issues of balancing paid and unpaid care work as well as sustainable work are of paramount importance and as such are the focus of the next two chapters.

The changing world of work will take decades to fully play out and will lead to sharp changes in the path of history and human development

Chapter **4**

Imbalances in paid and unpaid work

Infographic: Progress in gender equality on select dimensions: 1995 and 2015

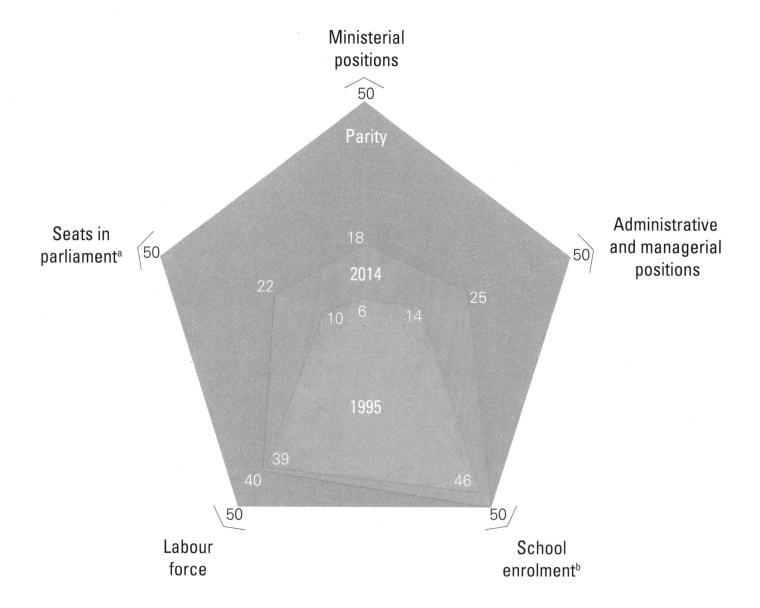

Ministerial positions

Administrative and managerial positions

Seats in parliament[a]

School enrolment[b]

Labour force

Parity

50

50

50

50

50

18

2014

22

25

10 6 14

1995

39

40

46

Note: Data are in percentages.

a. For countries with a bicameral legislative system, seats in parliament is calculated based on both houses.

b. Includes primary through tertiary education.

4.

Imbalances in paid and unpaid work

Both paid and unpaid work contribute to the realization of human potential—as preceding chapters have indicated. Indeed, most individuals need to accommodate both in their daily lives. However, there continue to be pronounced imbalances between men and women in how this balance is struck and in the degree of freedom available to make choices in this respect.

In these two domains of work, men's and women's roles are generally very different, reflecting societal contexts, norms and values, as well as perceptions, attitudes and historical gender roles. These lead to markedly different opportunities and outcomes for human development. For example, the total time spent on work by women tends to exceed that by men: An analysis of time use surveys representing 69 percent of the world's adult population shows that women account for 52 percent of total hours worked, men 48 percent (figure 4.1; see also table A4.1 at the end of the chapter).

Of the 59 percent of work that is paid, mostly outside the home, men's share is nearly twice that of women—38 percent versus 21 percent. The picture is reversed for unpaid work, mostly within the home and encompassing a range

of care responsibilities: Of the 41 percent of work that is unpaid, women perform three times more than men—31 percent versus 10 percent.

These numbers are only one dimension of imbalances that persist at many levels. The 1995 Human Development Report observed that women work more hours than men.[1] The data available at that time indicated gender inequalities in representation at decisionmaking levels in the private and public sectors and in participation in the labour force and school enrolment (see infographic at the beginning of the chapter). As the comparison with current data shows, at the global level there is a perceptible move towards equality on all these fronts but at different paces: The labour force participation rate, a measure of those already employed or actively seeking work, has moved only a little, while the gains towards parity in representation at decisionmaking levels have been much more pronounced, providing an indication of where constraints may be more entrenched.

Indeed, although women work more hours than men, their relatively limited participation in the labour force is symptomatic of the imbalanced sharing of care responsibilities and expectations that serves to circumscribe options. Expanding women's opportunities, choices and freedoms requires a range of actions that on the one hand remove barriers to their full and equal participation in the realm of paid work and on the other hand require the recognition, reduction and redistribution of the care burden that they disproportionately carry.

This general situation is changing in many regions, if slowly, as attitudes and norms shift and some positive public policies are pursued. More women are receiving higher levels of education, and in a globalized work environment with many new technologies, opportunities

Both paid and unpaid work contribute to the realization of human potential

FIGURE 4.1

Men dominate the world of paid work, and women the world of unpaid work

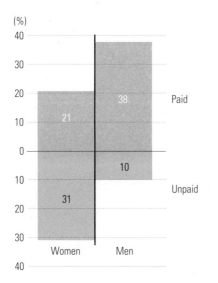

Note: Data are a female and male population–weighted average of 63 countries representing 69 percent of the world's adult (ages 15 and older) population.
Source: Human Development Report Office calculations based on Charmes (2015).

are opening in paid work. Women's representation in parliaments, in senior government positions and in senior business management has improved around the world: The glass ceiling has not been broken, but a few cracks are appearing. At the same time, there is also a move towards a more equitable sharing of the burden of unpaid care work between men and women. Nevertheless, the pace of these changes is slow, and much remains to be done.

Disparities in the world of paid work

Even though women carry more than half the burden of global work, they are disadvantaged in the world of work—in both paid and unpaid work. Women are less engaged in waged work, earn less than men, remain underrepresented in senior decisionmaking positions, encounter uneven barriers to entrepreneurship and, in many countries, are more likely to be in vulnerable employment.

Women are less engaged in the workforce

Conventional measures of work and employment do not take into account the engagement of individuals in unpaid care work, much of which takes place in the household: However, these measures indicate that women are less

likely to be working for pay or even looking for such work (figure 4.2). Both time use surveys and macroeconomic indicators such as the labour force participation rate support this point.

The labour force participation rate—the proportion of the working-age population in paid employment or looking for such work—differs for men and women. While a portion of the difference may be due to an undercounting of those working only sporadically or in home-based activities—kinds of employment where women, especially in the rural areas of developing countries are over-represented relative to men[2]—this effect is not large enough to completely account for the divergence. Two trends, in particular, stand out.

First, the labour force participation rate is consistently much lower for women than for men, both globally and by human development grouping (see figure 4.2). In many regions the gap has remained constant for decades. In 2015 the participation rate was around 77 percent for men and around 50 percent for women.[3]

Second, globally participation rates for women have fallen slightly in recent years, as have men's. The drop in the rate for women is due mainly to reductions in India (from 35 percent in 1990 to 27 percent in 2013) and China (from 73 percent in 1990 to 64 percent in 2013).[4] But low and very high human development countries have begun to see convergence in female and male rates, as the former has risen, while the latter has remained static or declined.

Women are less engaged in waged work

FIGURE 4.2

Women are less likely to be engaged in paid work, as shown by the labour force participation rate

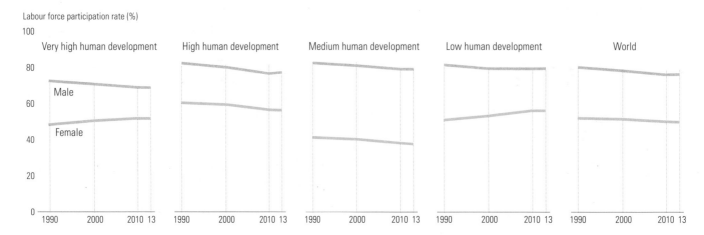

Source: Human Development Report Office calculations based on ILO (2015e).

These differences reflect economic, social and cultural factors that affect men and women differently, albeit with variations across countries. For example, over the ages 15–49 each birth is estimated to take a woman out of the labour force by an average of almost two years.[5] Falling fertility rates—observed in almost all regions —should therefore free up women's time, enabling greater participation in the labour force, provided other constraints are removed and suitable opportunities exist. Fertility rates are expected to continue to drop in the coming decades, gradually moving from the current global average of about 2.5 births per woman[6] to about 2.1 in 2075 and 1.99 in 2100.[7]

The decline in fertility rates is expected to increase female labour force participation. In a sample of 97 countries (covering about 80 percent of the world's population), it is estimated that 91 million additional women will join the labour force due to this effect over 2010–2030.[8]

Another factor that can contribute to higher labour force participation by women is improvement in their educational attainment, should opportunities exist (although there can be temporary declines if women are choosing higher education over labour market entry). This factor can be large: In a sample of 97 countries, increased education is estimated to boost female labour force participation by nearly 3 percentage points over 2010–2030.[9] However, such gains can be hard to realize if there is disparity in accessing labour market opportunities, or continuing imbalances in the sharing of the care burden or the meeting of social expectations. Issues such as these can make the transition to paid work more difficult for women. The declining disparities in educational attainment between women and men need to be matched by complementary actions that will facilitate a commensurate parity in the labour market.

A similar picture prevails with regard to employment: In 2015, 72 percent of working-age (ages 15 and older) men were employed, compared with only 47 percent of women (figure 4.3).

Achieving higher engagement in paid work confers multiple benefits not just to women but to societies and economies at large. It is well recognized that a higher female labour force participation rate boosts economic growth. For example, raising the rate in Japan from the current 66 percent to 80 percent (still 5 percentage points lower than the male participation rate) could boost the country's output 13 percent.[10] Annual economic losses due to gender gaps in effective labour (labour force participation rate and years of schooling) have been estimated at $60 billion in Sub-Saharan Africa.[11]

Women earn less

Even when working for a salary, earnings between women and men diverge. Globally, women earn 24 percent less than men.[12] The gap is due partly to the fact that women tend to be under-represented at higher levels of pay and in higher paid occupations. But even when doing similar work, women often earn less— with the gap generally greatest for the highest paying professions. In the United States female financial specialists earn only 66 percent of what their male counterparts earn. For dentists the proportion is 74 percent, and for accountants 76 percent.[13] In Latin America female top managers average only 53 percent of male top managers' salaries, and female scientists 65 percent.[14]

FIGURE 4.3

In 2015, 72 percent of working-age (ages 15 and older) men were employed, compared with 47 percent of women

(% of working-age population)

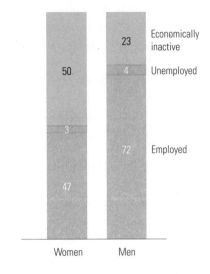

Source: UN 2015b.

Higher engagement in paid work confers multiple benefits not just to women but to societies

Some of the wage gaps can be explained by differences in education, skill levels and years of experience—often related to women's care responsibilities—but in most cases a large part remains unexplained (figure 4.4). In Estonia the wage gap is around 29 percent, but only a quarter is explained by differences in observable characteristics such as those related to human capital or the nature of the occupation—even if these differences were removed, women would still earn considerably less. As in many other countries, in Estonia too the unexplained part is larger in upper income deciles. In Denmark and Poland the explained gap is negative, meaning that women should be earning more than men after taking into account various characteristics.[15] One explanation for the unexplained portion of the wage gap is discrimination—whether overt or covert—at work.

Wage gaps are not only an economic issue; they also affect—and are affected by—power relations. On the one hand, earnings make for economic independence, a critical factor towards individual autonomy, voice and agency in households and the community. On the other hand, an unequal distribution of care responsibilities in the household may require one parent to take time off more frequently than the other, reducing the former parent's current and prospective earnings and perpetuating divergences. Equal pay for equal work is not just a matter of social justice; it also holds substantial social value, especially for empowering women in households and the community.

The glass ceiling continues to skew representation across genders

Wage gaps are but one indication of the lower representation of women in positions of leadership and seniority in the private sector. Globally, women hold only 22 percent of senior leadership positions, and 32 percent of businesses have no female senior managers.[16] The regional picture is quite varied (figure 4.5), and at the country level representation of women in senior management positions in businesses ranges from 8 percent in Japan to 40 percent in the Russian Federation.[17] The picture also varies within industries: Women hold only 19 percent of senior technology roles.[18] The situation is even starker at the top of enterprises, with only 9 percent of companies having a female chief executive officer in 2014. This vertical segregation at work accompanies a divide across occupations as well: Women's participation varies from 12 percent in mining and quarrying to 24 percent in professional services to 41 percent in education and social services.[19] Such occupational segregation has been pervasive over time and

Wage gaps are not only an economic issue

FIGURE 4.4

A large part of the wage gap between men and women is unexplained

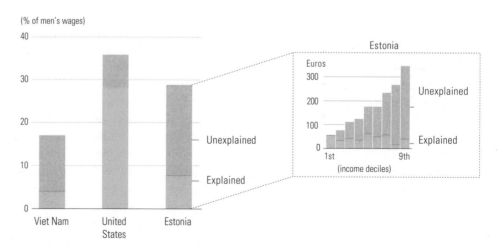

Note: Data are for the most recent year available.
Source: ILO 2015b

across levels of economic prosperity—in both advanced and developing countries men are over-represented in crafts, trades, plant and machine operations, and managerial and legislative occupations, and women in mid-skill occupations, such as clerks, service workers and shop and sales workers.

In public service, too—where equal pay for equal work is easier to realize—positions of leadership continue to be disproportionately occupied by men. That women hold about 25 percent or less of senior political and judicial positions likely influences the gender balance of laws and regulations and the way they are implemented (figure 4.6).

Asymmetric barriers to entrepreneurship

Women become entrepreneurs for many of the same reasons as men—to earn livelihoods, support their families, enrich their careers and gain a degree of independence.[20] For both women and men the highest rate of early-stage entrepreneurship (businesses less than 3.5 years old) is in Africa, followed by Latin America and the Caribbean (left panel of figure 4.7). However, in all regions women are less likely than men to initiate their own enterprises. Particularly in Latin America and the Caribbean, a disproportionately large number of female-headed

In public service positions of leadership continue to be disproportionately occupied by men

FIGURE 4.5

Women are underrepresented in senior business management across all regions, 2015

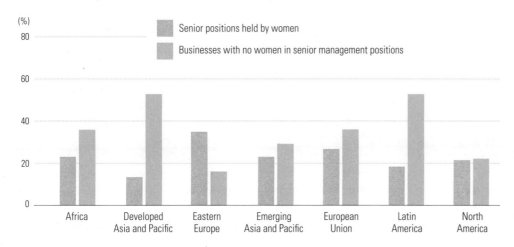

Source: Grant Thornton 2015.

FIGURE 4.6

Women are less likely to occupy positions of leadership in public service, 2014

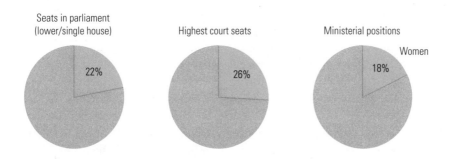

Note: For countries with a bicameral legislative system, seats in parliament are calculated based on both houses.
Source: Human Development Report Office, based on IPU (2015) and UNDP (2015b).

FIGURE 4.7

Fewer women than men are represented among both early-stage and established entrepreneurs, 2014

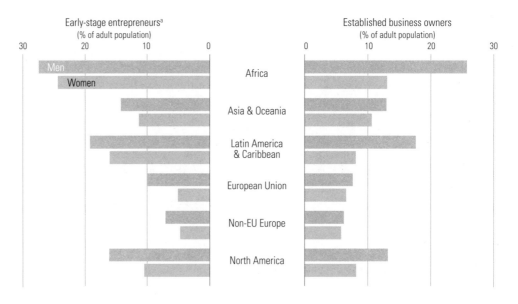

a. Refers to individuals who started a business within the last 3.5 years.
Source: Singer, Amorós and Moska Arreola 2015.

enterprises do not survive to become established businesses (right panel of figure 4.7).[21]

Why are there fewer female entrepreneurs? Reasons range from limited financial opportunities to unequal use of technology:

- *Unequal access to finance.* Some 42 percent of women worldwide did not have a bank account in 2014, and the proportion was even higher in developing countries (50 percent). And in 38 countries, including India, Mexico, Pakistan and Uganda, more than 80 percent of women are unbanked.[22] By contrast, in Japan and the Republic of Korea more than 90 percent of women have bank accounts.[23]
- *Legal restrictions and discriminatory practices.* In 22 countries covered by the Global Entrepreneurship and Development Index, married women do not enjoy the same legal rights as married men, and in 8 countries women do not enjoy the same legal access to property as men. In a third of these countries women's access to public spaces is restricted through either legal provisions or discriminatory practices.[24]
- *Unequal access to and use of technology.* As of 2013, only 39 percent of women in India were Internet users, compared with

61 percent of men. In China the percentage was 44 percent for women and 56 percent for men, and in Turkey only 44 percent of women were, compared with 64 percent of men.[25] Few women operate businesses in manufacturing, transport and advanced technology.[26]

Uneven presence in vulnerable work

While disparities within formal wage employment are rife, such work does confer some degree of economic security and contractual and legal protections for all workers. But working in the informal sector or being self-employed (but not an employer) can be marked by low and variable earnings, poor working conditions, inadequate voice and limited scope for collective actions, fostering insecurity and dependence.

One way to measure this is by enumerating workers in vulnerable employment, classified by labour statisticians as those working in a business operated by a family member (contributing family workers), or self-employed but not engaging any employee to work for them on a continuous basis (own-account workers). Another way is by counting workers in the informal sector, who are likely to remain outside

Women do not enjoy the same legal rights as men

the reach of the protections afforded by other forms of formal employment.

Among countries with data covering 84 percent of the global workforce, more than 48 percent of men and 41 percent of women are self-employed; in low-income countries the female rate (90 percent) is higher than the male rate (83 percent).[27] Globally, nearly 50 percent of employed women are in vulnerable employment, compared with 44 percent of men. In low-income countries the percentage is 86 percent for women and 77 percent for men.[28]

When grouped by human development categories, women are more likely to be in vulnerable employment in low and medium human development countries. Based on the available data, which have limited coverage in some cases, a similar situation appears to hold for geographic regions as well: Women are more likely to be in vulnerable employment in most regions (figure 4.8). The exception is Latin America and the Caribbean, where there is some parity, with vulnerable employment accounting for about a third of the working population for both women and men.[29]

Examples of such work can be found in many sectors, but two stand out as engaging especially large numbers of women: agriculture and paid domestic work.

Agriculture

In South Asia agriculture, primarily informal, accounts for almost 62 percent of female employment but less than 42 percent of male employment (figure 4.9).

Over and above this, the nature and returns to work may vary for men and women. For example, both men and women may be working on small landholdings, engaged principally in subsistence work,[30] but, as in Malawi, a greater proportion of women may be working unpaid on family farms (figure 4.10).

The share of the agricultural labour force that is female has either remained static or increased in all regions over the past 20 years, due to a number of factors, including the migration of men seeking better paid opportunities. Between 1980 and 2010 the female share in the agricultural labour force rose from 30 percent to 45 percent in the subregions of Near East and North Africa. This trend reflects a greater absolute number of women entering the agricultural labour force as well, implying continuing exposure to vulnerable employment.[31]

> Women are more likely to be in vulnerable employment

FIGURE 4.8

Across most parts of the world women are more likely to be in vulnerable employment, 2013

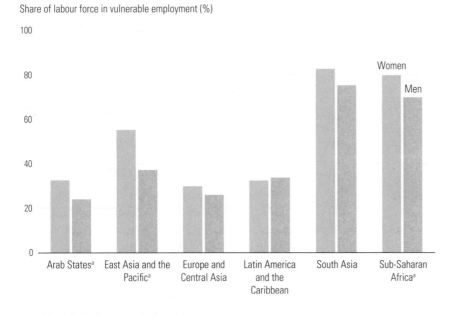

Share of labour force in vulnerable employment (%)

a. Data coverage is below the criteria used for regional aggregation.
Source: Human Development Report Office calculations based on ILO (2015e).

FIGURE 4.9

In South Asia agriculture accounts for more than 60 percent of female employment

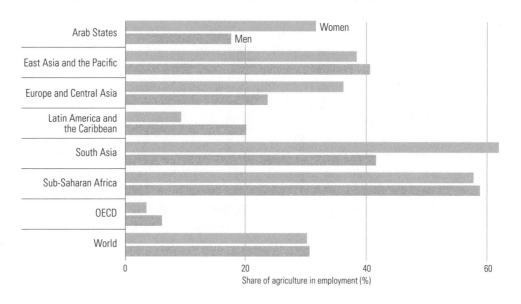

Note: Regional data are drawn from country data from the most recent year in 2005–2013. Excludes China because sex-disaggregated data are not available.
Source: ILO 2015e.

FIGURE 4.10

In rural Malawi women are more frequently unpaid, even when working outside the home, 2008

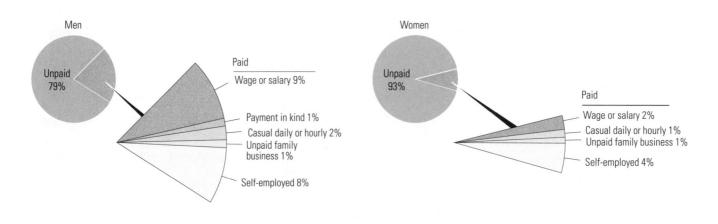

Source: FAO 2011a.

Paid domestic work

With rising household incomes, and increasing numbers of people working outside the home, the demand for paid domestic workers has risen. In 2010 an estimated 53 million people worldwide ages 15 and older were in domestic work. And 83 percent of them were women—accounting for 7.5 percent of female employees.[32] Many were migrants, both within their own countries and across borders.

Domestic work can be a critical source of income but holds limited potential for developing into high-quality employment. It encompasses different tasks and can engage both men and women in large numbers. For example, in 2004/05 India had 4.2 million domestic workers—1 percent of total employment (table 4.1)—about a quarter of them men. More than 70 percent were women working predominantly as housemaids or servants.[33] Some were full-time live-in domestic workers —they might have relatively stable employment and reasonable living conditions but little freedom. Others worked part time, living at home but visiting their employer's house once or twice a day. They could thus work in more than one household, with somewhat greater choices and exit options.[34]

The demand for domestic work sustains migration within countries—such as from rural to urban areas—or across national borders. For example, in many Latin American countries domestic migrant workers are from a neighbouring country—Bolivians in Argentina, Peruvians in Chile[35]—often reflecting lower travel costs, less restrictive entry requirements and established networks.

But others travel longer distances, from lower to higher income countries. Latin Americans work in the United States, and Ukrainian care workers find jobs in Italy.[36] Workers from the Philippines or Indonesia may move to Singapore, where one household in five employs a domestic worker.[37] Some of the biggest employers of such workers are the countries in the Gulf Cooperation Council. They employ about 2.4 million migrants, many of whom are female domestic workers (table 4.2).[38] Between 2001 and 2010 the number of migrant domestic workers dramatically increased.[39]

While migrant workers who travel to other countries may obtain jobs and earn more than they would at home, they can also find themselves trapped in abusive and exploitative settings behind closed doors with limited recourse to assistance. For example, their employer may withhold hold their passports and wages, not allowing them to resign their jobs before their contract ends. Work hours may be inordinately long, and employees may have limited or no access to services to reinforce their rights, becoming victims of exploitation and even physical and sexual violence.[40]

Many national workers reject these low-paid, low-skill and often low-prestige jobs, even though global demand for domestic work is rising. For example, in the United States at least 1.8 million workers are domestic workers in US homes, and most of them (95 percent)

are female or foreign born.[41] Migrant caregivers often leave their families and children back home, who themselves need to be cared for, often by grandparents or other relatives or hired local helpers, thus resulting in a global care chain parallel to the global flow of domestic workers.

Women's economic empowerment (or at least co-determination in resource distribution within the family) is greater when they are employers, or in waged and salaried work that is protected through legislation and services that guarantee their rights. Informal work, while providing much needed income, is inherently more exposed to risks and less likely to be effectively protected through safeguards such as legislation or organized collective action.

The advances towards greater equity that have taken place are not permanent or immutable—they are subject to both gains and reversals, where public policy can play a

Migrant caregivers often leave their families and children back home

TABLE 4.1

Domestic workers in India by sector, 2004–05 (thousands)

Type of domestic work	Female	Male	Total
Housemaid or servant	2,011	301	2,312
Cook	89	34	123
Gardener	4	15	19
Gatekeeper, *chowkidar* or guard	7	129	136
Governess or babysitter	63	25	88
Others	781	748	1,528
Total	2,955	1,252	4,206

Source: ILO 2013b.

TABLE 4.2

Migrant domestic workers in Gulf Cooperation Council countries (thousands)

Country	Women	Total
Bahrain (2011[a])	52	83
Kuwait (2010)	310	570
Oman (2009)	69	95
Qatar (2009)	48	80
Saudi Arabia (2009)	507	777

a. First quarter only.
Source: Rakkee and Sasikumar 2012.

major role. The special contribution by Roza Otunbayeva, former president of Kyrgyzstan, highlights the progress made and challenges ahead for women's empowerment in Central Asia, with a focus on Kyrgyzstan (signed box).

Imbalances in unpaid work

Unpaid work is apt to be overlooked in economic valuations but holds great worth for individuals and society (box 4.1) and can be a

---

SPECIAL CONTRIBUTION

Central Asia: emerged region, emerging challenges and opportunities for women

The Central Asian women's movement is approaching its centennial. Starting with the emancipation of Soviet women in the workforce, driven partly by the need to stand in for men who were fighting at the World War II front and partly by the post-war Soviet industrialization, women of Central Asia operated heavy machinery in the fields, managed power plants and piloted airplanes. They helped develop oil and gas fields and whole industries.

Much has changed since independence in the 1990s. De-industrialization, unfair privatization of public assets as a result of sharp social stratification, and high internal and external migration dramatically reduced living standards. Gradual collapse of social infrastructure and insignificant numbers of women among new business elites led to the virtual absence of women's economic capital and to their vulnerability to poverty and unemployment.

Central Asia, where two-thirds of the population resides in rural areas, is facing the consequences of climate change, loss of water resources and land degradation. Drug trafficking from Afghanistan and labour migration have boosted HIV prevalence in the region. High levels of corruption, low quality of public services and weak rule of law have led traditional and religious institutions to try to serve as substitutes for state governance institutions in rural areas. Outdated patriarchal values such as the superiority of men over women, early marriage, polygamy and the reinstatement of women's economic activity at home now compete against the apparent secular social norms.

These problems and their solutions were discussed at the 2015 National Women's Forum dedicated to the 90th anniversary of the Women's Movement of Kyrgyzstan, the 20th anniversary of the Beijing Platform and the 15th anniversary of the UN Security Council resolution on the active participation of women in conflict resolution and peace building. The 1995 Beijing Conference on the advancement of women and girls provided Central Asian delegates with both a powerful inspiration and a framework for national gender policies. Over the past 20 years we have formed national women's movements and the institutional mechanisms and national legislative and policy framework on gender equality.

Today, women make up 70 percent of Kyrgyzstan's civil society leaders. We head the country's most active human rights organizations and strong media resources, conduct judicial reform, fight corruption and work to improve the quality of public services. We have been at the forefront of all significant changes in our country, including the country's transition to a parliamentary form of government. Women actively participate in public supervisory boards at each ministry and at the Extractive Industries Transparency Initiative. The country's microcredit movement, in which women play a prominent role, has expanded significantly in the past 20 years, helping to feed the village and the urban poor. However, few women sit on the boards of directors of public and state companies. Sharp challenges face women entrepreneurs who wish to expand their businesses. We need business incubators and commercial and legal training. Instead of giving in to religious and conservative perspectives, we should strengthen the position of women in the creative industries and sports. We should build on decades of progress in which women have realized their aspirations and reached new heights.

The countries of Central and Eastern Europe have now undergone two decades of democratic development. While not long, this experience is increasingly embedded in our sociopolitical and historical memory. "Genes of democracy" have been life-giving agents of change at key historical junctures and are bringing these countries into the group of countries whose welfare and prosperity is directly related to prospects for gender equality. I am certain that the countries of Central Asia—having passed through Soviet emancipation and having achieved 100 percent literacy among women—will be able to cope with the problems of the transition period, with emerging and traditional challenges for gender equality. We must invest in women and girls and change gender stereotypes through inclusive, sustainable and equitable social and economic growth. If our state treats gender equality as an integral part of national development, we will be able to successfully modernize by helping all women and girls to achieve their full potential and aspirations.

Roza Otunbayeva
Former President of Kyrgyzstan

BOX 4.1

Monetary valuation of unpaid care work

The unpaid time that people devote to the care of family, friends and neighbours clearly contributes to economic living standards, social well-being and the development of human capabilities. At the same time it enables individuals to engage in the various forms of paid work in the economy. Yet while paid work is assigned a monetary value and features in national aggregates such as GDP, unpaid work remains largely unmeasured by such a metric and consequently invisible in discussions of economic policy.

Things are changing—due at least partly to UN resolutions insisting on improving the visibility of women's unpaid work—and many countries now administer nationally representative time use surveys asking individuals to recall their activities during the previous day. Over 2000–2010 at least 87 such surveys were conducted, more than over 1900–2000. Estimates of hours worked in providing various household services provide a basis for imputing a market value by asking how much it would cost to procure an equivalent number of hours of similar services from the market. This is valuation by a replacement cost—other approaches are possible, but this one is widely applied. The value of unpaid work contributions cannot always—or fully—be captured in market terms, but estimates of its monetary value, like efforts to estimate the value of unpriced environmental assets and services, can provide important insights.

Unpaid household work that leads to the production of goods (such as food for own consumption or collection of firewood or water necessary for the household) is considered part of production by the System of National Accounts, and most estimates of GDP include approximations of the value of this work. However, unpaid care work, such as meal preparation, housecleaning, laundry, water and firewood fetching and care of children, are explicitly excluded. Although time use surveys do not capture all forms of care work, they are useful for better estimating the total number of hours devoted to unrecognized and undervalued forms of work.

Valuation efforts, often inserted into "satellite accounts" that revolve around the conventional estimates, have gradually been gaining ground in national income accounting and are illuminating. Across a range of countries the results show that time devoted to unpaid care is not optional or idiosyncratic but carefully structured to meet routine family needs, especially the care needs of children, older people and those who are disabled or sick.

Among all countries that are attempting to measure the value of unpaid care work, estimates range from 20 percent to 60 percent of GDP.[1] In India unpaid care is estimated at 39 percent of GDP, in South Africa 15 percent.[2] Among Latin American countries the value is estimated at 26–34 percent of official GDP for Guatemala and 32 percent for El Salvador.

In 2008 the Organisation for Economic Co-operation and Development published estimates of household production in 27 countries using a replacement cost approach, which highlighted that the value of household production as a share of GDP varies considerably across countries. It is above 35 percent in Australia, Japan and New Zealand and below 20 percent in the Republic of Korea and Mexico, countries with lower GDP.

A greater acceptance of such valuations can help direct national policies. For example, in developing countries women spend a substantial amount of time tending to basic family needs, and access to clean water and modern energy services would greatly improve their productivity. However, estimates of the payoff to public investments typically does not take the value of such nonmarket work into account. Doing so could change how resources are allocated and projects prioritized for implementation.

Notes
1. Antonopoulos 2009. 2. Budlender 2010.
Source: Folbre 2015.

source of joy and fulfilment for many. Within the household and community, activities of this nature include various services related to the care of oneself and others. Some of this work is related to the everyday functioning of all households—cleaning, cooking and fetching water and firewood. But a substantial portion relates to the care of others—about 2 billion children for example.[42] Many adults may also need to be looked after to some degree—older people (people over age 80 now number around 120 million[43]), those living with disabilities (estimated at about a billion[44]) and the sick (for example, many of the 37 million living with HIV/AIDS).[45]

Such work is indispensable to maintaining and advancing capabilities and human development. It is essential to the functioning of societies and economies and makes possible much of the observed kinds of paid work. However, it is

Unpaid work is apt to be overlooked in economic valuations

unevenly shared between men and women—it is predominantly women who undertake much of this unpaid work (figure 4.11).

There are variations in how unpaid care work is shared between men and women across countries and over time. For example, in Argentina in 2013 an estimated 50 percent of men did unpaid domestic work, averaging 2.4 hours a day. For women the figures were 87 percent and 3.9 hours.[46] The same year the proportions were 30 percent for men and 54 percent for women in Bogotá, with women taking greater responsibility for food preparation, cleaning and maintenance.[47] How the burden is shared has also been evolving—in the United States women spent more than 240 minutes a day on housework and men less than 40 minutes in 1965; by 2012 women's involvement had fallen to around 140 minutes a day, and men's had risen to over 80 minutes. Despite the shift, the burden is still unequal.[48]

According to time use surveys in developing countries, women are typically responsible for more than 75 percent of the time their household dedicates to unpaid care. In low-income households that adds up to many more hours than in middle- or high-income households, which generally have better access to basic services and can afford to hire help or buy labour-saving technology.[49] In Africa alone women average 200 million hours a day collecting water.[50] Even when the burden of this work is ameliorated, it remains labour intensive and poses impediments to pursuing other activities such as education, paid work, participation or leisure.

Disparities in discretionary free time

Women work more than men, even if a large part is relatively invisible because it takes place in unpaid care activities. As a result, women have less discretionary free time than men do. In a sample of 62 countries men average 4.5 hours a day of leisure and social activities, compared with women's 3.9 hours.[51] The gap is wider at lower levels of human development: 29 percent (relative to men) in low human development countries, compared with 12 percent in very high human development countries (figure 4.12). In Sub-Saharan Africa women show both a high labour force participation rate and a high burden in care work, heavily restricting their free time—in Tanzania women have less than two hours of leisure a day.

In Denmark, Germany and New Zealand, where leisure time is more than five hours a day, the gap between men and women is very small

> **Women have less discretionary free time than men do**

FIGURE 4.11

Women take the major burden of unpaid care work, most recent year available

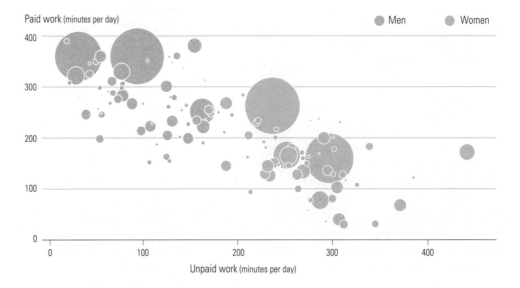

Note: Each bubble represents a country; the size of the bubble reflects the country's population.
Source: Human Development Report Office calculations based on Charmes (2015).

FIGURE 4.12

Men have more time for leisure and social activities than women do across all human development groupings, most recent year available

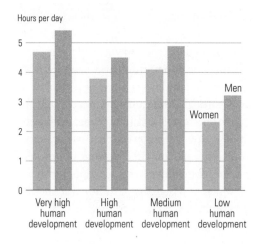

Note: Data are simple averages for each human development grouping and cover 62 countries.
Source: Human Development Report Office calculations based on Charmes (2015).

or nonexistent. These countries have strong public policies that promote gender equality and women's participation in the labour market as well as facilitative social norms. Although data are missing for some developing countries, the gap is 18 percent in Eastern Europe and Central Asia, 16 percent in the Arab States, 15 percent in Asia and the Pacific, 13 percent in South Asia and 7 percent in Latin America and the Caribbean.[52]

Sharing responsibilities for care work

The sources of caregiving are changing, and in many households men are taking on more domestic responsibilities, as with child care. This is happening even as the total time invested in the care of children within the family is growing. For example in the United States fathers spent on average 55 minutes a day on child care in 2010–2014, up from about 20 minutes in 1965. In comparison, an American mother spent about 100 minutes on child care in 2010–2014, up from about 90 minutes in 1965.[53] The increase relative to the 1965 value was smaller for women (roughly a tenth) than for men (more than double). Changes in family structure and the shift from extended families to more mobile nuclear families have also

contributed to this change, as have evolving social standards and the increasing engagement of women in paid work.

Grandparents, if available, often spend time caring for grandchildren (figure 4.13). In some cases there may be very little choice. In China, where parents from rural areas work in urban areas as migrant workers, grandparents provide care to 19 million children whose parents are both away.[54] Worldwide, as of 2013, approximately 18 million children under age 18 had lost one or both parents due to AIDS-related causes. Many of them are cared for by their grandparents.[55]

Emerging challenges— care gaps, health shocks and climate change

Unpaid care work carries within it an overriding human and social imperative. However, such a pressing imperative can also result in limiting the choices of those who are expected to provide it. As the need for care evolves and alternatives such as state-provided services do not keep up, these expectations and traditional roles can further circumscribe options, unless structural shifts take place towards a more equitable sharing of such work. As illustrative examples, this section presents three emerging

The sources of caregiving are changing

FIGURE 4.13

Grandparents often spend time caring for grandchildren, 2006–2007

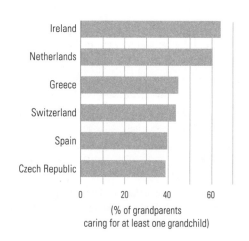

Source: OECD 2015a.

challenges—care gaps, health shocks and climate change.

Care gaps

Different age groups have different care needs, and as populations age, the nature of the services needed also changes. Traditionally, the care needs of infants and children have predominated, and economic, public, social and cultural institutions have evolved to meet them. While institutional arrangements change—for example, through the provision of parental leave or changes in what is expected of men and women—the broad contours of the kind of care services needed, how they will be provided, and money and time commitments are well understood. As fertility rates fall in most of the world, the number of children needing care is likely to decrease, although how the effort is distributed across the various actors will change and—hopefully—move to a more equitable distribution between men and women.

By contrast, as the number of children needing care falls, the care needs of older people will become increasingly important. The economic provisioning for this at the aggregate level is captured to some extent by the economic dependency ratio—the ratio of those ages 65 and older (not in the labour force) to those ages 15–64 (in the labour force). However, as with children, a substantial part of what is needed comes from care provided within families, which is not captured by conventional economic measures.

Recent estimates indicate that there is a global shortage of 13.6 million care workers, causing extreme deficits in long-term care services for those over age 65.[56] The total need for elder care increases with the number of older people and the frequency with which their inability to perform ordinary activities goes up. Some 110–190 million people worldwide experience major difficulties in functioning and need long-term care in their daily lives.[57] These needs can be met partly through paid care services acquired from outside the household (either market based or publicly provided); however, a considerable part comes from the unpaid care work of household and family members, provided predominantly by women. It is here that traditional gender roles may combine with increasing longevity, shrinking household sizes and limited access to alternatives to lead to a higher and disproportionately shared burden of care borne by women, restricting further the choices available in engaging in other forms of work.

While this is already a concern in several developed countries—notably Japan, where 26 percent of the population was over age 65 in 2014 and the cost of caring for them is expected to more than double by 2050,[58] in spite of the availability of alternative mechanisms (box 4.2)—it is also imminent in others. For instance, in the United Sates the burden of elder care relative to its value in 2010 rises relatively slowly—by about a sixth by 2030 and about a fifth by 2080 (figure 4.14). But in China the rise is much more rapid—by about two-fifths by 2030 and doubling by about 2050.[59]

FIGURE 4.14

The burden of elder care will rise much more quickly in China than in the United States

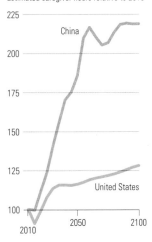

Estimated caregiver hours relative to 2010

Source: Mukherjee and Nayyar 2015.

BOX 4.2

Credits for long-term elder care in Japan

Japan has developed a system for providing care for older people through exchanges of "time credits." The mutual support networks of Fureai Kippu ("ticket for a caring relationship") emerged in urban grassroots mutual-help groups such as the Help of Daily Living Association in Tokyo and the Kobe Life Care Association.

Under the system people offer care to older people or disabled people in their own homes for meals or bathing or provide help shopping or preparing food. In return, the caregivers receive electronic tickets paid into a computerized savings account. They can save these Fureai Kippu for their own use in the future or transfer them to someone of their choice, typically a parent or family member. In some cases they can convert Fureai Kippu into a cash equivalent.

The system helps older people avoid or postpone moving to an expensive retirement home. It can also reduce the time they spend in hospital after a medical problem. Above all, it improves their quality of life.

Source: Hayashi 2012.

Regardless of where and when the need will become apparent, the care of older people is an increasingly urgent issue. Should conventional gender roles and a lack of public alternatives continue to prevail, women may find their choices to be increasingly constrained.

Health shocks

Care work becomes even more critical during severe health crises, such as those caused by HIV/AIDS, avian flu and Ebola. In countries with weak health services the burden often falls on caregivers working silently within the home. In these circumstances people are expected to sacrifice themselves for their families and communities, even putting their own well-being at risk. Traditional gender-differentiated roles then result in a further restriction of options for women, often greatest for the poor, who can least afford them.

For example, across Southern Africa unpaid, voluntary, informal networks of care providers have emerged as a critical vanguard in the provision of care to sick people and orphans in crises such as HIV/AIDS. As public health systems across the region face a barrage of challenges, family and community care providers, albeit with minimal support, fill in a health care gap left by governments.

The gender dimensions of this are evident. As seen in the recent Ebola outbreak, women were especially affected. Confirmed cases are overall equitably distributed between men and women, with women's numbers of cases just slightly higher. But the gendered dimensions of the outbreak go beyond infection rates. Evidence points to women's expected social role as carers of the sick for their increased vulnerability to infection and for greater calls on their time due to the sickness within the family and school closures. These responsibilities result in a decreased participation in economic activity, compounded further by road and market closures. Finally, in a vicious cycle, as the economy shrinks and revenues fall, cuts in public spending redistribute care costs to those who can least afford to pay.[60]

Climate change

Studies have already demonstrated that part of the unpaid work that women do is related to fetching water and gathering fuel and firewood—close to 2 billion people in developing countries use traditional biomass fuels as their primary source of energy.[61] A 2003 study highlighting examples from Sub-Saharan Africa noted that women spent more time than men in these activities, with variations driven by residence in rural or urban areas.[62] Other factors that impact the time spent include infrastructure and access to services, where household income is also a factor. Overall, the burden is likely to be the highest for the poor, especially in rural areas where access to modern cooking fuels and improved water resources is especially weak.

Biodiversity decline and deforestation have meant that wood—the most widely used solid fuel—is located farther from the places where people live. A similar result holds for groundwater. Meeting these pressing needs means that women and girls have less time than ever to take part in alternative activities, paid or unpaid.

These pressures are likely to be accentuated by climate change. The Intergovernmental Panel on Climate Change reported in its 2014 assessment that it is virtually certain that climate change will reduce renewable water resources substantially in most dry, subtropical regions, resulting in scarcity of drinking water and biomass-based fuels.[63] Many of these regions lie in Sub-Saharan Africa and other less developed parts of the world, where women and girls already spend considerable time each day in meeting these requirements for their homes. Should there continue to be a lack of relevant infrastructure or more equitable sharing between men and women, climate change will further reduce choices for women.

Towards a rebalancing—expanding choices, empowering people

The work that women do—paid and unpaid—has major human development implications for themselves and for others. Paid work provides economic autonomy along with opportunities for participation and social interaction, as well as for enhancing skills and capabilities, helping boost self-esteem and confidence. But unpaid

Part of the unpaid work that women do is related to fetching water and gathering fuel and firewood

care and community work are vital for human well-being and have both individual and social value. Unpaid caregivers seldom get to exercise choice in fulfilling their responsibilities and are often motivated by cultural, family and societal constraints. When people undertake care responsibilities and stay out of the labour force, they make large sacrifices, perhaps missing the chance to expand their capabilities in the workplace. They also often compromise their economic independence and personal autonomy, which can be crucial for them and their children.

Yet such caregiving makes a major contribution to human development, particularly by offering personalized care for family members —strengthening family bonds and boosting the physical and mental well-being of other family members, especially children and older people. Mothers who are able to breastfeed their babies, crucial for children's health,[64] and offer care for the first 15 months have positive effects on children's education performance.[65] And workplaces that provide enabling conditions and facilities are best able to support such caregiving benefits. Carework for the community enhances community well-being and improves social cohesion.

Growing evidence also shows that daughters of mothers in paid employment have advantages later in life. A recent study based on a sample of 50,000 adults in 25 countries concluded that daughters of working mothers who completed more years of schooling were more likely to be employed, especially in a supervisory role, and to earn a higher income. In the United States, where daughters of working mothers earned 23 percent more than daughters of stay-at-home mothers, some of these effects were stronger. The careers of sons appear not to be significantly influenced by having a working mother, not surprising as generally men are expected to work, but sons of working mothers appear to spend more time on child care and housework.[66]

An educated and professionally active mother can inspire and stimulate her children intellectually and serve as a positive role model for her children. However, in many countries public provision for family support services and facilities has become more difficult due to cutbacks in public social services, threatening women's participation in activity outside the home.

The questions become: How can society create an enabling environment where women can make empowered choices? What would be needed to translate such choices into an equitable balance between men and women in roles, responsibilities and outcomes for both paid and unpaid work?

Such interventions are discussed further in chapter 6 but need to be made along four axes, engaging a variety of actors and impacting all men and women:

- *Reducing and sharing the burden of unpaid care work*—including universal access to clean water, modern energy services for household needs, quality public services (including those related to health and care), workplace arrangements that accommodate flexible schedules without penalizing professional advancement and a shift in mindsets about gender-specific roles and responsibilities.
- *Expanding opportunities for women to engage in paid work*—including access to quality higher education in all fields, proactive recruitment efforts and reducing barriers to entrepreneurship.
- *Improving outcomes at work*—including legislative measures such as those related to workplace harassment and equal pay, mandatory parental leave, equitable opportunities to expand knowledge and expertise and measures to eliminate the attrition of human capital and expertise when engaged in care work (box 4.3).
- *Changing norms*—including promoting women in visible positions of seniority, responsibility and decisionmaking in both public and private spheres and encouraging the engagement of men in traditionally female-dominated professions.

Conclusions

Beyond economic contribution, paid and unpaid work has social value with considerable human development implications.

There have been positive developments in policies, social norms and attitudes and changing roles of men and women in various societies that should lead to a rebalancing in the world

An educated and professionally active mother can inspire and stimulate her children

BOX 4.3

Paid parental leave, including mandatory paternity leave

Paid parental leave is one of the most important benefits of flexibility extended to working parents, especially when mandatory paternity leave policies are implemented. The correct mix of policies can help ensure high rates of female labour force participation and motivated, satisfied workers with a good work–life balance.

Paid maternity leave benefits are essential for women to return to work after child bearing. About 85 percent of countries provide at least 12 weeks of maternity leave. Of 185 countries surveyed by the International Labour Organization, all but 2 (Papua New Guinea and the United States) allow mothers to receive at least some leave—paid for by the state, by employers or by some combination of both. Though only a third of countries meet the recommended minimum of at least 14 weeks off for new mothers, paid at least at two-thirds their salary and funded publicly, the picture is broadly good in developed countries and is improving in developing countries.[1]

But career breaks due to maternity leave alone have unintended consequences that may adversely impact women's careers. And such consequences are greater when leave is lengthy. For example, when women return to work after maternity leave, evidence suggests that as with other aspects of flexibility, they are penalized for taking paid maternity leave, particularly when the leave is generous. Time away from the labour market may reduce women's earning power and pension benefits as they miss chances to gain experience and win promotion. Moving into senior management becomes particularly hard.

In Germany each year of maternity leave a woman takes lowers her earnings upon resuming work by 6–20 percent. In France each year of absence is estimated to lower earnings by 7 percent.[2] The effect is magnified when lengthy maternity leave is combined with policies to encourage part-time work, which tempt more women back into the labour force but keep them in junior positions. In fact, it has been argued that the Elternkarenz, which pays parents to stay home for up to three years in Germany and up to two years in Austria, is effectively destroying career prospects for women who use it.[3]

Despite paid maternity leave, motherhood in the United Kingdom comes with a pay penalty: 60 percent of working mothers with children either at nursery or primary school work part time, as do half those with older, secondary school–age children. Just 10 percent of fathers work part time. Women who work part time average a third less an hour than men who work full time, and 40 percent of part-time female workers earn less than the living wage.[4]

A 1981 child care law in Chile was intended to increase the percentage of women who work, which is below 50 percent. It requires that companies with 20 or more female workers provide and pay for child care for women with children under age 2 in a nearby location where the women can go to feed them. It eased the transition back to work and helped children's development, but it also led to a 9–20 percent decline in women's starting salaries.[5]

Policies that effectively define women by their role as mothers affect workplace outcomes for all women, not just those who take long leaves. During the paid absence of a young parent, no employer may hire a permanent replacement. Since women are more likely than men to take long parental leaves, employers have a strong incentive to hire men. Europe's equal employment laws make such discrimination illegal, but evidence shows that employers differentiate anyway.[6]

The issue of paternity leave has received attention as the role of fathers in childrearing and sharing the care burden has been emphasized. Many countries now offer paternity leave. One approach that promotes balance is parental leave, to be split between mothers and fathers. Several European countries, as well as Australia and New Zealand, already have such a system. The downside is that because child rearing is traditionally seen as a mother's job, fathers tend not to take any leave, unless it is made mandatory.

In Austria, the Czech Republic and Poland, where all parental leave is transferrable, only about 3 percent of fathers use it. Countries are thus pursuing different approaches to overcome the problem. In Chile, Italy and Portugal paternal leave is compulsory. Fathers may be induced to use more paternal leave through such incentives as relaxing the gender-neutral approach and granting a bonus to parents who share parental leave more equally. For example, through similar measures, Germany saw the share of fathers taking time off rise from 3 percent in 2006 to 32 percent in 2013.[7]

> How can society create an enabling environment where women can make empowered choices?

Notes
1. *The Economist* 2015c. 2. *The Economist* 2015c. 3. Munk and Rückert 2015. 4. The Pregnancy Test 2014. 5. Villena, Sanchez and Rojas 2011; Prada, Rucci and Urzúa 2015. 6. Munk and Rückert 2015. 7. *The Economist* 2015c.
Source: Human Development Report Office.

of work. Education, social policies and modernizing societies have played a part.

Yet we are far from the desired results. Governments can promote measures to foster enabling conditions for men and women to make empowered choices through policy, but government measures can go only part of the way. The ultimate focus should be on sharing responsibilities and mutual contributions by men and women in overcoming imbalances in paid and unpaid work, critical not only in a rapidly changing and ageing world but also for sustainable work—a theme taken up in chapter 5.

Time use

Country	Survey year		Total paid work		Total unpaid work		Learning		Social life and leisure		Personal care and maintenance		Other (including travel)	
			Female	Male	Female	Male	Female	Male	Female	Male	Female	Male	Female	Male
			(minutes per day)		(minutes per day)		(minutes per day)		(minutes per day)		(minutes per day)		(minutes per day)	
Albania	2010–2011		117	257	314	52	56	57	163	226	699	733	83	108
Algeria	2012		30	198	312	54	42	30	240	330	798	768	18	72
Armenia	2004		101	291	312	63	43	36	189	255	749	733	46	61
Australia	2006		128	248	311	172	26	25	284	308	666	649	25	39
Austria	2008–2009		160	264	269	146	39	44	300	335	673	654	5	5
Belgium	2005		94	155	214	128	41	36	313	371	696	662	83	89
Belgium	1999		96	167	224	128	38	36	310	347	697	674	75	89
Benin	1998	Urban	240	237	199	63	68	110	144	243	723	722	31	49
Benin	1998	Rural	308	285	210	66	31	70	133	215	746	758	38	59
Benin	1998	Total	284[a]	268[a]	206[a]	65[a]	44[a]	84[a]	137[a]	225[a]	738[a]	745[a]	35[a]	55[a]
Bulgaria	2009–2010		137	190	298	164	24	27	241	296	730	750	37	26
Cambodia	2004		270	390	188	18	18	36	222	258	742	738	0	0
Canada	2010		180	255	257	170	37	34	309	346	656	634	0	0
Canada	2005		186	282	252	156	36	30	324	348	648	624	18	24
China	2008		263	360	237	94	31	34	215	251	696	704	0	0
Colombia	2012–2013		151[b]	311[b]	239[b]	67[b]	52[b]	56[b]	309[b]	338[b]	706[b]	690[b]	0[b]	0[b]
Costa Rica	2004		122[b]	352[b]	385[b]	105[b]	39[b]	37[b]	278[b]	289[b]	662[b]	653[b]	0[b]	0[b]
Denmark	2001		147	211	243	186	29	20	325	346	673	643	22	33
Ecuador	2012		150	306	273	78	83	87	174	190	877	875	0	0
El Salvador	2010		192	346	228	43	52	73	266	305	577	573	0	0
Estonia	2009–2010		161	197	261	169	30	36	267	314	670	666	51	60
Estonia	1999–2000		167	234	302	179	41	51	243	283	647	648	40	45
Ethiopia	2013	Urban	177	318	245	66	101	124	136	177	781	754	0	0
Ethiopia	2013	Rural	206	296	304	141	45	55	89	130	795	819	0	0
Ethiopia	2013	Total	200	301	291	125	58	70	99	140	792	805	0	0
Finland	2009		162	202	211	139	67	53	301	362	649	633	56	58
Finland	1999		183	267	221	130	69	51	284	323	638	621	52	56
Finland	1987		197	269	222	128	70	61	291	308	622	625	47	57
Finland	1979		187	258	226	120	80	73	266	297	632	634	56	69
France	2010		126	199	234	148	34	33	269	309	713	696	64	55
France	1999		120	207	267	151	36	43	249	285	731	716	37	38
France	1986		196	347	307	127	0	0	208	249	685	672	44	45
Germany	2001–2002		134	222	269	164	28	26	330	355	656	636	24	38
Ghana	2009		230	288	220	68	81	110	169	254	732	709	0	0
Greece	2013–2014		78	152	277	107	38	39	318	395	711	719	18	29
Hungary	1999–2000		171	261	268	127	33	34	256	304	683	681	28	32
India	1998–1999		160	360	297	31	0	0	241	277	736	765	0	0
Iran (Islamic Republic of)	Winter 2009	Urban	42	298	322	76	78	72	206	213	792	776	2	5
Iran (Islamic Republic of)	Summer 2009	Urban	39	274	284	77	28	22	226	242	862	823	1	1
Iran (Islamic Republic of)	Autumn 2008	Urban	38	276	316	80	81	82	215	218	790	780	1	5
Iran (Islamic Republic of)	Average of three surveys	Urban	40	283	307	78	62	59	216	224	815	793	1	4
Iraq	2007	Governorates centers	36	245	330	56	58	61	268	335	716	702	32	40
Iraq	2007	Other urban	26	231	340	62	48	63	284	340	705	701	38	44
Iraq	2007	Rural	40	226	362	58	30	55	249	334	720	713	41	55
Iraq	2007	Total	31	246	345	56	46	58	264	328	713	700	48	116
Ireland	2005		142	280	296	129	35	38	296	346	629	602	42	44
Italy	2008–2009		103	223	305	108	26	25	250	310	685	683	70	90
Italy	2002–2003		108	252	347	126	26	25	241	305	691	692	27	39
Italy	1988–1989		92	241	362	87	27	31	235	285	696	715	44	81
Japan	2011		165	330	254	77	57	59	247	276	673	660	46	41
Japan	2006		167	333	220	43	42	47	214	238	652	638	144	139
Japan	2001		162	330	219	35	44	50	218	249	652	637	145	139
Korea (Republic of)	2009		145	246	188	39	71	79	270	294	653	652	112	131
Korea (Republic of)	2004		154	260	194	36	69	79	234	259	634	635	156	172
Korea (Republic of)	1999		167	279	204	32	82	94	232	257	618	618	136	160
Kyrgyzstan	2010		163	267	275	100	69	76	288	348	644	647	1	2
Latvia	2003		234	337	277	143	15	11	233	273	653	646	27	30
Lithuania	2003		231	313	308	166	12	10	210	270	656	653	24	28
Madagascar	2001	Urban	189	320	234	65	112	133	164	191	783	779	13	16
Madagascar	2001	Rural	253	363	217	44	66	64	107	141	811	812	15	19

Time use (continued)

Country	Survey year		Total paid work (minutes per day)		Total unpaid work (minutes per day)		Learning (minutes per day)		Social life and leisure (minutes per day)		Personal care and maintenance (minutes per day)		Other (including travel) (minutes per day)	
			Female	Male	Female	Male	Female	Male	Female	Male	Female	Male	Female	Male
Madagascar	2001	Total	234 c	350 c	222 c	50 c	80 c	85 c	124 c	156 c	803 c	802 c	14 c	18 c
Mali	2008	Urban	138	282	235	27	71	104	249	330	729	658	20	43
Mali	2008	Rural	262	322	245	18	40	57	112	178	718	713	74	154
Mali	2008	Total	217	308	241	21	51	74	161	233	722	693	55	114
Mauritius	2003		116	296	277	73	65	67	290	345	709	695	0	0
Mexico	2009		172	381	442	155	72	75	150	174	604	584	0	71
Mexico	2002		122	327	385	88	77	116	165	186	596	583	95	140
Mongolia	2011		238	348	286	131	44	44	168	206	700	703	0	0
Morocco	2011–2012		81	325	300	43	.. d	.. d	.. d	.. d	.. d	.. d	.. d	.. d
Netherlands	2005–2006		146	279	254	133	37	42	297	308	657	619	49	61
New Zealand	2009–2010		143	254	247	141	36	41	311	311	664	650	39	42
New Zealand	1998–1999		136	253	250	138	42	48	295	305	657	645	60	49
Norway	2010		181	250	230	180	28	27	338	347	624	599	38	38
Norway	2000		179	274	236	161	27	22	348	353	611	586	39	43
Norway	1990		168	270	262	156	33	29	329	349	610	598	37	39
Norway	1980		143	280	286	146	31	30	333	340	619	611	28	33
Norway	1970		116	329	355	133	17	23	278	296	636	619	36	40
Oman	2007–2008		58	187	274	115	58	72	317	374	720	691	0	0
Pakistan	2007		78	322	287	28	58	82	194	243	824	767	0	0
Palestine, State of	2012–2013		36	249	293	55	81	76	337	361	693	697	0	1
Palestine, State of	1999–2000		32	307	301	54	97	96	342	374	685	649	6	3
Panama	2011	Urban	201	359	301	128	40	38	166	157	734	748	1	1
Peru	2010		183 e	361 e	339	136	75	78	127	122	715	742	1	1
Poland	2003–2004		136	234	295	157	39	39	285	332	658	642	27	35
Portugal	1999		178	298	302	77	34	40	175	255	686	685	57	79
Qatar	2012–2013		120 f	229 f	199 f	110 f	70 f	66 f	303 f	332 f	748 f	703 f	0 f	0 f
Romania	2011–2012		100	163	264	125	25	28	280	335	727	735	42	55
Serbia	2010–2011		129	227	301	148	25	19	305	341	673	665	22	28
Slovenia	2000–2001		169	236	286	166	41	36	287	339	630	632	26	30
South Africa	2010		129	214	229	98	64	71	259	307	758	750	0	0
South Africa	2000		116	190	216	83	96	109	276	330	734	727	0	0
Spain	2009–2010		128	205	263	126	39	39	271	326	686	693	52	53
Spain	2002–2003		119	243	280	101	43	42	265	321	681	684	51	50
Sweden	2010–2011		201	245	240	194	18	13	285	305	640	621	57	61
Sweden	2000–2001		180	265	254	183	22	16	279	295	638	611	67	70
Tanzania (United Republic of)	2006		205	276	212	73	75	87	103	148	846	858	0	0
Thailand	2009		268	360	188	55	65	64	198	233	719	733	0	0
Thailand	2004		281	372	174	49	75	70	182	209	724	733	0	0
Tunisia	2005–2006		108	298	326	54	47	47	244	312	692	702	16	26
Turkey	2006		68	267	371	88	20	24	254	286	672	672	55	103
United Kingdom	2005		145	233	232	131	14 g	14 g	296	328	673	652	84	82
United Kingdom	2000		140	248	261	153	15	16	266	300	678	649	78	75
United States	2013		166	252	252	163	29	28	318	349	658	634	21	16
United States	2012		176	250	249	156	28	32	321	361	656	631	15	14
United States	2011		177	254	249	165	27	29	318	346	656	631	18	17
United States	2010		176	245	252	167	29	28	313	349	653	632	22	20
United States	2009		171	256	258	166	30	26	321	352	649	631	16	14
United States	2008		179	271	259	162	31	25	319	347	644	630	14	10
United States	2007		188	271	263	169	26	25	310	344	643	625	11	12
United States	2006		181	272	263	164	32	27	307	342	649	628	13	12
United States	2005		180	266	262	163	25	27	311	342	649	631	11	10
United States	2004		178	265	268	165	29	29	313	347	642	626	9	8
United States	2003		173	274	268	168	29	26	315	340	643	622	12	11
Uruguay	2013				309 h	133 h								

NOTES

a Results are aggregated at the national level based on the distribution of the population between urban and rural areas at the time of the survey: 36% urban and 64% rural.

b Data are not based on a diary, so the total number of reported hours in the day is more than 24 hours because of simultaneous activities.

c Results are aggregated at the national level based on the distribution of the population between urban and rural areas at the time of the survey: 30% urban and 70% rural.

d Time use for learning, social life and leisure, personal care and maintenance and other is not disaggregated by gender. The corresponding total times are 29 minutes for learning, 400 minutes for social life and leisure, 636 minutes for personal care and maintenance and 59 for other (including travel).

e Calculated as the difference between the total number of minutes per day (1,440) and the total of minutes for unpaid and other activities.

f Refers to Qatari citizens only.

g Includes recreational study.

h Survey focused on unpaid work only.

DEFINITIONS

Total paid work: Working time in paid employment (corporations, quasi-corporations, nonprofit institutions and government), primary production activities, nonprimary production activities, construction activities and providing services for income.

Total unpaid work: Working time in providing unpaid domestic services for own final use, providing unpaid caregiving services to household members and providing community services and help to other households.

Learning: Time spent in learning activities, including attending classes at all levels of instruction (pre-primary, primary, secondary, technical and vocational, higher education, extra or makeup classes); attending literacy and other special programmes for handicapped children, adults and other groups who have no opportunity to attend school; completing homework assignments; conducting private studies and research; studying for examinations in courses; attending short-term courses, seminars and the like related to professional development and travelling to and from classes and school activities.

Social life and leisure: Time spent in socializing and community participation; attending cultural, entertainment and sports events; participating in hobbies, games and other pastime activities; participating in sports and outdoor activities and using mass media.

Personal care and maintenance: Time spent in personal care and maintenance, including activities related to biological needs (sleeping, eating, resting and the like); performing own personal and health-care and maintenance or receiving this type of care; participating in spiritual and religious care and activities; doing nothing; resting; relaxing; meditating; thinking and planning.

Other (including travel): Everything else, including travel, so that the total time sums to approximately 1,440–1,445 minutes.

MAIN DATA SOURCE

Columns 1–14: Charmes (2015); see http://hdr.undp.org for the full list of sources by country.

Chapter **5**

Moving to sustainable work

Infographic: The matrix of sustainable work

Increasing sustainability →

Greatest

← Decreasing human development

Increasing human development →

Limiting opportunities for the future but advancing human potential in the present

(for example, traditional water- and fertilizer-intensive agriculture)

Growing opportunities for the future; advancing human potential in the present

(for example, poverty-reducing solar power; volunteer-led reforestation)

Degrading opportunities for the future; destroying opportunities for the present

(for example, forced labour on deep-sea fishing vessels; trafficked workers clearing rainforest)

Supporting opportunities for the future but limiting human potential in the present

(for example, recycling without worker safeguards; removing contaminants without protective gear).

Least

← Decreasing sustainability

5.

Moving to sustainable work

The previous chapters have shown how the trends that are transforming the world of work—driven by market forces, but also encouraged and moderated by public actions and societal transitions—can deliver substantial future gains for human development. Another great imminent shift for how we live and work is the move towards sustainable development. It seems to have begun in some countries but is generally piecemeal. At the same time, it comes with a sombre imperative and urgency: If we collectively fail to make the move in a timely fashion, the thwarted human potential of current and future generations could prove catastrophic.

The move thus requires a deliberate, planned departure from business as usual. It calls for explicit collaboration and coordination across countries and regions and across public and private actors. And it brings to the fore some of the global and intergenerational ramifications of work.

Setting the context: goals for sustainable development

The Millennium Development Goals—anchored in the Millennium Declaration[1] of 2000—set quantified targets to reduce deprivations in the basic dimensions of human development by 2015. Since 2013 a process has been under way to help define the post-2015 development agenda and to extend and expand the work under the Millennium Development Goals. As part of this process the Member States of the United Nations formulated, in late 2014, goals for sustainable development.

Eliminating human deprivations and disparities in human progress is at the heart of the Sustainable Development Goals, a commitment to improving the living conditions of and opportunities for every individual in the world by 2030. Achieving the Sustainable Development Goals would transform the status of human development (box 5.1). Indeed the efforts of UN Member States to frame and adopt these goals signal the political intent to make the collective shift towards accelerating and sustaining human development.

The Sustainable Development Goals encompass three elements:
- They assert the primacy of poverty eradication within the framework of sustainable development.

- They are universal, covering all countries and individuals.
- They set time-bound targets with progress to be monitored and reported nationally, regionally and globally.

The Sustainable Development Goals thus continue to advance the practical realization of the human development approach: putting people at the centre of development to enhance the possibility of fully realizing their potential.

Sustainability in the human development framework

This universal approach—making development matter for all people in the world, now and in the future—is also at the core of the human development paradigm. The real foundation of human development is universalism in acknowledging the life claims of everyone. The universalism of life claims is the common thread that binds the demands of human development today with the exigencies of development tomorrow, especially environmental preservation and regeneration.

The strongest argument for protecting the environment, from a human development perspective, is to guarantee future generations a diversity and richness of choices similar to those enjoyed by previous generations. Human development and sustainability are thus essential components of the same ethic of the universalism of life claims.[2] This is also consistent with sustainable development—conventionally defined by the World Commission on Environment and Development (1987) as, "development that meets the needs of the present without compromising the ability of future generations to meet their own needs."[3]

> Human development and sustainability are essential components of the same ethic of the universalism of life claims

BOX 5.1

The Sustainable Development Goals and human development

One of the principal outcomes of the Rio+20 Conference in 2012 was the agreement by Member States to develop a set of Sustainable Development Goals that would transition the world from the Millennium Development Goals to the post-2015 development agenda. It was decided to establish an "inclusive and transparent intergovernmental process on Sustainable Development Goals that is open to all stakeholders, with a view to developing global sustainable development goals to be agreed by the United Nations General Assembly."[1]

Accordingly, an intergovernmental Open Working Group was created and met 13 times over 2013–2014, receiving inputs and information from academics, civil society representatives, technical experts and various entities from the multilateral system.

The group agreed on 17 goals and 169 targets that constitute the Sustainable Development Goals. Since January 2015 the UN General Assembly has been discussing them, leading to their adoption by the heads of states and governments at the UN Sustainable Development Summit in September 2015. They address five areas of critical importance for humanity and the planet.

Simultaneously, a process is under way at the United Nations Statistical Commission to agree on the statistical indicators to monitor progress towards the goals and targets at the global, regional and national levels. The indicators are expected to be finalized in 2016.

The Sustainable Development Goals address five areas of critical importance for humanity and the planet

Achieving the Sustainable Development Goals would transform the status of human development

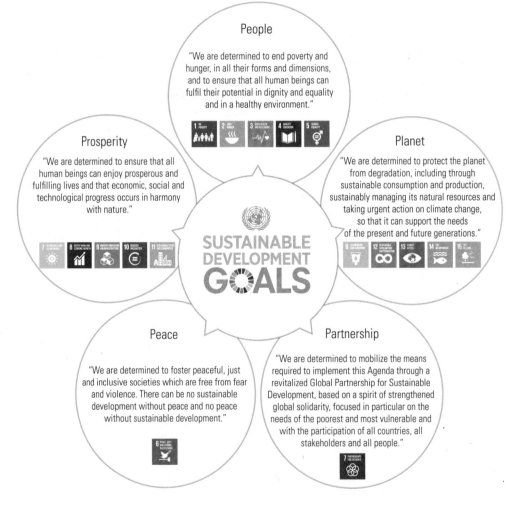

People

"We are determined to end poverty and hunger, in all their forms and dimensions, and to ensure that all human beings can fulfil their potential in dignity and equality and in a healthy environment."

Prosperity

"We are determined to ensure that all human beings can enjoy prosperous and fulfilling lives and that economic, social and technological progress occurs in harmony with nature."

Planet

"We are determined to protect the planet from degradation, including through sustainable consumption and production, sustainably managing its natural resources and taking urgent action on climate change, so that it can support the needs of the present and future generations."

SUSTAINABLE DEVELOPMENT GOALS

Peace

"We are determined to foster peaceful, just and inclusive societies which are free from fear and violence. There can be no sustainable development without peace and no peace without sustainable development."

Partnership

"We are determined to mobilize the means required to implement this Agenda through a revitalized Global Partnership for Sustainable Development, based on a spirit of strengthened global solidarity, focused in particular on the needs of the poorest and most vulnerable and with the participation of all countries, all stakeholders and all people."

Note: Icons for the Sustainable Development Goals were developed in collaboration with Trollbäck + Company.
1. UN 2012b.
Source: Human Development Report Office; UN 2015c.

Simply stated, ensuring sustainability for human development has three requirements, which the Sustainable Development Goals seek to ensure:

- The number and range of life-enhancing opportunities—at home, at work and in the community—available to all individuals over their lifetimes and that of their descendants does not diminish.
- The capabilities to take advantage of these opportunities—over all individuals' lifetimes and across generations—continue to be enhanced.
- Social, cultural, political or physical barriers that may inhibit individuals from accessing opportunities that best meet their capabilities—whether due to lack of participation, security, empowerment or infrastructure—are eliminated.

But achieving an equitable and enduring sharing of the fruits of development, within and across generations, is not straightforward. On the one hand, the collective resources available for making the necessary advances have never been greater. Global wealth and income are higher than ever before: Global wealth increased $20 trillion over 2013, to $263 trillion in 2014, while world GDP reached about $78 trillion.[4] Human ingenuity continues to generate innovative solutions that build on our increasing understanding of the world around us, thus providing grounds for optimism about enhancing opportunities and capabilities.

On the other hand, human activity, if it continues along a business-as-usual path, imperils the continuity of the gains so far, as well as the potential for advances in the future, both for ourselves and our descendants. These threats, driven most immediately by the unsustainable use of natural resources, as well as by the difficulty in resolving social barriers, vary both in scale—from the global to the local—and in scope. If unchecked or unmitigated, these threats will also limit the ability to continue to pursue some forms of work as natural resources degrade.

At the local level these connections have been clear for some time, and in many cases local solutions have been devised. For example, traditional societies in many parts of the world developed systems to sustainably harvest and use rainwater for agricultural work. Such systems, managed at the community level, are the theme of a special contribution made by His Excellency Maithripala Sirisena, President of Sri Lanka (signed box).

Other more recent examples abound. In 1962 Rachel Carson's Silent Spring spotlighted the harm caused to natural systems and humans through the overuse of dichlorodiphenyltrichloroethane (commonly known as DDT) as a pesticide in agriculture. Later the adverse local impacts of indoor and outdoor air pollution, land degradation, loss of forest cover and unsustainable patterns of water use, especially in agriculture, all sparked corrective action in many parts of the world.

Over time the scope of such measures has expanded from beyond the immediate community of producers and consumers to the provincial and national levels, where costs and benefits may be more widely dispersed. A common framework for estimating impacts, assessing trade-offs and making decisions is cost–benefit analysis.[5] Such analysis can include estimates of lost employment; however, not all job losses are equivalent. On the one hand a large unemployment figure could eventually have only a minor effect if all the workers are quickly rehired by another firm or in another industry. On the other hand a much smaller total number of jobs lost could be very harmful if concentrated in a one-factory town or if the affected workers have few transferable skills and prospects for future employment.

For an example of a cost–benefit analysis, take the 1990 Clean Air Act Amendments in the United States, which drastically cut emissions of air pollutants through regulations affecting many industries such as power plants, motor vehicles and other dispersed sources such as dry cleaners and commercial bakeries. Apart from the associated benefits to the environment, it is also estimated that millions of cases of respiratory problems and hundreds of thousands of serious illnesses and deaths were averted by 2010. These potentially prevented 13 million lost working days and 3.2 million lost school days.[6]

But the average earnings for workers in regulated sectors fell more than 5 percent in the three years after the new regulations were enacted. Earnings began to recover only five years after the policy change. The average worker

Achieving an equitable and enduring sharing of the fruits of development, within and across generations, is not straightforward

Community participation for improving rural livelihoods—lessons from the past

Centuries of irrigated agriculture and rice farming have been embedded in Sri Lanka's cultural traditions, enriching the lives of her people in all aspects. The relevance of these traditional occupations to the demands of a modern economy has to be defined now.

Strengthening the national economy for reducing inequality and eliminating poverty is a key objective of the new government of Sri Lanka. Improving the productivity of the agriculture sector is one main means of strengthening the national economy envisaged by the government. Since ancient times agriculture has received special attention as a means for ensuring food security and as a source of employment. Despite attempts to diversify the rural economy, the importance of agriculture as a source of employment has persisted over time, though in recent decades improving productivity has been a main objective for the sector.

The agriculture sector is faced with challenges on many fronts. Development of new technologies for controlling pests and weeds and improving the productivity of crops to keep abreast of competition is one challenge. It is connected to overuse, pollution and serious health issues, reducing labour productivity. Sustainable means of supplying water and other resources for agriculture is a second challenge. A third is the population growth, industrialization and urbanization that have made water and land more scarce and land less productive. Fourth, the unplanned use of forest land for development purposes has dried the groundwater table, reducing the availability of water for agriculture. Fifth, such developments have exposed hillsides to acute weather conditions. The topsoils of the slopes unprotected by trees are washed off. This makes land less arable. The washed-off soil sediments settle in rivers and irrigation systems. The irrigation system then needs regular maintenance and cleaning; otherwise siltation reduces the capacity of tanks and waterways to store and carry water. Sixth, changing weather patterns have reduced rainfall and increased the need for irrigated water.

These issues of maintaining rural agricultural ecosystems for sustaining agriculture have been a challenge for farmers for centuries. But collective action by farmer societies has ensured not only the protection of agricultural ecosystems, but also the collection and storage of periodic rainwater for use throughout the year.

Rural farmers in Sri Lanka who commenced their civilization in the dry zone overcame these challenges through collective action. Ancient farmers knew the importance of protecting the independent natural resource systems of the villages for sustaining agriculture and maximizing the use of scarce resources. They worked together to create irrigation systems that were able to collect periodic or seasonal rainwater and store them in village tanks to meet the livelihood needs of their communities. There are estimated to be more than 18,000 village tanks in Sri Lanka. They are strategically located or constructed to collect rainwater. Moreover, a collection of these small tanks creates a system of water collection and storage to ensure that farmers get maximum benefit from periodic or seasonal rains.

Such irrigation systems can be sustained only with proper management. First the forests and land in the catchment areas of these tanks need to be preserved to ensure proper water collection. The canals and waterways that connect the village tanks have to be removed of silt and cleaned to ensure proper flow of water. Localized governance of village tank systems has ensured that they are maintained regularly. Because the farmers who owned and managed the irrigation systems were also the main beneficiaries of these systems, they had to ensure timely management of the systems.

The protection of forests in the catchment areas to the village tanks were also ensured by the farmer societies. These self-governance strategies were both effective and low cost. Most important, farmers had to manage these common properties so that the benefits to the community were maximized rather than accruing to individuals. This need to manage common properties for the benefit of all brought together farmers to form societies in rural Sri Lanka. Farmers today benefit from the efforts of their predecessors who built the village irrigation resource management systems.

The agriculture sector today is different from the one that existed 50 years ago. Today's farmers are not subsistence farmers. They grow crops for commercial purposes. Some focus on integrated agricultural production enterprises aimed at adding value. As such, directly replicating the governance structures adopted by village farming societies may not be totally appropriate in today's context. But self-governance, knowledge of environmental resources and their protection as well as solutions for dealing with resource scarcity are all lessons we can learn from the practices of village farming communities of the past. Such collective action may enable rural farmers in Sri Lanka to protect their environment, adapt to changing weather patterns and improve the productivity of agricultural practices. Finally, with commercialized business acumen farmers can become bigger contributors to the economy. If they so succeed, our government's key objective of minimizing inequality and eliminating poverty will bear fruit.

Maithripala Sirisena
President of Sri Lanka

in a regulated sector is estimated to have lost the equivalent of 20 percent of pre-regulatory earnings.[7] But the precise cost depended on demographics, industry structure and the state of the economy,[8] and these estimates of forgone earnings were far below the associated health benefits that accrued to a wider range of individuals—even when measured in purely economic terms.[9]

Such effects now transcend local, national or even regional boundaries and affect the entire planet—winners and losers can be far removed by geography or time, presenting a notable challenge to how impacts are managed. One prominent example is climate change, associated with a long-term increase in average temperatures around the world, with changes in precipitation patterns, with a rise in the mean sea level and with a higher frequency of extreme weather events.

The calamitous impacts of climate change on human development—if not dealt with—have been put forward in many fora, including the 2011 Human Development Report.[10] These include changes that will greatly affect work and the conditions under which people—especially the poor—labour.

For example, in the Lower Mekong Basin in South-East Asia—occupied by primarily riverine and agrarian communities—climate change will seriously affect the lives and livelihoods of more than 42 million people. There will be major shifts in crop suitability, increased mortality of livestock, declining aquaculture yields, greater heat stress following working in the fields

and higher incidence of various diseases.[11] To adapt to these challenges, people will have to change how they work and manage risks.

At the global level current estimates indicate that if greenhouse gas emissions—the immediate driver of climate change—are unchecked, global surface temperature increases will be between 3.7°C and 4.8°C above pre-industrial levels by 2100.[12] To limit this to the agreed-upon target of 2°C requires a (generally accepted) reduction of 40–70 percent (relative to 2010) in total emissions by 2050. Such a reduction, while feasible, would evidently affect the hundreds of millions at work in the sectors that contribute these emissions (table 5.1).

These sectors tend to be connected to human development through multiple pathways, and how the mitigation efforts are undertaken has important consequences. For example, in developing countries there is a strong positive association between energy consumption and the Human Development Index (figure 5.1). This is also supported by abundant empirical evidence for access to modern energy services, improving health, reducing poverty, raising living standards and facilitating positive gender outcomes.

In 2012, 1.3 billion people lacked electricity, and 2.7 billion relied on traditional biomass as household fuel.[13] Expanding access to electricity would be an essential part of any programme to enhance human development, yet a business-as-usual approach would be difficult to reconcile with the global goal for containing climate change, even if consistent with local sustainability considerations.

> In developing countries there is a strong positive association between energy consumption and the Human Development Index

TABLE 5.1

Greenhouse gas emissions and employment by sector

Sector	Share of increase in greenhouse gas emissions over 2000–2010[a] (%)	People directly employed (millions)
Energy, including electricity and heat	34.6	30
Agriculture, forestry and other land use	24.0	1,044
Industry	21.0	200[b]
Transport	14.0	88
Buildings	6.4	110

a. Represents a composite measure of the total annual anthropogenic greenhouse gas emissions (carbon dioxide, methane, nitrous oxide, fluorinated gases) based on IPCC (2014b). According to IPCC (2014b), global greenhouse gas emissions caused by human activities rose from 2000 to 2010 by 1 gigatonne of carbon dioxide equivalent (2.2 percent) a year, reaching 49 gigatonnes of carbon dioxide equivalent a year.
b. The actual number is larger. The value reported is for resource-intensive manufacturing only, likely to be more important from a sustainability standpoint.
Source: IPCC 2014b; Poschen 2015.

FIGURE 5.1

There is a strong positive association between energy consumption and the Human Development Index for developing countries

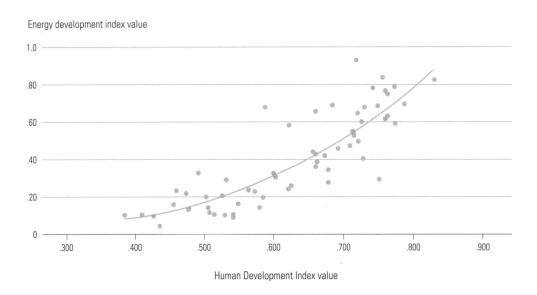

Source: Human Development Report Office calculations based on IEA (2012).

Work and sustainable development

As earlier chapters have made clear, work can have unintended consequences that go well beyond the people directly engaged in it—extending also to those far removed by geography or time. The consequences need not always be negative—the activities tied to such work could be changed to maintain human progress and produce positive benefits for many people not directly associated with them.

Sustainable work

Most effects of work on environmental sustainability and human development run along a continuum, encompassing positive and negative effects (see infographic at the beginning of the chapter). Work that promotes human development while reducing or eliminating negative side effects and unintended consequences is termed "sustainable work."

Sustainable work (in the top-right square of the matrix in the infographic) takes place in developed and developing economies, but it can differ in scale, in the conditions of work, in the links to human development and in the implications for policy. It is similar to "green jobs," broadly defined as "any decent job that contributes to preserving or restoring the quality of the environment, be it in agriculture, industry, services or administration," but differs in two important ways.[14]

First, sustainable work need not be undertaken in the context of earning a living and therefore encompasses the efforts of volunteers, artists, advocates and others. Second, it carries an imperative for advancing human development by channelling work towards the top-right in the matrix. For example, retrofitting car engines to reduce greenhouse gas emissions or replacing traditional biomass-burning cook stoves with solar cookers are both "green" activities, but the latter can have greater meaning for human development if it frees women and girls from having to collect firewood and enables them instead to take part in economic activities or go to school.[15]

Each country differs on the relative distribution of work within the matrix. There may also be trade-offs across the matrix. For example, volunteers afforesting degraded land that is lying fallow deliver benefits for sustainability and for human development. But if such

> Sustainable work promotes human development while reducing or eliminating negative side effects and unintended consequences

afforestation is accompanied by lost economic opportunities from not expanding agriculture into those areas, the trade-off would need to be resolved through the expression of a social preference (say, by demarcating a protected area to preserve biodiversity) or some alternative system of generating a livelihood.

For sustainable work to become more widely prevalent, three developments are needed in parallel, each requiring specific actions by national and international policymakers, industry and other private sector actors, civil society and individuals:

- *Termination.* Some existing work will have to end, and the workers displaced will need to be accommodated in other occupations (bottom-left square of the matrix in the infographic).
- *Transformation.* Some existing work will need to be transformed in order to be preserved through a combination of investment in adaptable new technologies and retraining or skill upgrading (top-left and bottom-right of the matrix in the infographic).
- *Creation.* Some work will be largely novel, benefiting both sustainability and human development, but will emerge from outside the current set of occupations (top-right square of the matrix in the infographic).

The move towards sustainable work departs from the business-as-usual path; it is also driven by an urgent timeline. Planetary boundaries, such as those related to greenhouse gas emissions or biosphere integrity, have already been transgressed, presaging an irreversible shift to a far less hospitable environment. So there is no time to lose. And as natural resources essential for livelihoods and life (such as the quality of topsoil or water) get depleted or degraded due to unsustainable use and climate change, the human development gains of local populations could be reversed, possibly accompanied by conflict or mass migrations.

Public policy can stimulate efforts by the private sector, social institutions, scientific bodies, advocacy groups, worker associations and individual leaders. For example, a review of planned and implemented environmental policy measures in nine developed and emerging countries found that appreciable gains in net employment were possible when complementary policies and incentives, including tax credits,

subsidies and worker training, were in place.[16] Even when a particular area of work is inherently green and arises spontaneously in response to market forces, additional efforts are needed to secure the most gains for human development, as suggested in a special contribution by Nohra Padilla, the recipient of the 2013 Goldman Environmental Prize (signed box).

Some of the issues linked to termination, transformation and creation are detailed below. Many sectors are involved (and many examples are possible), but we highlight those that impact the lives of hundreds of millions of the poor and that will be pivotal to accelerating the pace of human development.

Termination—managing losses

Some occupations can be expected to be preserved and even gain importance by moving to sustainable work. For example, as countries invest in mass transit systems to manage urbanization and commuter traffic, railway technicians are likely to be in demand. Others will not be so lucky.

The workers who might find themselves terminated can be expected to predominate in sectors that draw heavily on natural resources or emit greenhouse gases or other pollutants. About 50 million people are employed globally in such sectors (7 million in coal mining, for example). In Organisation for Economic Co-operation and Development countries the "seven most polluting industries account for over 80 percent of emissions while employing only about 10 percent of the workforce."[17] China expects to lose more than 800,000 jobs in obsolete power and steel plants over 2005–2020 due to national policies for pollution abatement and energy efficiency.[18] Indeed, such measures are at least partly responsible for the declining fortunes of the coal industry— the market capitalization of the four largest US coal companies was $1.2 billion in May 2015, down from $22 billion in 2010.[19]

Such changes are not new. For example, unprofitable coal mines were shut down in Poland in the 1990s. The collapse of the population of North Atlantic cod led to the closure of the cod fishing industry in Canada and Norway. Table 5.2 presents examples of industrywide closures, the measures taken to assist workers

The move towards sustainable work departs from the business-as-usual path

Recyclers: from waste pickers to global agents of sustainable development

The Millennium Development Goals and the post-2015 Agenda for Sustainable Development have drawn attention to the world's 20 million recyclers. Given the international nature of these platforms, this attention will likely lead to local and national policies to bring them dignified living conditions and decent work for them, as well as recognition of the benefits of the work that they do and resolution of the problems caused by capitalism, social injustice and economic inequality.

Many recyclers like us were born and grew up—and some have died —in the dump sites of developing countries across the globe. Certainly, none of us 30 years ago would have imagined that our work would be a source of recognition instead of an object of contempt and rejection. Many families displaced by conflicts and massive layoffs in factories—including immigrants and poor people excluded from production systems and from the benefits of social and human development, with limited opportunities for decent life and work—were forced to take on this humble work, rummaging through piles of garbage looking for recyclable waste.

Garbage was paradise for millions of families of "waste pickers," providing them with their daily food, blankets with which to cover themselves and the roofs of the shanties that they carved out for themselves in corners of the city. Even now, although they collect tonnes of materials that are recycled to produce new goods consumed by people throughout the world, regardless of social rank, many of today's recyclers still find in the trash much of what they need to survive.

The increasing use of recycling in production activities opened the doors to a magical world where recyclers discovered the importance of their work and the need to change their relationships with society and the market. Recyclers began to demand that society recognize, acknowledge and support their work. Their first gains were in organization: Cooperatives and associations were set up to protect their work at the local level and were followed by the formation of networks for regional, continental and global action. These organizations have fought for and have achieved policies that promote recycling as a component of public sanitation services and that reallocate public budgets accordingly. As a consequence of the demands of the recyclers, who have become increasingly organized and have assumed a role in defining the rules to protect the environment and human needs, governments in many countries have developed measures that strengthen recycling.

The numbers on recycling are impressive: millions of tonnes of recycled materials and new products on the market, millions in profits and savings for businesses and the public treasury, livelihoods for millions of people, reduced demand for natural resources and millions of square metres of land not used for garbage disposal. However, many recyclers are still working informally and very little money generated reaches the pockets of these workers, who are at the lowest rung of the recycling chain.

Cooperatives and associations that have successfully shifted from individual negotiations to collective bargaining for the benefit of the entire trade have pushed negotiations further, obtaining better market prices and additional income from services and waste recycling. However, much remains to be done so that more recyclers in the world can, in addition to gaining a livelihood, provide education opportunities and new prospects for their children. The children of this new generation of recyclers will be able to choose, thanks to their parent's work, between continuing in the family trade and entering another profession.

Leaders with vision are decisive in promoting policies and processes that benefit and recognize the Recyclers' Trade Union and that draw attention to the value chain in order to strengthen its activity. Organizational, commercial and business processes around recycling are not easy; they require transparent and collaborative relationships as well as government support programmes, policy measures and markets. In sum, they require a comprehensive system that addresses many pillars that, while difficult to build, are essential for recyclers to grow and strengthen their role for the benefit of society as a whole.

The world is entering a new phase, and all societies, through its representative institutions such as governments and grassroots organizations, should take the responsibility of ensuring a path of social progress. This must be achieved in particular for the millions of people who still do not enjoy fundamental human rights and minimum living standards. Those without access to water and basic sanitation services—both of which should be provided as a public good and not as a private for-profit service—should have access to a clean environment, to decent work and to dignity. But above all, we need to work together to overcome poverty, inequality and inequity so that social development instruments in the world take into consideration the hopes of millions of families.

Nohra Padilla
Director of the National Association of Recyclers (Colombia) and of the Association of Recyclers of Bogota, Recipient of the Goldman Environmental Prize in 2013

TABLE 5.2

Coping with industrywide closures

Closure	Companies affected	Workers affected	Measures for the displaced workers	Re-employment measures	Total cost of all measures (2015 US$)
Unprofitable coal mines in Poland (1990–2006)	37 mines	269,000	Soft loans for establishing business in other sectors Social benefit payment up to 24 months while searching for a job One-time payment based on multiple of average wage	54–65 percent found new jobs outside mining 33 percent of interviewees changed profession	$10.9 billion Debt forgiveness totaling $6.7 billion
Cod fisheries in Canada (1992–2001)	800 fish processing plants	30,000	Several billion-dollar relief package to coastal communities Government financial aid, retirement and retraining programmes Aid package for unemployed with weekly payments of $225–$406 About 28,000 received income support benefits	Most found employment in the shellfish industry	$3.7 billion
Restructuring forestry industries in Canada (2004–2014)	20 mills	118,000	Wage freeze Transition plans, including provincewide pension to encourage older mill workers to retire early Subsidizing mills to draw out closures Employment insurance for part-time work	Curtailing production and temporarily laying off workers, later rehiring Enabling workers to upgrade education or training in a different field	Estimated various resource-related "potential" projects: $140 billion
Restructuring forestry industry in China (2015–2017)	400 firms	100,000	Some of the state-owned forestry firms will work on conservation; others will operate like private businesses; aim to absorb workers or train them	Redeploy laid-off workers in forest management and protection	Natural forest protection programme with $12 billion budget $120.7 million compensation fund for land conversion
Coal mines in the United Kingdom (1984–2016)	167 mines	222,000	Monthly unemployment benefits for up to three years Employees' benefit trust Pension protection fund Protection on accrued benefits for employees, including pensions accrued prior to privatization Job creation agency set up by state	Offset by an estimated increase of 132,400 jobs in other industries and services in the same areas Increase in unemployment and incapacitated benefit claimants as well as early retirees	At least $768 million from government and EU sources

Source: Human Development Report Office calculations based on Suwala (2010); Schrank (2003); Liu, Yang and Li (2013); Beatty and others (2007); and Bennett and others (2000).

and the costs of those measures. In most cases governments played a major role in managing the aftermath of the closure, either directly or indirectly through existing safety nets.

One of the best documented experiences is Poland, where economic conditions led to the closure of 37 coal mines and the layoff of 269,000 workers over 16 years. Support measures included a lump-sum payment equivalent to several months' wages—larger if the worker left voluntarily; a monthly payment for up to 24 months while the worker looked for a job; and soft loans for setting up an enterprise, amounting to over $9 billion (not including debt forgiveness and other measures for the firms in excess of $5 billion). About 60 percent of the workers ultimately found new jobs. The Polish support package rehabilitated a large majority of workers; however, at a cost of about $35,000 per worker[20] it did not come cheap.

At particular risk in such terminations are older and often more experienced workers with limited geographic or occupational mobility as well as those in sectors where closures are unanticipated. As countries move towards sustainable work, special efforts may be needed to plan for the orderly winding down of activities in unviable sectors along with a package of diversified assistance, including training, job placement, business development, and income and health support.

Transformation—changing the nature of work

In many occupations how products are produced needs to change. For example, many workers worldwide are employed in work related to recycling, though the connection to workers' human development is not always straightforward. Ship breaking is a prime example in this regard—sustainable work can be promoted by implementing standards (box 5.2)

Work in agriculture—including farming, fisheries and forestry—engages more than a billion people worldwide, including most of those living on less than $1.25 a day.[21] The sector is responsible for a large proportion of greenhouse gas emissions and for crossing the planetary boundaries of the nitrogen and phosphorous cycles.[22] It is associated with unsustainable patterns of water and soil use. It is linked to deforestation and loss of biodiversity. And it is especially susceptible to the disruptions of climate change. Its transition to sustainable work is thus crucial, underscored by Sustainable Development Goal targets that aim to sustainably eradicate hunger and poverty by 2030.

Increasing agricultural production is still critical, given growing populations and changing consumption patterns, such as rising per capita consumption of animal-based proteins. World demand for agricultural products is expected to rise 1.1 percent a year through 2050.[23] Many workers continue to be engaged in the agricultural sector, particularly in Asia and Africa (figure 5.2). However, the three traditional ways of increasing production—yields, cropping intensity and area under cultivation—may all face limits. Increasing yields is possible if water and fertilizer are used more efficiently. Increasing cropping intensity and the area under cultivation are constrained by the need to protect forested areas and to manage (already degraded) soil quality. Matters will only get

Sustainable work can be promoted by implementing standards

BOX 5.2

Transforming ship breaking: promoting sustainable work by implementing standards

Ship breaking is a multibillion-dollar industry in which large commercial vessels are recycled at the end of their operational lives, typically at 25–30 years. Rather than leave them to gradually weather away and release large volumes of harmful pollutants to the environment, there is an active business in recycling all fuel, oils, engine parts and fixtures, followed by the steel hull. The top five ship-breaking countries are Bangladesh, China, India, Pakistan and Turkey. The industry provides a recycling solution to a gigantic problem, generates thousands of jobs for low-skilled workers (directly and in ancillary enterprises) and produces large volumes of steel, which can be especially useful for non-steel-producing countries.

All of these can be expected to lead to favourable outcomes for both sustainability and, at one level, for human development. However, the latter link is more tenuous: In many countries, environmental and labour safeguards for the industry are weak or not implemented. Workers toil in physically hazardous conditions and are exposed to toxic chemicals including asbestos, polychlorinated biphenyls and a range of heavy metals. Reports of children being employed are also frequent, and pollutants released during the process contaminate the local environment, accumulating to dangerous levels over time.

Regulation and appropriate policy measures can make the difference. In fact, a UN convention on ship breaking exists and is supposed to maintain environmental and labour standards.[1] However, implementation is uneven, though a number of studies show that an appropriate balancing of interests—strengthening the gains for human development—is possible. Such balancing would become even more important as the volume of the industry is set to grow as many more ships reach the end of their usable lives over the next decade.

Note
1. IMO 2009.
Source: Human Development Report Office.

FIGURE 5.2

The largest share of global employment in agriculture is in East Asia and South Asia

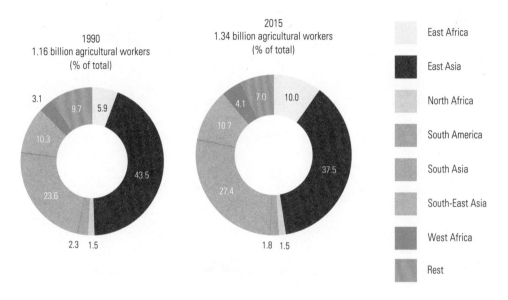

1990
1.16 billion agricultural workers
(% of total)

2015
1.34 billion agricultural workers
(% of total)

East Africa

East Asia

North Africa

South America

South Asia

South-East Asia

West Africa

Rest

Note: Economically active population in agriculture encompasses those engaged in or seeking work in agriculture, hunting, fishing or forestry. This includes all those whose livelihoods depend on agriculture (such as unpaid family members).
Source: FAO 2015.

worse with the precipitation changes associated with climate change.

Transforming the way farmers grow and process crops is thus crucial. Technologies and farming methods exist that can make the difference, but they need to be adopted faster. For example, about a third of total food production is lost or wasted, with cereals making up the largest portion of wasted food.[24] Technologies exist that can prevent this, including community-based mud silos (as in Ghana) and family-based storage units and hermetically sealed bags. Demonstrable, immediate gains are likely for the individual farmer as well. But adoption rates in developing countries are low, and major efforts are needed to expand them and to create new products for industrial or artisanal manufacture.[25]

Improved crop varieties, resistant to the effects of climate change, can help—as can better tilling, agroforestry and water harvesting (to conserve soil nutrients and promote multiple cropping), and smarter application of water and fertilizer (such as microdosing and precision agriculture). But adopting them is slow because of financial constraints, cultural and traditional norms, and limited awareness and training. These constraints can be alleviated

in particular contexts, through microfinance, demonstrations by peer farmers[26] and cell phone–based information services.[27]

Agricultural extension services are now provided by nongovernmental organizations and private sector workers as well as the more traditional government employees. The number of extension workers varies widely across countries, with more than 600,000 in China. Coverage tends to be uneven—with four extension workers per 1,000 family farms in Brazil and Ethiopia, but less than one per 1,000 in India.[28] Their reach among female farmers tends to be especially poor—a concern because women typically have a substantial role in agriculture in developing countries but have less access to market information, inputs and finances. To transform the way farmers cultivate and process crops, the coverage and quality of extension services have to increase.

Fertilizer use varies widely, and opportunities exist for redistribution across regions (map 5.1) —an occasion for cross-national collaboration. To curb excessive fertilizer use—limiting subsidies, introducing sustainable labelling schemes and insuring farmers against unanticipated losses—global policies need to integrate regional, national and even subnational practices.

Transforming
the way farmers
grow and process
crops is crucial

MAP 5.1

Fertilizer use varies widely

Phosphorous

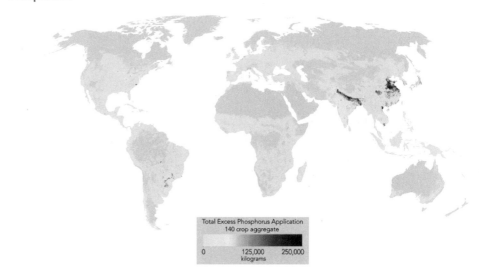

Nitrogen

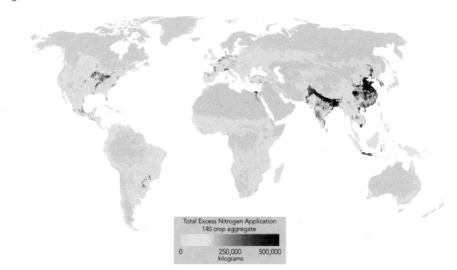

Note: Based on aggregation of 140 crops.
Source: Earthstat.org based on West and others (2014).

In turn, global conventions—where they exist—need to be translated into effective actions by local actors. Indeed, ship breaking (see box 5.2) is just one example of global recycling chains—which now include solar photovoltaic panels, in addition to cars, smartphones, tablets and other consumer durables—with many similar issues. As we move towards sustainability along one set of parameters, care must also be paid to the burgeoning problem of poorly regulated disposal and contamination of land, water and air with novel and toxic pollutants.

Creation—moving to new areas of work

Solar photovoltaic technologies, which convert sunlight to electricity, are an important part of many countries' renewable energy strategies. Their potential impact on human development differs radically depending on whether they replace grid-based electricity (which is generated by conventional means), as in many developed countries, or expand off-grid energy access, as in many developing countries. Solar photovoltaic technologies could become a key vehicle for achieving Sustainable Development Goal 7.1, to ensure universal access to affordable, reliable and modern energy services by 2030. Some countries are thus pursuing renewable energy sources to meet at least part of their energy demand (box 5.3).

By providing energy through photovoltaic technologies, developing countries can improve human development in several ways. In many parts of Africa and Asia solar home systems provide access to electricity in rural areas (table 5.3). Photovoltaic technologies also generate work for self-employed field assistants with basic technical and vocational skills who sell and install systems and provide regular maintenance. In India results of an industry survey suggested that off-grid solar photovoltaic technologies systems generated about 90 direct (involved in the actual manufacture of solar panels) and indirect (through employment with dealers, manufacture of products such as solar lanterns, home-lighting kits and the like) jobs per megawatt.[29]

Several countries in South Asia have demonstrated that such work is a viable option for women, which allows them to balance work with family care responsibilities. Renewable energy also stimulates small enterprises such as lantern rentals, phone and battery recharging and so on. And replacing kerosene-based lighting reduces carbon dioxide emissions. There are also reports of children spending longer hours studying.[30] Such positive impacts can be more widely distributed through skill development and

training (including for those at the higher end of the value chain), credit for purchases and better technical capacity for indigenous adaptation and innovation, as in Bangladesh. Many developing countries are installing—and therefore gaining expertise—in this technology (see table 5.3). Given the large number of people still without access to electricity, photovoltaic technologies represent an opportunity to leapfrog to a state that advances human development and sustainability simultaneously—an objective that can be fostered by the UN global technology transfer mechanism.

TABLE 5.3

Number of home solar systems installed and people without access to electricity (selected countries), various years

Region and country	Home solar systems installed (thousands)	People without access to electricity (%)
Africa		
Kenya	320	77.0
South Africa	150	14.6
Zimbabwe	113	59.5
Asia		
Bangladesh	3,800	40.4
India	892	21.3
Indonesia	264	4.0
Nepal	229	23.7
Sri Lanka	132	11.3

Source: IRENA 2013; World Bank 2015f; UNDESA 2012.

> Renewable energy also stimulates small enterprises

BOX 5.3

Implementing Tajikistan's National Programme on Application of Renewable Energy Sources

In Tajikistan the adverse impact of energy use on the environment has decreased over the past 20 years. Greenhouse gas emissions have decreased by more than 10 times as a result of a stimulated decrease in the use of organic fuels for energy production. The technologies for using renewable energy sources (solar, wind, geothermal and certain types of biomass) do not yet meet technical and economic parameters that could make their wider use possible (especially in rural areas, small social sites, for the production of greenhouse crops and in everyday life), but progress is being made. All these have implications for job creation.

In 2008 a government resolution led the Tajikistan Academy of Sciences to establish the Centre for Research and Use of Renewable Energy Sources. In implementing the Comprehensive Target Programme for Widespread Use of Renewable Energy Sources, research was conducted on all types of renewable energy (from small rivers, sun, wind, biomass and the like). This allowed for the analysis of the technical and economic potential of renewable resources in the country, including that of job creation. There are many completed and ongoing projects regarding the introduction of renewable energy sources with implications for people's lives and work.

Source: UNDP 2012d.

Sustainable Development Goals revisited—what they imply for work

Many of the Sustainable Development Goals and targets (see table 3 in the overview) have implications for sustainable work.[31] The most direct is goal 8 (promote sustained, inclusive and sustainable economic growth, full and productive employment, and decent work for all) and its associated targets. For example, target 8.7 seeks to take immediate and effective measures to eradicate forced labour, end modern slavery and human trafficking, and secure the prohibition and elimination of the worst forms of child labour, including recruitment and use of child soldiers, and by 2025 end child labour in all its forms.

Target 8.9—to devise and implement policies to promote sustainable tourism that creates jobs and promotes local culture and products by 2030—advocates for a particular kind of (sustainable) work. Target 8.8—to protect labour rights and promote safe and secure working environments for all workers, including migrant workers, in particular female migrants, and those in precarious employment—aims to strengthen the human development outcomes of workers, avoiding a race to the bottom.

Target 3.a—to strengthen the implementation of the World Health Organization Framework Convention on Tobacco Control[32] in all countries as appropriate—seeks to reduce work associated with tobacco production and distribution while improving the health of workers. Target 9.4—to upgrade, by 2030, infrastructure and retrofit industries to make them sustainable, with increased resource-use efficiency and greater adoption of clean and environmentally sound technologies and industrial processes, with all countries taking action in accordance with their respective capabilities—implies a specific direction towards upgrading skills and possibly towards new areas of work.

One cross-cutting aspect is voluntary work. Another is big data. Even as the wide availability of new digital technologies and communication platforms shape a new world of work, they also generate a massive pool of data to measure and assess various aspects of work, in turn improving analysis, policymaking and impact.

Supporting these efforts is a broader coalition. As mentioned in chapter 1, the United Nations has called for a data revolution[33] as an essential part of the post-2015 development agenda and the Sustainable Development Goals, with big data expected to be important (box 5.4).

Big data can indeed complement and enhance more conventional methods of data collection. It can be both timely and more frequent, and it offers the capacity to measure at a range of levels of aggregation from the individual to the global. Yet there are limits on reach and possibly in the robustness of the relationships inferred between online behaviour and broader socioeconomic indicators.

Some of these concerns can be addressed through further studies to establish the broader external validity of results. But baseline maps and other representations of the collected data must be constructed regularly to provide points of reference for policy action. Processes have to be in place for using the data systematically.

Terminating, transforming and creating work through the Sustainable Development Goals

A large number of the Sustainable Development Goal targets intend to reduce work that has negative implications for human development. Achieving target 8.7 would improve the lives of 168 million child labourers, about 36 million people in modern slavery[34] and 21 million people in forced labour.[35] Target 5.2 would assist 4.4 million sexually exploited women and girls,[36] while target 3.a would affect an estimated 100 million workers in tobacco—mostly in Brazil, China, India, Indonesia, the United States and Zimbabwe.[37] In these cases active policies and programmes will need to support the individuals formerly engaged in these areas of work.

Other goals and targets involve the transformation of current modes of work and the introduction of new approaches. Goal 2—end hunger, achieve food security and improved nutrition and promote sustainable agriculture—has the potential to transform how the very large number of people engaged in agriculture carry out their activities.

Much of the work associated with a move towards environmental sustainability (Target 9.4)

A large number of the Sustainable Development Goals targets intend to reduce work that has negative implications for human development

BOX 5.4

Big data: some applications for work and the Sustainable Development Goals

Online content—listening to individuals

The text of Google searches has been used in the United States to predict the number of people applying for unemployment benefits for the first time in a given week.[1] This sensitive indicator of the number of people laid off in a given week is important to policymakers, markets and others. Traditional methods use information collected administratively and are therefore available only with a delay.

However, when unemployment is anticipated by individuals, it is expected that search queries using terms such as "jobs," "unemployment office" and "apply for unemployment" would increase. Using the aggregate numbers of these searches in real time[2] improves the accuracy of predicting how fast layoffs are taking place by around 16 percent. Germany and Israel have seen similar improvements in forecasting accuracy.[3]

In yet another application, researchers in Ireland "listened online" to "conversations" in blogs, forums, websites and user posts on social media over June 2009–June 2011 to depict a qualitative picture that could complement official employment statistics[4] during the global economic crisis. Some 28,000 online documents were retrieved, each assigned a mood based on its content—anxiety, confidence, hostility, uncertainty, energy or happiness. Several leading and lagging relationships were found in the data. For example, anxiety peaked about five months before a spike in unemployment. Chatter about moving to worse housing situations increased about eight months after unemployment increased.[5] Such studies indicate the potential of using indicators mined from online information to improve predictions of large disruptions at work—and to refine social protection policies.

Mapping—using the data exhaust

Datasets generated by mobile phone use produce maps of human population while preserving the anonymity of users. These maps outperform other population mapping methods, are often more accurate, can map remote areas and can be produced and updated frequently and inexpensively.[6] They could also track changes in population movements in real time that indicate changes in livelihoods or coping patterns.[7] In Senegal it has been possible to map seasonal migration activity in 13 livelihood zones.[8]

Another mapping exercise used the individual digital fingerprints left by Twitter users in Spain to study work-related behaviors and indicators at the subnational level. Almost 20 million geolocated tweets from November 2012 to June 2013 revealed that in communities where a greater proportion of people worked, a larger share of daily tweets was sent in the morning hours (8–11 a.m.) on work days.[9] But the ratio of total Twitter users to the total population tended to be lower.[10]

Monitoring—deliberate and active collection

A different kind of impact on work can be expected when devices and sensors are used to monitor and report on indicators in real time. Since 2008, Beijing residents have been monitoring local air quality through home-based sensors, and they have shared and aggregated the information to pressure the municipality to take action.[11] Citizens are, in a sense, collecting new real-time data. Indeed, as citizen monitoring of this kind becomes more widespread, the collected data can be expected to accelerate the move towards sustainable work.

Notes
1. Choi and Varian 2009. 2. Google Trends provides daily and weekly reports on the volume of queries in different categories, including those related to jobs, welfare and employment. 3. Askitas and Zimmerman 2009; Suhoy 2009. 4. Global Pulse and SAS 2011. 5. The same study also examined 430,000 documents for the United States. Similar patterns were found between unemployment and online user-generated content. 6. Deville and others 2014. 7. Glass and others 2013; Bharti and others 2013. 8. Zufiria and others 2015. 9. Llorente and others 2014. 10. By contrast, some of the relationships found using other kinds of online content—for example, specific terms related to jobs—were found to not be closely associated with unemployment rates. 11. Lu and others 2015.
Source: Human Development Report Office.

will involve infrastructure and construction. In many developed and emerging economies, retrofitting for energy efficiency can have a substantial effect—in Germany it mobilized an estimated €100 billion over 2006–2013, supporting 300,000 jobs in the building industry.[38] And energy projects (goal 7) can drive long- and short-term jobs, directly and indirectly, when they enable other industries to grow and flourish. In 2014 renewable energy (excluding large hydro) employed an estimated 7.7 million directly and indirectly.[39] Large hydro was estimated to employ roughly 1.5 million direct jobs in the sector.[40] In renewable energy solar photovoltaic is the largest employer worldwide with 2.5 million jobs. These data are largely consistent with the estimates of the employment potential of renewable energy in 10 countries (map 5.2).

In 2014 renewable energy (excluding large hydro) employed an estimated 7.7 million directly and indirectly

MAP 5.2

The employment potential of renewable energy is considerable

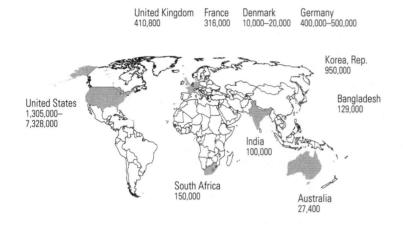

United Kingdom
410,800

France
316,000

Denmark
10,000–20,000

Germany
400,000–500,000

Korea, Rep.
950,000

United States
1,305,000–
7,328,000

Bangladesh
129,000

India
100,000

South Africa
150,000

Australia
27,400

Note: Refers to estimated employment potential by 2015 in Bangladesh, Denmark and the United Kingdom, by 2022 in India and by 2030 in the Republic of Korea and the United States.
Source: Strietska-Ilina and others 2011.

> Women are traditionally underrepresented in science and technology in many countries

In another example a global comprehensive plan for clean energy includes an option for a global climate stabilization programme with an annual investment of 1.5 percent of GDP for 20 years (two-thirds clean renewable energy, one-third energy efficiency).[41] This huge investment effort generates correspondingly large savings in the costs of adapting to climate change and could reduce carbon dioxide emissions 40 percent. The plan also appears good for economic development and job creation (table 5.4). The six countries in table 5.4 would gain 13.5 million net jobs, with the 13.5 million jobs lost in the fossil fuel sector more than offset by

27 million jobs created in clean energy. These numbers are conditional on a good transition plan and the capacity of countries to increase investment in new energies.

In moving to sustainable work, policymakers need to be aware of existing patterns of disadvantage and deprivation and target efforts accordingly. Women are traditionally underrepresented in science and technology in many countries and can thus be trapped in a cycle of skills that do not match market demands, which can lead to poorer outcomes for them. Deliberate targeting for solar photovoltaic technicians or for providing ecotourism facilities, while removing some traditional barriers such as access to finance, would help. Those already working at the lower end of green industries—for example, women are believed to constitute a large proportion of the estimated 15–20 million informal waste pickers in developing countries[42]—may need additional support to fully realize the human development potential of their work.

Building skills and capabilities through the Sustainable Development Goals

By strengthening health and education outcomes, especially for children, the Sustainable Development Goals can set the basis for people acquiring the skills to move to occupations that involve sustainable work. Target 4.1—to ensure all girls and boys complete free, equitable and quality primary and secondary education leading to relevant and effective learning outcomes

TABLE 5.4

Jobs generated from clean energy

	Total clean energy jobs created by investing 1.5 percent of GDP	Net clean energy jobs created after subtracting fossil fuel job losses	Clean energy job creation as share of total labor force (%)	
			Total jobs	Net jobs
India	12.0 million	5.7 million	2.6	1.4
China	11.4 million	6.4 million	1.5	0.6
United States	1.5 million	650,000	1.0	0.5
Indonesia	954,000	203,000	0.8	0.6
Brazil	925,000	395,000	0.9	0.4
South Africa	252,000	126,000	1.4	0.7

Source: Pollin 2015.

by 2030—will be pivotal in this respect, but success requires more and better teachers.

While alternative delivery models for education may somewhat modify this estimate (table 5.5), the 8.4 million additional trained teachers—3.3 million in primary education and 5.1 million in lower secondary education—needed through 2030[43] is a lower bound for the additional workers needed, as the number of managers, administrators, preschool teachers and trainers of teachers would grow as well.

Strengthening health outcomes will similarly require more trained health workers. The current stock of physicians, nurses and midwifes was around 34 million in 2012, only 3.6 percent of whom are in Sub-Saharan Africa, which has 12 percent of the world population.[44]

Estimates of the numbers of additional workers needed can vary. A conservative needs-based estimate (using a ratio of 3.45 health professionals per 1,000 people) indicates an increase of 10.1 million health professionals by 2030 (37 percent in Sub-Saharan Africa; table 5.6). However, the actual demand—estimated in a model incorporating income per capita growth and ageing as the main drivers—indicates that the required expansion could be up to 45 million health professionals at the global level. This latter model is arguably more realistic in estimating future demand based on past trends; however, it indicates that, notwithstanding the rapid population growth in Sub-Saharan Africa, the rate of formation of health professionals in the region would not be large enough to even meet the conservative requirement of 3.45 per 1,000. Clearly business-as-usual will not produce the required number of health professionals where they are most needed.

Recent experience shows that it can be done, but care is needed to ensure and maintain quality. Take Ghana: After it adopted a National Health Insurance Scheme in 2003 to move the country towards universal healthcare, the government doubled the number of trained nurses by 2009, using the human resource allocation quota system and establishing regional nurse training schools. Malawi, too, has set goals to fill more positions in healthcare. In response to a 65 percent vacancy rate, nursing education institutions increased the annual intake of student nurses and midwives 22 percent between 2004 and 2010,[45] which contributed to successes in health outcomes and progress towards the associated Millennium Development Goals.

Sustainable work is critical not only for sustaining the planet, but also for ensuring work that enables human development

TABLE 5.5

The demand for teachers

	Demand for teachers to reach universal coverage (thousands)							
	Primary education				Lower secondary education			
			Change				Change	
	Actual 2011	Required 2030	Absolute	Percent	Actual 2011	Required 2030	Absolute	Percent
Arab States	1,931	2,385	454	23.5	1,198	1,781	583	48.7
Central and Eastern Europe	1,127	1,238	111	9.8	1,570	1,901	331	21.1
Central Asia	340	385	45	13.2	406	473	67	16.5
East Asia and the Pacific	10,378	10,468	90	0.9	5,833	6,063	230	3.9
Latin America and the Caribbean	3,102	3,140	38	1.2	2,160	2,282	122	5.6
North America and Western Europe	3,801	4,103	302	7.9	2,555	2,725	170	6.7
South and West Asia	5,000	5,196	196	3.9	2,460	3,500	1,040	42.3
Sub-Saharan Africa	3,190	5,290	2,100	65.8	1,096	3,637	2,541	231.8
World	28,869	32,205	3,336	11.6	17,278	22,362	5,084	29.4

Source: Human Development Report Office estimates based on UNESCO (2014).

TABLE 5.6

The demand for health workers

	Demand for physicians, nurses and midwifes (thousands)							
	Demand-based				Needs-based[a]			
			Change				Change	
	Actual 2012	Model 2030	Absolute	Percent	Actual 2012	SDG 2030	Absolute	Percent
East Asia & Pacific	9,350	36,679	27,329	292.3	9,350	11,368	2,018	21.6
Europe & Central Asia	9,773	14,259	4,486	45.9	9,773	9,773	—	0.0
Latin America & Caribbean	3,723	5,964	2,241	60.2	3,723	4,151	427	11.5
Middle East & North Africa	1,629	3,443	1,814	111.4	1,629	2,069	440	27.0
North America	4,246	7,959	3,713	87.5	4,246	4,246	—	0.0
South Asia	3,443	6,875	3,432	99.7	3,443	6,924	3,482	101.1
Sub-Saharan Africa	1,229	3,585	2,356	191.8	1,229	4,986	3,758	305.8
World	33,989	79,360	45,371	133.5	33,989	44,114	10,125	29.8

a. Based on a threshold of 3.45 health workers per 1,000 people.
Source: Human Development Report Office estimates based on WHO (2014) and World Bank (2014b).

Conclusions

Sustainable work is critical not only for sustaining the planet, but also for ensuring work that continues to advance human development for future generations. Combining the two, sustainable work protects choices for the future while retaining them for the present. Enhancing human development through work—both for present as well as future generations—will require well thought out and well formulated policy options, and that is the theme of the concluding chapter.

Chapter **6**

Enhancing human development through work

Infographic: Policies for enhancing human development through work

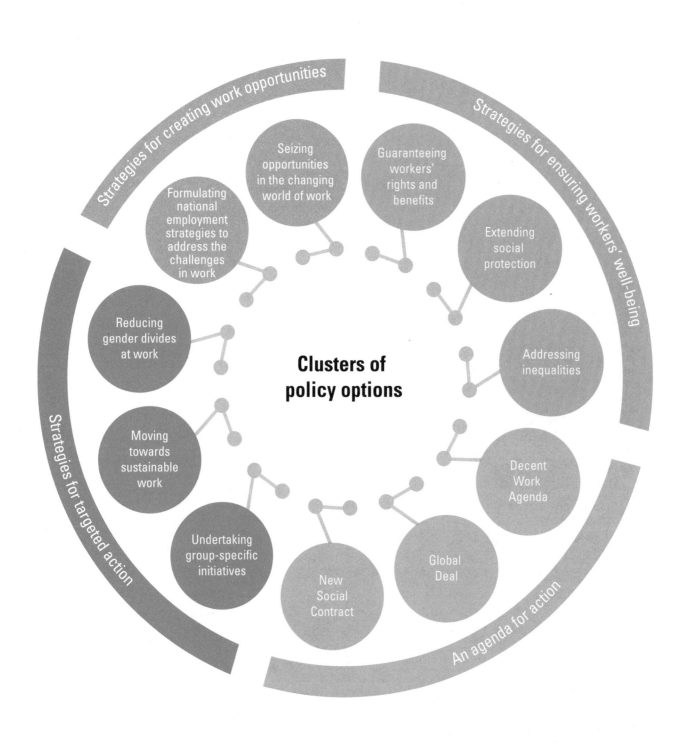

Strategies for creating work opportunities

Strategies for ensuring workers' well-being

Strategies for targeted action

An agenda for action

Clusters of policy options

Seizing opportunities in the changing world of work

Formulating national employment strategies to address the challenges in work

Guaranteeing workers' rights and benefits

Extending social protection

Reducing gender divides at work

Addressing inequalities

Moving towards sustainable work

Decent Work Agenda

Undertaking group-specific initiatives

New Social Contract

Global Deal

6.

Enhancing human development through work

The preceding chapters have analysed various aspects of the work–human development nexus—how work enhances and damages human development, how unpaid care work and paid work need to be made more equitable and how work needs to be sustainable in the changing world of work. As a response to these challenges, this chapter is oriented towards actions: providing policy options and making policy recommendations as to how human development can be enhanced through work.

The chapter groups policy options into three broad dimensions—strategies for creating work opportunities, strategies for ensuring workers' well-being and strategies for targeted action—each with a cluster of policy actions (see infographic at the beginning of the chapter). Beyond the policies, it also presents an overarching agenda for action based on three pillars—developing a New Social Contract, pursuing a Global Deal and implementing the Decent Work Agenda.

Strategies for creating work opportunities

Work for human development is about more than just jobs. It is also about expanding people's choices and making sure that opportunities are available. This includes ensuring that adequate and quality paid work opportunities are available and accessible for those who need and want paid work. In a world with high youth unemployment, financial volatility and large numbers of working poor, quality job creation becomes critical. For women, economic empowerment through paid work requires jobs to be available. This section focuses on policies to expand people's paid work choices.

Strategies for creating paid work opportunities have a two-pronged cluster of actions: formulating national employment strategies to address the challenges in the availability of quality paid work and enabling people and countries to seize opportunities in the changing world of work. The idea is to address this issue both from the demand as well as the supply side.

Formulating national employment strategies to address the challenges in work

The challenges in work—joblessness or poor-quality work—face many societies. Given the enormity of the problem, as discussed in earlier chapters, it is clear that a piecemeal approach to promoting work in its various dimensions is unlikely to bear fruit. More comprehensive national employment strategies are required, with a basic focus on creating more and better quality work for women and men. Such an approach, which places the needs of people at the core of economic policy, could be the centrepiece of a country's national development strategy (box 6.1). Some policy choices are as follows.

Setting an employment target

Employment targeting—re-orienting national monetary and fiscal policies towards generating employment—merits its own focus—a fact widely recognized by the international community.[1] That is why the proposed Sustainable Development Goals have a goal on productive employment and decent work for all. Interventions include:

- *Adding employment targeting to the national policy matrix.* More than a dozen countries have employment targets of some form (for example, creating 80,000 jobs over 2010–2014 in Honduras and reducing unemployment from 7.6 percent to 5–6 percent in Indonesia).[2] Such national commitments to work set the stage for job creation, prompting work initiatives in the public and private sectors. For example, providing credit to small and medium-sized

> The challenges in work—joblessness or poor-quality work—face many societies

BOX 6.1

National employment strategies

About 27 developing countries have adopted a national employment strategy, mostly since the 2008 global financial crisis. Another 18 are doing so, and 5 are revisiting their policies to better respond to new employment challenges. Some have attempted to directly integrate growth and employment—for example, the Growth and Employment Strategy in Cameroon and the Strategy for Accelerated Growth and Employment Promotion in Togo. Sri Lanka has integrated human resources and employment strategies in its National Human Resources and Employment Strategy, which began in 2014.

The employment–population ratio increased from 63 percent to 65 percent in Cameroon and from 70 percent to 75 percent in Togo after the countries adopted an employment strategy. Growth in labour productivity in Cameroon switched from a negative value to a positive 1.3 percent after five years. In Sri Lanka annual job growth is 12 percent.

Since the Arab Spring, Jordan and Tunisia have pursued comprehensive national employment strategies, broadening their focus from youth-centred active labour market policies to address many other employment challenges.

Source: ILO 2015a.

---

enterprises on favourable terms or directing commercial banks to have quotas to ensure adequate numbers of loans to sectors (such as agriculture) where most poor people work would help create jobs. Similarly, employment-friendly monetary policies, fiscal policies such as tax credits to small and medium-sized enterprises and subsidies for enterprises set up in less developed areas may accelerate job creation.

- *Pursuing dual targeting by central banks.* Job creation may require expanding central banks' traditional focuses from controlling inflation to generating employment. Inflation targeting, even if it does have a modest impact on reducing inflation and inflation expectations, does not seem to be associated with substantial improvements in real economic growth.[3] In fact, inflation targeting in developing countries has not been found to have a positive impact on growth.[4] Further, given the critical challenge of joblessness and the crucial importance of job creation for poverty reduction and human development, central banks may pursue dual targeting of inflation and employment.[5]
- *Considering specific monetary policy instruments to increase investments that grow jobs.* These include policies that increase financial support to business activities and investments in the real economy. For example, credit allocation mechanisms drawing on asset-based reserve requirements which make it more expensive for banks to hold reserves, loan

guarantees that reduce the risk of borrowing for entrepreneurs, support for pooling and underwriting small loans, use of the discount window to release funds from central banks for work-generating investments and capital management techniques. A study of several economies (Chile, Colombia, India, Malaysia and Singapore) found that these techniques —when supported by complementary factors such as sound assets, state capacity and central bank policy independence—can help create employment.[6] These instruments can also be made gender sensitive. For example, asset-based reserve requirements could be designed to generate more work for women, and central banks could give preferential access to the discount window for financial institutions investing in or lending to entities that will generate more and better work for women.

Formulating an employment-led development strategy

Employment can no longer be considered a derivative of economic growth. It has been viewed as such for too long, in a trickle-down notion that economic growth automatically leads to job creation.[7] But over the years the world has seen jobless growth, where economic growth does not go hand in hand with the creation of new jobs. It is crucial to ensure that the complementarity between labour and capital is retained and strengthened in the process of development.

> Job creation may require expanding central banks' traditional focuses from controlling inflation to generating employment

Some interventions in the context of employment-led development would entail:

- *Removing barriers critical to employment-led development.* For example, small and medium-sized enterprises often face biases in market entry and access to credit, and entrepreneurs may lack access to information and marketing skills. Women in particular face barriers in access to capital, technology and credit. Removing these barriers requires multiple levels of support to improve the productivity and income of these enterprises from all types of regulatory regimes and public and private institutions. In addressing credit, finance, training and skill development, some countries have, for example, formulated self-employment programmes as a critical part of their national employment plans (box 6.2).
- *Strengthening links between small and medium-sized enterprises (typically labour intensive) and large (typically international) capital-intensive firms.* Such relationships can help change patterns of growth by shifting resources to sectors with greater value added and job creation potential. This can be accomplished, for example, through industrial clusters supported by public investment. These networks can increase access to capital and technology and promote the transfer of skills.
- *Upgrading workers' skills over the lifecycle.* Success here could lead to a virtuous circle of employment–growth–employment, helping generate growth and more productive and better quality jobs while instilling individuals with more advanced work capabilities.
- *Focusing on sectors where the poor live and work.* In particular, a focus on poor people in rural areas who work in agriculture is important. Policy measures to protect and generate jobs in agriculture should aim to improve productivity without displacing jobs through intensive cultivation, changed cropping patterns, integrated input packages, better marketing and the like.
- *Designing and implementing a conducive legal and regulatory framework to tackle informal work.* Informal workers are among the most vulnerable and insecure. Regulatory frameworks for their work are critical from the perspective of incentives for enhancing productivity and value added and protection to reduce insecurity. Such frameworks could support innovation in the informal sector through low-cost technology adaptation, could be sensitive to the gender dimensions of informal work, could support informal workers in making their voices heard and give them recognition as citizens and entrepreneurs (box 6.3).
- *Adjusting the distribution of capital and labour in public spending to create jobs.* Public spending can support job creation through the types of technology used and the sectors in which spending takes place, with an eye to enhancing human development—for example,

Informal workers are among the most vulnerable and insecure

BOX 6.2

Self-employment programme of the Former Yugoslav Republic of Macedonia

A self-employment programme was an important element of The Former Yugoslav Republic of Macedonia's first national employment plan, launched in 2007. By 2015, 6,700 people had found secure and sustainable employment under the programme by creating their own companies or formalizing existing businesses. Self-employment opportunities in the country cover a broad spectrum, including dentistry studios, nursery gardens, hairdressing, eco-tourism, fashion design and the like.

Over the past eight years the state has invested more than $33 million in the programme. Some 70 percent of entrepreneurs who received training and equipment through the programme over the past five years have remained in business—well above the roughly 50 percent global success rate for new companies. In 2014 the programme was responsible for nearly 13 percent of all new private sector jobs and 14 percent of all new businesses created.

Women started 35 percent of the newly established businesses, and young entrepreneurs, 30 percent. The programme is a major achievement and an important milestone in promoting entrepreneurship and enabling young people and women to take action and improve their livelihoods. The programme has also specifically reached out to help single mothers and victims of domestic violence.

Source: UNDP 2013d.

Labour standard to tackle informal economy—a new milestone

Millions of workers are engaged in the informal economy, which is marked by low productivity, denial of workers' rights, a lack of adequate opportunities for quality work, insufficient social protection and an absence of social dialogue. These characteristics are also obstacles to developing sustainable enterprises. Consensus has been growing among governments, workers and employers that the sector needs incentives as well as protection of workers.

The International Labour Organization recently adopted new international labour standards—the first ever standards aimed specifically at tackling the informal economy—which are expected to help hundreds of millions of workers and economic units. The standards can be vital in protecting workers' fundamental rights, ensuring opportunities for income security, livelihoods and entrepreneurship; and promoting the creation, preservation and sustainability of enterprises and decent jobs in the informal economy. A huge step forward for millions of workers, there is now an international framework of guidance to help countries address the issues of the informal economy.

Source: ILO 2015c.

---

through spending in health, education and other social services. When this approach is taken, the public sector can also have a demonstration effect, guiding and signalling to the rest of the economy the availability and ways of using more labour-intensive versus capital-intensive technologies.

Moving to financial inclusion

An inclusive financial system is essential for structural transformation and the creation of work. More than 2.5 billion people around the world, roughly half the world's adults, are unbanked.[8] Women are disproportionately affected by a lack of access to finance. About three-quarters of people living on less than $2 a day do not have a bank account.[9] Yet the poorer a household is, the greater its need for protection against vulnerability, especially when out of work.

In developing countries lack of access to finance is a major hindrance to the operation and growth of enterprises. Access to finance was among the top five problems for 77 percent of businesses surveyed in a sample of developing countries.[10] In Sub-Saharan Africa, South Asia and Latin America and the Caribbean more than 30 percent of surveyed firms cited access to finance as a major constraint.[11] Policy options to expand access to finance, which can ultimately increase work opportunities, might encompass:

- *Extending banking services to the disadvantaged and to marginalized groups including women.* After pro-banking measures were adopted in Ecuador, the share of the population with a bank account increased from 29 percent in 2005 to 83 percent in 2011.[12]
- *Steering credit towards unserved, remote areas and targeted sectors.* In Argentina, Brazil, the Republic of Korea and Malaysia investment banks have played a central role in directing credit to targeted sectors.[13]
- *Lowering interest rates and providing credit guarantees and subsidized credit to small and medium-sized enterprises and export-oriented sectors.* These measures can help promote productivity and employment growth and can bear fruit quickly. In Rwanda a credit guarantee scheme enabled the country to become a major exporter of specialty coffee.[14]
- *Harnessing modern technology to promote financial inclusion.* M-Pesa in Kenya is a prime example of how mobile phone technology can be deployed to reach the unbanked (box 6.4).

Building a supportive macroeconomic framework

Maintaining macrofinancial stability is a precondition for a successful employment strategy. However, macroeconomic policies should go beyond ensuring economic stabilization, which is an important but not sufficient condition to creating work, and ensure that the environment is also conducive to generating jobs. Policies in this area need to address exchange rate management, capital account measures

An inclusive financial system is essential for structural transformation and the creation of work

BOX 6.4

M-Pesa—an innovative approach to financial inclusion

M-Pesa is a small-value money transfer system that was launched in Kenya in 2007. Its reach is extensive, offering a large segment of the population basic financial services. In 2012 the number of active users reached 15 million, or more than 60 percent of the country's adults and roughly 30 percent of the country's population. In 2014 there were 81,000 M-Pesa agent outlets in the country, up from 15,000 in 2013.

M-Pesa is used not only for standard money transfers and airtime purchases, but also to pay salaries, utility and other bills and to buy goods and services from both online and physical merchants. The transfer system is used by the government, businesses and nongovernmental organizations alike. The key drivers centre on a very supportive regulatory regime, innovative business models and technological advances in mobile phones. The approach is being replicated in other countries, as with bKash in Bangladesh.

Source: Nuzhat 2015.

and government fiscal space. Some options include:

- *Keeping the real exchange rate stable and competitive.* Financial volatility can make work environments and jobs less secure and reduce investments in the real economy. A stable exchange rate is an economic fundamental that can stimulate growth and employment. There seems to be a consensus that currency appreciation has a negative impact on employment by deteriorating international competitiveness.[15]
- *Assuring prudent capital account management.* This is important if the exchange rate policy is to promote job creation, as in Chile.[16] Transparent and widely agreed rules for capital controls should be pursued to reduce the volatility of capital flows in and out of national economies and thus the volatility of investments in productive work activities.
- *Restructuring budgets to allocate resources to job-creating sectors, to enhance human capabilities and to develop infrastructure.* This would entail reviewing existing expenditure patterns and reallocating resources towards employment generation (jobs for teachers and nurses in the social sectors, public works

programmes, operation and maintenance of physical infrastructure and the like).

- *Building fiscal space through comprehensive tax administration and efficiency in resource use.* Public spending can create jobs, but fiscal space for public spending requires a well structured, transparent and efficient plan and system for taxation and spending. Establishing sound fiscal measures during booms and normal periods of growth and drawing on a dedicated stabilization fund during recessions are also good options for supporting public investments. Overcoming leakages of resources would also enhance the use efficiency, equivalent to mobilizing new resources.
- *Promoting an enabling business environment.* Given that the private sector is often the main driver for job creation, an enabling business environment can be put in place by removing binding constraints in access to finance, infrastructure and regulation. Other elements include providing business incentives through tax relief. Business activities that create quality work can be encouraged. For example, by subsidizing inputs for businesses that set up in less developed areas, that provide job opportunities to disadvantaged groups and that use labour-intensive technologies. Direct transfers to employers in the form of wage subsidies or reductions in their taxes may also encourage them to hire more workers or to maintain existing jobs.
- *Ensuring high-quality infrastructure, including roads, electricity and telecommunications.* Mobile phones for poor female entrepreneurs have revolutionized their work and created more jobs. Expanding telecommunications and Internet infrastructures, as well as transportation arteries and access to clean energy, can immensely improve work options for people in rural areas and expand business activities and investment.
- *Adopting a regulatory framework that encourages competition, enhances efficiency and ensures transparency and accountability for business.* Cutting red tape, battling corruption, outlawing political use of licences and permits and accelerating decisionmaking in the public sector all help develop businesses and attract domestic and external investment, frequently boosting the number of jobs.

Financial volatility can make work environments and jobs less secure

Seizing opportunities in the changing world of work

There is an urgent need for policy attention to help people thrive in today's work environment. From a human development perspective this requires finding ways to harness new technologies in order to expand the choices people have in their work and lives. The changing world of work can contribute to human development by creating new and better work and by improving people's lot—all the more so if they are equipped with the skills, knowledge and competencies to capitalize on emerging opportunities. To be so equipped, it will take far more than business as usual or small, incoherent policy steps.

Heading off a race to the bottom

Given the potential (and realized) benefits of global production, a race to the bottom—ever lower wages and worsening working conditions—is not the only possible outcome. This scenario is due mainly to a misplaced emphasis on a static and narrow interpretation of competitiveness measured by financial costs per unit of production—a focus that might help companies' bottom lines today but is unsustainable in the long run. Ensuring decent wages, maintaining workers' safety and protecting their rights are universal issues that, if tackled, can pre-empt such a race. Some policy options include:
- *Focusing on the conditions in which goods are produced (beyond the economic aspects of costs and competitiveness).* Working conditions

are critical (box 6.5). And when labour is scarce, businesses that offer better work conditions will attract the most skilled labour. Consumers are also becoming increasingly aware of and sensitive to variations in work conditions, and buying intermediaries are facing growing pressures for ethical considerations to become more central, as illustrated by the global outcries caused by multiple accidents in the readymade garment industry in Bangladesh. With this pressure, major retailers that source their imports from Bangladeshi factories have initiated remedial plans to improve work conditions and are providing (some) financial support to workers.[17] Although these initiatives have limitations, they indicate that sweatshop conditions are not acceptable in global value chains.
- *Ensuring fair trade can help forestall the race to the bottom, as it is a critical consideration in many consumers' minds.* With consumers asking more questions about the products they buy, the single-minded pursuit of cost efficiency is likely to be short sighted and a poor business strategy. Cambodia's experience illustrates the ability of a country to maintain its export market without compromising the fairness of its wages and the quality of its work conditions (box 6.6).

Providing workers with new skills and education

The requirements of advanced skills and education will be pronounced in the future. Higher

> Sweatshop conditions are not acceptable in global value chains

BOX 6.5

Remaining competitive by improving working conditions

Some countries in global value chains that previously relied on low-wage employment now face labour shortages, giving workers more bargaining power. China has enacted several laws that hold promise for individual worker rights, enhanced employment security, reduced informal employment and broader access to social insurance. These include the Labour Contract Law (2008), the Labour Dispute Mediation and Arbitration Law (2008), the Employment Promotion Law (2008) and the

Social Insurance Law (2011).[1] Some have argued that the passage of these laws would place China's labour regulations third among Organisation for Economic Co-operation and Development members by strictness.[2] In 2013 Viet Nam also passed a new labour code that could enhance work conditions by regulating working time, extending maternity leave and providing more space for collective bargaining and a reformed wage system.[3]

Notes
1. Friedman and Kuruvilla 2015. 2. Gallagher and others 2014. 3. ILO 2014a.
Source: Human Development Report Office.

BOX 6.6

Cambodia—a success story in the globalized world of work

On entering readymade garment production for export in the 1990s, Cambodia was keen to earn a reputation for maintaining good labour standards, which offered advantages. In 1999 the government signed an agreement with the United States to adhere to the rights of workers in return for which the United States agreed to increase its annual import quota.

Not only did Cambodia benefit directly, but it also sustained that success after the quota system ended. In short, low wages and compromises over working conditions are not essential for maintaining competitiveness in export markets.

Source: Islam 2015.

and more specific skills will be needed for science and engineering jobs. Technology increasingly requires more than basic code skills and demands aptitudes in creativity, innovation and problem solving. Manufacturing activities are more skill intensive as routine activities are automated, and agriculture increasingly requires education and training to further boost productivity. Education, flexibility, adaptability and work-related skills are vital for workers to secure their livelihoods. This is the point emphasized in a special contribution by His Excellency Benigno S. Aquino III, President of the Philippines (signed box).

The shift from industrial to information-based knowledge economies is changing how we live, think, work and learn, pointing to four broad categories of skills that may be required:[18] ways of thinking, which include creativity, innovation, critical thinking, problem solving, decisionmaking and learning; ways of working, which refer to communication, collaboration and teamwork; tools for working, most of which are based on new information and communications technologies and information age literacy, including capabilities to learn and work through digital social networks; and ways of living in the world, such as a sense of global and local citizenship, a view towards life and career development, and a commitment to personal and social responsibility.

Concrete policy measures could focus on:

- *Designing and implementing a fit-for-future learning system.* High-quality education

with a curriculum that teaches students how to learn is an essential aspect of schooling, and so a learning paradigm shift may be in order—one that goes beyond the goal of universal access to schooling towards "access plus learning" with a focus on improving learning outcomes rather than just access.[19] Learning in a slew of foundational domains could be prioritized in early childhood, primary and lower secondary education, with space for specific skills development in subsequent years. Emphasis should also be on educating students in mathematics and the natural sciences, on teaching them to write and communicate persuasively and on giving them opportunities to develop skills in team cooperation and leadership. Education systems need to foster talent at all levels.

- *Increasing access to tertiary education.* In any society the system of higher education needs to reach more students, including those from poorer backgrounds. Therefore, classroom instruction cannot be divorced from the demands of the labour market. Some students will benefit from targeted training in industry-specific skills. In developed countries tertiary education attainment has increased 10 percent since 2000.[20] However, given the necessity for advanced skills, even higher rates of attainment will be required to meet future demand.

There are also concerns about a mismatch between the types of tertiary degrees that are in demand—science, technology, engineering and mathematics, in particular—and the number of people graduating with degrees in these fields. In the United States only 11 percent of college graduates were in science, technology, engineering and mathematics disciplines in 2008, compared with 42 percent in China, 35 percent in the Republic of Korea and 28 percent in Germany.[21] Women are substantially underrepresented in these fields, which could perpetuate wage gaps between men and women.

- *Upgrading or reorienting skills.* This may help displaced workers—whose livelihoods are threatened by the changing world of work—with training. Such workers can transition to new jobs at similar or higher wages after training and with wage subsidies and temporary income support (among other

In any society the system of higher education needs to reach more students, including those from poorer backgrounds

Building skills and protecting workers for inclusive growth

We believe: only inclusive growth can unlock the vast potential of the Filipino people. This principle has been at the core of our strategy, which is why we have made massive investments in our people. As a matter of fact, funding for social services currently takes up the lion's share of the budget, at 36.6 percent. In this way, we capacitate Filipinos to take part in growth: to ensure that their development and work are part of a virtuous cycle that yields not only individual, but also national success.

This virtuous cycle we speak of is largely dependent on ensuring that our people have the necessary knowledge and skills to achieve their goals. For the long-term, we have undertaken basic education reform: Filipino students will now spend 12 years of basic education in school, allowing sufficient time for the mastery of a curriculum that is at par with international standards. Our flagship poverty reduction program, the Conditional Cash Transfer Program, also contributes to this goal: For families to receive cash assistance, the primary requirement is to ensure that children attend school. Other investments have also been made, most prominently through our Technical Education and Skills Development Authority, which throughout our term has equipped 7.8 million graduates from various courses with skills relevant to different industries.

Throughout all this, we remain cognizant of the need to monitor and respond to trends in employment relevant to our communities. For example, our Commission on Higher Education has been working continuously to eliminate substandard and noncompliant programs, all while developing programs in high-demand and emerging fields. Further proof can be found in the way that we have addressed challenges—for instance, in helping out-of-school youth access economic opportunities. The Conditional Cash Transfer Program mentioned previously has already done much in this regard. Research conducted by the Philippine Institute for Development Studies shows that there has been a decrease in the number of out-of-school youth: from 2.9 million children in 2008 to 1.2 million in 2013. That is a staggering 1.7 million decrease.

The Department of Education's Abot-Alam Program, which translates to Knowledge Within Reach, aims to reduce this number even further. Launched in September 2014, this is the first initiative to map out-of-school

youth nationwide and match them with appropriate interventions in education, skills training or entrepreneurship. Our Department of Labor and Employment also helps at-risk youth to be job-ready in terms of acquiring the required skill sets by employers—that is, life skills, technical and company internship—through our JobStart Program.

The targeted manner in which we have addressed challenges and created opportunities for our youth can likewise be observed in other areas—for instance, in the informal sector. Domestic workers make up a significant part of the informal labour force, whether in the Philippines or elsewhere in the world. The nature of their employment makes it difficult to regulate; even so, my country has demonstrated extraordinary leadership in this regard.

In 2012 the Philippines became the second country to ratify International Labour Organization Convention No. 189 on Decent Work for Domestic Workers—the first international instrument recognizing domestic workers as having human and employment rights entitled to protection under the law. Our country has also signed bilateral agreements with the Kingdom of Saudi Arabia, Jordan and Lebanon for the protection of overseas Filipino domestic workers.

Here within our own shores, RA 10361, or An Act Instituting Policies for the Protection and Welfare of Domestic Workers—more commonly known as the Kasambahay Law—was finally enacted into law under my administration. Under this law our 1.9 million domestic workers are recognized as members of the formal sector, with all the appropriate rights, benefits, training and competency assessments.

These may be just a sampling of the measures we have implemented, but whether in education, youth employment or domestic work the message is clear: inclusivity. We are centred on our people: They are the end-all and be-all of everything we do. Thus, we will exert every effort to empower them, give them the same rights and protections and maximize their opportunities—regardless of social status, welfare and yes, even the nature of work. In this way each and every one of our people can see their work as meaningful, dignified and productive, as a means for their own development and success.

Benigno S. Aquino III
President of the Philippines

services). One particular example is the Trade Adjustment Assistance Programme under the North American Free Trade Agreement. Differences in the programmes available, variations in state and local implementation and the heterogeneity of participants in skills and experience produced mixed results in post-training skill gains, reemployment and wage increases, but having such a programme in place can help with job transitions.[22]

Lifelong learning and training are essential for upgrading skills and training for new types of work, and much learning takes place outside formal education. Workers, including those performing low-skilled tasks, must be prepared to learn and retrain throughout their working life.

- *Using adult education programmes for continued learning.* Northern European countries are particularly successful at lifelong learning, with over 60 percent of the adult population participating in adult education in Denmark, Finland, the Netherlands, Norway and Sweden.[23] Apprenticeships and industry-targeted training programmes as well as on-the-job training programmes help prepare young people for work and enable career shifts later in life. Both employers and governments may organize such programmes.

Policies that encourage training by employers, such as tax credits or preferential access to government contracts, are also needed. Women and girls require special attention, since they are often excluded from training opportunities for social or cultural reasons. In the informal sector women need equal access to information to anticipate the demands of markets and to understand how to be strategic in deciding what goods to produce. They also need access to global markets, whether through technology or other means.

On-the-ground learning is important, but training can be scaled up, taking advantage of communications technology, online learning platforms and global collaborations. For example, WeConnect International provides online learning to a global network of women-owned businesses in Chinese, English and Spanish, teaching entrepreneurial, business and leadership skills.[24] The nongovernmental organization Education for Employment designed a programme with McGraw-Hill and business leaders in Morocco to help fill skill gaps in Morocco's workforce, focusing on the professional skills needed in banking, retail and manufacturing.[25]

Innovating so that all income groups benefit

It has been suggested that neither workers nor employers will be the real winners in the economy of the future. Instead, a third party—people who can create new products, services and business models—will prosper immensely. This view highlights the potential challenge of delivering an acceptable standard of living for everyone outside this innovator class.[26] It also underscores the critical importance of a thriving, innovative business sector. If advances in digital technology can be harnessed to create new and better businesses, workers will have a better shot at sharing the growing prosperity. But if entrepreneurship declines, new technology will not guarantee overall social welfare gains, which is one reason the policy environment should be conducive for start-ups and social enterprises. Policy considerations (leaving the skill sets for future generations aside) include:

- *Reinventing work organizations.* There has never been a worse time to be competing with machines, but there has never been a better time to be a talented entrepreneur. But are there enough opportunities for all these entrepreneurs? Are we running out of innovations? When businesses are based on bits instead of atoms (a common characteristic in the digital age), each new product adds to the set of building blocks open to the next entrepreneur, instead of depleting the stock of ideas the way minerals or farmlands are depleted in the physical world.[27] Because innovation often relies heavily on combining and recombining previous innovations, the broader and deeper the pool of accessible ideas and individuals, the more opportunities there are for innovation, primarily to leverage ever-advancing technology and human skills.

- *Adopting complementary innovations.* General-purpose technologies, such as steam power, electricity and the internal combustion engine, not only got better over time, but also led to complementary innovations in the processes, companies and industries that used them. This led to a cascade of benefits both broad and deep. Many of the instruments of the current technological revolution (such as computers) are the general-purpose technologies of our era, already combined with networks and labelled information and communications technology. However, they have to be tied to complementary innovations that can have broad spillover effects on more of the population.

There has never been a worse time to be competing with machines, but there has never been a better time to be a talented entrepreneur

Using migration

Economic opportunities induce people to move to other lands to work and have a better life. Push factors such as droughts and conflicts also cause people to migrate. Migrant workers bring new knowledge, skills, creativity, innovation and experience, making migration mutually beneficial for migrants, who gain access to work, and receiving countries, which gain access to skills and experience. Furthermore, for some countries migrant workers add to a labour force that is declining because of various demographic transitions. The critical issue is how the human potential embodied in migration can be used for the benefit of all concerned. Some policy options may be:

- *Pursuing well formulated schemes for migrant workers.* Such schemes may include expanding programmes for seasonal workers in such sectors as agriculture and tourism, creating secure space for more low-skilled workers and well targeted programmes to match skilled workers, including professionals, with jobs. These could all be developed in the context of agreements reached in destination countries through political processes based on public discussions that balance different interests, local priorities and demands, and discussions involving source countries, employers and trade unions. Essential in all of these processes is increasing the security, protection and rights of migrant workers.
- *Undertaking actions in source countries.* Source countries can undertake skill development and training initiatives for aspiring migrant workers, pursuing orientation programmes to educate migrants on the laws, rules and culture in destination countries and their rights and obligations.
- *Improving the global management of migration.* The existing Global Forum on Migration and Development, with more than 150 participating countries, provides a good platform for addressing the challenge of migration through common responses. The importance of such a platform becomes even more critical when the world faces a crisis like the current one in Europe. Using such a forum as a basis, the global management of migration can be improved through relevant reforms that increase the security and opportunities for migrants during the process of migration and while working in receiving countries.

Strategies for ensuring workers' well-being

Work should enhance human development, but as previous chapters have shown, this link is not always automatic. The choices individuals have regarding work are also subject to myriad influences, and the quality of work varies. The previous section discussed ways to increase work opportunities. These policies can be complemented by policies that improve workers' well-being and expand the choices people have in the types of work they undertake—paid or unpaid.

Workers' well-being encompasses workers' rights and benefits. On the rights side, workers' safety, voice and participation are crucial. On the benefits side, income and social protection are of prime importance. Workers' well-being is linked to the quality of their work as well as the quality of their lives.

Guaranteeing workers' rights and benefits

Workers' rights are also human rights. Guaranteeing these rights thus has intrinsic as well as instrumental value. On the one hand ensuring workers' rights provides people with protection and safety, while on the other it can help ensure that they maintain a fair share of income, security, creative opportunity, social interaction and other work-related benefits.

Guaranteeing the rights and benefits of workers is at the heart of strengthening the positive links between work and human development and weakening the negative links. Positive links ensure that workers' rights and benefits go beyond good wages and include an environment where workers are more productive, safe and empowered. Weakening negative links is about guaranteeing workers' rights and benefits and eliminating exploitation, abuse, unsafe work environments and the destruction of dignity.

> The critical issue is how the human potential embodied in migration can be used for the benefit of all concerned

From the top—setting legislation and regulation

Legislation and regulation are critical for the protection of millions of workers around the world who are engaged in activities that damage human development or who are involved in high-risk work, as described in chapter 1. Policy options can take various forms:

- *Having well designed unemployment and wage policies.* Unemployment insurance and minimum wages protect workers and empower them with economic security. The two need to be strengthened in many cases. In 2013 just over 30 percent of the world's labour force was legally covered by unemployment benefits (cash periodic), up from 18 percent in the 1990s.[28] And the minimum wage was 58 percent of the average wage in South Asia and only 18 percent in Sub-Saharan Africa.[29] Two policy options in this regard are worth considering: increasing the minimum wage and complementing it with tax credits so that a living income can be provided to the working poor.[30] Such policies need not discourage investment because, contrary to conventional belief, the evidence does not confirm that labour regulations are a primary constraint to job creation.[31]

- *Protecting workers' rights and ensuring their safety.* There are already guiding agreements in place to help improve work safety and security. Namely, the eight fundamental conventions on freedom of association, forced labour, discrimination, child labour and domestic workers, which together constitute the Charter of Workers' Rights (see figure 9 in the overview and table A6.1 at the end of the chapter). But they need to be pursued and enforced by countries, with the following aims:
  - *Ratifying and implementing the eight conventions and reporting on progress of implementation.* More than 170 countries have ratified the Equal Remuneration Convention of 1951 and the Discrimination (Employment and Occupation) Convention of 1958, but some have not (figure 6.1). By 2014, 65 countries had laws prohibiting discrimination in employment based on sexual orientation in at least part of their

FIGURE 6.1

More than 170 countries have ratified the Equal Remuneration Convention of 1951 and the Discrimination (Employment and Occupation) Convention of 1958

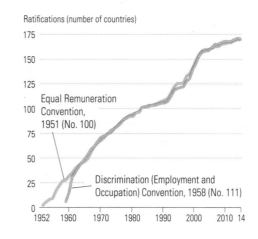

Source: ILO 2014d.

territory—more than triple the number 15 years ago (figure 6.2).[32] If support is needed, the global community should increase efforts for technical and financial assistance.
  - *Putting in legal frameworks and strengthening legislation to enhance and protect workers' rights and to remove all kinds of abuse* (box 6.7). Workers' rights, security, safety and working conditions may be part of such a framework. Globally, the same general labour law that covers other workers covers only 10 percent of paid domestic workers (figure 6.3).

Ensuring that people with disabilities can work

People with disabilities are just differently abled to do things. Indeed, if proper investments are made in accessibility and flexibility, people with disabilities can contribute considerably to work. The workplace could become an equalizing field for people with disabilities if the approach of employers was to change from one of charity and assistance to one of investment.[33] Some policy options include:

- *Providing an environment that is conducive to the productivity of people with disabilities.* For people with disabilities, finding and

Workers' rights are also human rights

FIGURE 6.2

By 2014, 65 countries had laws prohibiting discrimination in employment based on sexual orientation in at least part of their territory—more than triple the number 15 years ago

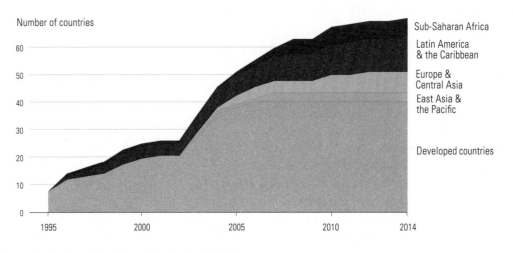

Source: Human Development Report Office calculations based on ILGA (2014).

BOX 6.7

Protecting the rights of sex workers

Protecting the rights of sex workers, ensuring their health and safety and protecting them against violence are basic policy priorities. In addition, bringing sex work into the framework of labour law allows for the same control and monitoring that other works are subjected to for forced labour.

In 2003 New Zealand adopted the Prostitution Reform Act, which has effectively decriminalized sex work in the country and introduced protections for the health and safety of sex workers and their clients. The act ensured sex workers' access to the legal system, empowering negotiations with and choices of clients, and relationships with police and health professionals have also improved. Evidence from the country suggests that support of sex workers enables them to organize and improve their ability to advocate and lobby on behalf of their community and to improve working conditions and workers' safety.[1]

Note
1. Barnett 2007.
Source: United Nations Development Programme's HIV and Health Group.

> People with disabilities are just differently abled to do things

sustaining work are difficult due to negative perceptions, cultural norms, transport problems, accessibility to resources and premises, and the like. Changes in information flows and infrastructures can enable people with disabilities to access work and enable employers to take advantage of a wealth of human ingenuity.

- *Encouraging behavioural shifts in favour of people with disabilities.* Changing social norms and perceptions to accept that people with disabilities are differently abled and to give them a fair chance in work is fundamental. This will have to be supplemented with legal frameworks that discourages discrimination against people with disabilities.

- *Enhancing capabilities and opportunities.* These can be done through skill and vocational training for people with disabilities, increasing access to productive resources such as finance for self-employment, and by providing information through appropriate mobile devices.

- *Ensuring accessibility.* Appropriate measurement must be taken to improve transportation to and from work and to improve access to premises and disability-friendly office space, workplaces and equipment.

FIGURE 6.3

Globally, the same general labour law that covers other workers covers only 10 percent of domestic workers, 2010

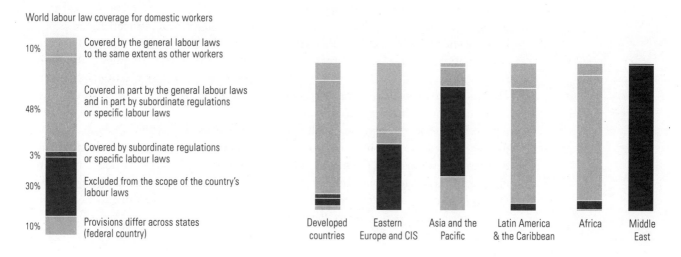

World labour law coverage for domestic workers

- Covered by the general labour laws to the same extent as other workers
- Covered in part by the general labour laws and in part by subordinate regulations or specific labour laws
- Covered by subordinate regulations or specific labour laws
- Excluded from the scope of the country's labour laws
- Provisions differ across states (federal country)

10% / 48% / 3% / 30% / 10%

Developed countries | Eastern Europe and CIS | Asia and the Pacific | Latin America & the Caribbean | Africa | Middle East

Source: ILO 2013b.

- *Adopting appropriate technology.* Technology can enhance the abilities of people with disabilities. Nevertheless, in many low-income countries only 5–15 percent of the people who require assistive devices and technologies have access to them.[34]
- *Pursuing affirmative action.* Ensuring that jobs are available for people with disabilities will require targeted interventions. In this regard, affirmative action (for example, the use of quotas) reserves jobs for people with disabilities and provides an opportunity for the rest of society to observe their capabilities and achievements, which may have a positive impact on changing social norms, biases and attitudes towards people with disabilities.
- *Using big data, but with caution.* Collecting and mobilizing data to monitor employment trends of people with disabilities can help inform policymaking. Big data has the potential to uncover discrimination in the labour market and, in doing so, trigger corrective policies. However, there are some risks. For example, the use of algorithms that recommend candidates to employers (based on historical interest or performance indicators, plus all other information available) might reproduce existing biases and prejudices on gender, race and social class (which can be inferred from social network information),

thereby introducing a liability for belonging to historically disadvantaged groups.

Making workers' rights and safety a cross-border issue

Cross-border action is fundamental in a globalized world where both work and workers move beyond borders. Measures may include:

- *Building a solid regulatory framework, based on global labour conventions, to facilitate the transborder movement of workers and help them reap the benefits of their work.* For example, for migrant workers, such a framework may streamline the migration of workers through legal means and ensure the efficient transfer of remittances, helping make effective use of these resources. It may also establish concrete guidelines and rules for working conditions, such as the working hours, pay and safety of these migrant workers, including those in paid domestic work. Such frameworks are often negotiated bilaterally, but a multicountry or subregional approach would establish some agreed-on guidelines and binding rules. If properly formulated, such frameworks can be effective regional or subregional public goods.

Such frameworks can be extended to cover economic migrants, who are undertaking all kinds of risks (for example, trying to

Cross-border action is fundamental in a globalized world where both work and workers move beyond borders

cross seas and oceans by overloaded and unseaworthy boats) to go from Sub-Saharan Africa to Europe and from South Asia to South-East Asia. As mobility increases and people continue to take risks, provisions can be made (such as prioritizing safety or introducing national quotas for migrants) in these frameworks.

- *Developing subregional remittance clearinghouses and remittance banks that have links with labour-sending countries.* Such institutions can ensure efficient and low-cost transfers of resources and the security of the migrant workers' hard-earned money.
- *Offering more support in source countries to educate migrant domestic workers about their rights and offering assistance in cases of abuse and exploitation.* In destination countries governments should implement formal contracts and labour protection laws for migrant domestic workers. They should also ease visa restrictions for family members and facilitate the integration of migrant children in national education systems to prevent the unnecessary separation of families.

From the bottom—promoting collective action and trade unionism

Collective action has strengthened the links between work and human development, including increases in compensation and social insurance and protection against health hazards at work. Collective action has been vital to forming shared values and worker solidarity, which enhances the agency and voice of individuals.

Globalization, the technological revolution and changes in labour markets are creating new ways of working, and it is clear that in this new and complex context policy alternatives for promoting collective action may need to be considered, such as:

- *Reforms for collective action.* Governance reforms that reaffirm labour standards, allow workers to bargain collectively and give all stakeholders—workers, managers and shareholders—a say in executive pay decisions can enhance worker voice and improve labour remuneration.
- *Emerging forms of collective action.* Different types of work require different types of workers' organizations. Informal worker bodies such as the Self-Employed Women's Association of India, transnational worker organizations such as the International Trade Union Confederation and bodies to protect immigrant workers and domestic workers such as the International Domestic Workers' Federation are a few of the existing institutions (box 6.8). But many workers remain without voice and influence, and there is much space for innovation in collective organizations that reflect the challenges and environments of the modern worker.
- *Innovative collective action for flexible workers.* Actions and institutions are needed to protect workers' rights and interests for those

BOX 6.8

The Self-Employed Women's Association—the world's largest trade union of informal workers

The Self-Employed Women's Association has nearly 2 million members, all working poor women, from multiple trades and occupations and from all religious and caste groups, in 10 states of India. It is also one of the most influential organizations of informal workers worldwide, having influenced policies, norms and practices at all levels.

It stresses self-reliance—individual and collective—and promotes organization around four sources of security: work, income, food and social security. It is primarily a trade union but engages in a wide range of interventions, including leadership development,

collective bargaining, policy advocacy, financial services (savings, loans and insurance), social services, housing and basic infrastructure services, and training and capacity building.

The association seeks to expand the voice of its members through representation at different levels by building their capacity and creating opportunities for them to participate in local councils; in municipal, state and national planning bodies; in tripartite boards; in minimum wage and other advisory boards; in sector-specific business associations; and in local, state and national labour federations.

Source: Chen, Bonner and Carré 2015.

who work in the "gig economy"—crowd workers and others. Take, the Freelancers Union, which has 250,000 independent contractors. In the United States there are 53 million freelancers, 40 percent of the country's workforce. About a tenth of the freelance population is temporary, working for one employer on a contract.[35] Although geographically dispersed, crowd workers are establishing digital versions of mutual and workplace solidarity.

- *Social movements through protest and demonstrations.* One form of worker agency can be observed through the rising number of social movements around the world, many of which are motivated by appeals for jobs, better labour conditions and higher wages. Unless employment and the needs of workers become policy priorities, prolonged and unsustainable periods of social unrest may become the reality in many countries.
- *New information and communications technology.* Technology can be used for mobilizing workers, allowing those with access to social networks to communicate and organize with others, wherever they are. This can mobilize support for workers and raise public awareness about working conditions, in particular publicizing individual cases and monitoring the activities of corporations. The Clean Clothes Campaign is one example of coalitions of local nongovernmental organizations and workers using the Internet.[36]
- *Stronger state action.* Tripartite consultations among governments, employers and workers have long characterized the regulation of work, but the balance in recent years has shifted towards dialogue between workers and employers with less state intervention. Given the decline of worker agency and increased social unrest, a greater state role is needed in tripartite arrangements.

Promotion of collective action through existing institutions (such as trade unions) as well as exploration of alternative institutional mechanisms are needed not only to secure workers' rights, voice and participation, but also for broader societal goals such as social cohesion, stability and development. If such mechanisms are absent or if existing institutions lose their strength, workers' well-being may decline, or political agitations may occur with adverse effects on societies. A study using panel data for 100 developed and developing countries for 1985–2002 shows that collective labour rights (as measured by the collective labour rights index) are linked to lower income inequality.[37] Therefore, in the interest of the broader health of society, all stakeholders should work towards strengthening institutions that uphold workers' well-being.

Extending social protection

Social protection is a more comprehensive concept than social security, social assistance or a social safety net—it combines all three systems and is critical for increasing workers' well-being and enhancing the choices people have in their work lives. Today, only 27 percent of the world's population is covered by a comprehensive social protection system. In other words 73 percent, about 5.2 billion people, do not have access to comprehensive social protection.[38] Most middle-income and some lower income countries have expanded their systems in recent years, although some of these advances are not fully legislated or do not have secured funding. In developed countries, despite growing demands for social protection in recent years, the 2008 financial crisis brought widespread tightening of social protection in its wake. For example, older people now get smaller pensions in at least 14 European countries.[39]

Moving towards greater social protection

Some policy options to expand social protection and enhance the links between work and human development include:

- *Pursuing well designed, appropriately targeted and well implemented social protection programmes.* A basic and modest set of social security guarantees through social transfers in cash and in kind can reasonably be provided for all citizens. The cost of setting such a floor with universal pension, basic health care, child benefits and employment schemes would range from about 4 percent of GDP in India to 11 percent of GDP in Burkina Faso.[40] Resources can be mobilized by adopting progressive taxes, restructuring expenditures, removing subsidies and extending contributory schemes by capturing

Today, only 27 percent of the world's population is covered by a comprehensive social protection system

more of the working population in the formal sector.

- *Combining social protection with appropriate work strategies.* Programmes would provide work to poor people while serving as a social safety net. Even though social protection may present disincentives for paid work, the consensus seems to be that social protection measures in themselves do not necessarily cause unemployment to increase —and they can, especially when combined with labour market policies, facilitate job creation. Creating work, reducing poverty and protecting people against shocks is an ideal outcome, as exemplified in the Rural Employment Opportunities for Public Assets in Bangladesh (box 6.9). Linking social protection (particularly unemployment benefits) to training and job search assistance also helps unemployed workers reintegrate into the labour force.

  Two types of actions are thus in order: compensatory measures in social protection and corrective measures in the labour market. Improving other labour market conditions for workers, participation and salary (including gaps for disadvantaged groups such as young and female workers) is vital for shaping future social protection systems.

- *Providing a living income.* A living income would provide a basic minimum income for all (a citizen's income), independent of the job market. The idea of a living income becomes more relevant in the current changing world of work, as automation may put many workers out of jobs, because of the changing nature of jobs and because many jobs may be at risk of disappearing (up to 50 percent of existing jobs may be at risk in the next 20 years).[41] A way forward, and one that would increase the ability of individuals to concentrate their time in forms of unpaid work that can be human development enhancing, would be to provide all citizens with an unconditional tax credit, which could be built up as the rewards from paid work fall. Two objections may be raised with regard to the notion of a living income—societies are too poor to afford it and it would be a disincentive to work. The first objection is not true for advanced economies, and the second one is irrelevant because the objective is not to enhance the incentive for paid work, but to enable people to live without paid work.

- *Tailoring successful social protection programmes to local contexts.* Programmes for cash transfers or conditional cash transfers have provided effective social protection, particularly in Latin America (for example, Bolsa Família in Brazil and Oportunidades, now called Prospera, in Mexico), and many have been replicated in other parts of the world (Sub-Saharan Africa). These programmes provide income support for poor families and build human capabilities by increasing funds for schooling and health

BOX 6.9

Rural Employment Opportunities for Public Assets in Bangladesh

Poverty affects millions of Bangladeshis, mostly in rural areas and households headed by women. Rural Employment Opportunities for Public Assets—a project supporting female-headed households—was launched in six food-insecure districts, with the support of Union Parishads, the lowest level of local government.

The project provided two years of employment for destitute women and employment for casual labourers during a lean period, offering them a safety net. Women also undertook training on social and legal issues, gender equality, human rights, primary health care, nutrition and income generation. Between 2008 and 2012, 25,000 women were employed for two years, and 500,000 work days for casual workers were created. More than 11,000 women received training.

From 2009 to 2012, 12,000 kilometres of vital rural eastern roads were repaired by women in crew groups in maintenance and post-flood repairs. Out of an individual daily wage of Taka 100 (around US$1.25), each woman makes a mandatory saving of Taka 30. During 2010 and 2011 each woman earned Taka 51,100 in cash and saved Taka 21,900, deposited in a local savings bank account under the project. In addition each woman received Taka 5,200 as an end of project bonus, to leave the project with total savings of Taka 27,100.

Source: EuropeAid 2012.

care for children. Conditional cash transfers in different forms have also been useful in addressing crisis-related labour market risks, as with Nicaragua's coffee price crisis.

- *Undertaking direct employment guarantee programmes.* Instead of cash transfers or conditional cash transfers, countries have also pursued employment guarantees. Jefes de Hogar in Argentina and the regional Karnali Employment Programme in Nepal are examples,[42] though the best known is the National Rural Employment Guarantee Scheme in India (box 6.10).

Targeting interventions for older people

Strengthening social protection for older people—particularly to provide the choice of retirement or cutting back on paid work—must be a priority. Policy choices include:

- *Expanding noncontributory basic social pensions systems.* Between 1990 and 2011 the number of countries adopting noncontributory basic and targeted pension systems more than doubled—from 10 to 21 for basic systems and from 20 to 46 for targeted ones.[43] Noncontributory regimes to minimize poverty for older people are only a first step: Programmes have to be well financed and endowed with regulations and institutions that enable efficient resource management.
- *Exploring fully funded contributory pension systems.* This modality—where pensions

draw on previous savings—has re-emerged in recent times. In 2011, 34 developing countries had such programmes, up from 5 in 1990, following the first programme launched in Chile in 1981.[44] After two decades challenges emerged in Chile, and the country introduced a comprehensive reform based on a new "solidarity pillar" (box 6.11). A few countries (Argentina and Bolivia) have also made radical changes in their systems.[45]

- *Financing social pensions for older people.* Such financing can be achieved through more contributions to pension systems and substantial improvements in the systems at reasonable cost, both of which are realistic. For example, in Latin America a noncontributory system equal to 10 percent of GDP per capita (eliminating poverty among older people), plus subsidies to supplement the contributory pillar, would cost an estimated 1.4–2.5 percent of GDP.[46]

Addressing inequalities

In the changing world of work, sustained human progress requires structural challenges, such as inequalities, to be tackled. As seen in chapter 3, workers are getting a smaller share of total income globally, while the proportion of returns to capital has risen sharply. There are also substantial inequalities in opportunities, as highlighted in chapter 1. Work seems to encompass myriad inequalities that may

> Strengthening social protection for older people must be a priority

---

BOX 6.10

National Rural Employment Guarantee Scheme in India—a milestone

The Mahatma Gandhi National Rural Employment Guarantee Act was mandated in 2005 to implement an ambitious, demand-driven employment-creation programme to benefit the rural poor through projects that improve agricultural productivity and alleviate land degradation. It guarantees rural households 100 days of unskilled manual work.

As the programme began to revamp prior employment programmes, evaluations have found that job creation accelerated from less than 1 billion working days among 20 million households in the act's first year of operation, 2006/2007, to 2.5 billion among 50 million households in 2010/2011. A simulation estimated that GDP would increase 0.02–0.03 percent, that labour

income would rise about 700 million rupees and that the welfare of the poorest households (as measured by Slutski-adjusted consumption relative to initial consumption) would increase up to 8 percent. People belonging to Scheduled Tribes or Scheduled Castes would also benefit.

Yet in evaluations, the programme's impacts are found to be asymmetrical between urban and rural dwellers, men and women, better-off and privileged population groups and more educated versus less educated groups. Its size has no precedent nationally or internationally, posing important design and management challenges.

Source: Zepeda and others 2013.

BOX 6.11

Chile's reforms to the reform: more solidarity, more contributions

In 1981 Chile pioneered fully funded contributory pension systems. But after two decades some problems became evident from the beneficiary side—too few contributors, a low level and density of contributions and wide gender imbalances. In response, the government adopted a comprehensive reform in 2008.

Some of the main points of the new architecture were a universal basic pension for those without substantial contributions (in time) and without large enough accumulation of resources at age 65 (extended to people with disabilities); a social security solidarity contribution that complements pension savings for those contributing to the system (for older people and people with disabilities); mandatory contributions from self-employed workers (to be completed by 2018),

enforced through the tax system; stronger mechanisms for complementary voluntary savings; and subsidies to pension contributions from young workers (to incentivize early participation in the system) and to young workers' employers (to incentivize hiring of young workers), targeting workers ages 18–35 with low salaries.

In 2009 an entitlement for women was added, to compensate for disparities in the labour market: Every woman will receive a bond for each child she has, deposited in her pension account, equivalent to 18 months of contributions based on the minimum wage. The public cost of the reformed system was estimated at 2.5 percent of GDP a year. In 2015 an International Presidential Commission presented an assessment of the system and proposals to address its weaknesses.

Source: Bosh, Melguizo and Pages 2013; Uthoff 2015.

---

be exacerbated without policy interventions. Widening inequalities are a threat to human development because they result in disparities in capabilities and choices.

Inequalities in outcome and opportunities could be reduced by focusing on the following:

- *Formulating and implementing pro-poor employment strategies.* A key problem is that inequality reproduces itself; it is thus important that employment strategies be pro-poor. Options include the creation of work in sectors where most poor people work; better access for poor households to such basic social services as health, education, safe water and sanitation; and access to such productive resources as inputs, credit and finance. Subsidies, targeted expenditures and pricing mechanisms can also be considered.
- *Providing complementary support.* This would come through marketing facilities, investments in physical infrastructure (particularly in rural areas, such as feeder roads), extension services and labour-intensive technologies. With the right incentives, the private sector can be induced to play a major role in building and running physical infrastructure. For example, in Brazil a $65 billion infrastructure package (about 3.5 percent of GDP) used concessions to the private sector to build 7,000 kilometres of highways, railways and ports, among other things.[47] Such

investments will immediately create work for low-skilled workers, with effects on poverty and inequality.

- *Regulating the financial sector to reduce the regressive effects of cycles.* For example, increases in physical investment produce sustained development for firms and workers, while increases in financial investment may be less stable and less likely to generate jobs. A more stable macroeconomic environment can favour a "productivist" rather than a "financierist" allocation of wealth-promoting investments in the real economy and create jobs.[48]
- *Removing asymmetries between the mobility of labour and of capital.* Labour mobility does not match that of capital, given intrinsic differences, but also as a matter of policy: Industrial countries promote capital mobility but discourage labour mobility. Nonetheless, regulating capital movement can reduce macroeconomic instability and middle-income traps in developing countries, as capital moves overseas when wages become too high. Migration policies discussed earlier in the chapter can at a minimum reduce the risks of migration.

Inequality also has a political dimension. Addressing inequality implies that existing shares of the fruits of development need to be rebalanced. Resistance from those who benefit

In the changing world of work, sustained human progress requires structural challenges, such as inequalities, to be tackled

from the current pattern of distribution is inevitable. But allowing inequality to grow could threaten the status quo of the system itself. Therefore, some forms of democratization, sharing and distributive policies in terms of transfers have to be pursued.

- *Democratizing education, particularly at the tertiary level, nationally and globally.* Countries place a high premium on tertiary education, which is not always evenly distributed among populations.[49] This is apparent within countries (as most workers with a tertiary education come from higher income families) and between countries (as the countries with greater levels and increases in tertiary education are industrial). In a world that demands skills for work, inequalities in tertiary education can reinforce inequalities in access to work and the related social and economic benefits. Montenegrins understand the need for equal access to tertiary education and support policies as such (box 6.12).
- *Pursuing profit sharing and employee ownership.* Profit sharing with labour and granting employees part ownership of enterprises may reduce inequality in income shares. Apart from reducing inequality, these programmes, combined with increased training and job security, may raise productivity and benefit workers.[50] Even so, profit sharing is the exception and not the rule. In the European Union fewer than 30 percent of companies have profit sharing, and fewer than 10 percent of workers own their own company's stocks.[51]
- *Adopting and enforcing distributive policies.* These could include progressive taxes on income and wealth, regulations to reduce rent extraction, stricter regulation (particularly of finance) and targeted public spending on the poor. In many countries (Senegal, for example) school lunch plans have helped deal with hunger and nutrition of children from poorer households and reduced some disparities in health outcomes.[52] Cash transfer programmes to poorer households (in South Africa, for example) have helped poorer households.[53] Conditional cash transfer programs have had a considerable positive impact on poverty and inequality (box 6.13). In developed countries better regulation of the financial sector would help narrow income inequality.

A 15-step agenda has been suggested to reduce inequality, some steps of which are captured in box 6.14.

Profit sharing with labour and granting employees part ownership of enterprises may reduce inequality in income shares

Strategies for targeted action

Strategies for targeted action are needed to complement strategies for employment creation

---

BOX 6.12

How Montenegrins value tertiary education

The density of universities in Montenegro has become comparable to or even higher than other countries in the region. Results of a 2011 United Nations Development Programme survey of citizens confirmed that Montenegrins highly value the completion of higher education. When asked, "In your opinion, what type of education would be the most appropriate for your child/grandchild?", about 60 percent of respondents considered higher education the most appropriate and 32 percent postgraduate and doctoral programmes.

The government emphasizes higher education in developing human capital. The overall budget for the sector has increased, and education strategies based on the premise of equal rights and the rights of students are being implemented. The fundamental goal of these measures is to ensure that the education system contributes to the creation of knowledge for personal and social development—that is, developing the skills necessary for economic progress and active participation in the democratic political community and for success in the world of fast, continuous and global changes.

What type of education is most appropriate?

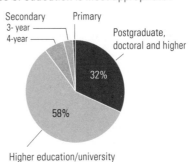

Source: UNDP 2013c.

BOX 6.13

Bolsa Família—Brazil's conditional cash transfer programme

Bolsa Família is one of the world's largest cash transfer programmes, administering benefits to 13.8 million households. It covered 26 percent of the population across all its municipalities in 2012 with a budget of $10.75 billion—0.53 percent of GDP—in 2013. It is fully financed from general government revenues through the social security budget.[1]

The programme's main goals are to reduce poverty, promote food security, break down the intergenerational cycle of poverty through human capital accumulation and increase access to public services, especially health, education and social assistance. Women are the programme recipients and are responsible for complying with the conditions: regular visits to health clinics, attendance of pregnant or breastfeeding women in scheduled prenatal and postnatal visits, attendance at educational activities on health and nutrition, full vaccination of children under age 7 and routine health checkups and growth monitoring.

Since the programme launched, 5 million Brazilians have exited extreme poverty, and by 2009 the programme had reduced the poverty rate by an estimated 8 percentage points.[2] The programme is credited with increasing enrolment 5.5 percentage points in grades 1–4 and 6.5 percentage points in grades 5–8, though its effect on dropout rates has not been as good. Over 2001–2009 Brazil's mean years of schooling rose from 6.8 to 8.3, while the Gini index of years of schooling fell from 0.347 to 0.288.[3]

Despite initial concerns that cash transfers to poor households could risk a decline in their labour supply and employment rate, the experience has been more encouraging. Bolsa Família has enabled an increase in the employment rate of the economically active population, a reduction of inactivity and informality rates, an increase in the proportion of workers contributing to social security and an increase in the average hourly wage for the primary occupation.[4]

Notes
1. Fultz and Francis 2013. 2. Soares 2012. 3. Glewwe and Kassouf 2008. 4. Machado and others 2011.
Source: Human Development Report Office.

BOX 6.14

Suggested measures for reducing inequalities

- Government should be more concerned with monopolies and competition policy.
- Trade unions should be bolstered to represent workers' interests.
- Government should provide public sector jobs at minimum wage to those who want them, in areas such as meals-on-wheels, care for older people, child care and the like.
- In addition to a minimum wage, there should be a framework to restrain pay at the highest levels. Some companies have voluntarily decreed that executive pay should be capped at 65 or 75 times the average pay in the firm.
- Personal income taxes should be made more progressive, with a maximum rate of 65 percent.
- Every child should get a "child benefit" payment, to help keep them out of poverty.

Source: Atkinson 2015.

and workers' well-being. Specific issue-based areas (such as sustainable work), specific groups (such as young people) and particular situations (such as conflict or post-conflict conditions) require special policy attention to enhance links between work and human development. Some of these issues (such as youth employment) may benefit from general policy actions (such as national employment strategies), but given the specific nature of some challenges, targeted actions are also required.

Reducing gender divides at work

> Gender divides at work—both paid and unpaid—are a manifestation of longstanding patterns of inequities

Gender divides at work—both paid and unpaid—are a manifestation of longstanding patterns of inequities. They can reinforce each other, trapping women and girls across generations in realms of limited choices and opportunities. Any re-balancing of work needs a coherent and simultaneous effort across many different dimensions. Policies that expand opportunities for women to engage in paid work, that enhance outcomes at work, that balance care work within households and that change gender norms regarding work can help reduce work-related gender inequalities. Policies could move in the following directions:

- *Expanding and strengthening gender-sensitive policies for female wage employment.*

Such strategies include improving access to higher education, particularly in math and sciences, for women to enhance skills, along with training that matches market demands. This could be accompanied by access to continuing professional development. Figure 6.4 shows a shallow U-shaped relationship for women between mean years of schooling and labour force participation. With few years of schooling, participation rates are high, likely because poor women with low education have to work to earn a living. It goes down as schooling years increase, but ultimately picks up with further increases in schooling.

- *Specific interventions.* Women can face harassment in the workplace and discrimination in hiring and compensation, access to finance and access to technology. Legislative measures are needed to reduce these inequalities in opportunity and to ensure that women and men receive equal pay for equal work. Proactive recruitment policies that encourage hiring female candidates in jobs, as well as programmes that reduce barriers to credit, finance and training, can increase opportunities in paid work. Reducing barriers to entrepreneurship can increase self-employment options. Regulations against workplace harassment are also imperative for workplace equality.

- *Focusing on both maternal and paternal parental leave.* Much can be achieved to balance the distribution of unpaid care work and to reduce wage gaps in paid work when fathers are included in generous parental leave policies—and even more when they have incentives to make use of them (box 6.15).

- *Enlarging care options, including day-care centres, after-school programmes, senior citizens' homes and long-term care facilities.* Employers can also offer child care onsite. Another alternative is to subsidize care work through instruments such as vouchers and tickets (box 6.16). The public provisioning of early childhood education can reduce care responsibilities as well as improve education and work outcomes later in the lifecycle.[54] Governments can require employers to offer leave for long-term care, as in Germany, where since 2015, employees have been able to take 10 days of leave to care for acutely ill family members, paid for by the social security system. If a family member needs longer term care, the employee can take more leave or reduce his or her working hours for up to six months. During that time he or she can obtain interest-free credit subsidized by the government. In some cases of hardship the debt can be cancelled. If the employee needs even more leave, he or she can reduce working time to 15 hours a week for up to

Women can face harassment in the workplace and discrimination in hiring and compensation, access to finance and access to technology

FIGURE 6.4

The relationship between mean years of schooling and labour force participation for women shows a shallow U shape

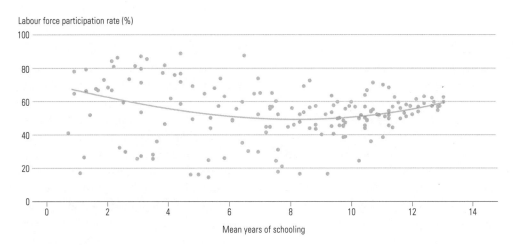

Labour force participation rate (%)

Mean years of schooling

Source: Human Development Report Office, calculations based on data from ILO (2015f) and UNESCO Institute for Statistics (2015).

Positive parental leave systems

For 40 years Sweden has had gender-neutral parental leave policies. Each parent can take at least two months of paid leave. Initially, few men made use of it, so they have been offered an incentive since 2002. If both parents take two months, the family gets one extra month of paid leave—an equality bonus. Now, 90 percent of Swedish fathers take paternity leave. They thus carry a greater proportion of childcare work and engage more in child-raising long after their paternity leave has ended. Perhaps not coincidentally, Sweden has one of the narrowest wage gaps in the world and one of the highest scores for female happiness.[1]

Similarly, a study on the parental insurance plan in Quebec, Canada, showed that the share of Québécois fathers taking paternity leave increased from about 10 percent in 2001 to more than 80 percent in 2010 and that fathers who took advantage of the leave devoted 23 percent more of their time to household chores—even one to three years after their leave ended.[2]

Notes
1. *The Economist* 2014b. 2. Patnaik 2015.
Source: Human Development Report Office.

Cash for care work

Governments can assist households by offering cash payments. Since the mid-1990s the Netherlands has had cash-for-care schemes. The benefits are based on a needs and income assessment, but the average is around €14,500 per person a year. There are few restrictions on how it can be used. Administrative costs are low, and evaluations have indicated that the cash is used effectively. Households consider the benefits adequate, and the scheme has proved popular.

Israel has a similar scheme. In 2008 the government introduced a pilot programme and in 2010 expanded it to cover 15 percent of the country. To be eligible for the cash benefit, an individual must receive medium- or high-intensity care by a caregiver who is not a family member. However, uptake is low and varies according to age, income and benefit level. Recipients are satisfied with the cash scheme but appear to benefit less than those receiving in-kind benefits.

In France beneficiaries can pay for long-term care services or directly hire a caregiver using the Chèque emploi services universel programme. They can then seek reimbursement from an accredited national organization. This system has the advantages of being transparent and of optimizing public expenditure.

Source: Colombo and others 2015.

two years, while retaining the right to return to the activities and hours covered.[55]

- *Pursuing proactive measures for increased representation of women in senior decision-making positions.* Proactive measures in human resource policies, gender requirements in selection and recruitment, and incentive mechanisms for retention could enhance women's representation in the public and private sectors. Policy priority should also consolidate progress on breaking the glass ceiling. There should be identical criteria for men and women for moving people into senior management positions, and it should be free of sex bias and based on equal pay for equal work. Women's representation can be increased through affirmative action measures. For example, introducing quotas for women on corporate boards, an initiative increasingly seen in the European Union.[56] Such efforts are even more effective when accompanied by policies that increase retention rates. Mentoring, coaching and sponsoring can empower women in the workplace—for example, by using successful senior female managers as role models and as sponsors.[57] All these approaches can help change norms and promote women to positions of seniority, responsibility and decisionmaking. A complementary approach for changing

Policy priority should also consolidate progress on breaking the glass ceiling

norms is to encourage the engagement of men in professions that have traditionally been dominated by women.

- *Encouraging flexible working arrangements, including telecommuting.* There should be sufficient incentives for women to return to work after giving birth. These may encompass the reservation of jobs for women on maternity leave for up to a year. Women could also be offered salary increases to return to work. Flexible work arrangements such as telecommuting or flexible hours can also allow women and men to balance paid and unpaid work.
- *Improving public services and infrastructures to reduce care work burdens.* Much time is spent in care work collecting water and fuel, cooking and performing other such tasks. Improved access to clean water and sanitation, energy services and public infrastructures, including for transportation, can greatly reduce the care work burden for families, making more time for paid work.
- *Valuing unpaid care work.* Such a valuation is not just cosmetic; it helps raise policy awareness, with implications for exploring options for remunerating such work. Valuation could also prompt the redistribution of care work between women and men. Various statistical methods (as described in

box 4.1 in chapter 4) can be used. But better methods for data collection will be needed as well.

- *Gathering better data on paid and unpaid work.* National statistical systems, using more female investigators and appropriate samples and questionnaires, can gather better data on unpaid work. With data, policies can be pursued that help balance paid and unpaid work—Norway has done it very successfully (box 6.17).

Moving towards sustainable work

It is possible to terminate, transform and create work in ways that will advance both human development and environmental sustainability. However, for this to occur coherence is needed in policies and actions across global, national and subnational levels, which can ensure that the most appropriate combination of skills, technologies, investments, regulations and social interventions are put into place.

There is a crisis in sustainability locally and globally, but there is also an opportunity for work to help move humanity to a more enduring and sustainable human development path. However, the timeframe within which key steps must be undertaken is tight, with delays making adverse impacts more likely.

Valuing unpaid care work is not just cosmetic; it helps raise policy awareness

BOX 6.17

Gender policies in Norway

Between 1970 and 2010 strong gender policies in Norway helped increase women's paid work and reduce their unpaid work. The government had already introduced paid maternity leave in 1956 but reformed the policy in 1993 to allow 49 weeks of fully paid parental leave for mothers and fathers. A series of legislative changes gradually increased the number of weeks of parental leave that are available only to the father, under the "paternal quota." A quota of four weeks was introduced in 1993 and extended (in stages) to 10 by 2009. This stimulated a huge increase in the share of fathers taking at least eight weeks of leave, from 8 percent in 1996 to 41 percent in 2010.[1]

Another important landmark was the 1979 Gender Equality Act. This prohibited any form of gender-based discrimination at the workplace, related to pregnancy,

birth or leave entitlements. It also stipulated that all public bodies should aim for gender equality when appointing members for councils, boards and committees. Between 2004 and 2006 rules were introduced governing the gender balance on the boards of publicly and privately owned public limited companies. In 2007 Norway implemented a legal entitlement to early child care, paid for partly by parents but with a maximum contribution of €300 per month. According to the 2014 World Economic Forum Global Gender Gap report, women's wages come closest to men's in Norway and Singapore—but women's wages are still only 80 percent of men's.[2] These policies have helped mitigate work–family trade-offs. They have also enlarged women's freedom of choice on family size and helped boost the birth rate.

Notes
1. Esther, Javorcik and Ulltveit-Moe 2015. 2. WEF 2014.
Source: Finland Ministry of Social Affairs and Health 2009.

Developing capabilities and skills

Moving towards sustainable work, as described in chapter 5, will require various transformations, including what is produced, how it is produced and where it is produced. A particularly important transformation is in the skill sets required for these shifts (for example, producing solar energy technicians in Nepal).

- *Identifying current and future skill needs.* Skill needs reflect both the current and anticipated needs of societies required, for example, to adopt more efficient and cleaner technologies in the immediate future, as well as the skills to support continuing innovation that will maintain the movement of work towards sustainability. Global data on current skill levels are not readily available, but their distribution is believed to be markedly uneven and may not correspond to those needed for sustainable work.
- *Developing skills to make transitions to sustainable work.* These would include technical and scientific skills that will enable the development, adaptation, installation and maintenance of sustainable solutions as well as literacy, numeracy, employability and entrepreneurship, enabling workers to learn and adopt new ways of working; and communication, training and education for all workers, facilitating the learning process.

Social sector policies must therefore incorporate this understanding of the dynamics of skill formation over the lifecycle and of the importance of education in children's early years for overcoming inequalities and producing marketable skills for the workforce. Private sector initiatives can also help (box 6.18).

- *Ensuring coherence and timeliness.* In developing skills for facilitating the move to sustainable work, there should be coherence and timeliness in the transition of skills mentioned above. Unless such measures are effective, the transition to sustainable work will face considerable practical, social and political hurdles.

Adopting different technologies and new investments

Adopting different technologies and new investments has to be contextualized in country situations. For example, poor countries, with weak infrastructures, low capabilities and inadequate resources for investments, will need support from the global community (for example, to increase technology transfer). But indigenous knowledge on many low-cost adaptive technologies exists, and many alternative technologies are in the public domain. South–South collaboration on lessons from various

---

BOX 6.18

Turkey's private sector initiative on overcoming skills mismatch at the local level

The Bursa Chamber of Commerce and Industry Education Foundation is a private sector–led and –owned initiative in Bursa, one of the most industrially advanced and trade-oriented cities in Turkey. Since 2009 the foundation has contributed to overcoming skills mismatches at the local level. It operates as a nonprofit vocational and technical training centre, with the mutually reinforcing goals of providing skilled human resources to industry and increasing employability of marginalized groups, particularly unemployed young people. The Bursa Chamber of Commerce and Industry is the owner of the foundation and funds all the activities of the training centre.

The overall goals of the initiative, along with the course and trainee selection process, take into account the skills needs of local sectors and insights from private sector leaders and industrialists in Bursa. Almost all the training and services offered are free. Private sector leadership, ownership and engagement have enabled the foundation to deliver market-relevant vocational and technical skills for disadvantaged people and to help direct them towards employment. The foundation has close and continuous ties to local industrialists, which allows for state-of-the-art training modalities and facilitates job placement for the trainees. It has the facilities and infrastructure to train more than 3,000 young people a year in a wide variety of areas, from textiles and mechatronics to automotive metals and hardware. The employment rate for successful trainees stands at 80 percent.

Source: UNDP 2014a.

experiments can help countries that are in need by spreading new ideas, scaling up implementation and encouraging replication.

- *Departing from business as usual.* Several alternative technologies are already in the public domain, such as climate-resistant crop varieties developed by public research institutions. Some are part of local indigenous knowledge systems (low-cost housing materials) or have been developed by practitioners and nongovernmental organizations (efficient cook stoves). In these cases the predominant challenges are identifying the technologies, adapting them to local contexts (if necessary) and scaling up their use.
- *Pursuing technology transfer.* As with renewable energy generation (hydro, solar and wind), adoption may depend on technology transfer, matched by greater investment. The average annual growth rate of renewable energy (including hydro, solar and wind) over 2010–2012 was 4 percent—too slow to reach the targets of Sustainable Energy for All by 2030. The annual growth rate should speed up to around 7.5 percent, and current annual investments of around $400 billion would need to triple to achieve 2030's targets.[58]
- *Leapfrogging to more sustainable work.* If investments are made in the most advanced and appropriate technologies, many countries may be able to leapfrog to more sustainable work—attaining and securing major gains in human development. While several sources of investment may be available—domestic resources, foreign direct investment, multilateral development banks—appropriate arrangements may also be needed for transferring technology through licensing or other arrangements.

International precedents exist for such transfers: Article 66.2 of the World Trade Organization's Agreement on Trade-Related Aspects of Intellectual Property Rights (1994) explicitly provides for an agreed-on mechanism to facilitate technology transfer to less developed countries.[59] In practice, the Montreal Protocol on substances that deplete the ozone layer (1989), the Clean Development Mechanism of the Kyoto Protocol (1997) and the technology mechanism of the United Nations Framework Convention on Climate Change (2012) have

all, to a greater or lesser extent, facilitated technology transfer. (The Rio + 20 outcome document recommended a "technology facilitation mechanism," which is also included in the Sustainable Development Goals.)

Incentivizing individual action, managing trade-offs and guarding against inequalities

Some of the solutions to advance sustainability will come from recognizing the positive externalities in people's work and incentivizing such actions (box 6.19). Others will require appropriate regulatory and macroeconomic policies to nudge actors in the right direction.

- *Pursue public policies for managing trade-offs.* Some workers will lose their jobs due to an end of activities in their industry or sector. Analysis of 21 country case studies identified the industries likely to expand under sustainability efforts—renewable energy, green buildings and retrofitting, transport, recycling, and waste and water management —and the industries likely to contract or substantially change, including agriculture and forestry, fisheries, extractive industries and fossil fuel generation, emission-intensive manufacturing, automotive production, shipbuilding and cement making.[60]

Another trade-off example comes from Sub-Saharan Africa, where rural poverty is widespread and much of the labour force lives in rural areas. Increasing farm productivity there has the potential to drive greater economic growth and poverty reduction, accelerating food security and human development. For example, since 2000 rural poverty in Ghana has fallen sharply, enabling it to become the first country in the region to meet Millennium Development Goal 1. Progress was driven in part by expanding the area under cocoa cultivation on labour-intensive smallholder farms.[61] But expansion and extensive cultivation have increased environmental stress. Trade-offs between the different and sometimes competing goals of sustainability and employment have to be reconciled to support sustainable work.

- *Implementing standards.* On promoting sustainable work, box 5.2 in chapter 5 discusses how implementing standards, regulation

> Some of the solutions to advance sustainability will come from recognizing the positive externalities in people's work

BOX 6.19

The "social wage" of work

A social wage compensates the effort by the worker in terms of the value of that effort to society

Helping conserve the environment or otherwise promote sustainability delivers benefits to society, and to future generations, that go far beyond the immediate benefit to the individual. This is a characteristic of several other forms of work as well, including some care work. The social value of these goods and services may diverge from their private (market) value—and will be underprovided in free market conditions.

This is especially relevant when the number or quality of workers engaged in the socially beneficial activity is inadequate to deliver the socially optimal level of the good or service. This can happen, for example, with a highly qualified worker who has a high reservation wage in an alternative occupation: however, the low market valuation of the more socially useful product restricts the amount he or she can be paid, and consequent low wages reduce the number or quality of people engaged in the activity.

This may—paradoxically—help attract highly qualified individuals who are volunteers (or part volunteers, accepting wages that are below what the private market would offer them) and are willing to work without pay because they believe strongly in the outcomes. Their commitment and abilities are important in correcting undersupply, as well as in motivating and raising public awareness so that ultimately the political system can take corrective action.

One approach could be to offer a social wage—a wage that compensates the effort by the worker in terms of the value of that effort to society. Incentives that correspond to this already exist in some areas, like hazard pay or a temporary promotion to public sector workers in conflict situations or dealing with an epidemic. A similar justification could be made for properly targeted subsidies or other incentives to promote, for example, more publicly available research on sustainability.

Source: Human Development Report Office.

and appropriate policy measures can make a difference in the context of the ship-breaking industry. It also rightly emphasizes the difficulties in implementing such standards, which are nonetheless imperative for quality sustainable work.

- *Addressing intergenerational inequalities.* The people most able to develop better remunerated skills for sustainable work are those who have the time and resources to access quality education and nutrition—the children of those at the upper end of the income and wealth distributions. Should the differential in earnings of these individuals (relative to lower skilled workers) be wide enough, their children in turn would be more likely to grasp the opportunity to develop more highly remunerated skills. Policies must level a playing field that is getting steeper over time, so that inequality in human development is not perpetuated across generations.

- *Manage and facilitate change.* Because moving to sustainable work will lead to jobs being lost, transformed (the majority) and created, public policies will be crucial in managing and facilitating change. Multiple stakeholders will have to work collaboratively—and

globally. And a mechanism is needed to translate the desired global outcomes into country actions (box 6.20).

Undertaking group-specific interventions

For some groups and situations, focused policy interventions will be needed because overall policy interventions may not adequately address specific challenges. Youth unemployment, older workers, work in conflict or post-conflict situations, and creative and voluntary work are some of the cases that may require special attention.

Youth unemployment

Earlier policy options, particularly for the changing world of work, relate to education and skills to prepare people for the future. These policies are especially relevant for addressing youth unemployment. But given the frequent severity of this challenge and its multi-dimensional (economic, social and political) impacts, it also needs targeted interventions. Exciting work opportunities for young people should be created so that they can unbridle

BOX 6.20

Possible measures at the country level for moving towards sustainable work

- Identifying appropriate technologies and investment options, including leapfrogging opportunities.
- Setting up regulatory and macroeconomic frameworks to facilitate adoption of sustainable policies.
- Ensuring that the population has the appropriate skills base—combining technical and high-quality skills with core abilities for learning, employability and communicating.
- Retraining and skills upgrading of large numbers of workers in informal sectors, such as agriculture. While some workers may be reached through the market, others will need the help of the public sector, nongovernmental organizations and others. These

opportunities can be a means to support women and other traditionally disadvantaged groups.
- Managing the adverse impacts of the transition through diversified packages of support and keeping the playing field level for breaking the transmission of intergenerational inequality.
- Continuing to build the skill base of the population. This will require a lifecycle approach that recognizes the cumulative nature of interventions that lead to learning. Large investments in the number and quality of health and education workers will be necessary, underscoring the continuing role of the public sector in transforming skills.

Source: Human Development Report Office.

their creativity, innovation and entrepreneurship in the new world of work.

- *Providing policy support to the sectors and entities that are creating new lines of work.* Such initiatives are ongoing, and new opportunities are being discovered every day, but they need policy support.
- *Investing in skills development, creativity and problem solving.* Special support should be extended to young women and men in apprenticeships, trade and vocational training, and on-the-job learning.
- *Providing supportive government policies to help young entrepreneurs.* Areas include advisory services for establishing businesses and initiatives, as well as better instruments and channels for financing. For example, in developing countries microfinance can provide small-scale community financing for youth; in more developed economies, particularly in information technology, venture capital can support start-ups and larger initiatives. More recently, crowdsourcing has emerged as an option to generate funds for small initiatives.[62]
- *Making tertiary learning more widely available through the Internet.* Massive open online courses are gaining followers among world-renowned academic institutions and among students. These Internet-driven advances are important to accelerate higher education, but interpersonal contacts as well as group work and problem solving are still necessary for gainful learning.

- *Using cash transfer programmes to provide employment for local young people and poor people.* In India and Uganda these programmes have provided resources for funding job searches and for supporting high-quality training and skills development.[63] They have also increased access to other sources of credit for entrepreneurship.

Older workers

Older people may want to continue working for two reasons: They want to remain active and engaged with their work or they cannot afford to retire. Older workers who leave paid employment can devote more time to care and volunteer work. They can thus contribute to society while maintaining a sense of social inclusion. In countries affected by HIV/AIDS, grandparents have served as foster parents to children orphaned by the disease.[64] Older relatives can also care for children whose parents have migrated for work. Nearly half the world's population will not receive a pension, and for the rest the pension may be inadequate.[65] Unless other family members support them, they must continue working, frequently in the informal sector.

Countries have identified older workers as assets and are making efforts to retain them, sometimes by removing laws on compulsory retirement or by increasing the pension age, as in France, Germany, Italy, Poland, Spain and

Exciting work opportunities for young people should be created

the United Kingdom.[66] This enables people to work longer and cuts the cost of pensions, but it may remove the choice for older people as to when to retire. Older workers benefit from greater access to part-time jobs and from more flexible work arrangements that allow them to ease into retirement. One option for flexibility is self-employment: Google has pledged to help first-time entrepreneurs over age 50.[67]

Work in conflict and post-conflict situations

It is important in conflict and post-conflict situations to focus on productive jobs that empower people, build agency, increase voice, offer social status and increase respect, cohesion, trust and people's willingness to participate in civil society. Such jobs can create economic and social ties and have the potential to build incentives to work across boundaries and resolve conflict. Some policy options are:

- *Creating new forms of work that emerge with the rise of conflict-specific needs.* For example, during combat, care for combatants (such as cooking), health care for the wounded and reconstruction may be extremely useful.
- *Supporting work in the health system may cover many goals.* In many conflict-afflicted countries the health system has collapsed, converting health services into a life-threatening challenge for helpers and the wounded. In this setting, international aid becomes indispensable, but local volunteers contribute substantially to providing crucial health services and saving lives.
- *Getting basic social services up and running.* Economic considerations apart, this has social and political benefits. Communities, nongovernmental organizations and public–private partnerships can be the drivers.
- *Initiating public works programmes.* Even emergency temporary jobs, cash for work and the like can provide much needed livelihoods and contribute to the building of critical physical and social infrastructures.
- *Formulating and implementing targeted community-based programmes.* Such programmes can yield multiple benefits, including stability. Economic activities can be jumpstarted by reconnecting people, reconstructing networks and helping restore the social fabric.

Creative and voluntary work

Creative work requires an enabling work environment, including financial support, and opportunities to collaborate and cross-fertilize ideas. This context is familiar in human development as a prerequisite for expanding human capabilities, including the capacity to be creative and contribute to the common good. Some key requirements for creativity and innovation to thrive are:

- *Innovating inclusively.* Here, new goods and services are developed for or by those living on the lowest incomes or by women, extending creative opportunities to groups that may be underrepresented.
- *Assuring democratic creativity.* Workplaces and online platforms can be organized in ways that encourage innovation at all levels.
- *Funding experimentation and risk.* This entails solving intractable social and environmental problems that may require foundations and public institutions to take funding risks on less proven approaches—for example, by supporting more basic research programmes. This may help innovators and creative workers patent their innovations.
- *Innovating for the public good.* Creativity and innovation can advance many objectives. Policies that direct innovation towards the greater social good, including volunteer work, can enhance human development.
- *Encouraging voluntary work.* This can also be done through various policy instruments. For example, fiscal policy instruments such as tax rebates, subsidies and public grants can help voluntary organizations and their work. At the political level public support to create and protect space for voluntary work can bring social benefits, particularly during emergencies like conflicts and natural disasters.

An agenda for action—three pillars

In addition to the policy options outlined in the three previous sections, a broader agenda for action, one that addresses the shifting global context of work, with three pillars may be pursued for making work enhance human development: a New Social Contract, a Global Deal and the Decent Work Agenda.

This is not an easy agenda. It will require political commitment, endorsement from the political process and support from top leadership. Some of the issues in the proposed action agenda are already part of the Post-2015 Development Agenda and the Sustainable Development Goals (such as the Decent Work Agenda); others have received political support but have not been adopted (such as a new Global Deal). So there is a momentum that needs to be galvanized. Some agenda actions (such as a New Social Contract) will require broad-based social dialogue at both the global and country levels.

Given the changing nature of the world with never-before-seen implications for work and human development, it is imperative that a serious dialogue start on these issues. In that spirit, the Report proposes the following agenda for action.

Developing a New Social Contract

During the 20th century social contracts among the state, employers and employees evolved in the industrial, service and public sectors, primarily in developed countries. This occurred more narrowly, and later, in developing countries, mostly in the formal and public sectors. Schematically, social contracts suggested arrangements whereby the state provided macroeconomic and monetary stability, development of the labour force through education and training (and quite a bit through military service) and legislation of labour rights, in return for negotiated contracts between employers and employees, often unionized, and labour market stability. As part of the implicit contract the state could levy corporate and personal taxes and additional labour charges to fund, initially, education services and some pensions and emergency health care. As societies prospered, services and social protection systems expanded further.

Social contracts served increasing segments of populations in developed countries and formal and public sector workers in developing countries as they decolonized. After the Second World War, when societies sought to rebuild and implement policies of "never again," education and social programmes (particularly social protection with unemployment insurance),

disability and old age pensions all expanded. Work opportunities also moved rapidly away from agriculture to urban centres in the formal economy. And in more recent years paid maternity leave and support for those at risk of marginalization also became part of the social protection system in many countries. Some systems also came to include the self-employed or those working in small businesses, often in the service industry.

In the new rapidly evolving world of work, participants are less likely to have long-term ties to a single employer or to be a member of a trade union. They may engage in crowdwork or freelance work for multiple employers and contractors. The traditional model that resulted in many of the social protection systems in place today is coming under strain as the world of work has become globalized, less organized and unionized, and more atomized. It does not fit the traditional arrangements for protection, and many freelancers are responsible for their own pensions and health care. How then does society fairly mobilize funds to cover a widening population that is not always in work, reach those working outside the formal sector, accommodate new labour market entrants (especially migrants) and cover those unable to find paid work?

Already there have been some initiatives, as with the Freelancers Union in the United States.[68] In Denmark security alongside reskilling and skills upgrading is provided in an increasingly flexible job market (box 6.21). But much more dialogue needs to take place at a much larger scale or countries risk losing the ground that was gained throughout the 20th century in terms of protecting workers and ensuring social welfare. It is time to open dialogue for a New Social Contract that addresses the challenges of the world of work today.

Pursuing a Global Deal

In an era of global production, national policies and social contracts may face additional difficulties. What works at home may not work in a competitive global environment. Further, true globalization rests on the ideas of sharing—we should share the task of finally taking responsibility for a "global working life."

Such moves to a Global Deal would require, apart from ratifying and implementing the

It is time to open dialogue for a New Social Contract that addresses the challenges of the world of work today

BOX 6.21

Flexicurity in Denmark

The Danish labour market has a lot of what is often called "flexicurity": coexistence of flexibility, in the form of low adjustment costs for employers and employees, and security, which is a by-product of Denmark's developed social safety net, ensuring high coverage and replacement rates.

The principal aim of flexicurity is to promote employment security over job security, meaning workers are protected, rather than their jobs. Consequently, employers benefit from all the advantages of a flexible labour force while employees can take comfort in a robust safety social net applied with active labour market policies.

Source: World Bank 2015b.

charters of workers' rights, mobilizing all partners—workers, businesses and governments —around the world, respecting workers' rights in practice and being prepared to negotiate agreements at all levels. This will not require new institutions, merely reoriented tasks and agendas in the strong international forums that the world already has.

A Global Deal can guide governments in implementing policies to meet the needs of their citizens. Without global agreements, national policies may respond to labour demands at home without accounting for externalities. This implies that a global–national compact is also necessary. International conventions such as the International Labour Organization

Convention Concerning Decent Work for Domestic Workers, which entered into force in September 2013, was a ground-breaking agreement that stands to establish global standards for the rights of paid domestic workers worldwide. This kind of agreement offers guiding principles to signatories but leaves space for national governments to implement policies that fit in different national contexts to meet commitments, motivated by global actions, but creating real change in local communities.

Implementing the Decent Work Agenda

The Decent Work Agenda refers to productive work for women and men in conditions of freedom, equity, security and human dignity. It involves opportunities for work that is productive and delivers a fair income; provides security in the workplace and social protection for workers and their families; offers prospects for personal development and social integration; gives people the freedom to express their concerns, to organize and to participate in decisions that affect their lives; and guarantees equal opportunities and equal treatment for all.[69] The four pillars of the agenda are in box 6.22.

The Decent Work Agenda and the human development framework are mutually reinforcing. Decent work enhances human development through each of its pillars. Employment creation and enterprise development provide income and livelihoods to people, crucial instruments for equity, and means

> A Global Deal can guide governments in implementing policies to meet the needs of their citizens

BOX 6.22

Four pillars of the Decent Work Agenda

- *Employment creation and enterprise development.* This requires acknowledging that a principal route out of poverty is jobs and that the economy needs to generate opportunities for investment, entrepreneurship, job creation and sustainable livelihoods.
- *Standards and rights at work.* People need representation opportunities to participate, to voice their views in order to obtain rights, and to earn respect. The International Labour Organization's normative work is key for compliance and measuring progress.

- *Social protection.* Fewer than 10 percent of people in the poorest countries have adequate social security protection. Basic social protection, such as health care and retirement security, is a foundation for participating productively in society and the economy.
- *Governance and social dialogue.* Social dialogue among governments, workers and employers can resolve important economic and social issues, encourage good governance, establish sound labour relations and boost economic and social progress.

Source: ILO 2008b.

for participation while facilitating self-esteem and dignity. Workers' rights help human development by upholding human rights, human freedom and labour standards. Social protection contributes to human development by ensuring safety nets and protecting people from risks and vulnerabilities. And social dialogue helps support human development through broad-based participation, empowerment and social cohesion.

Human development also contributes to the four pillars. Expanding human capabilities through human development enhances opportunities for employment and entrepreneurship. The participation aspect of human development helps enrich social dialogue. Human development also emphasizes the promotion of human rights, which boosts workers' rights and enhances human security. Given all these interlinks, implementing the Decent Work Agenda will help work enhance human development.

Conclusions

The world community has just agreed to a post-2015 development agenda and a set of Sustainable Development Goals. The agenda is global in scope, but its adaptation to national contexts and its implementation are likely to differ among countries. A unifying principle is likely to be "leave no one behind."

It is in this context of new global commitments and change that this report shows how human creativity, ingenuity, innovation and work can expand choices, enhance well-being and ensure freedom for every human being in an equitable and sustainable way—so that human progress, indeed, leaves no one behind.

Implementing the Decent Work Agenda will help work enhance human development

Status of fundamental labour rights conventions

Country	Freedom of association and collective bargaining		Elimination of forced and compulsory labour		Elimination of discrimination in employment and occupation		Abolition of child labour		
	C087: Freedom of Association and Protection of the Right to Organise Convention, 1948	C098: Right to Organise and Collective Bargaining Convention, 1949	C029: Forced Labour Convention, 1930	C105: Abolition of Forced Labour Convention, 1957	C100: Equal Remuneration Convention, 1951	C111: Discrimination (Employment and Occupation) Convention, 1958	C138: Minimum Age Convention, 1973		C182: Worst Forms of Child Labour Convention, 1999
	Entry into force: 4 July 1950	Entry into force: 18 July 1951	Entry into force: 1 May 1932	Entry into force: 17 January 1959	Entry into force: 23 May 1953	Entry into force: 15 June 1960	Entry into force: 19 June 1976		Entry into force: 19 November 2000
	Year of ratification	Year of ratification	Year of ratification	Year of ratification	Year of ratification	Year of ratification	Year of ratification	Age	Year of ratification
Afghanistan			1963		1969	1969	2010	14 years	2010
Albania	1957	1957	1957	1997	1957	1997	1998	16 years	2001
Algeria	1962	1962	1962	1969	1962	1969	1984	16 years	2001
Angola	2001	1976	1976	1976	1976	1976	2001	14 years	2001
Antigua and Barbuda	1983	1983	1983	1983	2003	1983	1983	16 years	2002
Argentina	1960	1956	1950	1960	1956	1968	1996	16 years	2001
Armenia	2006	2003	2004	2004	1994	1994	2006	16 years	2006
Australia	1973	1973	1932	1960	1974	1973			2006
Austria	1950	1951	1960	1958	1953	1973	2000	15 years	2001
Azerbaijan	1992	1992	1992	2000	1992	1992	1992	16 years	2004
Bahamas	2001	1976	1976	1976	2001	2001	2001	14 years	2001
Bahrain			1981	1998		2000	2012	15 years	2001
Bangladesh	1972	1972	1972	1972	1998	1972			2001
Barbados	1967	1967	1967	1967	1974	1974	2000	16 years	2000
Belarus	1956	1956	1956	1995	1956	1961	1979	16 years	2000
Belgium	1951	1953	1944	1961	1952	1977	1988	15 years	2002
Belize	1983	1983	1983	1983	1999	1999	2000	14 years	2000
Benin	1960	1968	1960	1961	1968	1961	2001	14 years	2001
Bolivia, Plurinational State of	1965	1973	2005	1990	1973	1977	1997	14 years	2003
Bosnia and Herzegovina	1993	1993	1993	2000	1993	1993	1993	15 years	2001
Botswana	1997	1997	1997	1997	1997	1997	1997	14 years	2000
Brazil		1952	1957	1965	1957	1965	2001	16 years	2000
Brunei Darussalam							2011	16 years	2008
Bulgaria	1959	1959	1932	1999	1955	1960	1980	16 years	2000
Burkina Faso	1960	1962	1960	1997	1969	1962	1999	15 years	2001
Burundi	1993	1997	1963	1963	1993	1993	2000	16 years	2002
Cabo Verde	1999	1979	1979	1979	1979	1979	2011	15 years	2001
Cambodia	1999	1999	1969	1999	1999	1999	1999	14 years	2006
Cameroon	1960	1962	1960	1962	1970	1988	2001	14 years	2002
Canada	1972		2011	1959	1972	1964			2000
Central African Republic	1960	1964	1960	1964	1964	1964	2000	14 years	2000
Chad	1960	1961	1960	1961	1966	1966	2005	14 years	2000
Chile	1999	1999	1933	1999	1971	1971	1999	15 years	2000
China					1990	2006	1999	16 years	2002
Colombia	1976	1976	1969	1963	1963	1969	2001	15 years	2005
Comoros	1978	1978	1978	1978	1978	2004	2004	15 years	2004
Congo	1960	1999	1960	1999	1999	1999	1999	14 years	2002
Costa Rica	1960	1960	1960	1959	1960	1962	1976	15 years	2001
Côte d'Ivoire	1960	1961	1960	1961	1961	1961	2003	14 years	2003
Croatia	1991	1991	1991	1997	1991	1991	1991	15 years	2001
Cuba	1952	1952	1953	1958	1954	1965	1975	15 years	
Cyprus	1966	1966	1960	1960	1987	1968	1997	15 years	2000
Czech Republic	1993	1993	1993	1996	1993	1993	2007	15 years	2001
Democratic Republic of the Congo	2001	1969	1960	2001	1969	2001	2001	14 years	2001
Denmark	1951	1955	1932	1958	1960	1960	1997	15 years	2000
Djibouti	1978	1978	1978	1978	1978	2005	2005	16 years	2005
Dominica	1983	1983	1983	1983	1983	1983	1983	15 years	2001
Dominican Republic	1956	1953	1956	1958	1953	1964	1999	14 years	2000
Ecuador	1967	1959	1954	1962	1957	1962	2000	14 years	2000
Egypt	1957	1954	1955	1958	1960	1960	1999	15 years	2002
El Salvador	2006	2006	1995	1958	2000	1995	1996	14 years	2000
Equatorial Guinea	2001	2001	2001	2001	1985	2001	1985	14 years	2001
Eritrea	2000	2000	2000	2000	2000	2000	2000	14 years	
Estonia	1994	1994	1996	1996	1996	2005	2007	15 years	2001
Ethiopia	1963	1963	2003	1999	1999	1966	1999	14 years	2003
Fiji	2002	1974	1974	1974	2002	2002	2003	15 years	2002
Finland	1950	1951	1936	1960	1963	1970	1976	15 years	2000
France	1951	1951	1937	1969	1953	1981	1990	16 years	2001

Country	Freedom of association and collective bargaining		Elimination of forced and compulsory labour		Elimination of discrimination in employment and occupation		Abolition of child labour		
	C087: Freedom of Association and Protection of the Right to Organise Convention, 1948	C098: Right to Organise and Collective Bargaining Convention, 1949	C029: Forced Labour Convention, 1930	C105: Abolition of Forced Labour Convention, 1957	C100: Equal Remuneration Convention, 1951	C111: Discrimination (Employment and Occupation) Convention, 1958	C138: Minimum Age Convention, 1973		C182: Worst Forms of Child Labour Convention, 1999
	Entry into force: 4 July 1950	Entry into force: 18 July 1951	Entry into force: 1 May 1932	Entry into force: 17 January 1959	Entry into force: 23 May 1953	Entry into force: 15 June 1960	Entry into force: 19 June 1976		Entry into force: 19 November 2000
	Year of ratification	Year of ratification	Year of ratification	Year of ratification	Year of ratification	Year of ratification	Year of ratification	Age	Year of ratification
Gabon	1960	1961	1960	1961	1961	1961	2010	16 years	2001
Gambia	2000	2000	2000	2000	2000	2000	2000	14 years	2001
Georgia	1999	1993	1993	1996	1993	1993	1996	15 years	2002
Germany	1957	1956	1956	1959	1956	1961	1976	15 years	2002
Ghana	1965	1959	1957	1958	1968	1961	2011	15 years	2000
Greece	1962	1962	1952	1962	1975	1984	1986	15 years	2001
Grenada	1994	1979	1979	1979	1994	2003	2003	16 years	2003
Guatemala	1952	1952	1989	1959	1961	1960	1990	14 years	2001
Guinea	1959	1959	1959	1961	1967	1960	2003	16 years	2003
Guinea-Bissau		1977	1977	1977	1977	1977	2009	14 years	2008
Guyana	1967	1966	1966	1966	1975	1975	1998	15 years	2001
Haiti	1979	1957	1958	1958	1958	1976	2009	14 years	2007
Honduras	1956	1956	1957	1958	1956	1960	1980	14 years	2001
Hungary	1957	1957	1956	1994	1956	1961	1998	16 years	2000
Iceland	1950	1952	1958	1960	1958	1963	1999	15 years	2000
India			1954	2000	1958	1960			
Indonesia	1998	1957	1950	1999	1958	1999	1999	15 years	2000
Iran, Islamic Republic of			1957	1959	1972	1964			2002
Iraq		1962	1962	1959	1963	1959	1985	15 years	2001
Ireland	1955	1955	1931	1958	1974	1999	1978	16 years	1999
Israel	1957	1957	1955	1958	1965	1959	1979	15 years	2005
Italy	1958	1958	1934	1968	1956	1963	1981	15 years	2000
Jamaica	1962	1962	1962	1962	1975	1975	2003	15 years	2003
Japan	1965	1953	1932		1967		2000	15 years	2001
Jordan		1968	1966	1958	1966	1963	1998	16 years	2000
Kazakhstan	2000	2001	2001	2001	2001	1999	2001	16 years	2003
Kenya		1964	1964	1964	2001	2001	1979	16 years	2001
Kiribati	2000	2000	2000	2000	2009	2009	2009	14 years	2009
Korea, Republic of					1997	1998	1999	15 years	2001
Kuwait	1961	2007	1968	1961		1966	1999	15 years	2000
Kyrgyzstan	1992	1992	1992	1999	1992	1992	1992	16 years	2004
Lao People's Democratic Republic			1964		2008	2008	2005	14 years	2005
Latvia	1992	1992	2006	1992	1992	1992	2006	15 years	2006
Lebanon		1977	1977	1977	1977	1977	2003	14 years	2001
Lesotho	1966	1966	1966	2001	1998	1998	2001	15 years	2001
Liberia	1962	1962	1931	1962		1959			2003
Libya	2000	1962	1961	1961	1962	1961	1975	15 years	2000
Lithuania	1994	1994	1994	1994	1994	1994	1998	16 years	2003
Luxembourg	1958	1958	1964	1964	1967	2001	1977	15 years	2001
Madagascar	1960	1998	1960	2007	1962	1961	2000	15 years	2001
Malawi	1999	1965	1999	1999	1965	1965	1999	14 years	1999
Malaysia		1961	1957	1958ª	1997	1997		15 years	2000
Maldives	2013	2013	2013	2013	2013	2013	2013	16 years	2013
Mali	1960	1964	1960	1962	1968	1964	2002	15 years	2000
Malta	1965	1965	1965	1965	1988	1968	1988	16 years	2001
Mauritania	1961	2001	1961	1997	2001	1963	2001	14 years	2001
Mauritius	2005	1969	1969	1969	2002	2002	1990	15 years	2000
Mexico	1950		1934	1959	1952	1961			2000
Moldova, Republic of	1996	1996	2000	1993	2000	1996	1999	16 years	2002
Mongolia	1969	1969	2005	2005	1969	1969	2002	15 years	2001
Montenegro	2006	2006	2006	2006	2006	2006	2006	15 years	2006
Morocco		1957	1957	1966	1979	1963	2000	15 years	2001
Mozambique	1996	1996	2003	1977	1977	1977	2003	15 years	2003
Myanmar	1955		1955						2013
Namibia	1995	1995	2000	2000	2010	2001	2000	14 years	2000
Nepal		1996	2002	2007	1976	1974	1997	14 years	2002
Netherlands	1950	1993	1933	1959	1971	1973	1976	15 years	2002
New Zealand		2003	1938	1968	1983	1983			2001

Status of fundamental labour rights conventions (continued)

Country	Freedom of association and collective bargaining		Elimination of forced and compulsory labour		Elimination of discrimination in employment and occupation		Abolition of child labour		
	C087: Freedom of Association and Protection of the Right to Organise Convention, 1948	C098: Right to Organise and Collective Bargaining Convention, 1949	C029: Forced Labour Convention, 1930	C105: Abolition of Forced Labour Convention, 1957	C100: Equal Remuneration Convention, 1951	C111: Discrimination (Employment and Occupation) Convention, 1958	C138: Minimum Age Convention, 1973		C182: Worst Forms of Child Labour Convention, 1999
	Entry into force: 4 July 1950	Entry into force: 18 July 1951	Entry into force: 1 May 1932	Entry into force: 17 January 1959	Entry into force: 23 May 1953	Entry into force: 15 June 1960	Entry into force: 19 June 1976		Entry into force: 19 November 2000
	Year of ratification	Year of ratification	Year of ratification	Year of ratification	Year of ratification	Year of ratification	Year of ratification	Age	Year of ratification
Nicaragua	1967	1967	1934	1967	1967	1967	1981	14 years	2000
Niger	1961	1962	1961	1962	1966	1962	1978	14 years	2000
Nigeria	1960	1960	1960	1960	1974	2002	2002	15 years	2002
Norway	1949	1955	1932	1958	1959	1959	1980	15 years	2000
Oman		1998	2005				2005	15 years	2001
Pakistan	1951	1952	1957	1960	2001	1961	2006	14 years	2001
Panama	1958	1966	1966	1966	1958	1966	2000	14 years	2000
Papua New Guinea	2000	1976	1976	1976	2000	2000	2000	16 years	2000
Paraguay	1962	1966	1967	1968	1964	1967	2004	14 years	2001
Peru	1960	1964	1960	1960	1960	1970	2002	14 years	2002
Philippines	1953	1953	2005	1960	1953	1960	1998	15 years	2000
Poland	1957	1957	1958	1958	1954	1961	1978	15 years	2002
Portugal	1977	1964	1956	1959	1967	1959	1998	16 years	2000
Qatar			1998	2007		1976	2006	16 years	2000
Romania	1957	1958	1957	1998	1957	1973	1975	16 years	2000
Russian Federation	1956	1956	1956	1998	1956	1961	1979	16 years	2003
Rwanda	1988	1988	2001	1962	1980	1981	1981	14 years	2000
Saint Kitts and Nevis	2000	2000	2000	2000	2000	2000	2005	16 years	2000
Saint Lucia	1980	1980	1980	1980	1983	1983			2000
Saint Vincent and the Grenadines	2001	1998	1998	1998	2001	2001	2006	14 years	2001
Samoa	2008	2008	2008	2008	2008	2008	2008	15 years	2008
San Marino	1986	1986	1995	1995	1985	1986	1995	16 years	2000
Sao Tome and Principe	1992	1992	2005	2005	1982	1982	2005	14 years	2005
Saudi Arabia			1978	1978	1978	1978	2014	15 years	2001
Senegal	1960	1961	1960	1961	1962	1967	1999 b	15 years	2000
Serbia	2000	2000	2000	2003	2000	2000	2000	15 years	2003
Seychelles	1978	1999	1978	1978	1999	1999	2000	15 years	1999
Sierra Leone	1961	1961	1961	1961	1968	1966	2011	15 years	2011
Singapore		1965	1965	1965 c	2002		2005	15 years	2001
Slovakia	1993	1993	1993	1997	1993	1993	1997	15 years	1999
Slovenia	1992	1992	1992	1997	1992	1992	1992	15 years	2001
Solomon Islands	2012	2012	1985	2012	2012	2012	2013	14 years	2012
Somalia	2014	2014	1960	1961		1961			2014
South Africa	1996	1996	1997	1997	2000	1997	2000	15 years	2000
South Sudan		2012	2012	2012	2012	2012	2012	14 years	2012
Spain	1977	1977	1932	1967	1967	1967	1977	16 years	2001
Sri Lanka	1995	1972	1950	2003	1993	1998	2000	14 years	2001
Sudan		1957	1957	1970	1970	1970	2003	14 years	2003
Suriname	1976	1996	1976	1976					2006
Swaziland	1978	1978	1978	1979	1981	1981	2002	15 years	2002
Sweden	1949	1950	1931	1958	1962	1962	1990	15 years	2001
Switzerland	1975	1999	1940	1958	1972	1961	1999 d	15 years	2000
Syrian Arab Republic	1960	1957	1960	1958	1957	1960	2001	15 years	2003
Tajikistan	1993	1993	1993	1999	1993	1993	1993	16 years	2005
Tanzania, United Republic of	2000	1962	1962	1962	2002	2002	1998	14 years	2001
Thailand			1969	1969	1999		2004	15 years	2001
The former Yugoslav Republic of Macedonia	1991	1991	1991	2003	1991	1991	1991	15 years	2002
Timor-Leste	2009	2009	2009						2009
Togo	1960	1983	1960	1999	1983	1983	1984	14 years	2000
Trinidad and Tobago	1963	1963	1963	1963	1997	1970	2004	16 years	2003
Tunisia	1957	1957	1962	1959	1968	1959	1995	16 years	2000
Turkey	1993	1952	1998	1961	1967	1967	1998	15 years	2001
Turkmenistan	1997	1997	1997	1997	1997	1997	2012	16 years	2010
Uganda	2005	1963	1963	1963	2005	2005	2003	14 years	2001
Ukraine	1956	1956	1956	2000	1956	1961	1979	16 years	2000
United Arab Emirates			1982	1997	1997	2001	1998	15 years	2001
United Kingdom	1949	1950	1931	1957	1971	1999	2000	16 years	2000
United States				1991					1999

Country	Freedom of association and collective bargaining		Elimination of forced and compulsory labour		Elimination of discrimination in employment and occupation		Abolition of child labour		
	C087: Freedom of Association and Protection of the Right to Organise Convention, 1948	C098: Right to Organise and Collective Bargaining Convention, 1949	C029: Forced Labour Convention, 1930	C105: Abolition of Forced Labour Convention, 1957	C100: Equal Remuneration Convention, 1951	C111: Discrimination (Employment and Occupation) Convention, 1958	C138: Minimum Age Convention, 1973		C182: Worst Forms of Child Labour Convention, 1999
	Entry into force: 4 July 1950	Entry into force: 18 July 1951	Entry into force: 1 May 1932	Entry into force: 17 January 1959	Entry into force: 23 May 1953	Entry into force: 15 June 1960	Entry into force: 19 June 1976		Entry into force: 19 November 2000
	Year of ratification	Year of ratification	Year of ratification	Year of ratification	Year of ratification	Year of ratification	Year of ratification	Age	Year of ratification
Uruguay	1954	1954	1995	1968	1989	1989	1977	15 years	2001
Uzbekistan		1992	1992	1997	1992	1992	2009	15 years	2008
Vanuatu	2006	2006	2006	2006	2006	2006			2006
Venezuela, Bolivarian Republic of	1982	1968	1944	1964	1982	1971	1987	14 years	2005
Viet Nam	2007				1997	1997	2003	15 years	2000
Yemen	1976	1969	1969	1969	1976	1969	2000	14 years	2000
Zambia	1996	1996	1964	1965	1972	1979	1976	15 years	2001
Zimbabwe	2003	1998	1998	1998	1989	1999	2000	14 years	2000

NOTES

a Not in force, denounced on 10 January 1990.

b The government declared in conformity with article 5, paragraph 2, of the convention that the provisions of the convention do not apply to traditional pastoral or rural work without remuneration that is carried out in a family setting by children under age 15 and that aims at better integrating them into their social surroundings and the environment.

c Not in force, denounced on 19 April 1979.

d Pursuant to article 3, the minimum age for underground work is 19 full years and 20 full years for apprentices.

DEFINITIONS

C087: Freedom of Association and Protection of the Right to Organise Convention, 1948: Says that workers and employers have the right to establish and join organizations of their own choosing without previous authorization, that workers' and employers' organizations have the right to establish and join federations and confederations and that any such organization, federation or confederation has the right to affiliate with international organizations of workers and employers.

C098: Right to Organise and Collective Bargaining Convention, 1949: Protects workers from anti-union discrimination, including requirements that a worker not join a union or relinquish trade union membership for employment, or dismissal because of union membership or participation in union activities and protects workers' and employers' organizations from interference by each other, in particular the establishment of workers' organizations under the domination of employers or employers' organizations or the support of workers' organizations by financial or other means in order to place such organizations under the control of employers or employers' organizations. The convention also enshrines the right to collective bargaining.

C029: Forced Labour Convention, 1930: Prohibits all forms of forced or compulsory labour—defined as work or service that is exacted from any person under the menace of penalty and for which the person has not offered himself or herself voluntarily. Exceptions are provided for work required by compulsory military service, normal civic obligations, as a consequence of a conviction in a court of law (provided that the work or service in question is carried out under the supervision and control of a public authority and that the person carrying it out is not hired to or placed at the disposal of private individuals, companies or associations), in cases of emergency and for minor communal services performed by the members of a community in the direct interest of the community.

C105: Abolition of Forced Labour Convention, 1957: Prohibits forced or compulsory labour as a means of political coercion or education or as a punishment for holding or expressing political views or views ideologically opposed to the established political, social or economic system; as a method of mobilizing and using labour for purposes of economic development; as a means of labour discipline; as a punishment for having participated in strikes and as a means of racial, social, national or religious discrimination.

C100: Equal Remuneration Convention, 1951: Ensures the principle of equal remuneration for men and women for work of equal value, with "remuneration" broadly defined to include the ordinary, basic or minimum wage or salary and any additional direct or indirect payment (in cash or in kind) from the employer.

C111: Discrimination (Employment and Occupation) Convention, 1958: Requires ratifying states to declare and pursue a national policy designed to promote equality of opportunity and treatment in respect of employment and occupation, with a view to eliminating any discrimination in these fields, including discrimination in access to vocational training, access to employment and to particular occupations, and terms and conditions of employment. Discrimination is defined as any distinction, exclusion or preference based on race, colour, sex, religion, political opinion, national extraction or social origin that nullifies or impairs equality of opportunity or treatment in employment or occupation.

C138: Minimum Age Convention, 1973: Sets the general minimum age for admission to employment or work at 15 years (13 for light work) and the minimum age for hazardous work at 18 (16 under certain strict conditions), with provisions for initially setting the general minimum age at 14 (12 for light work) where the economy and education facilities are insufficiently developed.

C182: Worst Forms of Child Labour Convention, 1999: Requires ratifying states to eliminate the worst forms of child labour (with child defined as a person under age 18), including all forms of slavery or practices similar to slavery, such as the sale and trafficking of children, debt bondage and serfdom and forced or compulsory labour, including forced or compulsory recruitment of children for use in armed conflict; child prostitution and pornography; using children for illicit activities, in particular for the production and trafficking of drugs; and work that is likely to harm the health, safety or morals of children. The convention also requires ratifying states to provide the necessary and appropriate direct assistance for the removal of children from the worst forms of child labour and for their rehabilitation and social integration and to ensure access to free basic education and, wherever possible and appropriate, vocational training for children removed from the worst forms of child labour.

MAIN DATA SOURCE

Columns 1–9: ILO (2015d).

Notes

Chapter 1

1   Vodopivec and Arunatilake 2008.
2   Nobel Media 2015.
3   Kabanda 2015.
4   Kretkowski 1998.
5   Tate 2013.
6   Miller 2015a.
7   OECD 2015c.
8   Kivimäki and others 2015.
9   UK Cabinet Office 2013.
10 Helliwell and Huang 2011a.
11 Clark and others 2008.
12 Beverly 2003; Gay 1994.
13 ILO 2014g.
14 ILO 2014g.
15 UN Volunteers 2011.
16 United States Equal Employment Opportunity Commission 2014.
17 Schifferes 2002.
18 ENAR 2013.
19 ILO 2011b.
20 Atal, Ñopo and Winder 2009.
21 WHO and World Bank 2011.
22 ILO 2013a.
23 Seckan 2013.
24 Chappell and Di Martino 2006.
25 ILO 2009.
26 World Bank 2011. This section draws on Stewart (2015) and Cramer (2015).
27 Kuehnast 2015.
28 UN 2000a.
29 Kuehnast 2015.
30 Kuehnast 2015.
31 UN Women 2012b.
32 ILO 2013c.
33 ILO 2013c.
34 ILO 2014e.
35 ILO 2014e.
36 ILO 2014e.
37 ILO 2014e.
38 Kaye 2006.
39 UNODC 2012.
40 Euronews 2015.
41 Human Rights Watch 2014a, 2014b.
42 Human Rights Watch 2014a, 2014b.
43 This section draws on a contribution by the HIV/AIDS Group of the United Nations Development Programme, which is gratefully acknowledged. UNAIDS 2012.
44 ILO 2010a, 2010b.
45 Shannon and others 2015.
46 Jana and others 2014.
47 ILO 2015f.
48 Shi 2008.
49 ILO 2013e.

Chapter 2

1   FAO 2014.
2   UNESCO 2014; WHO 2014; World Bank 2015f.
3   ILO 2015e.
4   Pollin 2015.
5   ILO 2013b.
6   Human Development Report Office calculation based on GERA (2015).
7   Kabanda 2015.
8   UNESCO and UNDP 2013.
9   Crow 2015.
10 Salamon, Sokolowski and Haddock 2011.
11 UNFPA and HelpAge International 2012.
12 UNDESA 2013a; World Bank 2015c.
13 World Bank 2015c.
14 World Bank 2015c, 2015e.
15 UN 2015b.
16 UN 2015b.
17 UN 2015b; UNAIDS 2015.
18 UN 2015b.
19 UN 2015b.
20 UNDP 2012b.
21 UNDP 2014d.
22 Hall 2015.
23 UNDP 2014e.
24 UN 2015b.
25 ILO 2015e.
26 UN 2015b.
27 The International Labour Organization defines the labour force participation rate as the proportion of a country's working-age population that engages actively in the labour market, either by working or looking for work. Depending on the structure of surveys, participation among certain groups of workers may be underestimated—particularly the number of employed people who work only a few hours in the reference period, are in unpaid employment or work near or in their home. The number of women tends to be underestimated to a larger extent than the number of men. See ILO (2015e) for more extensive definition and discussion.
28 UN 2015b.
29 IPU 2015; ILO 2015j.
30 Grant Thornton 2015.
31 BLS 2015b.
32 ILO 2015h.
33 ILO 2015i.
34 ILO 2015h.
35 UN 2015b.
36 WIEGO and ILO 2013.
37 UNDESA 2015.
38 UN 2015b.
39 ILO 2014b.
40 Eurostat 2015; OECD 2015b.
41 McKinsey Global Institute 2012b.
42 McKinsey Global Institute 2012b.
43 Ortiz and Cummins 2012.
44 OECD 2015b.
45 ILO 2012b.
46 ILO 2014e.
47 ILO 2013c.
48 UNDP 2013b.
49 Hellebrandt and Mauro 2015. The study estimates that global inequality has fallen—from a Gini coefficient of 0.69 in 2003 to 0.65 in 2013. For the global distribution of household incomes, the dominant forces are not changes in the distribution of income within countries, but changes in the relative average incomes of countries, weighted by population. Thus the extraordinary growth of China and, to less extent, India—which together account for almost 40 percent of the world's population—largely explain the decline in global household inequality.
50 Alvaredo and others 2011, 2013.
51 UNDP 2013b.
52 Oxfam 2015.
53 Oxfam 2015.
54 UNDP 2013b.
55 UNDP 2013b.
56 UNDESA 2013b.
57 Kharas and Gertz 2010.
58 Kharas and Gertz 2010.
59 UNDESA 2015.
60 Cortez 2012.
61 WEF 2015.
62 Anderlini 2015.
63 WMO 2014; Barbieri and others 2010.
64 World Bank 2015d.
65 WEF 2015.
66 Hawkins, Blackett and Heymans 2013.
67 Hawkins, Blackett and Heymans 2013.
68 UN 2015b.
69 UN 2015b.
70 IEP 2014.
71 UNDP 2014b.
72 Krug and others 2002.
73 UNESCO 2013b.
74 UNDP 2012c.
75 WHO 2013.
76 UN Women 2012a.
77 European Union Agency for Fundamental Rights 2014.
78 UN Women 2012, 2014.
79 Grosh, Bussolo and Freije 2014.
80 ILO 2015i.
81 WHO 2015b.
82 WHO 2015b.
83 WHO 2015a.
84 World Bank 2015a.
85 UN 2015b.
86 UN and others 2015.
87 UN 2015b.
88 UN 2015b.
89 World Bank 2002.
90 UN 2015b.
91 UNDP 2012a.
92 WHO 2003.
93 Guha-Sapir, Hoyois and Below 2014.
94 Norwegian Refugee Council and IDMC 2015.
95 "Big data" is a comprehensive concept that describes large and complex amounts of information. In contrast to traditional data, big data is characterized by the 5 V's (volume, velocity, variety, veracity and value). It is enormous in volume—on the order of zettabytes and brontobytes in a single dataset. Big data is also generated with high velocity (that is, the speed with which data must be stored and analyzed) and with large variety. Additionally, big data is unstructured and often takes qualitative information into account as well. The fifth V, value, accounts for the potential of big data to be used for development.
96 UN Global Pulse 2013.
97 A gigabyte equals 1 billion bytes, the basic unit of information.
98 Hsu and others 2014.

Chapter 3

1   World Bank 2015f; ILO 2014c.
2   World Bank 2015f.
3   Human Development Report Office calculation based on ILO 2014c.
4   FAO 2015. The Food and Agriculture Organization figures for the economically active population in agriculture are higher than the International Labour Organization figures on those employed in agriculture because the Food and Agriculture Organization defines the economically active population in agriculture (agricultural labour force) as the part of the economically active population engaged in or seeking work in agriculture, hunting, fishing or forestry, while the International Labour Organization definition for "employment by sector, agriculture" includes only employees (wage and salary earners) and excludes the self-employed and contributing family members. The employment share of the agricultural sector estimated by the International Labour Organization is lower than that estimated by the Food and Agriculture Organization.
5   FAO 2014.
6   FAO 2014.
7   FAO 2014.
8   ILO 2013d.
9   ILO 2014c.
10 Timmer and others 2014a.
11 McKinsey Global Institute 2012a.
12 International Federation of Robotics 2014.
13 WEF 2012a.
14 McKinsey Global Institute 2012a.
15 Rodrik 2015a.
16 ILO 2014c.
17 Warhurst and others 2012.
18 Timmer and others 2014a.
19 McKinsey Global Institute 2013.
20 Social Tech Guide 2015.
21 Cowen 2013.
22 Dobbs, Manyika and Woetzel 2015.

23 Ryder 2015.
24 Masters 2015.
25 Kagermann, Lukas and Wahlster 2011.
26 Kingsley-Hughes 2012.
27 Kearney, Hershbein and Boddy 2015.
28 Gordon 2014.
29 Milken Institute 2013.
30 ILO 2015i.
31 ILO 2015i.
32 Luce and others 2014.
33 Salazar-Xirinachs 2015.
34 ILO 2003a.
35 Bardhan, Jaffee and Kroll 2013.
36 Everest Research Institute 2008.
37 UNCTAD 2014.
38 OECD 2007.
39 OECD 2007.
40 Lippoldt 2012.
41 Andreoni 2015.
42 Elms and Low 2013.
43 ILO 2015i.
44 ILO 2015i.
45 OECD 2014.
46 Global trade is the equivalent of
    the sum of all exports and imports.
    Human Development Report Office
    calculations based on UNCTAD (2015).
47 McKinsey Global Institute 2014.
48 ITU 2015.
49 Gabre-Madhin 2012.
50 Atta, Boutraa and Akhkha 2011.
51 GSMA 2014.
52 Deloitte 2014b.
53 Deloitte 2014b.
54 Deloitte 2014b.
55 Aker and Mbiti 2010.
56 IDRC 2013.
57 Selim 2013.
58 CARE International website, www.
    lendwithcare.org.
59 Deloitte 2014b.
60 Twitter website, http://about.twitter.
    com/company.
61 Twitter website, http://about.twitter.
    com/company.
62 Wikipedia 2015.
63 McCarthy 2012.
64 Mandel 2013.
65 Selim 2013.
66 Calculated based on Global
    Entrepreneurship Monitor Database
    Accessed June 2015.
67 Boyde 2015.
68 Pooler 2014.
69 Amazon 2015.
70 The Economist 2014a.
71 McKinsey Global Institute 2014
72 Salazar-Xirinachs 2015.
73 McKinsey & Company 2014.
74 McKinsey & Company 2014.
75 Human Development Report Office
    calculations based on Wilson (2010).
76 UN 2012a.
77 Conti and Heckman 2010.
78 Conti and Heckman 2010.
79 Rodrik 2015a.
80 Dowdy 2014.
81 Global Workplace Analytics 2012.
82 Smeaton, Ray and Knight 2014. The
    data on flexible work are derived from

a range of sources, including case
studies, econometric secondary data
analyses, meta-analyses and primary
research using dedicated surveys that
elicit manager or employee views
of the costs or benefits of work–life
balance policies. Determining con-
clusive results therefore—whether
in terms of coverage, quality, diverse
indicators, methodology or relevance,
along with the interpretation and
generalizability of findings—remains
a challenge for this area of research.
83 Coenen and Kok 2014.
84 Brynjolfsson and McAfee 2014.
85 USPTO 2015.
86 USPTO 2015.
87 WIPO 2015.
88 Heyman 2015.
89 World Values Survey, sixth wave,
    2010–2014.
90 May 2007.
91 UN Volunteers 2014.
92 Clark 2013.
93 Lagesse 2015.
94 Ushahidi website, www.ushahidi.com.
95 Deloitte 2014a.
96 Yunus 2009.
97 Vissa 2015.
98 ITU 2013.
99 Maier 2008.
100 Maier 2008.
101 UNESCO 2015.
102 Mishkin 2014.
103 UNDESA 2015.
104 Jacobs 2015b.
105 Jacobs 2015b. The lump of labour fal-
    lacy is the contention that the amount
    of work available to labourers is fixed.
    It is considered a fallacy by most
    economists, who hold the view that
    the amount of work is not static.
106 CBRE Global 2014.
107 Autor 2014.
108 Cooper and Mishel 2015.
109 ILO 2015b.
110 Human Development Report Office
    calculations for 16 developed coun-
    tries based on EC (2015).
111 Stockhammer 2013.
112 Oxfam 2015.
113 Mishel and Davis 2014.
114 Stockhammer 2013.
115 Ffrench-Davis 2012.
116 Timmer and others 2014b.

Chapter 4

1   UNDP 1995.
2   Abdelali-Martin 2011.
3   UN 2015b.
4   ILO 2015e.
5   Bloom and others 2009.
6   UNDESA 2013b.
7   UNDESA 2013b.
8   Bloom and McKenna 2015.
9   This is purely the impact of higher
    female education on female labor force
    participation and does not account for
    the impacts of education on fertility.

Furthermore, increasing male education
can have an offsetting effect on female
labor force participation. The effect of
male education is about –2 percentage
points. Bloom and McKenna 2015.
10 Lewis 2015.
11 Bandara 2015.
12 UN Women 2015.
13 Miller 2014.
14 IADB 2012.
15 ILO 2015b.
16 Grant Thornton 2015.
17 Grant Thornton 2015.
18 Grant Thornton 2015.
19 Grant Thornton 2015.
20 Singer, Amorós and Moska Arreola
    2015.
21 Singer, Amorós and Moska Arreola
    2015.
22 Demirgüç-Kunt and others 2015;
    World Bank 2014a.
23 GEDI 2014.
24 GEDI 2014.
25 Statista 2014, 2015; Turkish
    Statistical Institute 2015.
26 Grant Thornton 2015.
27 ILO 2015h.
28 ILO 2015h.
29 Human Development Report Office
    calculations based on ILO (2015e).
30 FAO 2010.
31 FAO 2010, 2011b.
32 ILO 2013b.
33 ILO 2013b. Furthermore, as in ILO
    (2013b), India is characterized by
    high discrepancies between official
    estimates and estimates from other
    sources, and the results may have
    some margin of error. The conclusions
    here are based on the International
    Labour Organization's analysis of
    micro-data from the 2004/2005
    Employment and Unemployment
    Survey (61st round) by the National
    Sample Survey Organisation of India.
34 Raghuram 2001.
35 Tokman 2010. Due to the overlapping
    concept of domestic and care work,
    these statistics include both.
36 D'Cunha, Lopez-Ekra and Mollard
    2010.
37 TWC2 2011.
38 ILO 2013b.
39 Rakkee and Sasikumar 2012.
40 Human Rights Watch 2014b.
41 United Workers Congress n.d.
42 Human Development Report Office
    calculation of current global popula-
    tion under age 15; see also table 8 in
    Statistical annex.
43 UNFPA and HelpAge International
    2012
44 WHO and World Bank 2011.
45 UNAIDS 2015
46 INDEC 2014.
47 DANE 2014.
48 Sayer 2015.
49 Human Development Report Office
    calculations based on data from
    Charmes (2015).

50 Deen 2012.
51 Human Development Report Office
    calculations based on data from
    Charmes (2015). Values are the adult
    population–weighted average by sex.
52 Human Development Report Office
    calculations based on data from
    Charmes (2015).
53 BLS 2015c; Sayer 2015.
54 Ko and Hank 2013.
55 UN 2015b.
56 Scheil-Adlung 2015.
57 WHO and World Bank 2011.
58 The Japan Times 2015.
59 For a methodology for estimating
    the care burden, see Mukherjee and
    Nayyar (2015).
60 Elson, 2012.
61 UN WomenWatch 2009.
62 Charmes, 2006.
63 Jiménez Cisneros et al. 2014.
64 Baker and Milligan 2008.
65 Data are for educated mothers only.
    Liu and Skans 2010.
66 Miller 2015b.

Chapter 5

1   UN 2000b.
2   UNDP 1994.
3   World Commission on Environment
    and Development 1987.
4   Credit Suisse Research Institute 2014.
5   Arrow and others 1996.
6   EPA 2011.
7   Walker 2013.
8   Masur and Posner 2011.
9   Walker 2013.
10 UNDP 2011.
11 USAID 2013.
12 IPPC 2014a.
13 IEA 2014.
14 UNEP and others 2008.
15 Poschen (2015) further elaborates
    green jobs to be those that reduce
    consumption of energy and raw mate-
    rials, limit greenhouse gas emissions,
    minimize waste and pollution, protect
    and restore ecosystems, and enable
    enterprises and communities to adapt
    to climate change.
16 ILO 2013f.
17 UN 2015a.
18 IUES and CASS 2010.
19 The Economist 2015a.
20 Suwala 2010.
21 FAO 2015.
22 Steffen and others 2015.
23 FAO 2012.
24 WRI 2014.
25 Fuglie and Nin-Pratt 2012.
26 Yishay and Mobarak 2014.
27 Cole and Fernando 2012.
28 Swanson and Davis 2014.
29 Confederation of Indian Industry and
    India Ministry of New and Renewable
    Energy 2010.
30 IFC 2010.
31 UN 2015c.
32 WHO 2005.

33 UN 2014.
34 Walk Free Foundation 2015.
35 ILO 2012c.
36 Human Development Report Office calculation based on ILO (2014e) and UN Women (2014).
37 ILO 2003b.
38 ILO 2013f.
39 The 10 countries with the largest renewable energy employment were China, Brazil, the United States, India, Germany, Indonesia, Japan, France, Bangladesh and Colombia. IRENA 2015.
40 IRENA 2015.
41 Pollin 2015.
42 ILO 2013f.
43 UNESCO 2014.
44 WHO 2014; World Bank 2014b.
45 Columbia University 2013.

Chapter 6

1 ILO and others 2012.
2 ILO 2015a.
3 Epstein 2007a.
4 Epstein 2007a.
5 Derviş 2012; Krugman 2014; Rosengren 2013.
6 Derviş 2012; Krugman 2014; Rosengren 2013.
7 Jahan 2005.

8 World Bank 2013.
9 Demirgüç-Kunt and Klapper 2012.
10 WEF 2012b.
11 World Bank 2012.
12 Banco Central de Ecuador 2012.
13 Epstein 2007b.
14 ILO 2011a.
15 Galindo, Izquierdo and Montero 2006; Pratap, Lobato and Somuano 2003; Bleakley and Cowan 2002.
16 Epstein 2007a.
17 Islam 2015.
18 The University of Melbourne is leading a project on assessment and teaching of 21st century skills. Working with more than 250 research-ers across 60 institutions worldwide, the project categorized 21st century skills into four broad categories. See www.atc21s.org.
19 Force 2013.
20 OECD 2013a.
21 OECD 2013a.
22 Babcock and others 2012.
23 OECD 2013b.
24 WEConnect International website, www.weconnectinternational.org.
25 Education for Employment website, www.efe.org.
26 Brynjolfsson and McAfee 2014.
27 Brynjolfsson and McAfee 2011.

28 ILO 2015h.
29 Islam and Islam 2015.
30 Skidelsky 2015.
31 Evans 2015.
32 ILGA 2014.
33 Lamichhane 2015.
34 WHO and World Bank 2011.
35 Jacobs 2015a.
36 Clean Clothes Campaign website, www.cleanclothes.org.
37 Kerrissey 2015.
38 ILO 2014h.
39 ILO 2014h.
40 Islam and Islam 2015.
41 Skidelsky 2015.
42 Beazley 2014; Kostzer 2008; UNDP 2014d.
43 Based on Holzman (2012).
44 Holzman 2012.
45 Bosch, Melquizo and Pages 2013.
46 Bosch, Melquizo and Pages 2013.
47 Leahy 2015.
48 Ffrench-Davis 2010.
49 Montenegro and Patrinos 2014.
50 Blasi, Freeman and Krauss 2014.
51 Blasi, Freeman and Krauss 2014.
52 WFP 2015.
53 Arnold, Conway and Greenslade 2011.
54 UNDP 2014b.
55 German Federal Ministry of Labor and Social Affairs 2015.

56 Smale and Miller 2015.
57 Grant Thornton 2014.
58 The Sustainable Energy for All initia-tive has three goals, to be reached by 2030: ensuring universal access to modern energy services, doubling the rate of improvement of energy efficiency and doubling the share of renewables in the energy mix.
59 Article 66.2 of the Agreement on Trade-Related Aspects of Intellectual Property Rights (1994) reads "Developed country Members shall provide incentives to enterprises and institutions in their territories for the purpose of promoting and encourag-ing technology transfer to least-developed country Members in order to enable them to create a sound and viable technological base."
60 ILO 2011a.
61 UNDP 2013a.
62 Hazelhurst 2015.
63 Innovation for Poverty Action 2015.
64 Kasedde and others 2014.
65 ILO 2014h.
66 Eversheds 2014.
67 Jacobs 2015b.
68 Jacobs 2015b
69 ILO 2008a.

References

Abdelali-Martin, M. 2011. "Empowering Women in the Rural Labor Force with a Focus on Agricultural Employment in the Middle East and North Africa (MENA)." EGM/RW/2011/EP.9. Paper prepared for the Expert Group Meeting "Enabling Rural Women's Economic Empowerment: Institutions, Opportunities and Participation," 20–23 September, Accra. www.un.org/womenwatch/daw/csw/csw56/egm/Martini-EP-9-EGM-RW-Sep-2011.pdf. Accessed 21 August 2015.

ACE (Architects' Council of Europe). 2014. "The Architectural Profession in Europe 2014: A Sector Study." West Sussex, UK. www.ace-cae.eu/fileadmin/New_Upload/7._Publications/Sector_Study/2014/EN/2014_EN_FULL.pdf. Accessed 23 July 2015.

Agreement on Trade-Related Aspects of Intellectual Property Rights. 1994. Signed 15 April, Marrakesh, Morocco. www.wto.org/english/tratop_e/trips_e/t_agm0_e.htm. Accessed 15 June 2015.

Aker, J.C., and M.I. Mbiti. 2010. "Mobile Phones and Economic Development in Africa." Working Paper 211. Center for Global Development, Washington, DC.

Altindag, D., and N. Mocan. 2010. "Joblessness and Perceptions about the Effectiveness of Democracy." *Journal of Labor Research* 31(2): 99–123.

Alvaredo, F., A. Atkinson, T. Piketty, and E. Saez. 2011. The World Top Incomes Database. http://topincomes.g-mond.parisschoolofeconomics.eu/. Accessed 7 July 2015.

———. **2013.** "The Top 1 Percent in International and Historical Perspective." *Journal of Economic Perspectives* 27(3): 3–20. http://pubs.aeaweb.org/doi/pdfplus/10.1257/jep.27.3.3. Accessed 7 July 2015.

———. **2015.** 2015. "The World Top Incomes Database." http://topincomes.gmond.parisschoolofeconomics.eu/. Accessed 7 July 2015.

Amazon. 2015. "Working on HITs." www.mturk.com/mturk/welcome?variant=worker. Accessed 2 July 2015.

Anderlini, J. 2015. "China's Great Migration." *FT Magazine*, 30 April. www.ft.com/intl/cms/s/2/44096ed2-eeb0-11e4-a5cd-00144feab7de.html. Accessed 8 July 2015.

Andreoni, A. 2015. "Production as a Missing Dimension of Human Development." Background think piece for Human Development Report 2015. UNDP–HDRO, New York.

Antonopoulos, R. 2009. "The Unpaid Care Work-Paid Work Connection." Working Paper 86. International Labour Organization, Policy Integration and Statistics Department, Geneva.

Arnold, C., T. Conway, and M. Greenslade. 2011. "Cash Transfers: Literature Review." Policy Division, London.

Arrow, K., M. Cropper, G. Eads, R. Hahn, L. Lave, R. Noll, and others. 1996. "Is There a Role for Benefit-Cost Analysis in Environmental, Health, and Safety Regulation?" *Science* 272(5259): 221–22.

Askitas, N., and K. Zimmerman. 2009. "Google Econometrics and Unemployment Forecasting." *Applied Economics Quarterly* 55(2): 107–20.

Atal, J., H. Ñopo, and N. Winder. 2009. "New Century, Old Disparities: Gender and Ethnic Wage Gaps in Latin America." Working Paper 25. Inter-American Development Bank, Washington, DC. http://papers.ssrn.com/sol3/papers.cfm?abstract_id=1815933. Accessed 6 August 2015.

Atkinson A. 2015. *Inequality—What Can Be Done?* Cambridge, MA: Harvard University Press.

Atta, R., T. Boutraa, and A. Akhkha. 2011. "Smart Irrigation System for Wheat in Saudi Arabia Using Wireless Sensors Network Technology." *International Journal of Water Resources and Arid Environments* 1(6): 478–82.

Autor, D. 2014. "Polanyi's Paradox and the Shape of Employment Growth." Working Paper 20845. National Bureau of Economic Research, Cambridge, MA.

Babcock, L., W.J. Congdon, L.F. Katz, and S. Mullainathan. 2012. "Notes on Behavioral Economics and Labor Market Policy." *IZA Journal of Labor Policy* 1(1): 1–14.

Baker, M., and K. Milligan. 2008. "Maternal Employment, Breastfeeding and Health: Evidence from Maternity Leave Mandates." *Journal of Health Economics* 27(4): 871–87.

Banco Central de Ecuador. 2012. "De la Definicion de la Politica a la Practicia: Hacendo Inclusion Financiera." www.afa-global.org/library/publications. Quito.

Bandara, A. 2015. "The Economic Costs of Gender Gaps in Effective Labour: Africa's Missing Growth Reserve." *Feminist Economist* 21(2): 162–86. www.tandfonline.com/doi/pdf/10.1080/13545701.2014.986153. Accessed 23 July 2015.

Barbieri, A.F., E. Domingues, B.L. Queiroz, R.M. Ruiz, J.I. Rigotti, J.A.M. Carvalho, and M.F. Resende. 2010. "Climate Change and Population Migration in Brazil's Northeast: Scenarios for 2025–2050." *Population and Environment* 31(5): 344–70. http://link.springer.com/article/10.1007%2Fs11111-010-0105-1. Accessed 8 July 2015.

Bardhan, A., D.M. Jaffee, and C.A. Kroll, eds. 2013. *The Oxford Handbook of Offshoring and Global Employment.* Oxford, UK: Oxford University Press.

Barnett. T. 2007. "Decriminalizing Prostitution in New Zealand: The Campaign and the Outcome." http://myweb.dal.ca/mgoodyea/Documents/New%20Zealand/Decriminalising%20Prostitution%20in%20NZ.pdf. Accessed 10 July 2015.

Beatty, C., S. Fothergill, and R. Powell. 2007. "Twenty Years On: Has the Economy of the UK Coalfields Recovered?" *Environment and Planning* 39(7): 1654–75.

Beazley, R. 2014. "Social Protection through Public Works in Nepal: Improving the Karnali Employment Programme." Briefing Note. Oxford Policy Management, Oxford, UK.

Bennett, K., H. Beynon, and R. Hudson. 2000. *Coalfields Regeneration: Dealing with the Consequences of Industrial Decline.* Bristol, UK: The Policy Press.

Beverly, S. 2003. *Forces of Labor: Worker's Movements and Globalization since 1870.* Cambridge, UK: Cambridge University Press.

Bharti, N., X. Lu, L. Bengtsson, E. Wetter, and A. Tatem. 2013. "Rapid Assessment of Population Movements in Crises: The Potential and Limitations of Using Nighttime Satellite Imagery and Mobile Phone Data." Third conference on the Analysis of Mobile Phone Datasets and Networks, MIT (Media Lab), 1 May, Cambridge, MA.

Blasi, J., R. Freeman, and D. Krauss. 2014. *Citizen's Share: Reducing Inequality in the 21st Century?* New Haven, CT: Yale University Press.

Bleakley, H., and K. Cowan. 2002. "Corporate Dollar Debt and Depreciations: Much Ado about Nothing?" Working Paper 02-5. Federal Reserve Bank of Boston, MA.

Bloom, D.E., and M.J. McKenna. 2015. "Population, Labor Force, and Unemployment: Implications for the Creation of Jobs and of Decent Jobs, 1990-2030." Background think piece for Human Development Report 2015. UNDP–HDRO, New York.

Bloom, D.E., D. Canning, G. Fink, and J.E. Finlay. 2009. "Fertility, Female Labor Force Participation, and the Demographic Dividend." *Journal of Economic Growth* 14(2): 79–101. http://link.springer.com/article/10.1007%2Fs10887-009-9039-9. Accessed 11 June 2015.

BLS (United States Bureau of Labor Statistics). 2012. "International Comparisons of Manufacturing Productivity and Unit Labor Cost Trends, 2011." Economic News Release USDL-12-2365. Washington, DC. www.bls.gov/news.release/prod4.nr0.htm. Accessed 18 May 2015.

———. **2015a.** "Economic News Release: Table 1. Number and Percent of the U.S. Population who Were Eldercare Providers 1 by Sex and Selected Characteristics, Averages for the Combined Years 2011-12." Washington, DC. www.bls.gov/news.release/elcare.t01.htm. Accessed 27 July 2015.

———. **2015b.** "Economic News Release: Table A-12 Unemployment Persons by the Duration of Employment." Washington, DC.

———. **2015c.** "Economic News Release: American Time Use Survey." www.bls.gov/news.release/atus.toc.htm. Accessed 8 October 2015.

Bosch, M., A. Melguizo, and C. Pages. 2013. *Better Pensions, Better Jobs: Towards Universal Coverage in Latin America and the Caribbean.* Washington, DC: Inter-American Development Bank.

Boyde, E. 2015. "Asia Embraces the Start-up." *FT Wealth: Entrepreneurs*, May: 50–51.

Brynjolfsson, E., and A. McAfee. 2011. *Race against the Machine.* Lexington, MA: Digital Frontier Press.

———. **2014.** *The Second Machine Age: Work, Progress, and Prosperity in a Time of Brilliant Technologies.* New York: W.W. Norton and Company.

Budlender, D. 2010. "What Do Time Use Studies Tell Us about Unpaid Care Work?" In D. Budlender, ed., *Time Use Studies and Unpaid Care Work.* Geneva: United Nations Research Institute for Social Development.

Campbell, D. 2011. "Employment-led Growth and Growth-led Employment in the Recovery." In *The Global Crisis: Causes, Responses, and Challenges.* Geneva: International Labour Organization.

Carson, R. 1962. *Silent Spring.* Boston, MA: Houghton Mifflin.

CBRE Global. 2014. *Genesis Research Report: Fast Forward 2030: The Future of Work and the Workplace.* Beijing. www.cbre.com/o/international/AssetLibrary/CBRE_ Genesis_FAST_FORWARD_Workplace_2030_Exec_ Summary_E.pdf. Accessed15 June 2015.

Chappell, D., and V. Di Martino. 2006. *Violence at Work.* 3rd edition. Geneva: International Labour Organization. www.ilo.org/wcmsp5/groups/public/@dgreports/@dcomm /@publ/documents/publication/wcms_publ_9221108406_ en.pdf. Accessed 6 August 2015.

Charmes, J. 2006. "A Review of Empirical Evidence on Time Use in Africa from UN-Sponsored Surveys." In C.M. Blackden and Q. Wodon, *Gender, Time Use, and Poverty in Sub-Saharan Africa.* Working Paper 73. Washington, DC: World Bank.

————. **2015.** "Time Use across the World: Findings of a World Compilation of Time-Use Surveys." Working Paper. UNDP–HDRO, New York.

Chen, M., C. Bonner, and F. Carré. 2015. "Organizing Informal Workers: Benefits, Challenges and Successes." Background think piece for Human Development Report 2015. UNDP–HDRO, New York.

Choi, H., and H. Varian. 2009. "Predicting Initial Claims for Unemployment Benefits: Technical Report." Research at Google. http://research.google.com/archive/papers/ initialclaimsUS.pdf.

Cisco. 2015. "VNI Forecast Highlights." www.cisco.com/ web/solutions/sp/vni/vni_forecast_highlights/index.html. Accessed 15 June 2015.

Clark, A.E., E. Diener, Y. Georgellis, and R. Lucas. 2008. "Lags and Leads in Life Satisfaction: A Test of the Baseline Hypothesis." *Economic Journal*, 118(529): F222–F243.

Clark, L. 2013. "How the Red Cross and Digital Volunteers are Mapping Typhoon Haiyan to Save Lives." *Wired*, 14 November. www.wired.co.uk/news/archive/2013-11/14/ red-cross-typhoon-philippines. Accessed 4 May 2015.

Coenen, M., and R. Kok. 2014. "Workplace Flexibility and New Product Development Performance: The Role of Telework and Flexible Work Schedules." *European Management Journal* 32(4): 564–76. www.sciencedirect. com/science/journal/02632373/32/4. Accessed 4 September 2015.

Cole, S., and A.N. Fernando. 2012. "The Value of Advice: Evidence from Mobile Phone-based Agricultural Extension." Finance Working Paper 13-047. Harvard Business School, Cambridge, MA.

Colombo, F., A. Llena-Nozal, J. Mercier, and F. Tjadens. 2015. *Help Wanted? Providing and Paying for Long-term Care.* Paris. www.oecd.org/els/health-systems/help -wanted.htm. Accessed 8 June 2015.

Columbia University. 2013. "Council of Malawi to Launch Clinical Mentorship Initiative." http://icap.columbia.edu/ news-events/detail/icap-supports-the-nurses-and-midwives -council-of-malawi-to-launch-clinical. Accessed 27 May 2015.

Confederation of Indian Industry and India Ministry of New and Renewable Energy. 2010. "Human Resource Development Strategies for Indian Renewable Energy Sector." Hyderabad, India. http://mnre.gov.in/ file-manager/UserFiles/MNRE_HRD_Report.pdf. Accessed 20 July 2015.

The Conference Board. 2015. Total Economy Database. www.conference-board.org/data/economydatabase/. Accessed 15 May 2015.

Conti, G., and J.J. Heckman. 2010. "Understanding the Early Origins of the Education-Health Gradient: A Framework That Can Also Be Applied to Analyze Gene-Environment Interactions." *Perspectives on Psychological Science* 5(5): 585–605.

Cooper, D., and L. Mishel. 2015. "The Erosion of Collective Bargaining Has Widened the Gap between Productivity and Pay." Economic Policy Institute, Washington, DC. www.epi.org/publication/collective-bargainings-erosion -expanded-the-productivity-pay-gap/. Accessed 10 June 2015.

Cortez, A.L. 2012. "The International Development Strategy beyond 2015: Taking Demographic Dynamics into Account." Working Paper 122. United Nations Department of Economic and Social Affairs, New York. www.un.org/ esa/desa/papers/2012/wp122_2012. Accessed 4 June 2015.

Cowen, T. 2013. *Average Is Over: Powering America beyond the Age of Great Stagnation.* New York: Penguin.

Cramer, C. 2015. "Peace Work." Background think piece for Human Development Report 2015. UNDP–HDRO, New York.

Credit Suisse Research Institute. 2014. "Credit Suisse Global Wealth Report." Geneva. https://publications.credit -suisse.com/tasks/render/file/?fileID=60931FDE-A2D2 -F568-B041B58C5EA591A4. Accessed 15 June 2015.

Crow, D. 2015. "Doctors Hail New 'Pillars' in Fight against Cancer as Trial Data Back Latest Drug." *Financial Times*, 29 May.

D'Cunha, J., S. Lopez-Ekra, and B. Mollard. 2010. "Uncovering the Interfaces between Gender, Family, Migration and Development: The Global Care Economy and Chains." Background paper for Roundtable 2.2 on Migration, Gender and Family at the Global Forum for Migration and Development. www.gfmd.org/files/documents/gfmd_ mexico10_rt_2-2-annex_en.pdf. Accessed 30 June 2015.

DANE (Departamento Administrativo Nacional de Estadística de Colombia). 2014. "Encuesta nacional de uso del tiempo: Resultados para Bogotá." Bogota. www.dane.gov.co/files/investigaciones/boletines/ENUT/ Bol_ENUT_BTA_Ago2012_Jul2013.pdf. Accessed 29 June 2015.

Deen, T. 2012. "Women Spend 40 Billion Hours Collecting Water." Inter Press Service, 31 August. www.ipsnews. net/2012/08/women-spend-40-billion-hours-collecting -water/. Accessed 21 August 2015.

Deloitte. 2014a. "The Deloitte Millennial Survey: Big Demands and High Expectations: Executive Summary." London. http://www2.deloitte.com/content/dam/Deloitte/ global/Documents/About-Deloitte/2014_MillennialSurvey _ExecutiveSummary_FINAL.pdf. Accessed 10 July 2015.

————. **2014b.** "Value of Connectivity: Economic and Social Benefits of Expanding Internet Access." London. http:// www2.deloitte.com/content/dam/Deloitte/ie/Documents/ TechnologyMediaCommunications/2014_uk_tmt_value_ of_connectivity_deloitte_ireland.pdf. Accessed 10 July 2015.

Demirgüç-Kunt, A., and L. Klapper. 2012. "Measuring Financial Inclusion: The Global Findex Database" Policy Research Working Paper 6025. World Bank, Washington, DC.

Demirgüç-Kunt, A., L. Klapper, D. Singer, and P. Van Oudheusden. 2015. "The Global Findex Database 2014: Measuring Financial Inclusion around the World." Policy Research Working Paper 7255. World Bank, Development Research Group, Finance and Private Sector Development Team, Washington, DC.

Derviş, K. 2012. "Should Central Banks Target Employment?" 19 December. Brookings Institution. www. brookings.edu/research/opinions/2012/12/19-central -banks-employment-dervis. Accessed 18 June 2015.

Deville, P., C. Linard, S. Martin, M. Gilbert, F. Stevens, A. Gaughan, and others. 2014. "Dynamic Population Mapping Using Mobile Phone Data." *Proceedings of the National Academy of Sciences* 111(45): 15888–93.

Dishman, L. 2013. "Where Are All the Women Creative Directors?" *Fast Company*, 26 February. www.fastcompany. com/3006255/where-are-all-women-creative-directors. Accessed 23 July 2015.

Dobbs, R., J. Manyika, and J. Woetzel. 2015. *No Ordinary Disruption: The Four Global Forces Breaking All the Trends.* New York: Public Affairs.

Domestic Workers United and Data Center. 2006. "Home Is Where Work Is: Inside New York's Domestic Work Industry." Bronx, NY, and Oakland, CA. www.datacenter. org/reports/homeiswheretheworkis.pdf. Accessed 30 June 2015.

Donay, C. 2014. "The Positive Shock of the New." Briefing for Entrepreneur Summit 2014. Pictet Wealth Management, Geneva.

Dowdy, C. 2014. "Make Yourself More at Home in the Office." *Financial Times*, 24 November.

Easton, M. 2014. "Vicar or Publican: Which Jobs Make You Happy?" BBC News, 20 March. www.bbc.com/news/ magazine-26671221. Accessed 6 August 2015.

EC (European Commission). 2015. AMECO database. http://ec.europa.eu/economy_finance/ameco/user/serie/ SelectSerie.cfm. Accessed 1 July 2015.

***The Economist.* 2014a.** "Arrested Development." 4 October. http://www.economist.com/news/special-report/ 21621158-model-development-through-industrialisation -its-way-out-arrested-development. Accessed 1 July 2015.

————. **2014b.** "Why Swedish Men Take So Much Paternity Leave." 22 June. www.economist.com/blogs/economist -explains/2014/07/economist-explains-15. Accessed 22 June 2015.

————. **2015a.** "Black Moods." 6 June. www.economist. com/news/business/21653622-coals-woes-are-spreading -it-still-has-its-fans-black-moods. Accessed 15 June 2015.

————. **2015b.** "Made to Measure." 30 May. www. economist.com/news/technology-quarterly/21651925 -robotic-sewing-machine-could-throw-garment-workers -low-cost-countries-out. Accessed 15 June 2015.

————. **2015c.** "Parenting and Work: A Father's Place." 14 May. www.economist.com/news/international/21651203 -men-have-long-been-discouraged-playing-equal-role -home-last-starting. Accessed 23 July 2015.

Elms, D.K., and P. Low, eds. 2013. *Global Value Chains in a Changing World.* Geneva: World Trade Organization.

Elson, D. 2012. "Social Reproduction in the Global Crisis." In P. Utting, S. Razavi, and R. Varghese Buchholz, eds., *The*

Global Crisis and Transformative Social Change. London: Palgrave McMillan.

ENAR (European Network Against Racism). 2013. "ENAR Shadow Report 2012/13 on Racism in Europe: Key Findings on Racism and Discrimination in Employment." Brussels. www.enar-eu.org/IMG/pdf/key_findings_shadow_report_2012-13_layout.pdf. Accessed 1 July 2015.

EPA (United States Environmental Protection Agency). 2011. "The Benefits and Costs of the Clean Air Act from 1990 to 2010: Final Report – Rev. A." Washington, DC. www.epa.gov/cleanairactbenefits/feb11/fullreport_rev_a.pdf. Accessed 7 July 2015.

Epstein, G. 2007a. *Central Banks, Inflation Targeting and Employment Creation.* Economic and Labour Market Paper 2007/2. Geneva: International Labour Organization.

———. **2007b.** "Central Banks as Agents of Employment Creation." ST/ESA/2007/DWP/38. Working Paper 38. United Nations Department of Economic and Social Affairs, New York.

Esther A., B. Javorcik, and K. Ulltveit-Moe. 2015. "Globalization: A Woman's Best Friend? Exporters and the Gender Wage Gap." Discussion Paper. University of Oxford, UK. www.economics.ox.ac.uk/materials/papers/13874/paper743.pdf. Accessed 5 June 2015.

ETUI (European Trade Union Institute). 2015. "Job Quality Index (JQI)." www.etui.org/Topics/Labour-market-employment-social-policy/Job-quality-index-JQI. Accessed 1 July 2015.

Eurofound. 2013. "European Working Conditions Survey." www.eurofound.europa.eu/european-working-conditions-surveys-ewcs. Accessed 1 July 2015.

Euronews. 2015. "Greek Island of Samos Feels Strain of Migrant Influx." 21 May. www.euronews.com/2015/05/21/greek-island-of-samos-feels-strain-of-migrant-influx/. Accessed 2 July 2015.

EuropeAid. 2012. "Food Security: Rural Employment Opportunities for Public Assets (REOPA), Bangladesh." Brussels. http://ec.europa.eu/europeaid/documents/case-studies/bangladesh_food-security_reopa_en.pdf. Accessed 12 June 2015.

European Union Agency for Fundamental Rights. 2014. *Violence against Women: An EU-wide Survey.* Vienna. http://fra.europa.eu/sites/default/files/fra-2014-vaw-survey-main-results-apr14_en.pdf. Accessed 14 July 2015.

Eurostat. 2015. "Statistics Explained: Unemployment Statistics." http://ec.europa.eu/eurostat/statistics-explained/index.php/Main_Page. 1 June 2015.

Evans, P. 2015. "Expanding the Supply of Capability-Enhancing Jobs and Transforming Employment Structures: The Role of the Public Sector." Background think piece for Human Development Report 2015. UNDP–HDRO, New York.

Everest Research Institute. 2008. "Share of Market for Business Process Offshoring by Location." Dallas, TX.

Eversheds. 2014. "Compulsory Retirement: An International Comparison." Global Employment HR e-briefing, 11 April. www.eversheds.com/global/en/what/articles/index.page?ArticleID=en/Employment_and_labour_law/Global_Employment_HR_e-briefing-Compulsory_retirement_an_international_comparison. Accessed 6 July 2015.

FAO (Food and Agriculture Organization). 2009. *The State of Food Insecurity in the World 2009: Economic Crises: Impacts and Lessons Learned.* Rome. ftp://ftp.fao.org/docrep/fao/012/i0876e/i0876e.pdf. Accessed 14 July 2015.

———. **2010.** "Gender and Rural Employment Policy: Differentiated Pathways out of Poverty." Policy Brief 1–7. Rome. www.fao.org/docrep/013/i2008e/i2008e00.htm. Accessed 23 July 2015.

———. **2011a.** *Gender Inequalities in Rural Employment in Malawi: An Overview.* Rome. www.fao.org/docrep/016/ap092e/ap092e00.pdf. Accessed 23 July 2015.

———. **2011b.** *The State of Food and Agriculture 2010–11: Women in Agriculture: Closing the Gender Gap for Development.* Rome. www.fao.org/docrep/013/i2050e/i2050e.pdf. Accessed 23 July 2015.

———. **2012.** "World Agriculture towards 2030/2050: The 2012 Revision." ESA Working Paper 12-03. Rome. www.fao.org/docrep/016/ap106e/ap106e.pdf. Accessed 15 May 2015.

———. **2014.** *The State of Food and Agriculture 2014: Innovation in Family Farming.* Rome. www.fao.org/3/a-i4040e.pdf. Accessed 20 May 2015.

———. **2015.** FAOSTAT database. http://faostat3.fao.org/home/E. Accessed 10 June 2015.

Ffrench-Davis, R. 2010. "Macroeconomics for Development: From "Financierism" to "Productivism"." *CEPAL Review* 102: 7–26. www.cepal.org/publicaciones/xml/0/43000/rvi102ffrenchdavis.pdf. Accessed 15 June 2015.

———. **2012.** "Employment and Real Macroeconomic Stability: The Regressive Role of Financial Flows in Latin America." *International Labour Review* 151(1–2): 21–41.

Finland Ministry of Social Affairs and Health. 2009. "Promoting Children's Welfare in the Nordic Countries." *Our Schools/Our Selves*, Spring. www.policyalternatives.ca/sites/default/files/uploads/publications/National%20Office/2009/04/Promoting%20Children's%20Welfare%20in%20the%20Nordic%20Countries.pdf. Accessed 22 June 2015.

Folbre, N. 2015. "Valuing Non-Market Work." Background think piece for Human Development Report 2015. UNDP–HDRO, New York.

Force, L.M.T. 2013. *Toward Universal Learning: What Every Child Should Learn.*" Montreal, Canada: United Nations Educational, Scientific and Cultural Organization Institute for Statistics; Washington, DC: Center for Universal Education at Brookings. www.brookings.edu/~/media/Research/Files/Reports/2013/02/learning-metrics/LMTFRpt1TowardUnivrslLearning.pdf?la=en. Accessed 8 June 2015.

Frey, C., and M. Osborne. 2013. "The Future of Employment: How Susceptible Are Jobs to Computerisation?" Oxford Martin School, Oxford, UK.

Friedman, E., and S. Kuruvilla. 2015. "Experimentation and Decentralization in China's Labor Relations." *Human Relations* 68(2): 181–95.

Fuglie, K., and A. Nin-Pratt. 2012. "Agricultural Productivity: A Changing Global Harvest." In International Food Policy Research Institute, *Global Food Policy Report 2012.* Washington, DC. www.ifpri.org/publication/agricultural-productivity-changing-global-harvest. Accessed 3 June 2015.

Fultz, E., and J. Francis. 2013. "Cash Transfer Programmes, Poverty Reduction and Empowerment of Women: A Comparative Analysis: Experiences from Brazil, Chile, India, Mexico and South Africa." Working Paper 4/2013. International Labour Organization, Geneva. www.ilo.org/wcmsp5/groups/public/---dgreports/---gender/documents/publication/wcms_233599.pdf. Accessed 16 April 2015.

Gabre-Madhin, E. 2012. "A Market for Abdu: Creating a Commodity Exchange in Ethiopia." International Food Policy Research Institute, Washington, DC.

Galindo A., A. Izquierdo, and J. Montero. 2006. "Real Exchange Rates, Dollarization and Industrial Employment in Latin America." Working Paper 575. Inter-American Development Bank, Research Department, Washington, DC.

Gallagher, M., J. Giles, A. Park, and M. Wang. 2014. "China's 2008 Labor Contract Law: Implementation and Implications for China's Workers." *Human Relations.*

Gay, S. 1994. *Manufacturing Militance: Workers' Movements in Brazil and South Africa.* Berkeley, CA: University of California Press. Accessed 20 July 2015.

GEDI (Global Entrepreneurship and Development Institute). 2014. "The Gender Global Entrepreneurship and Development Index (GEDI): A 30-country Analysis of the Conditions That Foster High-potential Female Entrepreneurship." Washington, DC. http://cleancookstoves.org/binary-data/RESOURCE/file/000/000/299-2.pdf. Accessed 10 August 2015.

GERA (Global Entrepreneurship Research Association). 2015. Global Entrepreneurship Monitor Database. www.gemconsortium.org/data/sets. Accessed 15 June 2015.

German Federal Ministry of Labor and Social Affairs. 2015. "Social Security at a Glance 2015." Bonn. www.bmas.de/SharedDocs/Downloads/DE/PDF-Publikationen/a998-social-security-at-a-glance-total-summary.pdf?__blob=publicationFile. Accessed 15 June 2015.

Germany Trade and Invest. 2014. "Industrie 4.0: Smart Manufacturing for the Future." Berlin. www.gtai.de/GTAI/Content/EN/Invest/_SharedDocs/Downloads/GTAI/Brochures/Industries/industrie4.0-smart-manufacturing-for-the-future-en.pdf. Accessed 24 June 2015.

Glass, H., I. Kirkpatrick, and A. Schiff. 2013. "Analysing and Mapping Population Movements from Anonymous Cellphone Activity Data." Paper presented at the Third International Conference on the Analysis of Mobile Phone Datasets NetMob 2013 Special Session on the D4D Challenge, 1–3 May, Cambridge, MA. http://perso.uclouvain.be/vincent.blondel/netmob/2013/NetMob2013-program.pdf. Accessed 10 June 2015.

Glewwe, P., and A.L. Kassouf. 2008. "What Is the Impact of the Bolsa Familia Programme on Education?" OnePager 107. International Policy Centre for Inclusive Growth, Brasilia. www.ipc-undp.org/pub/IPCOnePager107.pdf. Accessed 10 June 2015.

Global Pulse and SAS Institute Inc. 2011. "Using Social Media and Online Conversations to Add Depth to Unemployment Statistics." Methodological White Paper. www.unglobalpulse.org/projects/can-social-media-mining-add-depth-unemployment-statistics. Accessed 27 May 2015.

Global Workplace Analytics. 2012. "Latest Telecommuting Statistics." http://globalworkplaceanalytics.com/telecommuting-statistics. Accessed 18 May 2015.

Goldin, C. 2014. "A Grand Gender Convergence: Its Last Chapter." *American Economic Review* 104(4): 1091–1119.

http://scholar.harvard.edu/files/goldin/files/goldin_aeapress_2014_1.pdf. Accessed 29 June 2015.

Goldin, C., and L. Katz. 2008. *The Race between Education and Technology*. Cambridge, MA: Belknap Press.

Gordon, R.J. 2014. "The Demise of U. S. Economic Growth: Restatement, Rebuttal, and Reflections." Working Paper 19895. National Bureau of Economic Research, Cambridge, MA.

Grant Thornton. 2014. *Women in Business: From Classroom to Boardroom*. Grant Thornton International Business Report 2014. London. www.grantthornton.at/files/GTI%20IBR/women-in-business-international-business-report.pdf. Accessed 14 June 2015.

———. 2015. *Women in Business: The Path to Leadership*. Grant Thornton International Business Report 2015. London. www.grantthornton.be/Resources/IBR-2015-Women-in-Business.pdf. Accessed 15 June 2015.

Grosh, M., M. Bussolo, and S. Freije, eds. 2014. *Understanding the Poverty Impact of the Global Financial Crisis in Latin America and the Caribbean*. Washington, DC: World Bank.

GSMA (Groupe Speciale Mobile Association). 2014. *The Mobile Economy: Sub-Saharan Africa 2014*. London. www.gsmamobileeconomyafrica.com/GSMA_ME_SubSaharanAfrica_Web_Singles.pdf. Accessed 15 June 2015.

Guha-Sapir, D., P. Hoyois, and R. Below. 2014. *Annual Disaster Statistical Review 2013: The Numbers and Trends*. Brussels: Centre for Research on the Epidemiology of Disasters. www.cred.be/sites/default/files/ADSR_2013.pdf. Accessed 14 July 2015.

Hall, K. 2015. "Child Poverty: Measurements, Trends and Policy Directions." Paper presented at a conference on measuring deprivation to promote human development organized by Academy of Science of South Africa, 9–10 June, Muldersdrift, South Africa.

Handa, S., and B. Davis. 2006. "The Experience of Conditional Cash Transfers in Latin America and the Caribbean." *Development Policy Review* 24(5): 513–36.

Hawkins, P., I. Blackett, and C. Heymans. 2013. *Poor-Inclusive Urban Sanitation: An Overview*. Washington, DC: World Bank. www.wsp.org/sites/wsp.org/files/publications/WSP-Poor-Inclusive-Urban-Sanitation-Overview.pdf. Accessed 20 July 2015.

Hayashi, M. 2012. "Japan's Fureai Kippu Time-banking in Elderly Care: Origins, Development, Challenges and Impact." *International Journal of Community Currency Research* 16(A): 30–44. http://ijccr.net/2012/08/16/japans-fureai-kippu-time-banking-in-%E2%80%A8elderly-care-origins-development-%E2%80%A8challenges-and-impact. Accessed 30 June 2015.

Hazelhurst, J. 2015. "The Search for Seed Capital." *FT Wealth: Entrepreneurs*, 8 May, p. 40–41.

Heckman. J. 2013. *Giving Kids a Fair Chance*. Cambridge, MA: The MIT Press.

Hellebrandt, T., and P. Mauro. 2015. "The Future of Worldwide Income Distribution." Working Paper 15-7. Peterson Institute for International Economics, Washington, DC. www.iie.com/publications/wp/wp15-7.pdf. Accessed 7 July 2015.

Helliwell, J., and H. Huang. 2011a. "New Measures of the Costs of Unemployment: Evidence from the Subjective Well-being of 2.3 Million Americans." Working Paper 2011-03. University of Alberta, Department of Economics, Edmonton, Canada. www.ualberta.ca/~econwps/2011/wp2011-03.pdf. Accessed 6 August 2015.

———. 2011b. "Wellbeing and Trust in the Workplace." *Journal of Happiness Studies* 12(5): 747–67.

Heyman, S. 2015. "A Museum at the Forefront of Digitization." *New York Times*, 13 May.

Holzman, R. 2012. "Global Pension Systems and Their Reform: Worldwide Drivers, Trends, and Challenges." Social Protection and Labor Discussion Paper 1213. World Bank, Washington, DC.

Hsu, A., O. Malik, L. Johnson, and D.C. Esty. 2014. "Development: Mobilize Citizens to Track Sustainability." *Nature* 508(7494): 33–35.

Human Rights Watch. 2014a. *Hidden Away: Abuses against Migrant Domestic Workers in the UK*. New York. www.hrw.org/node/124191. Accessed 6 August 2015.

———. 2014b. "'I Already Bought You.' Abuse and Exploitation of Female Migrant Domestic Workers in the United Arab Emirates." www.hrw.org/report/2014/10/22/i-already-bought-you/abuse-and-exploitation-female-migrant-domestic-workers-united. Accessed 6 August 2015.

IADB (Inter-American Development Bank). 2012. "La mujer latinoamericana y caribeña: más educada pero peor pagada." Washington, DC. www.iadb.org/es/noticias/articulos/2012-10-15/diferencia-salarial-entre-hombres-y-mujeres,10155.html. Accessed 3 July 2015.

IDRC (International Development Research Centre). 2013. "Growth and Economic Opportunities for Women." Ottawa. www.idrc.ca/EN/Documents/GrOW-Literature-ReviewEN.pdf. Accessed 3 June 2015.

IEA (International Energy Agency). 2012. *World Energy Outlook 2012*. Paris. www.iea.org/publications/freepublications/publication/WEO2012_free.pdf. Accessed 26 May 2015.

———. 2014. *World Energy Outlook 2014: Executive Summary*. Paris. www.iea.org/publications/freepublications/publication/WEO_2014_ES_English_WEB.pdf. Accessed 26 May 2015.

IEP (Institute for Economics and Peace). 2014. *Global Terrorism Index 2014: Measuring and Understanding the Impact of Terrorism*. New York. www.visionofhumanity.org/sites/default/files/Global%20Terrorism%20Index%202014_0.pdf. Accessed 4 June 2015.

IFC (International Finance Corporation). 2010. "Solar Lighting for the Base of the Pyramid: Overview of an Emerging Market." Washington, DC. www.ifc.org/wps/wcm/connect/a68a120048fd175eb8dcbc849537832d/SolarLightingBasePyramid.pdf?MOD=AJPERES. Accessed 1 June 2015.

ILGA (International Lesbian Gay Bisexual Trans and Intersex Association). 2014. "State- Sponsored Homophobia: A World Survey of Laws: Criminalization, Protection and Recognition of Same-sex Love." Brussels. http://old.ilga.org/Statehomophobia/ILGA_SSHR_2014_Eng.pdf. Accessed 20 July 2015.

ILO (International Labour Organization). 2003a. "Employment and Social Policy in Respect of Export Processing Zones." GB.286/ESP/3. Committee on Employment and Social Policy, Geneva.

———. 2003b. "Up in Smoke: What Future for Tobacco Jobs?" 18 September. www.ilo.org/global/about-the-ilo/newsroom/features/WCMS_071230/lang--en/index.htm. Accessed 15 May 2015.

———. 2008a. *Toolkit for Mainstreaming Employment and Decent Work*. Geneva. www.ilo.org/wcmsp5/groups/public/---dgreports/---exrel/documents/publication/wcms_172612.pdf. Accessed 15 June 2015.

———. 2008b. "ILO Declaration on Social Justice for a Fair Globalization." Geneva. www.ilo.org/wcmsp5/groups/public/---dgreports/---cabinet/documents/genericdocument/wcms_371208.pdf. Accessed 15 June 2015.

———. 2009. "Violence at Work in the European Union." Geneva. www.ilo.org/wcmsp5/groups/public/---ed_protect/---protrav/---safework/documents/publication/wcms_108536.pdf. Accessed 22 July 2015.

———. 2010a. "Fifth Item on the Agenda: HIV/AIDS and the World of Work: Report of the Committee on HIV/AIDS." Provisional Record 13(Rev.), 99th Session, Geneva. www.ilo.org/wcmsp5/groups/public/---ed_norm/---relconf/documents/meetingdocument/wcms_141773.pdf. Accessed 2 July 2015.

———. 2010b. "Recommendation 200: Recommendations Concerning HIV and AIDS and the World of Work Adopted by the Conference at Its Ninety-ninth Session." 17 June, Geneva. www.ilo.org/wcmsp5/groups/public/---ed_norm/---relconf/documents/meetingdocument/wcms_142613.pdf. Accessed 20 June 2015.

———. 2011a. *Efficient Growth, Employment and Decent Work in Africa: Time for a New Vision*. Geneva.

———. 2011b. *Equality at Work: The Continuing Challenge: Global Report under the Follow-up to the ILO Declaration on Fundamental Principles and Rights at Work*. Geneva. www.ilo.org/wcmsp5/groups/public/---ed_norm/---declaration/documents/publication/wcms_166583.pdf. Accessed 1 July 2015.

———. 2012a. *Decent Work Indicators: Concepts and Definitions*. 1st edition. Geneva. www.ilo.org/wcmsp5/groups/public/---dgreports/---integration/documents/publication/wcms_229374.pdf. Accessed 5 August 2015.

———. 2012b. *Global Employment Trends for Youth 2012*. Geneva. www.ilo.org/wcmsp5/groups/public/@dgreports/@dcomm/documents/publication/wcms_180976.pdf. Accessed 9 June 2015.

———. 2012c. *ILO Global Estimate of Forced Labour: Results and Methodology*. Geneva. www.ilo.org/wcmsp5/groups/public/---ed_norm/---declaration/documents/publication/wcms_182004.pdf. Accessed 4 June 2015.

———. 2013a. "Discrimination at Work on the Basis of Sexual Orientation and Gender Identity: Results of Pilot Research." GB.319/LILS/INF/1. Legal Issues and International Labour Standards Section Governing Body 319th Session, 16–31 October, Geneva. www.ilo.org/wcmsp5/groups/public/---ed_norm/---relconf/documents/meetingdocument/wcms_221728.pdf. Accessed 22 July 2015.

———. 2013b. "Domestic Workers across the World: Global and Regional Statistics and the Extent of Legal Protection." Geneva. www.ilo.org/wcmsp5/groups/public/---dgreports/---dcomm/---publ/documents/publication/wcms_173363.pdf. Accessed 23 July 2015.

———. 2013c. "Global Estimates and Trends of Child Labour 2000-2012." Geneva. www.ilo.org/wcmsp5/

groups/public/---ed_norm/---ipec/documents/publication/
wcms_221881.pdf. Accessed 1 July 2015.

———. 2013d. *Marking Progress against Child Labour: Global Estimates and Trends 2000-2012.* Geneva. www.ilo. org/wcmsp5/groups/public/---ed_norm/---ipec/documents /publication/wcms_221513.pdf. Accessed 6 July 2015.

———. 2013e. *Protecting Workplace Safety and Health in Difficult Economic Times: The Effect of the Financial Crisis and Economic Recession on Occupational Safety and Health.* Geneva. www.ilo.org/wcmsp5/groups/public/ ---ed_protect/---protrav/---safework/documents/ publication/wcms_214163.pdf. Accessed 2 July 2015.

———. 2013f. "Sustainable Development, Decent Work and Green Jobs: Report V." International Labour Conference, 102nd Session, Geneva. www.ilo.org/wcmsp5/groups/pub-lic/@ed_norm/@relconf/documents/meetingdocument/ wcms_207370.pdf. Accessed 3 June 2015.

———. 2014a. "Case Study: Better Work Vietnam Shows Path for Labour Law Reform." Geneva. http://betterwork. org/global/wp-content/uploads/ILO-Better-Work-Vietnam. LRWeb_.pdf. Accessed 15 June 2015.

———. 2014b. *Global Employment Trends 2014: Risk of a Jobless Recovery.* Geneva.

———. 2014c. Key Indicators of the Labor Market data-base. www.ilo.org/empelm/what/WCMS_114240/ lang--en/index.htm. Accessed 20 June 2015.

———. 2014d. NORMLEX database. www.ilo.org/dyn/ normlex/en/. Accessed 20 June 2015.

———. 2014e. *Profits and Poverty: The Economics of Forced Labour.* Geneva. www.ilo.org/wcmsp5/groups/ public/---ed_norm/---declaration/documents/publication/ wcms_243391.pdf. Accessed 22 July 2015.

———. 2014f. "Social Protection for Older Persons: Key Policy Trends and Statistics." Social Protection Policy Paper. Geneva.

———. 2014g. *World of Work Report 2014: Developing with Jobs.* Geneva. www.ilo.org/wcmsp5/groups/ public/---dgreports/---dcomm/documents/publication/ wcms_243961.pdf. Accessed 6 August 2015.

———. 2014h. *World Social Protection Report 2014/15: Building Economic Recovery, Inclusive Development and Social Justice.* Geneva.

———. 2015a. *1. What Is A National Employment Policy?* National Employment Policies: A Guide for Workers' Organization. Geneva.

———. 2015b. *Global Wage Report 2014/15: Wages and Income Inequality.* Geneva. www.ilo.org/wcmsp5/groups/ public/---dgreports/---dcomm/---publ/documents/ publication/wcms_324678.pdf. Accessed 20 July 2015.

———. 2015c. "ILO Adopts Historic Labour Standard to Tackle the Informal Economy." 12 June. www.ilo.org/ilc/ ILCSessions/104/media-centre/news/WCMS_375615/ lang--en/index.htm. Accessed 12 June 2015.

———. 2015d. ILO Social Protection Sector databases. www.ilo.org/protection/information-resources/databases. Accessed 15 May 2015.

———. 2015e. *Key Indicators of the Labour Market.* 8th edition. Geneva. www.ilo.org/empelm/what/ WCMS_114240/. Accessed 18 May 2015.

———. 2015f. "Mining: A Hazardous Work." www.ilo.org/ safework/areasofwork/hazardous-work/WCMS_124598/ lang--en/index.htm. Accessed 2 July 2015.

———. 2015g. "Overview and Topics of Labour Statistics." www.ilo.org/global/statistics-and-databases/statistics -overview-and-topics/lang--en/index.htm. Accessed 15 May 2015.

———. 2015h. *World Employment and Social Outlook: The Changing Nature of Jobs.* Geneva. www.ilo.org/wcmsp5/ groups/public/---dgreports/---dcomm/---publ/documents/ publication/wcms_368626.pdf. Accessed 6 July 2015.

———. 2015i. *World Employment and Social Outlook: Trends 2015.* Geneva. www.ilo.org/wcmsp5/groups/ public/---dgreports/---dcomm/---publ/documents/ publication/wcms_337069.pdf. Accessed 6 July 2015.

———. 2015j. ILOSTAT database. www.ilo.org/ilostat. Accessed 30 March 2015.

ILO (International Labour Organization), UNCTAD (United Nations Conference on Trade and Develop-ment), UNDESA (United Nations Department of Economic and Social Affairs), and WTO (World Trade Organization). 2012. "Macroeconomic Growth, Inclusive Growth and Employment." Thematic Think Piece. UN Sys-tem Task Team on the Post-2015 UN Development Agenda. New York. www.un.org/millenniumgoals/pdf/Think%20 Pieces/12_macroeconomics.pdf. Accessed 15 June 2015.

IMO (International Maritime Organization). 2009. "The Hong Kong International Convention for the Safe and Environmentally Sound Recycling of Ships." www.imo.org/ en/About/conventions/listofconventions/pages/the-hong -kong-international-convention-for-the-safe-and -environmentally-sound-recycling-of-ships.aspx. Accessed 17 June 2015.

INDEC (Instituto Nacional de Estadística y Censo de Argentina). 2014. "Encuesta sobre trabajo no remu-nerado y uso de tiempo." 10 July. Buenos Aires. www. indec.mecon.ar/uploads/informesdeprensa/tnr_07_14.pdf. Accessed 29 June 2015.

Innovation for Poverty Action. 2015. "Northern Uganda Social Action Fund–Youth Opportunities Program." www. poverty-action.org/project/0189. Accessed 1 June 2015.

International Federation of Robotics. 2014. Industrial Robot Statistics. www.ifr.org/industrial-robots/statistics/. Accessed 2 June 2015.

International Women's Media Foundation. 2011. *Global Report on the Status of Women in the News Media.* Washington, DC. www.iwmf.org/wp-content/ uploads/2013/09/IWMF-Global-Report-Summary.pdf. Accessed 29 June 2015.

IPCC (Intergovernmental Panel on Climate Change). 2014a. *Climate Change 2014: Mitigation of Climate Change.* Contribution of Working Group III to the Fifth Assessment Report of the Intergovernmental Panel on Climate Change. Geneva. www.ipcc.ch/report/ar5/wg3/. Accessed 26 May 2015.

———. 2014b. *Climate Change 2014: Synthesis Report.* Contribution of Working Groups I, II and III to the Fifth Assessment Report of the Intergovernmental Panel on Climate Change. Geneva. www.ipcc.ch/report/ar5/syr/. Accessed 26 May 2015.

IPU (Inter-Parliamentary Union). 2015. *Women in Politics: 2015.* New York. http://www.ipu.org/pdf/publications/ wmnmap15_en.pdf. Accessed 26 May 2015.

IRENA (International Renewable Energy Agency). 2013. *Renewable Energy and Jobs.* Abu Dhabi. www.irena.org/ rejobs.pdf. Accessed 27 May 2015.

———. 2015. *Renewable Energy and Jobs: Annual Review 2015.* Abu Dhabi. www.irena.org/DocumentDownloads/ Publications/IRENA_RE_Jobs_Annual_Review_2015.pdf. Accessed 27 May 2015.

Islam, R. 2015. "Globalization of Production, Work and Human Development: Is Race to the Bottom Inevitable?" Background think piece for Human Development Report 2015. UNDP–HDRO, New York.

Islam, R., and I. Islam. 2015. *Employment and Inclusive Development.* New York: Routledge.

ITU (International Telecommunication Union). 2013. "ICT Facts and Figures: The World in 2013." Geneva.

———. 2015. "ICT Facts and Figures: The World in 2015." Geneva. www.itu.int/en/ITU-D/Statistics/Documents/ facts/ICTFactsFigures2015.pdf. Accessed 15 June 2015.

IUES (Institute for Urban and Environmental Studies) and CASS (Chinese Academy of Social Sciences). 2010. "Study on Low Carbon Development and Green Employment in China." Beijing.

Jacobs, E. 2014. "Women Earn Less than Men Even When They Set the Pay." *Financial Times,* 6 November. www. ft.com/cms/s/0/79a98b40-59d6-11e4-9787-00144feab 7de.html. Accessed 27 June 2015.

———. 2015a. "Workers of the Gig Economy." *Financial Times,* 13 March.

———. 2015b. "Working Older." *Financial Times,* 3 July.

Jahan, S. 2005. "Reorienting Development: Towards an Engendered Employment Strategy." Working Paper 5. United Nations Development Programme, International Poverty Centre, Brasilia.

Jana, S., B. Dey, S. Reza-Paul, and R. Steen. 2014. "Combatting Human Trafficking in the Sex Trade: Can Sex Workers Do It Better?" *Journal of Public Health* 36(4): 622–28.

***The Japan Times.* 2015.** "Seniors Grow to 26% of Population as Japan Shrinks for Fourth Year Straight." 17 April. www.japantimes.co.jp/news/2015/04/17/national/ seniors-grow-26-population-japan-shrinks-fourth-year -straight/. Accessed 23 July 2015.

Jaumotte, F., and C. Buitron. 2015. "Power from the People." *Finance & Development* 52(1): 29–31.

Jiménez Cisneros, B.E., Oki, T., Arnell, N.W., and others. 2014. "Freshwater Resources." In C.B. Field, V.R. Barros, D.J. Dokken, and others, eds., *Climate Change 2014: Impacts, Adaptation, and Vulnerability. Part A: Global and Sectoral Aspects.* Contribution of Working Group II to the Fifth Assessment Report of the Intergovernmental Panel on Climate Change. New York: Cambridge University Press.

Kabanda, P. 2015. "Work as Art: How the Linkages between Creative Work and Human Development Flow: Analytical Paper on the Linkages between Creative Work and Human Development." Background think piece for Human Development Report 2015. UNDP–HDRO, New York.

Kagermann, H., W.-D. Lukas, and W. Wahlster. 2011. "Industrie 4.0: Mit dem Internet der Dinge auf dem Weg zur 4. industriellen Revolution." www.ingenieur.de/ Themen/Produktion/Industrie-40-Mit-Internet-Dinge-Weg -4-industriellen-Revolution. Accessed 3 June 2015.

Kalil, T., and J. Miller. 2015. "Advancing U.S. Leadership in High-Performance Computing." White House Blog, 29 July. www.whitehouse.gov/blog/2015/07/29/

advancing-us-leadership-high-performance-computing. Accessed 14 September 2015.

Karabarbounis, L., and B. Neiman. 2014. "The Global Decline of the Labor Share." *Quarterly Journal of Economics* 129(1): 61–103.

Kasedde, S., A. Doyle, J. Seeley, and D. Ross. 2014. "They Are Not Always a Burden: Older People and Child Fostering in Uganda during the HIV Epidemic." *Social Science and Medicine* 113: 161–68.

Kaye, M. 2006. *Contemporary Forms of Slavery in Argentina.* Anti-Slavery International 2006, London. www.antislavery.org/includes/documents/cm_docs/2009/c/contemporary_forms_of_slavery_in_argentina.pdf. Accessed 2 July 2015.

Kearney, M.S., B. Hershbein, and D. Boddy. 2015. "The Future of Work in the Age of the Machine." Framing Paper. The Hamilton Project, Washington, DC. www.hamiltonproject.org/papers/future_of_work_in_machine_age/. Accessed 9 June 2015.

Kerrissey, J. 2015. "Collective Labour Rights and Income Inequality." *American Sociological Review* 80(3): 626–53.

Kharas, H., and G. Gertz. 2010. "The New Global Middle Class: A Crossover from West to East." In C. Li, ed., *China's Emerging Middle Class: Beyond Economic Transformation.* Washington, DC: Brookings Institution Press.

Kingsley-Hughes, A. 2012. "Apple Sells 645,000 Devices per Day in Q2." ZDNet, 25 April. www.zdnet.com/article/apple-sells-645000-devices-a-day-during-q2/. Accessed 18 May 2015.

Kivimäki, M., M. Jokela, S.T. Nyberg, A. Singh-Manoux, E.I. Fransson, L. Alfredsson, and others. 2015. "Long Working Hours and Risk of Coronary Heart Disease and Stroke: A Systematic Review and Meta-analysis of Published and Unpublished Data for 603,838 Individuals." *Lancet,* Online First, 9 August.

Ko, P., and K. Hank. 2013. "Grandparents Caring for Grandchildren in China and Korea: Findings from CHARLS and KLoSA." *Journal of Gerontology: Series B: Psychological Sciences and Social Sciences* 69(4): 646–51. http://psychsocgerontology.oxfordjournals.org/content/69/4/646.full.pdf+html. Accessed 23 July 2015.

Kostzer, D. 2008. "Argentina: A Case Study of the Plan Jefes y Jefas de Hogar Desocupados, or Employment Road to economic Recovery." Working Paper 534. Levy Economics Institute of Bard College, Annandale-on-Hudson, New York.

Kretkowski, P. 1998. "The 15 Percent Solution." *Wired,* 23 January. http://archive.wired.com/techbiz/media/news/1998/01/9858. Accessed 5 August 2015.

Krug, E.G., L.L. Dahlberg, J.A. Mercy, A.B. Zwi, and R. Lozano, eds. 2002. *World Report on Violence and Health.* Geneva: World Health Organization. http://whqlibdoc.who.int/publications/2002/9241545615_eng.pdf. Accessed 4 June 2015.

Krugman, P. 2014. "Inflation Targets Reconsidered." Draft paper for European Central Bank Forum on Central Banking, 21–23 May, Sintra, Portugal. https://2014.ecbforum.eu/up/artigos-bin_paper_pdf_0134658001400681089-957.pdf. Accessed 15 May 2015.

Kuehnast, K. 2015. "Gender and Peacebuilding: Why Women's Involvement in Peacebuilding Matters." United States Institute of Peace. www.buildingpeace.org/think-global-conflict/issues/gender-and-peacebuilding. Accessed 29 June 2015.

Kurtzleben, D. 2013. "CHARTS: New Data Show, Women, More Educated Doing Most Volunteering." U.S. News, 27 February. www.usnews.com/news/articles/2013/02/27/charts-new-data-show-women-more-educated-doing-most-volunteering. Accessed 29 June 2015.

Kynge, J., and J. Wheatley. 2015. "Emerging Markets: Redrawing the World Map." *Financial Times*, 3 August.

Lagesse, D. 2015. "Virtual Volunteers Use Twitter and Facebook to Make Maps of Nepal." NPR, 15 May. www.npr.org/sections/goatsandsoda/2015/05/05/404438272/virtual-volunteers-use-twitter-and-facebook-to-make-maps-of-nepal. Accessed 21 May 2015.

Lambert, K., and C. Driscoll. 2003. "Nitrogen Pollution: From the Sources to the Sea." Hubbard Brook Research Foundation, Hanover, NH.

Lamichhane, K. 2015. *Disability, Education and Employment in Developing Countries: From Disability to Investment.* New Delhi, India: Cambridge University Press.

Leahy, J. 2015. "Rouseff Unveils Plan to Recovery." *Financial Times*, 10 June.

Lewis, L. 2015. "Mind the Gap." *Financial Times*. Big Read, 7 July. https://soundcloud.com/ft-analysis/japan-mind-the-gap. Accessed 23 July 2015.

Lippoldt, D., ed. 2012. *Policy Priorities for International Trade and Jobs.* Paris: Organisation for Economic Co-operation and Development.

Littleton, C. 2014. "Employment of Women in Film Production Dips Below 1998 Levels." *Variety*, 14 January. http://variety.com/2014/film/news/employment-of-women-in-film-production-dips-below-1998-levels-1201055095/. Accessed 23 July 2015.

Liu, Q., and O.N. Skans. 2010. "The Duration of Paid Parental Leave and Children's Scholastic Performance." *B. E. Journal of Economic Analysis and Policy* 10(1).

Liu, Y., F. Yang, and X. Li. 2013. "Employment and Decent Work in China's Forestry Industry." Draft Working Paper. International Labour Organization, Sectoral Activities Department, Country Office for China and Mongolia, Geneva. http://apgreenjobs.ilo.org/resources/employment-and-decent-work-in-china2019s-forestry-industry-draft. Accessed 5 June 2015.

Llorente, A., M. Garcia-Herranz, M. Cebrian, and E. Moro. 2014. "Social Media Fingerprints of Unemployment." arXiv:1411.3140 [physics.soc-ph].

Lu, Y., N. Nakicenovic, M. Visbeck, and A. Stevance. 2015. "Five Priorities for the UN Sustainable Development Goals." *Nature* 520: 432–33. www.nature.com/polopoly_fs/1.17352!/menu/main/topColumns/topLeftColumn/pdf/520432a.pdf. Accessed 8 June 2015.

Luce, S., J. Luff, J.A. McCartin, and R. Milkman, eds. 2014. *What Works for Workers? Public Policies and Innovative Strategies for Low-Wage Workers.* New York: Russell Sage Foundation.

Machado A., G. Geaquinto Fontes, M. Furlan Antigo, R. Gonzalez, and F. Veras Soares. 2011. "Bolsa Família as Seen through the Lens of the Decent Work Agenda." OnePager 133. International Policy Centre for Inclusive Growth, Brasilia. www.ipc-undp.org/pub/IPCOnePager133.pdf. Accessed 14 July 2015.

Maddison, A. 2008. "Historical Statistics of the World Economy: 1-2008 AD." www.ggdc.net/maddison/Historical_Statistics/horizontal-file_02-2010.xls. Accessed 25 June 2015.

Maier, S. 2008. "Empowering Women through ICT-Based Business Initiatives: An Overview of Best Practices in E-Commerce/E-Retailing Projects." *Information Technologies and International Development* 4(2): 43–60.

Mandel, M. 2013. "752,000 App Economy Jobs on the 5th Anniversary of the App Store." Progressive Policy Institute, Washington, DC. www.progressivepolicy.org/slider/752000-app-economy-jobs-on-the-5th-anniversary-of-the-app-store/. Accessed 25 June 2015.

Masters, B. 2015. "The Nuts and Bolts of Robot-Human Working Relations." *Financial Times*, 3 July.

Masur, J., and E. Posner. 2012. "Regulation, Unemployment, and Cost-Benefit Analysis." *Virginia Law Review* 98(3): 579–634.

May, M. 2007. *The Elegant Solution: Toyota's Formula for Mastering Innovation.* New York: Simon and Schuster.

McCarthy, T. 2012. "Encyclopedia Britannica Halts Print Publication after 244 Years." *The Guardian*, 13 March. www.theguardian.com/books/2012/mar/13/encyclopedia-britannica-halts-print-publication. Accessed 26 May 2015.

McKinsey & Company. 2014. "Education to Employment: Getting Europe's Youth to Work." New York. www.mckinsey.com/insights/social_sector/converting_education_to_employment_in_europe. Accessed 25 June 2015.

McKinsey Global Institute. 2012a. "Manufacturing the Future: The Next Era of Global Growth and Innovation." New York.

———. **2012b.** "The World at Work: Jobs, Pay and Skills for 3.5 Billion People." New York.

———. **2013.** "Disruptive Technologies: Advances That Will Transform Life, Business, and the Global Economy." New York.

———. **2014.** "Global Flows in a Digital Age: How Trade, Finance, People, and Data Connect the World Economy." New York. www.mckinsey.com/insights/globalization/global_flows_in_a_digital_age. Accessed 18 June 2015.

Milken Institute. 2013. "In Tech We Trust? A Debate with Peter Thiel and Marc Andreessen." Milken Institute, Santa Monica, CA.

Miller, C.C. 2014. "Pay Gap Is Because of Gender, Not Jobs." *New York Times*, 23 April. www.nytimes.com/2014/04/24/upshot/the-pay-gap-is-because-of-gender-not-jobs.html. Accessed 3 July 2015.

———. **2015a.** "The 24/7 Work Culture's Toll on Families and Gender Equality." *New York Times*, 28 May. www.nytimes.com/2015/05/31/upshot/the-24-7-work-cultures-toll-on-families-and-gender-equality.html. Accessed 5 August 2015.

———. **2015b.** "Mounting Evidence of Advantages for Children of Working Mothers." *New York Times*, 15 May. www.nytimes.com/2015/05/17/upshot/mounting-evidence-of-some-advantages-for-children-of-working-mothers.html. Accessed 11 August 2015.

Mishel, L., and A. Davis. 2014. "CEO Pay Continues to Rise as Typical Workers Are Paid Less." Issue Brief 380. Economic Policy Institute, Washington, DC.

Mishkin, S. 2014. "Saudi Arabia to Use edX Web Courses to Train Unemployed." *Financial Times*, 15 July. www.ft.com/

intl/cms/s/0/67fe0cb8-0c3d-11e4-943b-00144feabdc0.
html. Accessed 22 July 2015.

Montenegro, C., and H. Patrinos. 2014. "Comparable
Estimates of Returns to Schooling around the World."
Policy Research Working Paper 7020. World Bank,
Washington, DC.

Mukherjee, S., and S. Nayyar. 2015. "Aging and the
Care Burden: Implications for Developed and Developing
Countries." Working Paper. UNDP–HDRO, New York.

Munk, M.R., and R. Rückert. 2015. "To Work or Not
Shouldn't Be a Question." *Science* 348(6233): 470.
www.sciencemag.org/cgi/pmidlookup?view=long&pm
id=25908825. Accessed 29 June 2015.

Nobel Media. 2015. "889 Nobel Laureates since 1901."
www.nobelprize.org/nobel_prizes/. Accessed 30 June
2015.

**Norwegian Refugee Council and IDMC (Internal
Displacement Monitoring Center). 2015.** *Global
Estimates 2015: People Displaced by Disasters.*
Geneva. www.internal-displacement.org/assets/library/
Media/201507-globalEstimates-2015/20150713-global
-estimates-2015-en-v1.pdf. Accessed 31 July 2015.

Nuzhat, N. 2015. "Mobile Technology towards Human
Development." Background think piece for Human
Development Report 2015. UNDP–HDRO, New York.

**OECD (Organisation for Economic Co-operation and
Development). 2007.** *Offshoring and Employment: Trends
and Impacts.* Paris.

———. **2013a.** *Education at a Glance 2013: OECD
Indicators.* Paris.

———. **2013b.** *OECD Skills Outlook 2013: First Results from
the Survey of Adult Skills.* Paris.

———. **2014.** *OECD Employment Outlook 2014.* Paris.

———. **2015a.** OECD Family database. Paris. www.oecd.
org/social/family/database.htm. Accessed 23 July 2015.

———. **2015b.** OECDStat. http://stats.oecd.org. Accessed
2 July 2015.

———. **2015c.** "Work-Life Balance." www.oecdbetterlife
index.org/topics/work-life-balance/. Accessed 22 July
2015.

Ortiz, I., and M. Cummins. 2012. "When the Global Crisis
and Youth Bulge Collide." Social and Economic Policy
Working Paper. United Nations Children's Fund, New York.

Oxfam. 2015. "Wealth: Having It All and Wanting More."
Issue Briefing. Oxford, UK. www.oxfam.org/sites/www.
oxfam.org/files/file_attachments/ib-wealth-having-all
-wanting-more-190115-en.pdf. Accessed 10 July 2015.

Patnaik, A. 2015. "Reserving Time for Daddy: The Short
and Long-Run Consequences of Fathers' Quotas." Cornell
University, Ithaca, NY. http://ssrn.com/abstract=2475970.
Accessed 18 June 2015.

**Pearlin, L.I., S. Schieman, E.M. Fazio, and S.C.
Meersman. 2005.** "Stress, Health, and the Life Course:
Some Conceptual Perspectives." *Journal of Health and
Social Behavior* 46(2): 205–19.

Pollin, R. 2015. *Greening the Global Economy.* Cambridge,
MA: MIT Press.

Pooler, M. 2014. "Crowdworkers Team Up on Pay and
Practices." *Financial Times*, 3 November.

Poschen, P. 2015. *Decent Work, Green Jobs and the
Sustainable Economy.* Geneva: International Labour

Organization. www.ilo.org/global/publications/books/
WCMS_373209/lang--en/index.htm. Accessed 1 June
2015.

Prada, M.F., G. Rucci, and S.S. Urzúa. 2015. "The Effects
of Mandated Child Care on Female Wages in Chile."
Working Paper 21080. National Bureau of Economic
Research, Cambridge, MA. www.nber.org/papers/w21080.
pdf. Accessed 29 June 2015.

Pratap, S., I. Lobato, and A. Somuano. 2003. "Debt
Composition and Balance Sheet Effects of Exchange Rate
Volatility in Mexico: A Firm Level Analysis." *Emerging
Markets Review* 4(4): 450–71.

The Pregnancy Test. 2014. "Ending Discrimination at Work
for New Mothers." Trade Union Congress, London. www.
tuc.org.uk/sites/default/files/pregnancytestreport.pdf.
Accessed 29 June 2015.

**Qiang C., S. Kuek, A. Dymond, and S. Esselaar.
2011.** "Mobile Applications for Agriculture and Rural
Development." World Bank, Washington, DC. http://
siteresources.worldbank.org/INFORMATIONAND
COMMUNICATIONANDTECHNOLOGIES/Resources/
MobileApplications_for_ARD.pdf. Accessed 4 June 2015.

Raghuram, P. 2001. "Caste and Gender in the Organisation
of Paid Domestic Work in India." *Work, Employment &
Society* 15(3): 607–17.

Rakkee, T., and S.K. Sasikumar. 2012. *Migration of
Women Workers from South Asia to the Gulf.* New Delhi:
V.V. Giri National Labour Institute and UN Women. www.
ucis.pitt.edu/global/sites/www.ucis.pitt.edu.global/files/
migration_women_southasia_gulf.pdf. Accessed 24 July
2015.

Rodrik, D. 2015a. "Premature Deindustrialization." Working
Paper 20935. National Bureau of Economic Research,
Cambridge, MA.

———. **2015b.** "Work and Human Development in a De-
industrializing World." Background think piece for Human
Development Report 2015. UNDP–HDRO, New York.

Rosengren, E. 2013. "Should Full Employment Be a
Mandate for Central Banks?" Speech at the Federal
Reserve Bank of Boston's 57th Economic Conference on
"Fulfilling the Full Employment Mandate – Monetary
Policy & The Labor Market," 12 April, Boston, MA.

Ryder, G. 2015. "Labor in the Age of Robots." Project
Syndicate, 22 January. www.project-syndicate.org/
commentary/labor-in-the-age-of-robots-by-guy-ryder
-2015-01. Accessed 15 June 2015.

**Salamon, L.M., S.W. Sokolowski, and M.A. Haddock.
2011.** "Measuring the Economic Value of Volunteer Work
Globally: Concepts, Estimates and a Roadmap to the
Future." *Annals of Public and Cooperative* 82(3): 217–52.
http://ccss.jhu.edu/wp-content/uploads/downloads/
2011/10/Annals-Septmeber-2011.pdf. Accessed 1 July
2015.

Salazar-Xirinachs, J. 2015. "Trends and Disruptions and
Their Implication for the Future of Jobs." World Economic
Forum, International Business Council, Geneva.

Sayer, L.C. 2015. "The Complexities of Interpreting Changing
Household Patterns." Council on Contemporary Families, 7
May. https://contemporaryfamilies.org/complexities-brief
-report/. Accessed 29 June 2015.

Schifferes, S. 2002. "Racism at Work: Workers Feel the
Effects of Prejudice." BBC. http://news.bbc.co.uk/hi/

english/static/in_depth/uk/2002/race/racism_at_work.
stm. Accessed 1 July 2015.

Schrank, W. 2003. "Introducing Fisheries Subsidies."
Fisheries Technical Paper 437. Food and Agriculture
Organization, Rome. www.fao.org/3/a-y4647e.pdf.
Accessed 10 June 2015.

Scott, J. 2008. "SLIDES: Threats to Biological Diversity:
Global, Continental, Local." Presentation at Summer
Conference on Shifting Baselines and New Meridians:
Water, Resources, Landscapes, and the Transformation of
the American West, 4–6 June, Boulder, CO. http://scholar.
law.colorado.edu/cgi/viewcontent.cgi?article=1014&
context=water-resources-and-transformation-of-American
-West. Accessed 10 June 2015.

Seckan, B. 2013. "Workplace Violence in America:
Frequency and Effects." *Journalist's Resources*, 12
May. http://journalistsresource.org/studies/economics/
workers/workplace-violence-america-frequency-effects.
Accessed 22 July 2015.

Selim, N. 2013. "Innovative Approaches to Job Creation."
Background note for World Development Report 2013.
World Bank, Washington, DC.

**Shannon, K., S.Z. Strathdee, S.M. Goldenberg, P.
Duff, P. Mwangi, M. Rusakova, and others. 2015.**
"Global Epidemiology of HIV among Female Sex Workers:
Influence of Structural Determinants." *Lancet* 385(9962):
55–71. www.thelancet.com/journals/lancet/article/
PIIS0140-6736(14)60931-4/abstract. Accessed 2 July
2015.

Scheil-Adlung, X. 2015. "Long-term Care Protection for
Older Persons: A Review of Coverage Deficits in 46
Countries." ESS Working Paper 50. International Labour
Organization, Geneva. www.ilo.org/wcmsp5/groups/
public/---ed_protect/---soc_sec/documents/publication/
wcms_407620.pdf. Accessed 8 October 2015.

Shi, L. 2008. "Rural Migrant Workers in China: Scenario,
Challenges and Public Policy." Working Paper 89.
International Labour Organization, Policy Integration and
Statistics Department, Geneva. www.ilo.org/wcmsp5/
groups/public/---dgreports/---integration/documents/
publication/wcms_097744.pdf. Accessed 6 August 2015.

Siemens AG. 2015. "Fact Sheet: Amberg Electronics Plant."
Munich, Germany. www.siemens.com/press/pool/de/
events/2015/corporate/2015-02-amberg/factsheet-amberg
-en.pdf. Accessed 3 June 2015.

Singer, S., J.E. Amorós, and D. Moska Arreola. 2015.
Global Entrepreneurship Monitor: 2014 Global Report.
London: Global Entrepreneurship Research Association.
www.gemconsortium.org/report. Accessed 18 June 2015.

Skarhed, A. 2010. "Förbud mot köp av sexuell tjänst
En utvärdering 1999–2008." Government of Sweden,
Stockholm.

Skidelsky, R. 2015. "Minimum Wage or Living Income"
Project Syndicate, 16 July. www.project-syndicate.org/
commentary/basic-income-tax-subsidies-minimum-wage
-by-robert-skidelsky-2015-07. Accessed 20 July 2015.

Smale, A., and C.C. Miller. 2015. "Germany Sets Gender
Quota in Boardrooms." *New York Times*, 6 March.

Smeaton, D., K. Ray, and G. Knight. 2014. "Costs and
Benefits to Business of Adopting Work Life Balance
Working Practices: A Literature Review." UK Department
for Business, Innovation and Skills, London. http://www.
psi.org.uk/images/uploads/bis-14-903-costs-and-benefits

-to-business-of-adopting-work-life-balance-working -practices-a-literature-review.pdf. Accessed 4 September 2015.

Soares, S. 2012. "Bolsa Família, Its Design, Its Impacts and Possibilities for the Future." Working Paper 89. International Policy Center for Inclusive Growth, Brasilia.

Social Tech Guide. 2015. "Project Details: 3D Hubs." http://socialtech.org.uk/projects/3d-hubs/. Accessed 4 June 2015.

Statista. 2014. "Statistics and Facts on Internet Usage in India." www.statista.com/topics/2157/internet-usage-in -india/. Accessed 2 June 2015.

———. 2015. "Distribution of Internet Users in China from December 2012 to December 2014, by Gender." www. statista.com/statistics/265148/percentage-of-internet -users-in-china-by-gender/. Accessed 25 June 2015.

Steffen, W., K. Richardson, J. Rockström, S. Cornell, I. Fetzer, E. Bennett, and others. 2015. "Planetary Boundaries: Guiding Human Development on a Changing Planet." *Science* 347(6223). www.sciencemag.org/content/ 347/6223/1259855.abstract. Accessed 26 May 2015.

Stevens, G. 2012. "Women in Architecture 1: First Thoughts." Architectural Blatherations blog. www.archsoc.com/kcas/ ArchWomen.html. Accessed 16 April 2012.

Stewart, F. 2015. "Employment in Conflict and Post-Conflict Situations." Background think piece for Human Development Report 2015. UNDP–HDRO, New York.

Stiglitz, J. 2015. *The Great Divide*. New York: W.W. Norton and Company.

Stockhammer, E. 2013. "Why Have Wage Shares Fallen? A Panel Analysis of the Determinants of Functional Income Distribution. Conditions of Work and Employment 35. International Labour Organization, Geneva.

Strietska-Ilina, O., C. Hofmann, M. Durán Haro, and S. Jeon. 2011. *Skills for Green Jobs: A Global View*. Geneva: International Labour Organization. www.ilo.org/wcmsp5/ groups/public/---dgreports/---dcomm/---publ/documents/ publication/wcms_159585.pdf. Accessed 28 May 2015.

Suhoy T. 2009. "Query Indices and a 2008 Downturn: Israeli Data." Discussion Paper 2009.06. Bank of Israel, Jerusalem. www.bankisrael.gov.il/deptdata/mehkar/ papers/dp0906e.pdf. Accessed 5 June 2015 .

Surowiecki, J. 2008. "The Open Secret of Success." *New Yorker*, 12 May.

Suwala, W. 2010. "Lessons Learned from the Restructuring of Poland's Coal-mining Industry." International Institute for Sustainable Development, Winnipeg, Canada.

Swanson, B., and K. Davis. 2014. "Status of Agricultural Extension and Rural Advisory Services Worldwide." Global Forum for Rural Advisory Services, Lindau, Switzerland. www.g-fras.org/images/wwes/GFRAS-Status_of_Rural_ Advisory_Services_Worldwide.pdf. Accessed 20 July 2015.

Tal, B. 2015. "Canadian Employment Quality Index: Employment Quality—Trending Down." Canadian Imperial Bank of Commerce. Toronto, Canada. http://research. cibcwm.com/economic_public/download/eqi_20150305. pdf. Accessed 1 July 2015.

Tate, R. 2013. "Google Couldn't Kill 20 Percent Time Even If It Wanted To." *Wired*, 21 August. www.wired. com/2013/08/20-percent-time-will-never-die/. Accessed 3 June 2015.

Tcherneva, P. 2014. "The Social Enterprise Model for a Job Guarantee in the United States." Policy Note Archive 2014/1. Levy Economics Institute of Bard College, Annandale-on-Hudson, New York.

Timmer, M.P, G.J. de Vries, and K. de Vries. 2014a. "Patterns of Structural Change in Developing Countries." Research Memorandum 149. Groningen Growth and Development Centre, Groningen, The Netherlands.

Timmer, M.P., A.A. Erumban, B. Los, R. Stehrer, and G.J. de Vries. 2014b. "Slicing Up Global Value Chains." *Journal of Economic Perspectives* 28(2): 99–118.

Tokman, V. 2010. "Domestic Workers in Latin America: Statistics for new policies." Working Paper (Statistics) 17. Women in Informal Employment Globalizing and Organizing, Cambridge, MA. http://wiego.org/sites/wiego. org/files/publications/files/Tokman_WIEGO_WP17.pdf. Accessed 30 June 2015.

Turkish Statistical Institute. 2015. "Information and Communication Technology (ICT) Usage in Households and by Individuals." Ankara. www.turkstat.gov.tr/PreTablo.do ?alt_id=1028. Accessed 20 June 2015.

TWC2 (Transient Workers Count Too). 2011. "Fact Sheet: Foreign Domestic Workers in Singapore (Basic Statistics)." https://twc2.org.sg/2011/11/16/fact-sheet-foreign-domestic -workers-in-singapore-basic-statistics/. Accessed 30 June 2015.

UK Cabinet Office. 2013. "Subjective Wellbeing and Employment: Analysis of the Annual Population Survey (APS) Wellbeing Data, Apr-Oct 2011." London. www.gov. uk/government/uploads/system/uploads/attachment_ data/file/225510/subjective_wellbeing_employment.pdf. Accessed 6 August 2015.

UN (United Nations). 2000a. "Resolution 1325 (2000)." S/ RES/1325, 31 October. http://daccess-dds-ny.un.org/doc/ UNDOC/GEN/N00/720/18/PDF/N0072018.pdf. Accessed 29 June 2015.

———. 2000b. "United Nations Millennium Declaration." Resolution 55/2 adopted by the General Assembly, 8 September, New York. www.un.org/millennium/declaration/ ares552e.htm. Accessed 16 June 2015.

———. 2012a. "Education and Skills for Inclusive Sustainable Development beyond 2015." Thematic Think Piece. UN System Task Team on the Post-2015 UN Development Agenda. New York. www.un.org/ millenniumgoals/pdf/Think%20Pieces/4_education.pdf. Accessed 1 July 2015.

———. 2012b. "The Future We Want." Resolution 66/288 adopted by the General Assembly, 27 July. New York. www.un.org/ga/search/view_doc.asp?symbol=A/ RES/66/288&Lang=E. Accessed 1 July 2015.

———. 2014. *A World That Counts: Mobilising the Data Revolution for Sustainable Development*. New York. www. undatarevolution.org/report/. Accessed 25 June 2015.

———. 2015a. "Achieving Sustainable Development through Employment Creation and Decent Work for All." Background Note. New York. www.un.org/en/africa/osaa/ pdf/events/20150330/backgrounder.pdf. Accessed 25 June 2015.

———. 2015b. *The Millennium Development Goals Report 2015*. New York. www.un.org/millenniumgoals/2015_ MDG_Report/pdf/MDG%202015%20rev%20(July%201). pdf. Accessed 30 July 2015.

———. 2015c. "Transforming Our World: The 2030 Agenda for Sustainable Development." New York.

UN (United Nations), World Bank, European Union, and African Development Bank. 2015. *Recovering from the Ebola Crisis*. Geneva. www.ilo.org/wcmsp5/groups/ public/---dgreports/---dcomm/---publ/documents/ publication/wcms_359364.pdf. Accessed 20 July 2015.

UN Global Pulse. 2013. "Big Data for Development: A Primer." www.unglobalpulse.org/sites/default/files/ Primer%202013_FINAL%20FOR%20PRINT.pdf. Accessed 20 June 2015.

UN Volunteers. 2011. *State of the World's Volunteerism Report: Universal Values for Global Well-being*. New York: United Nations Development Programme. www.unv.org/ fileadmin/docdb/pdf/2011/SWVR/English/SWVR2011_ full.pdf. Accessed 20 July 2015.

———. 2014. "UNV Online Volunteering Service." www. onlinevolunteering.org/en/vol/about/index.html. Accessed 20 July 2015.

UN Women. 2012a. *Estimating the Costs of Domestic Violence against Women in Viet Nam*. Hanoi. www. unwomen.org/~/media/headquarters/attachments/ sections/library/publications/2013/2/costing-study-viet -nam%20pdf.pdf. Accessed 4 June 2015.

———. 2012b. "Women's Participation in Peace Negotiations: Connections between Presence and Influence." In *UN Women Sourcebook on Women, Peace and Security*. New York.

———. 2014. "Facts and Figures: Ending Violence against Women: A Pandemic in Diverse Forms." www.unwomen. org/en/what-we-do/ending-violence-against-women/ facts-and-figures. Accessed 4 June 2015.

———. 2015. *Progress of the World's Women 2015-2016: Transforming Economies, Realizing Rights*. New York. http://progress.unwomen.org/en/2015/pdf/UNW_ progressreport.pdf. Accessed 20 July 2015.

UN WomenWatch. 2009. "Fact Sheet: Women, Gender Equality and Climate Change." www.un.org/womenwatch/ feature/climate_change/. Accessed 16 October 2015.

UNAIDS (Joint United Nations Programme on HIV and AIDS). 2012. *UNAIDS Guidance Note on HIV and Sex Work*. Geneva. www.unaids.org/sites/default/files/ sub_landing/files/JC2306_UNAIDS-guidance-note-HIV -sex-work_en.pdf. Accessed 2 July 2015.

———. 2015. *How AIDS Changed Everything—MDG6: 15 Years; 15 Lessons of Hope from the AIDS Response*. Geneva. www.unaids.org/en/resources/documents/2015/ MDG6_15years-15lessonsfromtheAIDSresponse. Accessed 28 May 2015.

UNCTAD (United Nations Conference on Trade and Development). 2014. *Services: New Frontier for Sustainable Development*. Geneva.

———. 2015. UNCTADStat database. http://unctadstat. unctad.org/wds/ReportFolders/reportFolders.aspx. Accessed 2 June 2015.

UNDESA (United Nations Department of Economic and Social Affairs). 2012. "World Population 2012." New York. www.un.org/en/development/desa/population/ publications/pdf/trends/WPP2012_Wallchart.pdf. Accessed 20 May 2015.

———. 2013a. *International Migration Report 2013*. New York. www.un.org/en/development/desa/population/

publications/pdf/migration/migrationreport2013/Full_Document_final.pdf. Accessed 20 August 2015.

———. 2013b. *World Population Prospects: The 2012 Revision. Volume II: Demographic Profiles*. New York. http://esa.un.org/wpp/Documentation/pdf/WPP2012_Volume-II-Demographic-Profiles.pdf. Accessed 4 June 2015.

———. 2015. *World Population Prospects: The 2015 Revision*. New York. http://esa.un.org/unpd/wpp/Publications/Files/Key_Findings_WPP_2015.pdf Accessed 30 July 2015.

UNDP (United Nations Development Programme). 1994. *Human Development Report 1994*. New York.

———. 1995. *Human Development Report 1995*. New York.

———. 2011. *Human Development Report 2011: Sustainability and Equity: A Better Future for All*. New York.

———. 2012a. *Africa Human Development Report 2012: Towards a Food Secure Future*. New York. www.undp.org/content/dam/undp/library/corporate/HDR/Africa%20HDR/UNDP-Africa%20HDR-2012-EN.pdf. Accessed 4 June 2015.

———. 2012b. *Malaysia National Human Development Report 2012*. Kuala Lumpur.

———. 2012c. *Somalia National Human Development Report: Empowering Youth for Peace and Development*. Nairobi. http://hdr.undp.org/sites/default/files/reports/242/somalia_report_2012.pdf. Accessed 31 July 2015.

———. 2012d. *National Human Development Report: Tajikistan: Institutions and Development*. Dushanbe.

———. 2013a. *Accelerating Progress: Sustaining Results*. New York.

———. 2013b. *Humanity Divided: Confronting Inequality in Developing Countries*. New York. www.undp.org/content/dam/undp/library/Poverty%20Reduction/Inclusive%20development/Humanity%20Divided/HumanityDivided_Full-Report.pdf. Accessed 25 June 2015.

———. 2013c. *National Human Development Report 2013: People Are the Real Wealth of the Country: How Rich Is Montenegro?* Podgorica. www.me.undp.org/content/dam/montenegro/docs/publications/NHDR/NHDR2013/NHDR%202013%20ENG.pdf. Accessed 8 June 2015.

———. 2013d. "Self-Employment Success Stories." Republic of Macedonia, Employment Service Agency, Podgorica. www.mk.undp.org/content/dam/the_former_yugoslav_republic_of_macedonia/docs/SuccessStories_fullPreview.pdf. Accessed 8 June 2015.

———. 2014a. "How the Private Sector Develops Skills Lessons from Turkey." Istanbul International Center for Private Sector in Development, Ankara.

———. 2014b. *Human Development Report 2014: Sustaining Human Progress: Reducing Vulnerabilities and Building Resilience*. New York. http://hdr.undp.org/sites/default/files/hdr14-report-en-1.pdf. Accessed 16 July 2015.

———. 2014c. *National Human Development Report 2014: Good Corporate Citizens: Public and Private Goals Aligned for Human Development*. Chisinau. http://hdr.undp.org/sites/default/files/engleza_final.pdf. Accessed 21 July 2015.

———. 2014d. *Nepal Human Development Report 2014*. Kathmandu. http://hdr.undp.org/sites/default/files/nepal_nhdr_2014-final.pdf. Accessed 21 July 2015.

———. 2014e. *Roman Poverty from a Human Development Perspective*. New York. www.eurasia.undp.org/content/dam/rbec/docs/roma%20poverty%20from%20a%20human%20development%20perspective.pdf. Accessed 31 July 2015.

———. 2015a. *National Human Development Report 2014: Ethiopia: Accelerating Inclusive Growth for Sustainable Human Development in Ethiopia*. http://hdr.undp.org/sites/default/files/nhdr2015-ethiopia-en.pdf. Accessed 7 July 2015.

———. 2015b. "Integrated Results and Resources Framework." Annex II in *Report of the Administrator on the Strategic Plan: Performance and Results for 2014*. Presented in Annual Session, June. www.undp.org/content/dam/undp/library/corporate/Executive%20Board/2015/Annual-session/English/dp2015-11_Annexes%20I%20II%20and%20III.docx. Accessed 16 October 2015.

UNECE (United Nations Economic Commission for Europe) Expert Group on Measuring Quality of Employment. 2012. "Statistical Framework for Measuring Quality of Employment." Draft revised after the Expert Group meeting on 22–23 November. www.unece.org/fileadmin/DAM/stats/documents/ece/ces/ge.12/2013/Statistical_framework_for_measuring_quality_of_employment.pdf. Accessed 1 July 2015.

UNEP (United Nations Environment Programme), ILO (International Labour Organization), IOE (International Organization of Employers), and ITUC (International Trade Union Confederation). 2008. *Green Jobs: Towards Decent Work in a Sustainable, Low-carbon World*. Nairobi. www.unep.org/PDF/UNEPGreenjobs_report08.pdf. Accessed 18 May 2015.

UNESCO (United Nations Educational, Scientific and Cultural Organization). 2013a. "Literacy for All Remains an Elusive Goal, New UNESCO Data Shows." UNESCO Media Services, 5 September. www.unesco.org/new/en/media-services/single-view/news/literacy_for_all_remains_an_elusive_goal_new_unesco_data_shows/#.VawSjflVikp. Accessed 20 May 2015.

———. 2013b. "UNESCO: Half of All Out-of-school Children Live in Conflict-affected Countries." UNESCO Media Services, 11 July. www.unesco.org/new/en/media-services/single-view/news/unesco_half_of_all_out_of_school_children_live_in_conflict_affected_countries/#.VXCzU89Viko. Accessed 20 May 2015.

———. 2014. *EFA Global Monitoring Report 2013/4: Teaching and Learning: Achieving Quality for All*. Paris. http://unesco.nl/sites/default/files/dossier/gmr_2013-4.pdf. Accessed 17 June 2015.

———. 2015. "A Complex Formula: Girls and Women in Science, Technology, Engineering and Mathematics in Asia." Bangkok. http://unesdoc.unesco.org/images/0023/002315/231519e.pdf. Accessed 17 June 2015.

UNESCO (United Nations Educational, Scientific and Cultural Organization) Institute for Statistics. 2015. UIS.Stat. http://data.uis.unesco.org. Accessed 17 June 2015.

UNESCO (United Nations Educational, Scientific and Cultural Organization) and UNDP (United Nations Development Programme). 2013. *Creative Economy Report 2013 Special Edition: Widening Local Development Pathways*. Paris.

UNFPA (United Nations Population Fund) and HelpAge International. 2012. *Ageing in the Twenty-First Century: A Celebration and a Challenge*. New York and London. www.unfpa.org/sites/default/files/pub-pdf/Ageing%20report.pdf. Accessed 11 August 2015.

United States Equal Employment Opportunity Commission. 2014. "Charge Statistics. FY 1997 through FY 2014." http://eeoc.gov/eeoc/statistics/enforcement/charges.cfm. Accessed 1 July 2015.

United Workers Congress. n.d. "Domestic Workers." www.unitedworkerscongress.org/domestic-workers.html. Accessed 30 June 2015.

UNODC (United Nations Office on Drugs and Crime). 2012. *Global Report on Trafficking in Persons 2012*. Vienna. www.unodc.org/documents/data-and-analysis/glotip/Trafficking_in_Persons_2012_web.pdf. Accessed 6 August 2015.

USAID (United States Agency for International Development). 2013. *USAID Mekong ARCC Climate Change Impact and Adaptation Study for the Lower Mekong Basin: Main Report*. Bangkok: Regional Development Mission for Asia. http://mekongarcc.net/sites/default/files/mekong_arcc_main_report_printed_-_final.pdf. Accessed 15 June 2015.

USPTO (United States Patent and Trademark Office). 2015. Statistical database. www.uspto.gov/learning-and-resources/statistics. Accessed 9 June 2015.

Uthoff, A. 2015. "Reforma al Sistema de Pensiones Chileno." Serie Financiamiento para el Desarrollo 240. http://repositorio.cepal.org/bitstream/handle/11362/5221/S1100849_es.pdf?sequence=1. Accessed 23 January 2015.

Vanek, J., M.A. Chen, F. Carré, J. Heintz, and R. Hussmanns. 2014. "Statistics on the Informal Economy: Definitions, Regional Estimates and Challenges." Working Paper (Statistics) 2. Women in Informal Employment Globalizing and Organizing, Cambridge, MA. http://wiego.org/sites/wiego.org/files/publications/files/Vanek-Statistics-WIEGO-WP2.pdf. Accessed 29 June 2015.

Villena, M.G., R. Sanchez, and E. Rojas. 2011. "Unintended Consequences of Childcare Regulation in Chile: Evidence from a Regression Discontinuity Design." Munich Personal RePEc Archive. http://mpra.ub.uni-muenchen.de/62096/1/MPRA_paper_62096.pdf. Accessed 29 June 2015.

Vissa, B. 2015. "Entrepreneurial Skills for Social Good." *FT Wealth*, 5 July 2015.

Vodopivec, M., and N. Arunatilake. 2008. "Population Ageing and the Labor Market: The Case of Sri Lanka." SP Discussion Paper 821. World Bank, Washington, DC. http://siteresources.worldbank.org/SOCIALPROTECTION/Resources/SP-Discussion-papers/Labor-Market-DP/0821.pdf. Accessed 5 August 2015.

Walk Free Foundation. 2015. *The Global Slavery Index 2014*. Dalkeith, Australia. http://d3mj66ag90b5fy.cloudfront.net/wp-content/uploads/2014/11/Global_Slavery_Index_2014_final_lowres.pdf. Accessed 18 June 2015.

Walker, R. 2013. "The Transitional Costs of Sectoral Reallocation: Evidence from the Clean Air Act and the Workforce." *Quarterly Journal of Economics* 128(4): 1787–1835.

Warhurst, C., F. Carré, P. Findlay, and C. Tilly. 2012. *Are Bad Jobs Inevitable?* Basingstoke, UK: Palgrave Macmillan.

WEF (World Economic Forum). 2012a. "The Future of Manufacturing: Opportunities to Drive Economic Growth." Geneva. http://www3.weforum.org/docs/WEF_MOB_FutureManufacturing_Report_2012.pdf. Accessed 6 July 2015.

———. 2012b. *The Global Competitiveness Report 2012-2013.* Geneva.

———. 2014. *Global Gender Gap Report 2014.* Geneva. http://reports.weforum.org/global-gender-gap-report-2014/. Accessed 4 June 2015.

———. 2015. *Global Risks 2015, 10th Edition.* Geneva. www.weforum.org/risks. Accessed 20 July 2015.

West, P., J. Gerber, P. Engstrom, N. Mueller, K. Brauman, K. Carlson, and others. 2014. "Leverage Points for Improving Global Food Security and the Environment." *Science* 345: 325–28.

WFP (World Food Programme). 2015. "Senegal: Building Sustainable School Meal Programmes." 6 May. www.wfp.org/stories/cash-vouchers-changing-students-lives-senegal. Accessed 2 June 2015.

WHO (World Health Organization). 2003. "Climate Change and Human Health: Risks and Responses." www.who.int/globalchange/summary/en/. Accessed 18 June 2015.

———. 2005. "The WHO Framework Convention on Tobacco Control: An Overview." www.who.int/fctc/about/WHO_FCTC_summary_January2015.pdf. Accessed 12 June 2015.

———. 2013. *Global and Regional Estimates of Violence against Women.* Geneva. http://apps.who.int/iris/bitstream/10665/85239/1/9789241564625_eng.pdf. Accessed 4 June 2015.

———. 2014. "Health Workforce 2030: Towards a Global Strategy on Human Resources for Health." Global Health Workforce Alliance Synthesis Paper of the Thematic Working Groups. Geneva. www.who.int/workforcealliance/media/news/2014/public_consultations_GHWA_Synthesis_Paper_Towards_GSHRH_21Jan15.pdf. Accessed 4 June 2015.

———. 2015a. "Household Air Pollution and Health." Fact Sheet 292. Geneva. www.who.int/mediacentre/factsheets/fs292/en/. Accessed 8 July 2015.

———. 2015b. "Noncommunicable Diseases." Fact sheet. www.who.int/mediacentre/factsheets/fs355/en/. Accessed 8 July 2015.

WHO (World Health Organization) and World Bank. 2011. *World Report on Disability.* Geneva. www.who.int/disabilities/world_report/2011/report.pdf. Accessed 30 June 2015.

WIEGO (Women in Informal Employment Globalizing and Organizing) and ILO (International Labour Organization). 2013. *Women and Men in the Informal Economy: A Statistical Picture.* 2nd Edition. Cambridge, MA, and Geneva. www.ilo.org/wcmsp5/groups/public/---dgreports/---stat/documents/publication/wcms_234413.pdf. Accessed 29 June 2014.

Wikipedia. 2015. "Wikipedia:About." https://en.wikipedia.org/wiki/Wikipedia:About. Accessed 2 June 2015.

Wilson, R.A. 2010. *Skills Supply and Demand in Europe: Medium-term Forecast up to 2020.* Luxembourg: Publications Office of the European Union.

WIOD (World Input-Output Database). 2014. World Input-Output Database www.wiod.org/new_site/data.htm Accessed 26 June 2015.

WIPO (World Intellectual Property Organization). 2015. Statistics database. www.wipo.int/ipstats/en/. Accessed 3 June 2015.

WMO (World Meteorological Organization). 2014. *The Impact of Climate Change: Migration and Cities in South America.* Geneva. www.wmo.int/bulletin/en/content/impact-climate-change-migration-and-cities-south-america. Accessed 12 May 2015.

World Bank. 2002. "Improving Livelihoods on Fragile Lands." In *World Development Report 2003: Sustainable Development in a Dynamic World: Transforming Institutions, Growth, and Quality of Life.* Washington, DC. http://elibrary.worldbank.org/doi/abs/10.1596/0821351508_Chapter4. Accessed 30 June 2015.

———. 2011. *World Development Report 2011: Conflict, Security and Development.* Washington, DC. http://siteresources.worldbank.org/INTWDRS/Resources/WDR2011_Full_Text.pdf. Accessed 6 August 2015.

———. 2012. "Enterprise Surveys." www.enterprisesurveys.org. Accessed 22 June 2015.

———. 2013. "Three Quarters of the World's Poor are Unbanked." 19 April. http://go.worldbank.org/72MAKHBAM0. Accessed 20 May 2015.

———. 2014a. Global Findex Database. http://datatopics.worldbank.org/financialinclusion/. Accessed 10 August 2015.

———. 2014b. World Development Indicators database. http://data.worldbank.org/data-catalog/world-development-indicators. Accessed 5 June 2015.

———. 2015a. *Ending Poverty and Hunger by 2030: An Agenda for the Global Food System.* Washington, DC. http://www-wds.worldbank.org/external/default/WDSContentServer/WDSP/IB/2015/06/03/090224b082eed2bb/2_0/Rendered/PDF/Ending0poverty0e0global0foodsystem.pdf. Accessed 8 July 2015.

———. 2015b. *Global Monitoring Report 2014/2015: Ending Poverty and Sharing Prosperity.* Washington, DC. www.worldbank.org/content/dam/Worldbank/gmr/gmr2014/GMR_2014_Full_Report.pdf. Accessed 15 June 2015.

———. 2015c. "Migration and Remittances: Recent Development and Outlook: Special Topic: Financing for Development." Migration and Development Brief 24. Washington, DC. http://siteresources.worldbank.org/INTPROSPECTS/Resources/334934-1288990760745/MigrationandDevelopmentBrief24.pdf. Accessed 17 September 2015.

———. 2015d. "Overview." www.worldbank.org/en/topic/urbandevelopment/overview. Accessed 8 July 2015.

———. 2015e. "Remittances Growth to Slow Sharply in 2015, as Europe and Russia Stay Weak; Pick Up Expected Next Year." Press release, 13 April. Washington, DC. www.worldbank.org/en/news/press-release/2015/04/13/remittances-growth-to-slow-sharply-in-2015-as-europe-and-russia-stay-weak-pick-up-expected-next-year. Accessed 22 July 2015.

———. 2015f. World Development Indicators database. http://data.worldbank.org/data-catalog/world-development-indicators. Accessed 15 June 2015.

World Commission on Environment and Development. 1987. *Our Common Future.* New York: Oxford University Press.

World Values Survey. 2014. Wave 6. www.worldvaluessurvey.org/wvs.jsp. Accessed 12 June 2015.

WRI (World Resources Institute). 2014. *Creating a Sustainable Food Future.* Washington, DC. www.wri.org/sites/default/files/wri13_report_4c_wrr_online.pdf. Accessed 20 June 2015.

Wyne, J. 2014. "The Next Step: Breaking Barriers to Scale for MENA's Entrepreneurs." Wamda Research Lab, Beirut.

Yang, S., and L. Zheng. 2011. "The Paradox of De-coupling: A Study of Flexible Work Programs and Workers' Productivity." *Social Science Research* 40(1): 299–311. www.sciencedirect.com/science/article/pii/S0049089X10000761. Accessed 29 June 2015.

Yishay, A., and A. Mobarak. 2014. "Social Learning and Communication." Working Paper 20139. National Bureau of Economic Research, Cambridge, MA. www.nber.org/papers/w20139. Accessed 18 May 2015.

Yunus, M. 2009. *Creating a World without Poverty: Social Business and the Future of Capitalism.* New York: Public Affairs.

Zabriskie, R.B., and B.P. McCormick. 2001. "The Influences of Family Leisure Patterns on Perceptions of Family Functioning." *Family Relations* 50(3): 281–89.

Zepeda, E., S. McDonald, M. Panda, and G. Kumar. 2013 *Employing India. Guaranteeing Jobs for the Rural Poor.* Washington, DC: Carnegie Endowment for International Peace.

Zufiria, P., D. Pastor-Escuredo, L. Úbeda-Medina, M. Hernández-Medina, I. Barriales-Valbuena, A. Morales, and others. 2015. "Mobility Profiles and Calendars for Food Security and Livelihoods Analysis." D4D Challenge proceedings, Netmob 2015. www.unglobalpulse.org/sites/default/files/UNGP%20Case%20Study_D4D%20Mobility_2015.pdf. Accessed 15 July 2015.

Statistical annex

Readers guide

The 16 statistical tables in this annex as well as the statistical tables following chapters 2, 4 and 6 provide an overview of key aspects of human development. The first seven tables contain the family of composite human development indices and their components estimated by the Human Development Report Office (HDRO). The remaining tables present a broader set of indicators related to human development.

Unless otherwise specified in the notes, tables use data available to the HDRO as of 15 April 2015. All indices and indicators, along with technical notes on the calculation of composite indices and additional source information, are available online at http://hdr.undp.org/en/data.

Countries and territories are ranked by 2014 Human Development Index (HDI) value. Robustness and reliability analysis has shown that for most countries the differences in HDI are not statistically significant at the fourth decimal place.[1] For this reason countries with the same HDI value at three decimal places are listed with tied ranks.

Sources and definitions

Unless otherwise noted, the HDRO uses data from international data agencies with the mandate, resources and expertise to collect national data on specific indicators.

Definitions of indicators and sources for original data components are given at the end of each table, with full source details in *Statistical references*.

Gross national income per capita in purchasing power parity terms

In comparing standards of living based on income across countries, the income component of the HDI uses gross national income (GNI) per capita converted into purchasing power parity (PPP) terms to eliminate differences in national price levels.

The International Comparison Programme (ICP) survey is the world's largest statistical initiative that produces internationally comparable price levels, economic aggregates in real terms and PPP estimates. Estimates from ICP surveys conducted in 2011 and covering 190 countries were used to compute the 2014 HDI values.

Methodology updates

The 2015 Report retains all the composite indices from the family of human development indices—the HDI, the Inequality-adjusted Human Development Index, the Gender Development Index, the Gender Inequality Index and the Multidimensional Poverty Index. The methodology used to compute these indices is the same as one used in the 2014 Report. For details see *Technical notes 1–5* at http://hdr.undp.org.

Comparisons over time and across editions of the Report

Because national and international agencies continually improve their data series, the data—including the HDI values and ranks—presented in this Report are not comparable to those published in earlier editions. For HDI comparability across years and countries, see table 2, which presents trends using consistent data.

Discrepancies between national and international estimates

National and international data can differ because international agencies harmonize national data using a consistent methodology and occasionally produce estimates of missing data to allow comparability across countries. In other cases international agencies might not have access to the most recent national data. When HDRO becomes aware of discrepancies, it brings them to the attention of national and international data authorities.

Country groupings and aggregates

The tables present weighted aggregates for several country groupings. In general, an aggregate is shown only when data are available for at least half the countries and represent at least two-thirds of the population in that classification. Aggregates for each classification cover only the countries for which data are available.

Human development classification

HDI classifications are based on HDI fixed cut-off points, which are derived from the quartiles of distributions of component indicators. The cut-off points are HDI of less than 0.550 for low human development, 0.550–0.699 for medium human development, 0.700–0.799 for high human development and 0.800 or greater for very high human development.

Regional groupings

Regional groupings are based on United Nations Development Programme regional classifications. Least Developed Countries and Small Island Developing States are defined according to UN classifications (see www.unohrlls.org).

Developing countries

Aggregates are provided for the group of countries classified as developing countries.

Organisation for Economic Co-operation and Development

Aggregates are presented for the 34 members of the Organisation of Economic Co-operation and Development, 31 of which are developed countries and 3 of which are developing countries. Aggregates refer to all countries from the group for which data are available.

Country note

Data for China do not include Hong Kong Special Administrative Region of China, Macao Special Administrative Region of China or Taiwan Province of China.

Symbols

A dash between two years, as in 2005–2014, indicates that the data are from the most recent year available during the period specified. A slash between years, as in 2005/2014, indicates average for the years shown. Growth rates are usually average annual rates of growth between the first and last years of the period shown.

The following symbols are used in the tables:

..	Not available
0 or 0.0	Nil or negligible
—	Not applicable

Statistical acknowledgements

The Report's composite indices and other statistical resources draw on a wide variety of the most respected international data providers in their specialized fields. HDRO is particularly grateful to the Centre for Research on the Epidemiology of Disasters; Economic Commission for Latin America and the Caribbean; Eurostat; Food and Agriculture Organization; Gallup; ICF Macro; Internal Displacement Monitoring Centre; International Labour Organization; International Monetary Fund; International Telecommunication Union; Inter-Parliamentary Union; Luxembourg Income Study; Organisation for Economic Co-operation and Development; United Nations Children's Fund; United Nations Conference on Trade and Development; United Nations Department of Economic and Social Affairs; United Nations Economic and Social Commission for West Asia; United Nations Educational, Scientific and Cultural Organization Institute for Statistics; Office of the United Nations High Commissioner for Refugees; United Nations Office on Drugs and Crime; United Nations World Tourism Organization; World Bank; and World Health Organization. The international education database maintained by Robert Barro (Harvard University) and Jong-Wha Lee (Korea University) was another invaluable source for the calculation of the Report's indices.

Statistical tables

The first seven tables relate to the five composite human development indices and their components.

Since the 2010 *Human Development Report*, four composite human development indices—the HDI, the Inequality-adjusted Human Development Index, the Gender Inequality Index and the Multidimensional Poverty Index—have been calculated. Last year's Report introduced the Gender Development Index, which compares the HDI calculated separately for women and men and is included again in this year's Report.

The remaining tables present a broader set of human development–related indicators and provide a more comprehensive picture of a country's human development. Three of these tables are presented as annexes to chapters 2, 4 and 6.

Table 1, Human Development Index and its components, ranks countries by 2014 HDI value and details the values of the three HDI components: longevity, education (with two indicators) and income. The table also presents the difference in rankings by HDI and GNI.

Table 2, Human Development Index trends, 1990–2014, provides a time series of HDI values allowing 2014 HDI values to be compared with those for previous years. The table uses the most recently revised historical data available in 2015 and the same methodology applied to compute the 2014 HDI. Along with historical HDI values, the table includes the change in HDI rank over the last five years and the average annual HDI growth rates across four different time intervals, 1990–2000, 2000–2010, 2010–2014 and 1990–2014.

Table 3, Inequality-adjusted Human Development Index, contains two related measures of inequality—the IHDI and the loss in HDI due to inequality. The IHDI looks beyond the average achievements of a country in health, education and income to show how these achievements are distributed among its residents. The IHDI can be interpreted as the level of human development when inequality is accounted for. The relative difference between the IHDI and HDI is the loss due to inequality in distribution of the HDI within the country. The table also presents the coefficient of human inequality, which is an unweighted average of inequalities in three dimensions. In addition, the table shows each country's difference in rank on the HDI and the IHDI. A negative value means that taking inequality into account lowers a country's rank in the HDI distribution. The table also presents three standard measures of income inequality: the ratio of the top and the bottom quintiles; the Palma ratio, which is the ratio of income of the top 10 percent and the bottom 40 percent; and the Gini coefficient.

Table 4, Gender Development Index, measures disparities in HDI by gender. The table contains HDI values estimated separately for women and men; the ratio of which is the GDI. The closer the ratio to 1, the smaller the gap between women and men. Values for the three HDI components—longevity, education (with two indicators) and income—are also presented by gender. The table also includes country groupings by absolute deviation from gender parity in HDI values.

Table 5, Gender Inequality Index, presents a composite measure of gender inequality using three dimensions: reproductive health, empowerment and the labour market. Reproductive health is measured by two indicators: the maternal mortality ratio and the adolescent birth rate. Empowerment is measured by the share of parliamentary seats held by women and the share of population with at least some secondary education. And labour market is measured by participation in the labour force. A low Gender Inequality Index value indicates low inequality between women and men, and vice-versa.

Table 6, Multidimensional Poverty Index: developing countries, captures the multiple deprivations that people face in their education, health and living standards. The MPI shows both the incidence of nonincome multidimensional poverty (a headcount of those in multidimensional poverty) and its intensity (the relative number of deprivations poor people experience at the same time). Based on intensity thresholds, people are classified as near multidimensional poverty, multidimensionally poor or in severe poverty, respectively. The contributions of deprivations in each dimension to overall poverty are also included. The table also presents measures of income poverty—population living on less than PPP $1.25 per day and population living below the national poverty line. This year's Multidimensional Poverty Index estimations use the revised methodology that was introduced in the 2014 Report. The revised methodology includes some modifications to the original set of 10 indicators: height-for-age replaces weight-for-age for children under age 5 because stunting is a better indicator of chronic malnutrition, a child death is considered a health deprivation only if it happened in the five years prior to the survey, the minimum threshold for education deprivation was raised from five years of schooling to six to reflect the standard definition of primary schooling used in the Millennium Development Goals and in international measures of functional literacy and the indicators for household assets were expanded to better reflect rural as well as urban households.

Table 7, Multidimensional Poverty Index: changes over time, presents estimates of Multidimensional Poverty Index values and its components for two or more time points for countries for which consistent data were available in 2015. Estimation is based on the revised methodology introduced in the 2014 Report.

Table 8, Population trends, contains major population indicators, including total population, median age, dependency ratios and total fertility rates, which can help assess the burden of support that falls on the labour force in a country. Deviations from the natural sex ratio at birth have implications for population replacement levels, suggest possible future social and economic problems and may indicate gender bias.

Table 9, Health outcomes, presents indicators of infant health (percentage of infants who are exclusively breastfed for the first six months of life, percentage of infants who lack immunization for DTP and measles, and infant mortality rate), child health (child mortality rate and percentage of children under age 5 who are stunted) and adult health (adult mortality rates by gender, deaths due to malaria and tuberculosis, HIV prevalence and life expectancy at age 60). Two indicators of quality of health care are also included: number of physicians per 10,000 people and public health expenditure as a share of GDP.

Table 10, Education achievements, presents standard education indicators along with indicators of education quality, including average test scores on reading, mathematics and science for 15-year-old students. The table provides indicators of educational attainment—adult and youth literacy rates and the share of the adult population with at least some secondary education. Gross enrolment ratios at each level of education are complemented by primary school dropout rates. The table also includes two indicators of education quality—primary school teachers trained to teach and the pupil–teacher ratio—as well as an indicator on public expenditure on education as a share of GDP.

Table 11, National income and composition of resources, covers several macroeconomic indicators such as gross domestic product (GDP), gross fixed capital formation and taxes on income, profit and capital gain as percentage of total tax revenue. Gross fixed capital formation is a rough indicator of national income that is invested rather than consumed. In times of economic uncertainty or recession, gross fixed capital formation typically declines. General government final consumption expenditure (presented as a share of GDP and as average annual growth) and research and development expenditure are indicators of public spending. In addition, the table presents three indicators of debt—domestic credit provided by the banking sector, external debt stock and total debt service, all measured as a percentage of GDP or gross national income (GNI). The consumer price index is a measure of inflation; two indicators related to the price of food are presented as well—the price level index and the price volatility index.

Table 12, Environmental sustainability, covers environmental vulnerability and effects of environmental threats. The table shows the proportion of fossil fuels and renewable energy sources in the primary energy supply, levels and annual growth of carbon dioxide emissions per capita and measures of ecosystem and natural resources preservation (natural resource depletion as a percentage of GNI, forest area and change in forest area and fresh water withdrawals). The table contains the under-five mortality rates due to outdoor and indoor air pollution and to unsafe water, unimproved sanitation or poor hygiene. The table also presents an indicator of the direct impacts of natural disasters (average annual population affected per million people).

Table 13, Work and employment, contains indicators on three components: employment, unemployment and labour productivity. Two key indicators related to employment are highlighted: the employment to population ratio and the labour force participation rate. The table also reports employment in agriculture and services and the change since 1990. Also presented is the percentage of the labour force with tertiary education, which is associated with the high-skilled labour force. The table brings together indicators related to vulnerable employment and different forms of unemployment. And labour productivity is accounted for by output per worker and hours worked per week.

Table 14, Human security, reflects the extent to which the population is secure. The table begins with the percentage of registered births, followed by the number of refugees by country of origin and number of internally displaced persons. It shows the size of the homeless population due to natural disasters, orphaned children population and prison population. Indicators on homicide and suicide (by gender) are provided.

And the table includes the depth of food deficit and an indicator on violence against women.

Table 15, International integration, provides indicators of several aspects of globalization. International trade is measured as share of GDP. Financial flows are represented by net inflows of foreign direct investment and private capital, official development assistance and inflows of remittances. Human mobility is captured by the net migration rate, the stock of immigrants, the net number of tertiary students from abroad (expressed as a percentage of total tertiary enrolment in that country) and the number of international inbound tourists. International communication is represented by the share of the population that uses the Internet, the number of mobile phone subscriptions per 100 people and the percentage change in mobile phone subscriptions between 2009 and 2014.

Table 16, Supplementary indicators: perceptions of well-being, includes indicators that reflect individuals' opinions and self-perceptions of relevant dimensions of human development—quality of education, quality of health care, standard of living and labour market, personal safety and overall satisfaction with freedom of choice and life. The table also contains indicators reflecting perceptions of government policies on preservation of the environment and overall trust in the national government and judicial system.

Chapter 1 annex table, Work with exploitation, risks and insecurities, brings together indicators of work that represents risk to human development—child labour, domestic workers and working poor. The table also presents recent counts of occupational injuries. Three indicators—unemployment benefits, paid maternity leave and old age pension—indicate security stemming from employment.

Chapter 4 annex table, Time use, compiles data from more than 100 time use surveys conducted over the last 25 years with information on the time women and men spend daily on major activities—paid and unpaid work, learning, social life and leisure, personal care and maintenance, and other (unaccounted) activities.

Chapter 6 annex table, Status of fundamental rights conventions, shows when countries ratified key labour rights conventions. The eight selected conventions cover four key groups of rights and freedoms: freedom of association and collective bargaining, elimination of forced and compulsory labour, elimination of discrimination in respect of employment, and occupation and abolition of child labour.

Note

1. Aguna and Kovacevic (2011) and Høyland, Moene and Willumsen (2011).

Human development indices

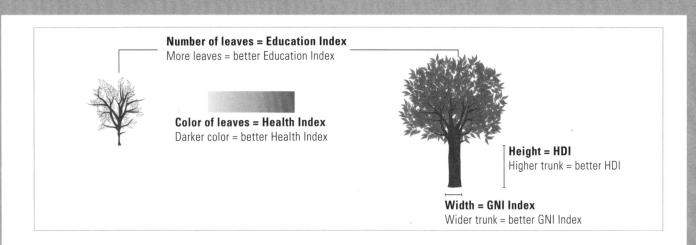

Number of leaves = Education Index
More leaves = better Education Index

Color of leaves = Health Index
Darker color = better Health Index

Height = HDI
Higher trunk = better HDI

Width = GNI Index
Wider trunk = better GNI Index

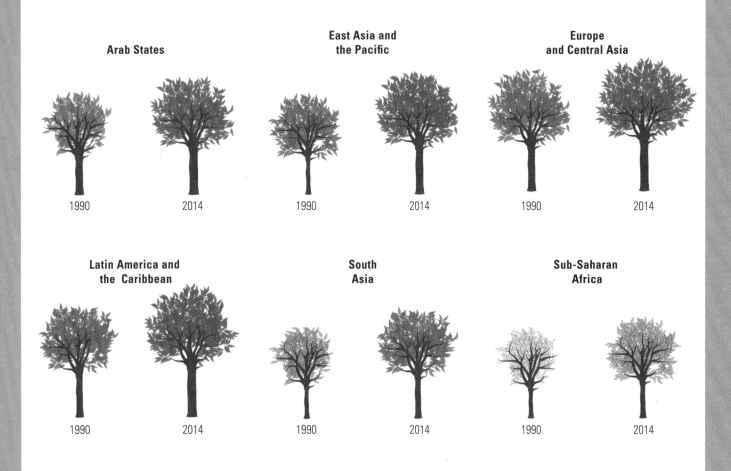

Arab States

1990 · 2014

East Asia and the Pacific

1990 · 2014

Europe and Central Asia

1990 · 2014

Latin America and the Caribbean

1990 · 2014

South Asia

1990 · 2014

Sub-Saharan Africa

1990 · 2014

Note: Infographic was inspired by the work of Jurjen Verhagen, the winner of the 2015 Cartagena Data Fest visualization contest.

TABLE 1

Human Development Index and its components

TABLE
1

HDI rank	Human Development Index (HDI) Value 2014	Life expectancy at birth (years) 2014	Expected years of schooling (years) 2014[a]	Mean years of schooling (years) 2014[a]	Gross national income (GNI) per capita (2011 PPP $) 2014	GNI per capita rank minus HDI rank 2014
VERY HIGH HUMAN DEVELOPMENT						
1 Norway	0.944	81.6	17.5	12.6[b]	64,992	5
2 Australia	0.935	82.4	20.2[c]	13.0	42,261	17
3 Switzerland	0.930	83.0	15.8	12.8	56,431	6
4 Denmark	0.923	80.2	18.7[c]	12.7	44,025	11
5 Netherlands	0.922	81.6	17.9	11.9	45,435	9
6 Germany	0.916	80.9	16.5	13.1[d]	43,919	11
6 Ireland	0.916	80.9	18.6[c]	12.2[e]	39,568	16
8 United States	0.915	79.1	16.5	12.9	52,947	3
9 Canada	0.913	82.0	15.9	13.0	42,155	11
9 New Zealand	0.913	81.8	19.2[c]	12.5[b]	32,689	23
11 Singapore	0.912	83.0	15.4[f]	10.6[e]	76,628[g]	−7
12 Hong Kong, China (SAR)	0.910	84.0	15.6	11.2	53,959	−2
13 Liechtenstein	0.908	80.0[h]	15.0	11.8[i]	79,851[g,j]	−10
14 Sweden	0.907	82.2	15.8	12.1	45,636	−1
14 United Kingdom	0.907	80.7	16.2	13.1[d]	39,267	9
16 Iceland	0.899	82.6	19.0[c]	10.6[e]	35,182	12
17 Korea (Republic of)	0.898	81.9	16.9	11.9[e]	33,890	13
18 Israel	0.894	82.4	16.0	12.5	30,676	16
19 Luxembourg	0.892	81.7	13.9	11.7	58,711	−11
20 Japan	0.891	83.5	15.3	11.5[e]	36,927	7
21 Belgium	0.890	80.8	16.3	11.3[d]	41,187	0
22 France	0.888	82.2	16.0	11.1	38,056	4
23 Austria	0.885	81.4	15.7	10.8[d]	43,869	−5
24 Finland	0.883	80.8	17.1	10.3[e]	38,695	0
25 Slovenia	0.880	80.4	16.8	11.9	27,852	12
26 Spain	0.876	82.6	17.3	9.6	32,045	7
27 Italy	0.873	83.1	16.0	10.1[d]	33,030	4
28 Czech Republic	0.870	78.6	16.4	12.3	26,660	10
29 Greece	0.865	80.9	17.6	10.3	24,524	14
30 Estonia	0.861	76.8	16.5	12.5[e]	25,214	12
31 Brunei Darussalam	0.856	78.8	14.5	8.8[e]	72,570[k]	−26
32 Cyprus	0.850	80.2	14.0	11.6	28,633	3
32 Qatar	0.850	78.2	13.8[l]	9.1	123,124[g]	−31
34 Andorra	0.845	81.3[h]	13.5[f]	9.6[m]	43,978[n]	−18
35 Slovakia	0.844	76.3	15.1	12.2[d]	25,845	5
36 Poland	0.843	77.4	15.5	11.8	23,177	10
37 Lithuania	0.839	73.3	16.4	12.4	24,500	7
37 Malta	0.839	80.6	14.4	10.3	27,930	−1
39 Saudi Arabia	0.837	74.3	16.3	8.7[d]	52,821	−27
40 Argentina	0.836	76.3	17.9	9.8[d]	22,050[k]	11
41 United Arab Emirates	0.835	77.0	13.3[o]	9.5[d]	60,868	−34
42 Chile	0.832	81.7	15.2	9.8	21,290	11
43 Portugal	0.830	80.9	16.3	8.2	25,757	−2
44 Hungary	0.828	75.2	15.4	11.6[d]	22,916	3
45 Bahrain	0.824	76.6	14.4[p]	9.4[b]	38,599	−20
46 Latvia	0.819	74.2	15.2	11.5[d]	22,281	4
47 Croatia	0.818	77.3	14.8	11.0	19,409	11
48 Kuwait	0.816	74.4	14.7[l]	7.2	83,961[g]	−46
49 Montenegro	0.802	76.2	15.2	11.2	14,558	27
HIGH HUMAN DEVELOPMENT						
50 Belarus	0.798	71.3	15.7	12.0[q]	16,676	14
50 Russian Federation	0.798	70.1	14.7	12.0	22,352	−1
52 Oman	0.793	76.8	13.6	8.0	34,858	−23
52 Romania	0.793	74.7	14.2	10.8	18,108	10
52 Uruguay	0.793	77.2	15.5	8.5	19,283	7
55 Bahamas	0.790	75.4	12.6[r]	10.9	21,336	−3
56 Kazakhstan	0.788	69.4	15.0	11.4[e]	20,867	−1
57 Barbados	0.785	75.6	15.4	10.5[q]	12,488	27
58 Antigua and Barbuda	0.783	76.1	14.0	9.2[r]	20,070	−1
59 Bulgaria	0.782	74.2	14.4	10.6[d]	15,596	13
60 Palau	0.780	72.7[h]	13.7	12.3[f]	13,496	18
60 Panama	0.780	77.6	13.3	9.3	18,192	1

HDI rank		Human Development Index (HDI)	Life expectancy at birth	Expected years of schooling	Mean years of schooling	Gross national income (GNI) per capita	GNI per capita rank minus HDI rank
		Value	(years)	(years)	(years)	(2011 PPP $)	
		2014	2014	2014[a]	2014[a]	2014	2014
62	Malaysia	0.779	74.7	12.7[l]	10.0	22,762	−14
63	Mauritius	0.777	74.4	15.6	8.5	17,470	0
64	Seychelles	0.772	73.1	13.4	9.4[r]	23,300	−19
64	Trinidad and Tobago	0.772	70.4	12.3[l]	10.9	26,090	−25
66	Serbia	0.771	74.9	14.4	10.5	12,190	20
67	Cuba	0.769[s]	79.4	13.8	11.5[q]	7,301[t]	47
67	Lebanon	0.769	79.3	13.8	7.9[l]	16,509	−1
69	Costa Rica	0.766	79.4	13.9	8.4	13,413	10
69	Iran (Islamic Republic of)	0.766	75.4	15.1	8.2[e]	15,440	4
71	Venezuela (Bolivarian Republic of)	0.762	74.2	14.2	8.9[d]	16,159	−2
72	Turkey	0.761	75.3	14.5	7.6	18,677	−12
73	Sri Lanka	0.757	74.9	13.7	10.8[b]	9,779	29
74	Mexico	0.756	76.8	13.1	8.5	16,056	−4
75	Brazil	0.755	74.5	15.2[u]	7.7	15,175	−1
76	Georgia	0.754	74.9	13.8	12.1[q]	7,164	40
77	Saint Kitts and Nevis	0.752	73.8[h]	12.9	8.4[r]	20,805	−21
78	Azerbaijan	0.751	70.8	11.9	11.2[l]	16,428	−11
79	Grenada	0.750	73.4	15.8	8.6[r]	10,939	14
80	Jordan	0.748	74.0	13.5	9.9	11,365	11
81	The former Yugoslav Republic of Macedonia	0.747	75.4	13.4	9.3[q]	11,780	9
81	Ukraine	0.747	71.0	15.1	11.3[e]	8,178	25
83	Algeria	0.736	74.8	14.0	7.6	13,054	−1
84	Peru	0.734	74.6	13.1	9.0	11,015	8
85	Albania	0.733	77.8	11.8[l]	9.3	9,943	14
85	Armenia	0.733	74.7	12.3	10.9[e]	8,124	22
85	Bosnia and Herzegovina	0.733	76.5	13.6	8.3[v]	9,638	19
88	Ecuador	0.732	75.9	14.2	7.6	10,605	7
89	Saint Lucia	0.729	75.1	12.6	9.3[q]	9,765	14
90	China	0.727	75.8	13.1	7.5[b]	12,547	−7
90	Fiji	0.727	70.0	15.7[l]	9.9	7,493	21
90	Mongolia	0.727	69.4	14.6	9.3[e]	10,729	4
93	Thailand	0.726	74.4	13.5	7.3	13,323	−13
94	Dominica	0.724	77.8[h]	12.7[w]	7.9[r]	9,994	4
94	Libya	0.724	71.6	14.0[l]	7.3[e]	14,911[k,x]	−19
96	Tunisia	0.721	74.8	14.6	6.8[q]	10,404	1
97	Colombia	0.720	74.0	13.5	7.3[d]	12,040	−9
97	Saint Vincent and the Grenadines	0.720	72.9	13.4[l]	8.6[r]	9,937	3
99	Jamaica	0.719	75.7	12.4	9.7[e]	7,415	13
100	Tonga	0.717	72.8	14.7	10.7[e]	5,069	32
101	Belize	0.715	70.0	13.6	10.5	7,614	9
101	Dominican Republic	0.715	73.5	13.1	7.6	11,883	−12
103	Suriname	0.714	71.1	12.7[l]	7.7[v]	15,617	−32
104	Maldives	0.706	76.8	13.0[l]	5.8[z]	12,328	−19
105	Samoa	0.702	73.4	12.9[f]	10.3[f]	5,327	24
MEDIUM HUMAN DEVELOPMENT							
106	Botswana	0.698	64.5	12.5	8.9[e]	16,646	−41
107	Moldova (Republic of)	0.693	71.6	11.9	11.2	5,223	23
108	Egypt	0.690	71.1	13.5	6.6[e]	10,512	−12
109	Turkmenistan	0.688	65.6	10.8	9.9[r]	13,066	−28
110	Gabon	0.684	64.4	12.5[l]	7.8[v]	16,367	−42
110	Indonesia	0.684	68.9	13.0	7.6[z]	9,788	−9
112	Paraguay	0.679	72.9	11.9	7.7[b]	7,643	−3
113	Palestine, State of	0.677	72.9	13.0	8.9	4,699[x]	21
114	Uzbekistan	0.675	68.4	11.5	10.9[aa]	5,567	10
115	Philippines	0.668	68.2	11.3	8.9[d]	7,915	−7
116	El Salvador	0.666	73.0	12.3	6.5	7,349	−3
116	South Africa	0.666	57.4	13.6	9.9	12,122	−29
116	Viet Nam	0.666	75.8	11.9[w]	7.5[e]	5,092	15
119	Bolivia (Plurinational State of)	0.662	68.3	13.2	8.2	5,760	4
120	Kyrgyzstan	0.655	70.6	12.5	10.6	3,044	29
121	Iraq	0.654	69.4	10.1	6.4[e]	14,003	−44
122	Cabo Verde	0.646	73.3	13.5	4.7[r]	6,094	−1
123	Micronesia (Federated States of)	0.640	69.1	11.7	9.7[f]	3,432	21

TABLE 1   Human Development Index and its components | 209

TABLE 1   HUMAN DEVELOPMENT INDEX AND ITS COMPONENTS

TABLE
1

HDI rank	Human Development Index (HDI) Value 2014	Life expectancy at birth (years) 2014	Expected years of schooling (years) 2014[a]	Mean years of schooling (years) 2014[a]	Gross national income (GNI) per capita (2011 PPP $) 2014	GNI per capita rank minus HDI rank 2014
124 Guyana	0.636	66.4	10.3	8.5[e]	6,522	−4
125 Nicaragua	0.631	74.9	11.5[l]	6.0[e]	4,457	12
126 Morocco	0.628	74.0	11.6	4.4[b]	6,850	−8
126 Namibia	0.628	64.8	11.3	6.2[e]	9,418	−21
128 Guatemala	0.627	71.8	10.7	5.6	6,929	−11
129 Tajikistan	0.624	69.4	11.2	10.4[v]	2,517	27
130 India	0.609	68.0	11.7	5.4[e]	5,497	−4
131 Honduras	0.606	73.1	11.1	5.5	3,938	7
132 Bhutan	0.605	69.5	12.6	3.0[q]	7,176	−17
133 Timor-Leste	0.595	68.2	11.7	4.4[y]	5,363[ab]	−6
134 Syrian Arab Republic	0.594	69.6	12.3	6.3[e]	2,728[k,x]	21
134 Vanuatu	0.594	71.9	10.6[l]	6.8[aa]	2,803	19
136 Congo	0.591	62.3	11.1	6.1[b]	6,012	−14
137 Kiribati	0.590	66.0	12.3	7.8[r]	2,434	21
138 Equatorial Guinea	0.587	57.6	9.0[l]	5.5[y]	21,056	−84
139 Zambia	0.586	60.1	13.5	6.6[e]	3,734	2
140 Ghana	0.579	61.4	11.5	7.0	3,852	−1
141 Lao People's Democratic Republic	0.575	66.2	10.6	5.0[q]	4,680	−6
142 Bangladesh	0.570	71.6	10.0	5.1[e]	3,191	5
143 Cambodia	0.555	68.4	10.9	4.4[y]	2,949	7
143 Sao Tome and Principe	0.555	66.5	11.3	4.7[y]	2,918	8
LOW HUMAN DEVELOPMENT						
145 Kenya	0.548	61.6	11.0	6.3[b]	2,762	9
145 Nepal	0.548	69.6	12.4	3.3[e]	2,311	16
147 Pakistan	0.538	66.2	7.8	4.7	4,866	−14
148 Myanmar	0.536	65.9	8.6	4.1[e]	4,608[k]	−12
149 Angola	0.532	52.3	11.4	4.7[y]	6,822	−30
150 Swaziland	0.531	49.0	11.3	7.1[b]	5,542	−25
151 Tanzania (United Republic of)	0.521	65.0	9.2	5.1[e]	2,411	8
152 Nigeria	0.514	52.8	9.0[l]	5.9[y]	5,341	−24
153 Cameroon	0.512	55.5	10.4	6.0[e]	2,803	−1
154 Madagascar	0.510	65.1	10.3	6.0[q]	1,328	24
155 Zimbabwe	0.509	57.5	10.9	7.3[e]	1,615	13
156 Mauritania	0.506	63.1	8.5	3.8[e]	3,560	−14
156 Solomon Islands	0.506	67.9	9.2	5.0[f]	1,540	16
158 Papua New Guinea	0.505	62.6	9.9[r]	4.0[e]	2,463	−1
159 Comoros	0.503	63.3	11.5	4.6[y]	1,456	16
160 Yemen	0.498	63.8	9.2	2.6[e]	3,519	−17
161 Lesotho	0.497	49.8	11.1	5.9[z]	3,306	−16
162 Togo	0.484	59.7	12.2	4.5[y]	1,228	17
163 Haiti	0.483	62.8	8.7[r]	4.9[y]	1,669	4
163 Rwanda	0.483	64.2	10.3	3.7	1,458	11
163 Uganda	0.483	58.5	9.8	5.4[e]	1,613	6
166 Benin	0.480	59.6	11.1	3.3[e]	1,767	0
167 Sudan	0.479	63.5	7.0	3.1[b]	3,809	−27
168 Djibouti	0.470	62.0	6.4	3.8[q]	3,276[k]	−22
169 South Sudan	0.467	55.7	7.6[r]	5.4	2,332	−9
170 Senegal	0.466	66.5	7.9	2.5	2,188	−8
171 Afghanistan	0.465	60.4	9.3	3.2[e]	1,885	−7
172 Côte d'Ivoire	0.462	51.5	8.9	4.3[b]	3,171	−24
173 Malawi	0.445	62.8	10.8	4.3[e]	747	13
174 Ethiopia	0.442	64.1	8.5	2.4	1,428	2
175 Gambia	0.441	60.2	8.8	2.8[e]	1,507	−2
176 Congo (Democratic Republic of the)	0.433	58.7	9.8	6.0	680	11
177 Liberia	0.430	60.9	9.5[l]	4.1[e]	805	7
178 Guinea-Bissau	0.420	55.2	9.0	2.8[r]	1,362	−1
179 Mali	0.419	58.0	8.4	2.0	1,583	−8
180 Mozambique	0.416	55.1	9.3	3.2[y]	1,123	1
181 Sierra Leone	0.413	50.9	8.6[l]	3.1[e]	1,780	−16
182 Guinea	0.411	58.8	8.7	2.4[y]	1,096	0
183 Burkina Faso	0.402	58.7	7.8	1.4[y]	1,591	−13
184 Burundi	0.400	56.7	10.1	2.7[e]	758	1
185 Chad	0.392	51.6	7.4	1.9	2,085	−22

HDI rank	Human Development Index (HDI) Value 2014	Life expectancy at birth (years) 2014	Expected years of schooling (years) 2014[a]	Mean years of schooling (years) 2014[a]	Gross national income (GNI) per capita (2011 PPP $) 2014	GNI per capita rank minus HDI rank 2014
186 Eritrea	0.391	63.7	4.1	3.9[r]	1,130	−6
187 Central African Republic	0.350	50.7	7.2	4.2[q]	581	1
188 Niger	0.348	61.4	5.4	1.5[e]	908	−5
OTHER COUNTRIES OR TERRITORIES						
Korea (Democratic People's Rep. of)	..	70.3	..	..	..	..
Marshall Islands	..	..	..	..	4,674	..
Monaco	..	..	..	..	..	..
Nauru	..	..	9.3	..	..	..
San Marino	..	..	..	..	..	..
Somalia	..	55.4	..	..	..	..
Tuvalu	..	..	..	..	5,278	..
Human development groups						
Very high human development	0.896	80.5	16.4	11.8	41,584	—
High human development	0.744	75.1	13.6	8.2	13,961	—
Medium human development	0.630	68.6	11.8	6.2	6,353	—
Low human development	0.505	60.6	9.0	4.5	3,085	—
Developing countries	0.660	69.8	11.7	6.8	9,071	—
Regions						
Arab States	0.686	70.6	12.0	6.4	15,722	—
East Asia and the Pacific	0.710	74.0	12.7	7.5	11,449	—
Europe and Central Asia	0.748	72.3	13.6	10.0	12,791	—
Latin America and the Caribbean	0.748	75.0	14.0	8.2	14,242	—
South Asia	0.607	68.4	11.2	5.5	5,605	—
Sub-Saharan Africa	0.518	58.5	9.6	5.2	3,363	—
Least developed countries	0.502	63.3	9.3	4.1	2,387	—
Small island developing states	0.660	70.1	11.4	7.9	6,991	—
Organisation for Economic Co-operation and Development	0.880	80.2	15.8	11.5	37,658	—
World	**0.711**	**71.5**	**12.2**	**7.9**	**14,301**	—

NOTES

a Data refer to 2014 or the most recent year available.

b Based on Barro and Lee (2013b).

c For the purpose of calculating the HDI value, expected years of schooling is capped at 18 years.

d Updated by HDRO based on data from UNESCO Institute for Statistics (2015) and Barro and Lee (2014).

e Based on Barro and Lee (2014).

f Based on data from the national statistical office.

g For the purpose of calculating the HDI value, GNI per capita is capped at $75,000.

h Value from UNDESA (2011).

i Calculated as the average of mean years of schooling for Austria and Switzerland.

j Estimated using the purchasing power parity (PPP) rate and projected growth rate of Switzerland.

k HDRO estimate based on data from World Bank (2015a) and United Nations Statistics Division (2015).

l Updated by HDRO based on data from UNESCO Institute for Statistics (2015).

m Assumes the same adult mean years of schooling as Spain.

n Estimated using the PPP rate and projected growth rate of Spain.

o Based on data from UNESCO Institute for Statistics (2011).

p Based on data on school life expectancy from UNESCO Institute for Statistics (2013).

q Based on data from United Nations Children's Fund (UNICEF) Multiple Indicator Cluster Surveys for 2005–2014.

r Based on cross-country regression.

s The 2013 HDI value published in the 2014 *Human Development Report* was based on miscalculated GNI per capita in 2011 PPP dollars, as published in the World Bank (2014). A more realistic value, based on the model developed by HDRO and verified and accepted by Cuba's National Statistics Office, is $7,222. The corresponding 2013 HDI value is 0.759 and the rank is 69th.

t Based on a cross-country regression model and projected growth rates from ECLAC (2014).

u HDRO calculations based on data from National Institute for Educational Studies of Brazil (2013).

v Updated by HDRO based on data from UNESCO Institute for Statistics (2015) and a UNICEF Multiple Indicator Cluster Survey.

w Based on data on school life expectancy from UNESCO Institute for Statistics (2012).

x Based on projected growth rates from UNESCWA (2014).

y Based on data from ICF Macro Demographic and Health Surveys for 2005–2014.

z Updated by HDRO based on data from UNESCO Institute for Statistics (2015), an ICF Macro Demographic and Health Survey and Barro and Lee (2014).

aa Updated by HDRO based on data from a UNICEF Multiple Indicator Cluster Survey.

ab Based on data from Timor-Leste Ministry of Finance (2015).

DEFINITIONS

Human Development Index (HDI): A composite index measuring average achievement in three basic dimensions of human development—a long and healthy life, knowledge and a decent standard of living. See *Technical note 1* at http://hdr.undp.org/en for details on how the HDI is calculated.

Life expectancy at birth: Number of years a newborn infant could expect to live if prevailing patterns of age-specific mortality rates at the time of birth stay the same throughout the infant's life.

Expected years of schooling: Number of years of schooling that a child of school entrance age can expect to receive if prevailing patterns of age-specific enrolment rates persist throughout the child's life.

Mean years of schooling: Average number of years of education received by people ages 25 and older, converted from education attainment levels using official durations of each level.

Gross national income (GNI) per capita: Aggregate income of an economy generated by its production and its ownership of factors of production, less the incomes paid for the use of factors of production owned by the rest of the world, converted to international dollars using PPP rates, divided by midyear population.

MAIN DATA SOURCES

Column 1: HDRO calculations based on data from UNDESA (2015), UNESCO Institute for Statistics (2015), United Nations Statistics Division (2015), World Bank (2015a), Barro and Lee (2014) and IMF (2015).

Column 2: UNDESA (2015).

Column 3: UNESCO Institute for Statistics (2015).

Column 4: UNESCO Institute for Statistics (2015), Barro and Lee (2014), UNICEF Multiple Indicator Cluster Surveys and ICF Macro Demographic and Health Surveys.

Column 5: World Bank (2015a), IMF (2015) and United Nations Statistics Division (2015).

Column 6: Calculated based on data in columns 1 and 5.

TABLE 1 Human Development Index and its components | 211

TABLE 2

TABLE 2

Human Development Index trends, 1990–2014

		Human Development Index (HDI)							HDI rank		Average annual HDI growth			
		Value								Change	(%)			
HDI rank		1990	2000	2010	2011	2012	2013	2014	2013	2009–2014[a]	1990–2000	2000–2010	2010–2014	1990–2014
VERY HIGH HUMAN DEVELOPMENT														
1	Norway	0.849	0.917	0.940	0.941	0.942	0.942	0.944	1	0	0.77	0.25	0.11	0.44
2	Australia	0.865	0.898	0.927	0.930	0.932	0.933	0.935	2	0	0.36	0.33	0.20	0.32
3	Switzerland	0.831	0.888	0.924	0.925	0.927	0.928	0.930	3	0	0.67	0.40	0.14	0.47
4	Denmark	0.799	0.862	0.908	0.920	0.921	0.923	0.923	4	1	0.76	0.53	0.41	0.61
5	Netherlands	0.829	0.877	0.909	0.919	0.920	0.920	0.922	5	0	0.56	0.36	0.34	0.44
6	Germany	0.801	0.855	0.906	0.911	0.915	0.915	0.916	6	3	0.66	0.58	0.26	0.56
6	Ireland	0.770	0.861	0.908	0.909	0.910	0.912	0.916	8	−2	1.12	0.54	0.21	0.72
8	United States	0.859	0.883	0.909	0.911	0.912	0.913	0.915	7	−3	0.28	0.28	0.18	0.26
9	Canada	0.849	0.867	0.903	0.909	0.910	0.912	0.913	8	1	0.22	0.41	0.28	0.31
9	New Zealand	0.820	0.874	0.905	0.907	0.909	0.911	0.913	10	−1	0.64	0.35	0.24	0.45
11	Singapore	0.718	0.819	0.897	0.903	0.905	0.909	0.912	11	11	1.33	0.92	0.41	1.00
12	Hong Kong, China (SAR)	0.781	0.825	0.898	0.902	0.906	0.908	0.910	12	2	0.55	0.85	0.32	0.64
13	Liechtenstein	..	..	0.902	0.903	0.906	0.907	0.908	13	−2	..	..	0.14	..
14	Sweden	0.815	0.897	0.901	0.903	0.904	0.905	0.907	14	−1	0.96	0.04	0.16	0.45
14	United Kingdom	0.773	0.865	0.906	0.901	0.901	0.902	0.907	15	−2	1.13	0.46	0.02	0.67
16	Iceland	0.802	0.859	0.892	0.896	0.897	0.899	0.899	16	−1	0.69	0.38	0.20	0.48
17	Korea (Republic of)	0.731	0.821	0.886	0.891	0.893	0.895	0.898	17	0	1.16	0.77	0.33	0.86
18	Israel	0.785	0.850	0.883	0.888	0.890	0.893	0.894	18	1	0.80	0.38	0.31	0.54
19	Luxembourg	0.779	0.851	0.886	0.888	0.888	0.890	0.892	19	−3	0.88	0.41	0.16	0.56
20	Japan	0.814	0.857	0.884	0.886	0.888	0.890	0.891	19	−3	0.51	0.31	0.18	0.37
21	Belgium	0.806	0.874	0.883	0.886	0.889	0.888	0.890	21	−2	0.81	0.10	0.21	0.41
22	France	0.779	0.848	0.881	0.884	0.886	0.887	0.888	22	−1	0.85	0.38	0.20	0.55
23	Austria	0.794	0.836	0.879	0.881	0.884	0.884	0.885	23	1	0.53	0.50	0.17	0.46
24	Finland	0.783	0.857	0.878	0.881	0.882	0.882	0.883	24	−1	0.90	0.25	0.13	0.50
25	Slovenia	0.766	0.824	0.876	0.877	0.878	0.878	0.880	25	−1	0.73	0.61	0.13	0.58
26	Spain	0.756	0.827	0.867	0.870	0.874	0.874	0.876	26	2	0.90	0.47	0.27	0.62
27	Italy	0.766	0.829	0.869	0.873	0.872	0.873	0.873	27	−1	0.79	0.47	0.13	0.55
28	Czech Republic	0.761	0.821	0.863	0.866	0.867	0.868	0.870	28	0	0.76	0.50	0.21	0.56
29	Greece	0.759	0.799	0.866	0.864	0.865	0.863	0.865	29	−2	0.51	0.81	−0.04	0.55
30	Estonia	0.726	0.780	0.838	0.849	0.855	0.859	0.861	30	3	0.73	0.71	0.69	0.71
31	Brunei Darussalam	0.782	0.819	0.843	0.847	0.852	0.852	0.856	31	1	0.46	0.29	0.37	0.38
32	Cyprus	0.733	0.800	0.848	0.852	0.852	0.850	0.850	32	−2	0.87	0.59	0.04	0.62
32	Qatar	0.754	0.809	0.844	0.841	0.848	0.849	0.850	33	−1	0.71	0.42	0.18	0.50
34	Andorra	..	..	0.823	0.821	0.844	0.844	0.845	34	..	..	..	0.66	..
35	Slovakia	0.738	0.763	0.827	0.832	0.836	0.839	0.844	36	3	0.34	0.82	0.48	0.56
36	Poland	0.713	0.786	0.829	0.833	0.838	0.840	0.843	35	1	0.99	0.53	0.41	0.70
37	Lithuania	0.730	0.754	0.827	0.831	0.833	0.837	0.839	37	−1	0.32	0.93	0.38	0.58
37	Malta	0.729	0.766	0.824	0.822	0.830	0.837	0.839	37	4	0.49	0.74	0.45	0.59
39	Saudi Arabia	0.690	0.744	0.805	0.816	0.826	0.836	0.837	39	10	0.76	0.79	1.00	0.81
40	Argentina	0.705	0.762	0.811	0.818	0.831	0.833	0.836	40	7	0.78	0.62	0.75	0.71
41	United Arab Emirates	0.726	0.797	0.828	0.829	0.831	0.833	0.835	40	−6	0.94	0.39	0.21	0.59
42	Chile	0.699	0.752	0.814	0.821	0.827	0.830	0.832	42	2	0.74	0.79	0.56	0.73
43	Portugal	0.710	0.782	0.819	0.825	0.827	0.828	0.830	43	0	0.97	0.47	0.33	0.65
44	Hungary	0.703	0.769	0.821	0.823	0.823	0.825	0.828	44	−4	0.90	0.67	0.21	0.69
45	Bahrain	0.746	0.794	0.819	0.817	0.819	0.821	0.824	45	−6	0.62	0.32	0.14	0.41
46	Latvia	0.692	0.727	0.811	0.812	0.813	0.816	0.819	47	−5	0.49	1.09	0.25	0.70
47	Croatia	0.670	0.749	0.807	0.814	0.817	0.817	0.818	46	−1	1.12	0.75	0.32	0.83
48	Kuwait	0.715	0.804	0.809	0.812	0.815	0.816	0.816	47	−3	1.18	0.06	0.23	0.55
49	Montenegro	..	..	0.792	0.798	0.798	0.801	0.802	49	1	..	..	0.32	..
HIGH HUMAN DEVELOPMENT														
50	Belarus	..	0.683	0.786	0.793	0.796	0.796	0.798	51	4	..	1.41	0.39	..
50	Russian Federation	0.729	0.717	0.783	0.790	0.795	0.797	0.798	50	8	−0.17	0.88	0.47	0.38
52	Oman	..	..	0.795	0.793	0.793	0.792	0.793	52	−4	..	..	−0.06	..
52	Romania	0.703	0.706	0.784	0.786	0.788	0.791	0.793	53	−1	0.04	1.06	0.26	0.50
52	Uruguay	0.692	0.742	0.780	0.784	0.788	0.790	0.793	54	4	0.70	0.50	0.40	0.57
55	Bahamas	..	0.778	0.774	0.778	0.783	0.786	0.790	55	2	..	−0.06	0.51	..
56	Kazakhstan	0.690	0.679	0.766	0.772	0.778	0.785	0.788	56	6	−0.15	1.20	0.73	0.56
57	Barbados	0.716	0.753	0.780	0.786	0.793	0.785	0.785	56	−3	0.50	0.36	0.18	0.39
58	Antigua and Barbuda	..	..	0.782	0.778	0.781	0.781	0.783	58	−6	..	..	0.03	..
59	Bulgaria	0.695	0.713	0.773	0.775	0.778	0.779	0.782	59	0	0.26	0.81	0.29	0.49
60	Palau	..	0.743	0.767	0.770	0.775	0.775	0.780	62	0	..	0.31	0.44	..
60	Panama	0.656	0.714	0.761	0.759	0.772	0.777	0.780	60	4	0.85	0.64	0.61	0.72
62	Malaysia	0.641	0.723	0.769	0.772	0.774	0.777	0.779	60	1	1.21	0.62	0.32	0.82

		Human Development Index (HDI)							HDI rank		Average annual HDI growth			
		Value								Change	(%)			
HDI rank		1990	2000	2010	2011	2012	2013	2014	2013	2009–2014[a]	1990–2000	2000–2010	2010–2014	1990–2014
63	Mauritius	0.619	0.674	0.756	0.762	0.772	0.775	0.777	62	6	0.86	1.15	0.68	0.95
64	Seychelles	..	0.715	0.743	0.752	0.761	0.767	0.772	68	8	..	0.39	0.97	..
64	Trinidad and Tobago	0.673	0.717	0.772	0.767	0.769	0.771	0.772	64	−4	0.63	0.74	0.01	0.57
66	Serbia	0.714	0.710	0.757	0.761	0.762	0.771	0.771	64	−1	−0.05	0.65	0.45	0.32
67	Cuba	0.675	0.685	0.778	0.776	0.772	0.768	0.769	66	−14	0.15	1.28	−0.28	0.54
67	Lebanon	..	..	0.756	0.761	0.761	0.768	0.769	66	1	..	..	0.43	..
69	Costa Rica	0.652	0.704	0.750	0.756	0.761	0.764	0.766	69	1	0.77	0.64	0.52	0.67
69	Iran (Islamic Republic of)	0.567	0.665	0.743	0.751	0.764	0.764	0.766	69	7	1.62	1.11	0.74	1.26
71	Venezuela (Bolivarian Republic of)	0.635	0.673	0.757	0.761	0.764	0.764	0.762	69	−4	0.59	1.17	0.18	0.76
72	Turkey	0.576	0.653	0.738	0.751	0.756	0.759	0.761	72	16	1.26	1.23	0.79	1.17
73	Sri Lanka	0.620	0.679	0.738	0.743	0.749	0.752	0.757	74	5	0.91	0.85	0.62	0.83
74	Mexico	0.648	0.699	0.746	0.748	0.754	0.755	0.756	73	−2	0.77	0.65	0.35	0.65
75	Brazil	0.608	0.683	0.737	0.742	0.746	0.752	0.755	74	3	1.18	0.76	0.60	0.91
76	Georgia	..	0.672	0.735	0.740	0.747	0.750	0.754	76	4	..	0.89	0.65	..
77	Saint Kitts and Nevis	..	..	0.739	0.741	0.743	0.747	0.752	79	..	..	..	0.44	..
78	Azerbaijan	..	0.640	0.741	0.742	0.745	0.749	0.751	77	−2	..	1.46	0.35	..
79	Grenada	..	..	0.737	0.739	0.740	0.742	0.750	82	..	..	..	0.43	..
80	Jordan	0.623	0.705	0.743	0.743	0.746	0.748	0.748	78	−8	1.25	0.53	0.17	0.77
81	The former Yugoslav Republic of Macedonia	..	..	0.738	0.742	0.743	0.744	0.747	81	−2	..	..	0.31	..
81	Ukraine	0.705	0.668	0.732	0.738	0.743	0.746	0.747	80	2	−0.54	0.92	0.51	0.24
83	Algeria	0.574	0.640	0.725	0.730	0.732	0.734	0.736	84	4	1.09	1.26	0.35	1.04
84	Peru	0.613	0.677	0.718	0.722	0.728	0.732	0.734	85	15	1.00	0.58	0.57	0.75
85	Albania	0.624	0.656	0.722	0.728	0.729	0.732	0.733	85	2	0.50	0.96	0.35	0.67
85	Armenia	0.632	0.648	0.721	0.723	0.728	0.731	0.733	87	1	0.24	1.08	0.41	0.62
85	Bosnia and Herzegovina	..	..	0.710	0.724	0.726	0.729	0.733	89	2	..	..	0.78	..
88	Ecuador	0.645	0.674	0.717	0.723	0.727	0.730	0.732	88	5	0.45	0.61	0.52	0.53
89	Saint Lucia	..	0.683	0.730	0.730	0.730	0.729	0.729	89	−5	..	0.66	−0.02	..
90	China	0.501	0.588	0.699	0.707	0.718	0.723	0.727	93	13	1.62	1.74	1.02	1.57
90	Fiji	0.631	0.678	0.717	0.720	0.722	0.724	0.727	91	1	0.72	0.56	0.36	0.59
90	Mongolia	0.578	0.589	0.695	0.706	0.714	0.722	0.727	95	14	0.18	1.68	1.11	0.96
93	Thailand	0.572	0.648	0.716	0.721	0.723	0.724	0.726	91	3	1.25	1.00	0.35	1.00
94	Dominica	..	0.694	0.723	0.723	0.723	0.723	0.724	93	−10	..	0.41	0.03	..
94	Libya	0.679	0.731	0.756	0.711	0.745	0.738	0.724	83	−27	0.75	0.34	−1.07	0.27
96	Tunisia	0.567	0.654	0.714	0.715	0.719	0.720	0.721	96	−1	1.43	0.88	0.26	1.00
97	Colombia	0.596	0.654	0.706	0.713	0.715	0.718	0.720	97	3	0.93	0.76	0.50	0.79
97	Saint Vincent and the Grenadines	..	0.674	0.711	0.713	0.715	0.717	0.720	98	−5	..	0.55	0.30	..
99	Jamaica	0.671	0.700	0.727	0.727	0.723	0.717	0.719	98	−23	0.42	0.38	−0.30	0.28
100	Tonga	0.650	0.671	0.713	0.716	0.717	0.716	0.717	100	−4	0.32	0.60	0.14	0.41
101	Belize	0.644	0.683	0.709	0.711	0.716	0.715	0.715	101	−7	0.59	0.38	0.19	0.43
101	Dominican Republic	0.596	0.655	0.701	0.704	0.708	0.711	0.715	103	0	0.95	0.68	0.50	0.76
103	Suriname	..	..	0.707	0.709	0.711	0.713	0.714	102	−5	..	..	0.24	..
104	Maldives	..	0.603	0.683	0.690	0.695	0.703	0.706	104	2	..	1.25	0.86	..
105	Samoa	0.621	0.649	0.696	0.698	0.700	0.701	0.702	105	−3	0.45	0.70	0.21	0.52
MEDIUM HUMAN DEVELOPMENT														
106	Botswana	0.584	0.561	0.681	0.688	0.691	0.696	0.698	106	1	−0.41	1.96	0.61	0.74
107	Moldova (Republic of)	0.652	0.597	0.672	0.679	0.683	0.690	0.693	107	2	−0.87	1.19	0.78	0.26
108	Egypt	0.546	0.622	0.681	0.682	0.688	0.689	0.690	108	−3	1.31	0.90	0.33	0.98
109	Turkmenistan	..	..	0.666	0.671	0.677	0.682	0.688	109	..	..	..	0.80	..
110	Gabon	0.620	0.632	0.663	0.668	0.673	0.679	0.684	111	1	0.20	0.48	0.76	0.41
110	Indonesia	0.531	0.606	0.665	0.671	0.678	0.681	0.684	110	3	1.34	0.92	0.71	1.06
112	Paraguay	0.579	0.623	0.668	0.671	0.669	0.677	0.679	113	−1	0.74	0.70	0.41	0.67
113	Palestine, State of	..	..	0.670	0.675	0.685	0.679	0.677	111	−4	..	..	0.29	..
114	Uzbekistan	..	0.594	0.655	0.661	0.668	0.672	0.675	114	0	..	0.98	0.77	..
115	Philippines	0.586	0.623	0.654	0.653	0.657	0.664	0.668	115	−1	0.61	0.50	0.52	0.55
116	El Salvador	0.522	0.603	0.653	0.658	0.662	0.664	0.666	115	0	1.46	0.79	0.50	1.02
116	South Africa	0.621	0.632	0.643	0.651	0.659	0.663	0.666	117	4	0.17	0.18	0.87	0.29
116	Viet Nam	0.475	0.575	0.653	0.657	0.660	0.663	0.666	117	1	1.92	1.29	0.47	1.41
119	Bolivia (Plurinational State of)	0.536	0.603	0.641	0.647	0.654	0.658	0.662	119	2	1.19	0.61	0.79	0.88
120	Kyrgyzstan	0.615	0.593	0.634	0.639	0.645	0.652	0.655	121	3	−0.37	0.68	0.84	0.26
121	Iraq	0.572	0.606	0.645	0.648	0.654	0.657	0.654	120	−2	0.58	0.62	0.34	0.56
122	Cabo Verde	..	0.572	0.629	0.637	0.639	0.643	0.646	122	2	..	0.96	0.66	..
123	Micronesia (Federated States of)	..	0.603	0.638	0.640	0.641	0.639	0.640	123	−2	..	0.56	0.06	..
124	Guyana	0.542	0.602	0.624	0.630	0.629	0.634	0.636	124	1	1.05	0.36	0.47	0.66
125	Nicaragua	0.495	0.565	0.619	0.623	0.625	0.628	0.631	125	1	1.34	0.91	0.51	1.02

TABLE 2   Human Development Index trends, 1990–2014 | 213

TABLE 2 HUMAN DEVELOPMENT INDEX TRENDS, 1990–2014

TABLE 2

		Human Development Index (HDI)							HDI rank		Average annual HDI growth			
		Value								Change	(%)			
HDI rank		1990	2000	2010	2011	2012	2013	2014	2013	2009–2014[a]	1990–2000	2000–2010	2010–2014	1990–2014
126	Morocco	0.457	0.528	0.611	0.621	0.623	0.626	0.628	126	5	1.44	1.48	0.69	1.33
126	Namibia	0.578	0.556	0.610	0.616	0.620	0.625	0.628	128	3	−0.39	0.94	0.70	0.35
128	Guatemala	0.483	0.552	0.611	0.617	0.624	0.626	0.627	126	0	1.35	1.03	0.65	1.10
129	Tajikistan	0.616	0.535	0.608	0.612	0.617	0.621	0.624	129	1	−1.39	1.28	0.68	0.06
130	India	0.428	0.496	0.586	0.597	0.600	0.604	0.609	131	6	1.49	1.67	0.97	1.48
131	Honduras	0.507	0.557	0.610	0.612	0.607	0.604	0.606	131	−4	0.95	0.91	−0.16	0.75
132	Bhutan	..	..	0.573	0.582	0.589	0.595	0.605	134	..	..	..	1.39	..
133	Timor-Leste	..	0.468	0.600	0.611	0.604	0.601	0.595	133	1	..	2.51	−0.22	..
134	Syrian Arab Republic	0.553	0.586	0.639	0.635	0.623	0.608	0.594	130	−15	0.58	0.88	−1.82	0.30
134	Vanuatu	..	..	0.589	0.590	0.590	0.592	0.594	135	1	..	..	0.19	..
136	Congo	0.534	0.489	0.554	0.560	0.575	0.582	0.591	138	2	−0.87	1.25	1.61	0.42
137	Kiribati	..	..	0.588	0.585	0.587	0.588	0.590	136	−1	..	..	0.09	..
138	Equatorial Guinea	..	0.526	0.591	0.590	0.584	0.584	0.587	137	−5	..	1.18	−0.18	..
139	Zambia	0.403	0.433	0.555	0.565	0.576	0.580	0.586	139	1	0.71	2.52	1.36	1.57
140	Ghana	0.456	0.485	0.554	0.566	0.572	0.577	0.579	140	−2	0.63	1.33	1.13	1.00
141	Lao People's Democratic Republic	0.397	0.462	0.539	0.552	0.562	0.570	0.575	141	2	1.51	1.56	1.62	1.55
142	Bangladesh	0.386	0.468	0.546	0.559	0.563	0.567	0.570	142	0	1.94	1.57	1.07	1.64
143	Cambodia	0.364	0.419	0.536	0.541	0.546	0.550	0.555	144	1	1.40	2.50	0.87	1.77
143	Sao Tome and Principe	0.455	0.491	0.544	0.548	0.552	0.553	0.555	143	−2	0.76	1.02	0.52	0.83
LOW HUMAN DEVELOPMENT														
145	Kenya	0.473	0.447	0.529	0.535	0.539	0.544	0.548	145	0	−0.58	1.70	0.92	0.62
145	Nepal	0.384	0.451	0.531	0.536	0.540	0.543	0.548	146	3	1.62	1.64	0.78	1.49
147	Pakistan	0.399	0.444	0.522	0.527	0.532	0.536	0.538	147	0	1.07	1.62	0.79	1.25
148	Myanmar	0.352	0.425	0.520	0.524	0.528	0.531	0.536	148	1	1.90	2.03	0.72	1.76
149	Angola	..	0.390	0.509	0.521	0.524	0.530	0.532	149	1	..	2.70	1.11	..
150	Swaziland	0.536	0.496	0.525	0.528	0.529	0.530	0.531	149	−5	−0.78	0.57	0.28	−0.04
151	Tanzania (United Republic of)	0.369	0.392	0.500	0.506	0.510	0.516	0.521	151	2	0.60	2.46	1.05	1.44
152	Nigeria	..	..	0.493	0.499	0.505	0.510	0.514	152	2	..	..	1.06	..
153	Cameroon	0.443	0.437	0.486	0.496	0.501	0.507	0.512	154	6	−0.13	1.07	1.32	0.61
154	Madagascar	..	0.456	0.504	0.505	0.507	0.508	0.510	153	−4	..	1.02	0.27	..
155	Zimbabwe	0.499	0.428	0.461	0.474	0.491	0.501	0.509	158	12	−1.53	0.75	2.50	0.08
156	Mauritania	0.373	0.442	0.488	0.489	0.498	0.504	0.506	156	1	1.71	0.98	0.92	1.28
156	Solomon Islands	..	0.446	0.494	0.501	0.504	0.505	0.506	155	−2	..	1.02	0.57	..
158	Papua New Guinea	0.353	0.424	0.493	0.497	0.501	0.503	0.505	157	−2	1.87	1.51	0.60	1.51
159	Comoros	..	..	0.488	0.493	0.499	0.501	0.503	158	−1	..	..	0.75	..
160	Yemen	0.400	0.441	0.496	0.495	0.496	0.498	0.498	160	−8	0.99	1.19	0.08	0.92
161	Lesotho	0.493	0.443	0.472	0.480	0.484	0.494	0.497	161	1	−1.05	0.62	1.30	0.03
162	Togo	0.404	0.426	0.459	0.468	0.470	0.473	0.484	167	3	0.52	0.76	1.29	0.75
163	Haiti	0.417	0.442	0.471	0.475	0.479	0.481	0.483	162	−3	0.58	0.62	0.67	0.61
163	Rwanda	0.244	0.333	0.453	0.464	0.476	0.479	0.483	163	5	3.16	3.13	1.61	2.89
163	Uganda	0.308	0.393	0.473	0.473	0.476	0.478	0.483	164	−2	2.47	1.86	0.51	1.89
166	Benin	0.344	0.392	0.468	0.473	0.475	0.477	0.480	165	−2	1.33	1.78	0.64	1.40
167	Sudan	0.331	0.400	0.465	0.466	0.476	0.477	0.479	165	−5	1.90	1.52	0.74	1.55
168	Djibouti	..	0.365	0.453	0.462	0.465	0.468	0.470	168	0	..	2.17	0.97	..
169	South Sudan	..	..	0.470	0.458	0.457	0.461	0.467	171	..	..	..	−0.15	..
170	Senegal	0.367	0.380	0.456	0.458	0.461	0.463	0.466	170	−3	0.36	1.83	0.55	1.00
171	Afghanistan	0.297	0.334	0.448	0.456	0.463	0.464	0.465	169	0	1.20	2.97	0.97	1.89
172	Côte d'Ivoire	0.389	0.398	0.444	0.445	0.452	0.458	0.462	172	0	0.23	1.12	0.98	0.72
173	Malawi	0.284	0.340	0.420	0.429	0.433	0.439	0.445	174	2	1.83	2.14	1.49	1.90
174	Ethiopia	..	0.284	0.412	0.423	0.429	0.436	0.442	175	2	..	3.78	1.78	..
175	Gambia	0.330	0.384	0.441	0.437	0.440	0.442	0.441	173	−2	1.55	1.38	−0.02	1.22
176	Congo (Democratic Republic of the)	0.355	0.329	0.408	0.418	0.423	0.430	0.433	176	3	−0.77	2.18	1.52	0.83
177	Liberia	..	0.359	0.405	0.414	0.419	0.424	0.430	177	1	..	1.20	1.50	..
178	Guinea-Bissau	..	..	0.413	0.417	0.417	0.418	0.420	178	−4	..	..	0.42	..
179	Mali	0.233	0.313	0.409	0.415	0.414	0.416	0.419	179	−3	2.97	2.73	0.61	2.47
180	Mozambique	0.218	0.300	0.401	0.405	0.408	0.413	0.416	180	0	3.25	2.96	0.94	2.74
181	Sierra Leone	0.262	0.299	0.388	0.394	0.397	0.408	0.413	182	0	1.32	2.63	1.59	1.91
182	Guinea	..	0.323	0.388	0.399	0.409	0.411	0.411	181	1	..	1.83	1.50	..
183	Burkina Faso	..	..	0.378	0.385	0.393	0.396	0.402	184	2	..	..	1.58	..
184	Burundi	0.295	0.301	0.390	0.392	0.395	0.397	0.400	183	0	0.20	2.62	0.66	1.28
185	Chad	..	0.332	0.371	0.382	0.386	0.388	0.392	186	1	..	1.12	1.37	..
186	Eritrea	..	..	0.381	0.386	0.390	0.390	0.391	185	−5	..	..	0.62	..
187	Central African Republic	0.314	0.310	0.362	0.368	0.373	0.348	0.350	187	0	−0.14	1.58	−0.84	0.45

	Human Development Index (HDI)							HDI rank		Average annual HDI growth			
	Value								Change	(%)			
HDI rank	1990	2000	2010	2011	2012	2013	2014	2013	2009–2014[a]	1990–2000	2000–2010	2010–2014	1990–2014
188 Niger	0.214	0.257	0.326	0.333	0.342	0.345	0.348	188	0	1.85	2.40	1.69	2.05
OTHER COUNTRIES OR TERRITORIES													
Korea (Democratic People's Rep. of)	..	..	..	..	..	..	..	..	..	..	..	..	..
Marshall Islands	..	..	..	..	..	..	..	..	..	..	..	..	..
Monaco	..	..	..	..	..	..	..	..	..	..	..	..	..
Nauru	..	..	..	..	..	..	..	..	..	..	..	..	..
San Marino	..	..	..	..	..	..	..	..	..	..	..	..	..
Somalia	..	..	..	..	..	..	..	..	..	..	..	..	..
Tuvalu	..	..	..	..	..	..	..	..	..	..	..	..	..
Human development groups													
Very high human development	0.801	0.851	0.887	0.890	0.893	0.895	0.896	—	—	0.61	0.42	0.26	0.47
High human development	0.592	0.642	0.723	0.730	0.737	0.741	0.744	—	—	0.81	1.20	0.71	0.95
Medium human development	0.473	0.537	0.611	0.619	0.623	0.627	0.630	—	—	1.28	1.29	0.78	1.20
Low human development	0.368	0.404	0.487	0.492	0.497	0.502	0.505	—	—	0.92	1.90	0.92	1.32
Developing countries	0.513	0.568	0.642	0.649	0.654	0.658	0.660	—	—	1.02	1.23	0.70	1.06
Regions													
Arab States	0.553	0.613	0.676	0.679	0.684	0.686	0.686	—	—	1.02	0.99	0.38	0.90
East Asia and the Pacific	0.516	0.593	0.686	0.693	0.702	0.707	0.710	—	—	1.39	1.48	0.87	1.34
Europe and Central Asia	0.651	0.665	0.731	0.739	0.743	0.746	0.748	—	—	0.22	0.94	0.59	0.58
Latin America and the Caribbean	0.625	0.684	0.734	0.738	0.743	0.745	0.748	—	—	0.91	0.70	0.47	0.75
South Asia	0.437	0.503	0.586	0.596	0.599	0.603	0.607	—	—	1.42	1.55	0.86	1.38
Sub-Saharan Africa	0.400	0.422	0.499	0.505	0.510	0.514	0.518	—	—	0.54	1.68	0.94	1.08
Least developed countries	0.348	0.399	0.484	0.491	0.495	0.499	0.502	—	—	1.39	1.95	0.92	1.54
Small island developing states	0.574	0.607	0.656	0.658	0.658	0.658	0.660	—	—	0.56	0.79	0.13	0.59
Organisation for Economic Co-operation and Development	0.785	0.834	0.872	0.875	0.877	0.879	0.880	—	—	0.61	0.44	0.24	0.48
World	**0.597**	**0.641**	**0.697**	**0.703**	**0.707**	**0.709**	**0.711**	**—**	**—**	**0.71**	**0.85**	**0.47**	**0.73**

TABLE
2

NOTES

a A positive value indicates an improvement in rank.

DEFINITIONS

Human Development Index (HDI): A composite index measuring average achievement in three basic dimensions of human development—a long and healthy life, knowledge and a decent standard of living. See *Technical note 1* (http://hdr.undp.org/en) for details on how the HDI is calculated.

Average annual HDI growth: A smoothed annualized growth of the HDI in a given period, calculated as the annual compound growth rate.

MAIN DATA SOURCES

Columns 1–7: HDRO calculations based on data from UNDESA (2015), UNESCO Institute for Statistics (2015), United Nations Statistics Division (2015), World Bank (2015a), Barro and Lee (2014) and IMF (2015).

Column 8: Calculated based on data in column 6.

Column 9: Calculated based on HDI data from HDRO and data in column 7.

Columns 10–13: Calculated based on data in columns 1, 2, 3 and 7.

TABLE 2   Human Development Index trends, 1990–2014 | 215

TABLE **3**

Inequality-adjusted Human Development Index

TABLE 3

	Human Development Index (HDI)	Inequality-adjusted HDI (IHDI)			Coefficient of human inequality	Inequality in life expectancy	Inequality-adjusted life expectancy index	Inequality in education[a]	Inequality-adjusted education index	Inequality in income[a]	Inequality-adjusted income index	Income inequality		
	Value	Value	Overall loss (%)	Difference from HDI rank[b]		(%)	Value	(%)	Value	(%)	Value	Quintile ratio	Palma ratio	Gini coefficient
HDI rank	2014	2014	2014	2014	2014	2010–2015[c]	2014	2014[d]	2014	2014[d]	2014	2005–2013[c]	2005–2013[c]	2005–2013[c]
VERY HIGH HUMAN DEVELOPMENT														
1 Norway	0.944	0.893	5.4	0	5.3	3.4	0.916	2.3	0.886	10.2	0.878	4.0	0.9	26.8
2 Australia	0.935	0.858	8.2	−2	7.9	4.2	0.920	1.9	0.914	17.7	0.752	5.8	1.3	34.0
3 Switzerland	0.930	0.861	7.4	0	7.3	3.9	0.931	5.7	0.816	12.3	0.839	5.2	1.2	32.4
4 Denmark	0.923	0.856	7.3	−1	7.1	4.0	0.889	3.0	0.897	14.4	0.787	4.0	0.9	26.9
5 Netherlands	0.922	0.861	6.6	3	6.5	3.9	0.911	4.1	0.858	11.6	0.817	4.5	1.0	28.9
6 Germany	0.916	0.853	6.9	0	6.7	3.7	0.902	2.4	0.871	14.1	0.790	4.7	1.1	30.6
6 Ireland	0.916	0.836	8.6	−3	8.5	3.7	0.902	5.4	0.858	16.3	0.756	5.3	1.2	32.1
8 United States	0.915	0.760	17.0	−20	15.7	6.2	0.853	5.3	0.842	35.6	0.610	9.8	2.0	41.1
9 Canada	0.913	0.832	8.8	−2	8.6	4.6	0.910	3.9	0.841	17.4	0.754	5.8	1.3	33.7
9 New Zealand	0.913	..	..	..	..	4.8	0.905	..	..	..	..	..	..	..
11 Singapore	0.912	..	..	..	..	2.8	0.942	..	..	..	..	..	..	..
12 Hong Kong, China (SAR)	0.910	..	..	..	..	2.8	0.957	..	..	..	..	..	..	..
13 Liechtenstein	0.908	..	..	..	..	..	..	..	..	..	..	..	..	..
14 Sweden	0.907	0.846	6.7	3	6.5	3.1	0.927	3.5	0.813	13.1	0.804	3.7	0.9	26.1
14 United Kingdom	0.907	0.829	8.6	−2	8.4	4.5	0.892	2.8	0.860	17.8	0.742	7.6	1.7	38.0
16 Iceland	0.899	0.846	5.9	4	5.8	2.8	0.936	2.4	0.832	12.2	0.777	3.8	0.9	26.3
17 Korea (Republic of)	0.898	0.751	16.4	−19	15.9	3.9	0.915	25.5	0.644	18.4	0.718	..	..	..
18 Israel	0.894	0.775	13.4	−9	12.9	3.8	0.923	9.9	0.776	25.0	0.649	10.3	2.2	42.8
19 Luxembourg	0.892	0.822	7.9	0	7.7	3.3	0.918	6.0	0.729	13.9	0.829	..	..	..
20 Japan	0.891	0.780	12.4	−5	12.2	3.2	0.945	19.8	0.649	13.5	0.773	5.4	1.2	32.1
21 Belgium	0.890	0.820	7.9	1	7.9	3.9	0.899	8.1	0.762	11.6	0.804	5.0	1.3	33.1
22 France	0.888	0.811	8.7	0	8.6	4.0	0.919	8.0	0.751	13.9	0.772	5.1	1.2	31.7
23 Austria	0.885	0.816	7.8	2	7.6	3.7	0.909	3.5	0.771	15.5	0.777	4.6	1.1	30.0
24 Finland	0.883	0.834	5.5	10	5.5	3.5	0.902	2.1	0.800	10.8	0.803	4.0	1.0	27.8
25 Slovenia	0.880	0.829	5.9	8	5.8	3.8	0.894	2.6	0.840	11.0	0.757	3.6	0.8	24.9
26 Spain	0.876	0.775	11.5	0	11.0	3.9	0.926	5.2	0.759	23.9	0.663	7.6	1.4	35.8
27 Italy	0.873	0.773	11.5	−1	11.3	3.4	0.938	10.6	0.700	19.8	0.702	6.9	1.4	35.5
28 Czech Republic	0.870	0.823	5.4	10	5.3	3.7	0.868	1.4	0.854	11.0	0.751	3.9	0.9	26.4
29 Greece	0.865	0.758	12.4	−5	12.1	4.0	0.899	11.6	0.735	20.6	0.660	6.4	1.4	34.7
30 Estonia	0.861	0.782	9.2	6	8.9	5.6	0.825	2.4	0.853	18.6	0.680	5.6	1.2	32.7
31 Brunei Darussalam	0.856	..	..	..	..	4.4	0.865	..	..	..	..	..	..	..
32 Cyprus	0.850	0.758	10.7	−2	10.6	3.7	0.892	13.1	0.673	15.0	0.727	..	..	..
32 Qatar	0.850	..	..	..	..	6.0	0.841	..	..	..	..	..	..	..
34 Andorra	0.845	..	..	..	..	..	..	..	..	..	..	..	..	..
35 Slovakia	0.844	0.791	6.2	9	6.2	5.6	0.818	1.5	0.813	11.3	0.744	4.1	0.9	26.6
36 Poland	0.843	0.760	9.8	2	9.6	5.7	0.833	5.6	0.778	17.5	0.678	5.2	1.3	32.8
37 Lithuania	0.839	0.754	10.1	−1	10.0	6.6	0.766	5.9	0.817	17.5	0.686	5.7	1.2	32.6
37 Malta	0.839	0.767	8.5	4	8.5	4.8	0.888	7.3	0.690	13.3	0.737	..	..	..
39 Saudi Arabia	0.837	..	..	..	..	8.7	0.762	..	..	..	..	..	..	..
40 Argentina	0.836	0.711	15.0	−8	14.5	9.3	0.786	8.1	0.759	26.3	0.601	10.6	2.3	43.6
41 United Arab Emirates	0.835	..	..	..	..	5.5	0.829	..	..	..	..	..	..	..
42 Chile	0.832	0.672	19.3	−13	18.2	5.9	0.893	12.6	0.655	36.0	0.519	12.6	3.3	50.8
43 Portugal	0.830	0.744	10.4	1	10.1	3.9	0.900	5.9	0.685	20.3	0.668	..	..	..
44 Hungary	0.828	0.769	7.2	10	7.1	5.4	0.803	3.2	0.789	12.6	0.717	4.5	1.0	28.9
45 Bahrain	0.824	..	..	..	..	6.3	0.816	..	..	..	..	..	..	..
46 Latvia	0.819	0.730	10.8	0	10.5	7.6	0.771	3.5	0.778	20.5	0.649	6.9	1.5	36.0
47 Croatia	0.818	0.743	9.1	3	8.9	5.2	0.836	4.3	0.745	17.2	0.659	5.3	1.4	33.6
48 Kuwait	0.816	..	..	..	..	7.2	0.776	..	..	..	..	..	..	..
49 Montenegro	0.802	0.728	9.2	1	9.2	7.6	0.799	7.4	0.735	12.6	0.658	4.7	1.1	30.6
HIGH HUMAN DEVELOPMENT														
50 Belarus	0.798	0.741	7.1	4	7.1	6.8	0.736	3.7	0.804	10.8	0.690	3.8	0.9	26.5
50 Russian Federation	0.798	0.714	10.5	1	10.3	9.8	0.695	2.3	0.788	18.7	0.664	7.3	1.8	39.7
52 Oman	0.793	..	..	..	..	7.0	0.813	..	..	..	..	..	..	..
52 Romania	0.793	0.711	10.3	2	10.2	8.8	0.768	4.7	0.719	17.1	0.651	4.1	0.9	27.3
52 Uruguay	0.793	0.678	14.5	−4	14.2	9.2	0.799	9.2	0.647	24.2	0.602	9.0	2.0	41.3
55 Bahamas	0.790	..	..	..	..	9.4	0.772	..	..	..	..	..	..	..
56 Kazakhstan	0.788	0.694	11.9	1	11.8	16.7	0.633	5.9	0.751	12.7	0.704	4.0	1.0	28.6
57 Barbados	0.785	..	..	..	..	8.1	0.786	5.5	0.734	..	..	..	..	..
58 Antigua and Barbuda	0.783	..	..	..	..	8.0	0.794	..	..	..	..	..	..	..
59 Bulgaria	0.782	0.699	10.5	3	10.4	7.9	0.768	5.5	0.710	17.8	0.627	6.4	1.4	34.3

TABLE
3

HDI rank	Human Development Index (HDI) Value	Inequality-adjusted HDI (IHDI) Value	Overall loss (%)	Difference from HDI rank[b]	Coefficient of human inequality	Inequality in life expectancy (%)	Inequality-adjusted life expectancy index Value	Inequality in education[a] (%)	Inequality-adjusted education index Value	Inequality in income[a] (%)	Inequality-adjusted income index Value	Income inequality Quintile ratio	Palma ratio	Gini coefficient
	2014	2014	2014	2014	2014	2010–2015[c]	2014	2014[d]	2014	2014[d]	2014	2005–2013[e]	2005–2013[e]	2005–2013[e]
60 Palau	0.780	..	..	..	..	..	..	12.0	0.696	23.0	0.570			
60 Panama	0.780	0.604	22.5	–20	21.7	12.1	0.779	16.6	0.567	36.5	0.499	17.6	3.6	51.9
62 Malaysia	0.779	..	..	..	..	4.9	0.800	..	..	..	..	11.3	2.6	46.2
63 Mauritius	0.777	0.666	14.2	–2	14.1	9.2	0.760	13.2	0.623	19.8	0.625	5.9	1.5	35.9
64 Seychelles	0.772	..	..	..	..	7.9	0.752	..	..	..	..	18.8	6.4	65.8
64 Trinidad and Tobago	0.772	0.654	15.2	–3	15.0	16.4	0.648	6.6	0.659	21.9	0.656	..	..	..
66 Serbia	0.771	0.693	10.1	5	10.1	8.5	0.773	8.1	0.688	13.5	0.627	4.6	1.1	29.7
67 Cuba	0.769	..	..	..	..	5.1	0.867	11.3	0.681	..	..	..	..	..
67 Lebanon	0.769	0.609	20.8	–15	20.2	6.7	0.852	24.1	0.491	30.0	0.540	..	..	..
69 Costa Rica	0.766	0.613	19.9	–11	19.1	7.3	0.847	15.5	0.561	34.3	0.486	12.8	2.9	48.6
69 Iran (Islamic Republic of)	0.766	0.509	33.6	–41	32.1	12.5	0.746	37.3	0.433	46.6	0.407	7.0	1.7	38.3
71 Venezuela (Bolivarian Republic of)	0.762	0.612	19.7	–11	19.4	12.2	0.732	17.6	0.570	28.4	0.550	11.7	2.4	44.8
72 Turkey	0.761	0.641	15.8	0	15.7	11.0	0.757	14.2	0.563	21.8	0.618	8.0	1.9	40.0
73 Sri Lanka	0.757	0.669	11.6	7	11.6	8.3	0.774	12.8	0.646	13.7	0.597	5.8	1.6	36.4
74 Mexico	0.756	0.587	22.4	–12	21.8	10.9	0.778	19.7	0.518	34.6	0.501	11.1	2.8	48.1
75 Brazil	0.755	0.557	26.3	–20	25.6	14.5	0.717	23.6	0.518	38.7	0.465	16.9	3.8	52.7
76 Georgia	0.754	0.652	13.6	5	13.2	12.9	0.736	3.3	0.761	23.4	0.494	8.9	2.0	41.4
77 Saint Kitts and Nevis	0.752	..	..	..	..	..	..	..	..	..	..	..	..	..
78 Azerbaijan	0.751	0.652	13.2	7	12.9	21.7	0.612	8.3	0.645	8.9	0.702	5.1	1.3	33.0
79 Grenada	0.750	..	..	..	..	8.4	0.753	..	..	..	..	..	..	..
80 Jordan	0.748	0.625	16.5	2	16.4	11.9	0.732	16.9	0.586	20.5	0.568	5.1	1.4	33.7
81 The former Yugoslav Republic of Macedonia	0.747	0.622	16.7	2	16.1	7.6	0.788	10.6	0.608	30.1	0.504	9.3	2.3	44.2
81 Ukraine	0.747	0.689	7.8	16	7.7	10.4	0.703	3.6	0.770	9.2	0.604	3.4	0.9	24.8
83 Algeria	0.736	..	..	..	..	16.7	0.702	..	..	..	..	..	..	..
84 Peru	0.734	0.563	23.4	–10	23.0	13.9	0.723	23.3	0.509	31.9	0.484	11.9	2.5	45.3
85 Albania	0.733	0.634	13.5	8	13.4	9.9	0.801	11.9	0.561	18.3	0.567	4.3	1.0	29.0
85 Armenia	0.733	0.658	10.2	14	10.1	12.7	0.735	3.7	0.679	13.9	0.572	4.5	1.1	30.3
85 Bosnia and Herzegovina	0.733	0.635	13.3	9	13.1	6.7	0.811	12.5	0.573	20.2	0.551	5.4	1.3	33.0
88 Ecuador	0.732	0.570	22.1	–4	21.8	13.4	0.745	21.1	0.510	30.9	0.487	12.0	2.7	46.6
89 Saint Lucia	0.729	0.613	15.9	5	15.5	9.9	0.764	9.2	0.600	27.4	0.502	..	..	..
90 China	0.727	..	..	..	..	9.8	0.774	..	..	29.5	0.514	10.1	2.1	37.0
90 Fiji	0.727	0.616	15.3	8	15.1	12.3	0.675	10.5	0.686	22.6	0.505	8.0	2.2	42.8
90 Mongolia	0.727	0.633	12.9	12	12.8	16.6	0.634	9.4	0.647	12.3	0.619	6.2	1.6	36.5
93 Thailand	0.726	0.576	20.6	1	19.9	9.8	0.755	16.1	0.519	34.0	0.488	6.9	1.8	39.4
94 Dominica	0.724	..	..	..	..	..	..	..	..	..	..	..	..	..
94 Libya	0.724	..	..	..	..	10.1	0.714	..	..	..	..	..	..	..
96 Tunisia	0.721	0.562	22.0	–2	21.4	10.6	0.753	34.6	0.415	18.9	0.569	6.4	1.5	35.8
97 Colombia	0.720	0.542	24.7	–10	24.1	13.5	0.718	21.3	0.489	37.4	0.453	17.5	4.0	53.5
97 Saint Vincent and the Grenadines	0.720	..	..	..	..	13.0	0.708	..	..	..	..	..	..	..
99 Jamaica	0.719	0.593	17.5	7	16.9	15.0	0.729	5.6	0.628	30.1	0.454	9.6	2.5	45.5
100 Tonga	0.717	..	..	..	..	13.7	0.701	..	..	..	..	..	..	..
101 Belize	0.715	0.553	22.6	–3	21.7	11.4	0.682	15.9	0.611	37.9	0.406	..	..	..
101 Dominican Republic	0.715	0.546	23.6	–6	23.4	16.9	0.684	22.9	0.474	30.3	0.503	10.3	2.5	45.7
103 Suriname	0.714	0.543	24.0	–5	23.3	13.6	0.680	19.0	0.492	37.3	0.478	..	..	..
104 Maldives	0.706	0.531	24.9	–6	23.8	8.1	0.803	40.0	0.333	23.2	0.559	6.8	1.6	37.4
105 Samoa	0.702	..	..	..	..	13.3	0.713	..	..	..	..	..	..	..
MEDIUM HUMAN DEVELOPMENT														
106 Botswana	0.698	0.431	38.2	–23	36.5	21.9	0.535	32.1	0.437	55.5	0.344	22.9	5.8	60.5
107 Moldova (Republic of)	0.693	0.618	10.8	20	10.8	11.0	0.707	7.3	0.651	14.0	0.514	4.6	1.1	30.6
108 Egypt	0.690	0.524	24.0	–5	22.8	13.4	0.681	40.9	0.351	14.2	0.604	4.4	1.2	30.8
109 Turkmenistan	0.688	..	..	..	..	26.0	0.519	..	..	..	..	..	..	..
110 Gabon	0.684	0.519	24.0	–6	24.0	28.0	0.492	23.5	0.465	20.4	0.613	8.3	2.1	42.2
110 Indonesia	0.684	0.559	18.2	6	18.2	16.4	0.629	20.8	0.486	17.3	0.573	5.7	1.5	38.1
112 Paraguay	0.679	0.529	22.1	–1	21.8	19.2	0.657	16.2	0.492	30.1	0.458	13.0	2.9	48.0
113 Palestine, State of	0.677	0.577	14.9	16	14.8	13.1	0.707	16.5	0.549	15.0	0.494	5.5	1.4	34.5
114 Uzbekistan	0.675	0.569	15.8	14	15.3	24.3	0.563	1.4	0.672	20.1	0.485	5.8	1.5	35.2
115 Philippines	0.668	0.547	18.1	7	17.8	15.2	0.629	11.6	0.539	26.8	0.483	8.4	2.2	43.0
116 El Salvador	0.666	0.488	26.7	–6	26.2	14.5	0.697	30.2	0.389	34.0	0.428	8.4	2.0	41.8
116 South Africa	0.666	0.428	35.7	–15	33.0	25.7	0.427	16.1	0.594	57.3	0.310	28.5	8.0	65.0
116 Viet Nam	0.666	0.549	17.5	9	17.4	12.1	0.755	18.0	0.474	22.0	0.463	6.1	1.5	35.6
119 Bolivia (Plurinational State of)	0.662	0.472	28.7	–5	28.4	24.5	0.561	24.7	0.480	36.1	0.391	15.2	2.7	46.6

TABLE 3 Inequality-adjusted Human Development Index | 217

TABLE 3 INEQUALITY-ADJUSTED HUMAN DEVELOPMENT INDEX

TABLE 3

HDI rank	Human Development Index (HDI) Value	Inequality-adjusted HDI (IHDI) Value	Overall loss (%)	Difference from HDI rank[b]	Coefficient of human inequality	Inequality in life expectancy (%)	Inequality-adjusted life expectancy index Value	Inequality in education[a] (%)	Inequality-adjusted education index Value	Inequality in income[a] (%)	Inequality-adjusted income index Value	Income inequality Quintile ratio	Palma ratio	Gini coefficient
	2014	2014	2014	2014	2014	2010–2015[c]	2014	2014[d]	2014	2014[d]	2014	2005–2013[e]	2005–2013[e]	2005–2013[e]
120 Kyrgyzstan	0.655	0.560	14.5	17	14.2	20.0	0.623	5.0	0.665	17.7	0.424	5.4	1.3	33.4
121 Iraq	0.654	0.512	21.8	2	21.5	17.6	0.626	30.6	0.342	16.1	0.626	4.4	1.1	29.5
122 Cabo Verde	0.646	0.519	19.7	5	19.4	12.0	0.722	18.2	0.433	28.0	0.447	8.7	2.3	43.8
123 Micronesia (Federated States of)	0.640	..	..	..	..	19.8	0.606	..	..	..	..	40.2	7.0	61.1
124 Guyana	0.636	0.520	18.3	8	18.1	19.2	0.577	10.5	0.511	24.4	0.477	..	..	..
125 Nicaragua	0.631	0.480	24.0	1	23.6	13.2	0.733	29.5	0.366	28.3	0.411	11.0	2.5	45.7
126 Morocco	0.628	0.441	29.7	–2	28.5	16.8	0.692	45.8	0.253	23.0	0.492	7.3	2.0	40.9
126 Namibia	0.628	0.354	43.6	–25	39.3	21.7	0.539	27.8	0.377	68.3	0.217	19.6	5.8	61.3
128 Guatemala	0.627	0.443	29.4	1	28.9	17.4	0.658	36.2	0.308	33.1	0.429	14.8	3.6	52.4
129 Tajikistan	0.624	0.515	17.5	10	17.0	29.3	0.537	6.5	0.615	15.0	0.414	4.7	1.1	30.8
130 India	0.609	0.435	28.6	1	27.7	25.0	0.554	42.1	0.292	16.1	0.508	5.0	1.4	33.6
131 Honduras	0.606	0.412	32.1	–7	30.7	17.0	0.678	26.4	0.361	48.6	0.285	23.5	5.0	57.4
132 Bhutan	0.605	0.425	29.8	–2	28.9	22.2	0.593	44.8	0.249	19.6	0.519	6.8	1.8	38.7
133 Timor-Leste	0.595	0.412	30.7	–4	29.4	22.8	0.572	47.6	0.247	17.8	0.494	4.4	1.2	30.4
134 Syrian Arab Republic	0.594	0.468	21.2	8	20.8	12.6	0.667	31.5	0.376	18.3	0.408	5.7	1.5	35.8
134 Vanuatu	0.594	0.492	17.2	12	17.2	15.4	0.675	17.5	0.430	18.5	0.410	..	..	..
136 Congo	0.591	0.434	26.6	6	26.2	36.0	0.417	21.5	0.402	21.2	0.488	8.2	1.9	40.2
137 Kiribati	0.590	0.405	31.5	–2	30.1	20.6	0.562	21.4	0.474	48.4	0.249	..	..	..
138 Equatorial Guinea	0.587	..	..	..	..	44.4	0.322	..	..	..	..	..	..	..
139 Zambia	0.586	0.384	34.4	–6	33.9	37.2	0.387	21.7	0.466	42.6	0.314	17.4	4.8	57.5
140 Ghana	0.579	0.387	33.1	–3	33.1	30.8	0.441	36.7	0.350	31.7	0.377	9.3	2.2	42.8
141 Lao People's Democratic Republic	0.575	0.428	25.6	7	25.3	21.5	0.558	34.1	0.304	20.3	0.463	5.8	1.6	36.2
142 Bangladesh	0.570	0.403	29.4	1	29.0	20.1	0.634	38.6	0.274	28.3	0.375	4.7	1.3	32.1
143 Cambodia	0.555	0.418	24.7	7	24.6	25.3	0.557	28.3	0.321	20.3	0.407	4.6	1.3	31.8
143 Sao Tome and Principe	0.555	0.418	24.7	6	24.7	26.9	0.523	21.4	0.368	25.8	0.378	5.6	1.3	33.9
LOW HUMAN DEVELOPMENT														
145 Kenya	0.548	0.377	31.3	–3	31.1	31.5	0.439	26.0	0.380	36.0	0.321	11.0	2.8	47.7
145 Nepal	0.548	0.401	26.8	3	25.9	21.1	0.602	41.4	0.266	15.1	0.403	5.0	1.3	32.8
147 Pakistan	0.538	0.377	29.9	0	28.6	29.9	0.498	44.4	0.208	11.6	0.519	4.1	1.1	29.6
148 Myanmar	0.536	..	..	..	..	27.1	0.515	19.4	0.303	..	..	..	..	..
149 Angola	0.532	0.335	37.0	–8	36.6	46.2	0.267	34.6	0.310	28.9	0.453	9.0	2.2	42.7
150 Swaziland	0.531	0.354	33.3	–2	33.1	35.0	0.290	26.8	0.404	37.6	0.379	14.0	3.5	51.5
151 Tanzania (United Republic of)	0.521	0.379	27.3	4	27.2	30.4	0.482	28.5	0.304	22.7	0.372	6.2	1.7	37.8
152 Nigeria	0.514	0.320	37.8	–9	37.5	40.8	0.299	43.3	0.254	28.4	0.430	9.1	2.2	43.0
153 Cameroon	0.512	0.344	32.8	–1	32.4	39.4	0.331	34.8	0.318	23.1	0.387	7.5	1.9	40.7
154 Madagascar	0.510	0.372	27.0	4	26.8	24.8	0.522	35.0	0.318	20.4	0.311	7.4	1.9	40.6
155 Zimbabwe	0.509	0.371	27.0	4	26.7	26.8	0.422	17.4	0.449	35.8	0.270	..	..	..
156 Mauritania	0.506	0.337	33.4	1	32.9	36.6	0.420	40.8	0.214	21.2	0.425	7.8	1.9	40.5
156 Solomon Islands	0.506	0.385	23.8	11	23.8	22.3	0.573	22.8	0.328	26.3	0.304	..	..	..
158 Papua New Guinea	0.505	..	..	..	..	26.5	0.481	11.5	0.360	..	..	..	..	..
159 Comoros	0.503	0.268	46.7	–18	46.0	34.2	0.438	47.6	0.248	56.0	0.178	26.7	7.0	64.3
160 Yemen	0.498	0.329	34.0	0	33.0	30.3	0.469	48.1	0.177	20.6	0.427	5.6	1.5	35.9
161 Lesotho	0.497	0.320	35.6	–2	34.9	33.5	0.305	24.3	0.383	47.0	0.280	20.4	4.3	54.2
162 Togo	0.484	0.322	33.4	1	33.1	36.8	0.386	38.9	0.299	23.5	0.290	10.7	2.6	46.0
163 Haiti	0.483	0.296	38.8	–7	38.2	27.9	0.475	38.3	0.249	48.4	0.219	26.6	5.5	59.2
163 Rwanda	0.483	0.330	31.6	4	31.6	30.2	0.475	29.4	0.289	35.2	0.262	11.0	3.2	50.8
163 Uganda	0.483	0.337	30.2	6	30.2	33.8	0.392	29.4	0.319	27.3	0.305	8.8	2.4	44.6
166 Benin	0.480	0.300	37.4	–2	37.1	37.0	0.384	44.8	0.230	29.4	0.306	8.2	2.2	43.5
167 Sudan	0.479	..	..	..	..	32.8	0.450	42.7	0.171	..	..	6.2	1.4	35.3
168 Djibouti	0.470	0.308	34.6	1	33.7	32.5	0.436	47.0	0.162	21.7	0.413	7.7	1.9	40.0
169 South Sudan	0.467	..	..	..	..	40.8	0.325	39.6	0.235	..	..	..	..	..
170 Senegal	0.466	0.305	34.4	1	34.0	29.5	0.504	44.7	0.168	27.7	0.337	7.8	1.9	40.3
171 Afghanistan	0.465	0.319	31.4	5	30.0	34.3	0.408	44.8	0.202	10.8	0.396	4.0	1.0	27.8
172 Côte d'Ivoire	0.462	0.287	38.0	–1	37.6	40.2	0.290	45.1	0.214	27.4	0.379	9.4	2.2	43.2
173 Malawi	0.445	0.299	32.9	2	32.6	40.0	0.395	30.2	0.308	27.7	0.220	9.7	2.6	46.2
174 Ethiopia	0.442	0.312	29.4	7	28.0	30.2	0.474	44.3	0.176	9.5	0.363	5.3	1.4	33.6
175 Gambia	0.441	..	..	..	..	34.8	0.403	..	..	26.9	0.299	11.0	2.8	47.3
176 Congo (Democratic Republic of the)	0.433	0.276	36.2	0	35.3	49.9	0.298	27.7	0.341	28.2	0.208	9.3	2.4	44.4
177 Liberia	0.430	0.280	34.8	2	34.1	33.1	0.421	46.4	0.215	22.7	0.243	7.0	1.7	38.2
178 Guinea-Bissau	0.420	0.254	39.6	–5	39.4	45.3	0.296	40.3	0.206	32.5	0.267	5.9	1.5	35.5
179 Mali	0.419	0.270	35.7	1	34.5	45.6	0.318	41.6	0.176	16.1	0.350	5.2	1.3	33.0

TABLE
3

HDI rank	Human Development Index (HDI) Value	Inequality-adjusted HDI (IHDI) Value	Overall loss (%)	Difference from HDI rank[b]	Coefficient of human inequality	Inequality in life expectancy (%)	Inequality-adjusted life expectancy index Value	Inequality in education[a] (%)	Inequality-adjusted education index Value	Inequality in income[a] (%)	Inequality-adjusted income index Value	Income inequality Quintile ratio	Palma ratio	Gini coefficient
	2014	2014	2014	2014	2014	2010–2015[c]	2014	2014[d]	2014	2014[d]	2014	2005–2013[e]	2005–2013[e]	2005–2013[e]
180 Mozambique	0.416	0.273	34.3	3	34.1	40.2	0.323	33.8	0.242	28.4	0.262	9.8	2.5	45.7
181 Sierra Leone	0.413	0.241	41.7	−4	40.0	51.2	0.232	49.6	0.171	19.2	0.351	5.6	1.5	35.4
182 Guinea	0.411	0.261	36.5	0	35.2	40.3	0.356	48.3	0.167	17.1	0.300	5.5	1.3	33.7
183 Burkina Faso	0.402	0.261	35.0	2	34.6	41.1	0.351	38.6	0.161	24.2	0.317	7.0	1.9	39.8
184 Burundi	0.400	0.269	32.6	5	31.5	43.6	0.318	36.9	0.234	14.1	0.263	4.8	1.3	33.3
185 Chad	0.392	0.236	39.9	−1	39.6	46.1	0.262	41.9	0.157	30.7	0.318	10.0	2.2	43.3
186 Eritrea	0.391	..	..	..	..	24.7	0.506	..	..	..	..	..	..	..
187 Central African Republic	0.350	0.198	43.5	−1	43.1	45.7	0.256	34.5	0.224	49.2	0.135	18.0	4.5	56.3
188 Niger	0.348	0.246	29.2	3	28.4	37.9	0.396	35.0	0.129	12.3	0.292	4.5	1.2	31.2
OTHER COUNTRIES OR TERRITORIES														
Korea (Democratic People's Rep. of)	..	..	..	..	..	15.4	0.655	..	..	..	..	..	..	..
Marshall Islands	..	..	..	..	..	..	..	..	..	..	..	..	..	..
Monaco	..	..	..	..	..	..	..	..	..	..	..	..	..	..
Nauru	..	..	..	..	..	..	..	..	..	..	..	..	..	..
San Marino	..	..	..	..	..	..	..	..	..	..	..	..	..	..
Somalia	..	..	..	..	..	42.1	0.316	43.5	..	..	..	..	..	..
Tuvalu	..	..	..	..	..	..	..	10.5	..	..	..	..	..	..
Human development groups														
Very high human development	0.896	0.788	12.1	—	11.8	4.9	0.885	8.0	0.782	22.5	0.706	—	—	—
High human development	0.744	0.600	19.4	—	19.0	10.7	0.757	16.8	0.542	29.4	0.527	—	—	—
Medium human development	0.630	0.468	25.8	—	25.5	21.9	0.584	34.7	0.348	19.8	0.503	—	—	—
Low human development	0.505	0.343	32.0	—	31.7	35.0	0.405	37.9	0.247	22.0	0.404	—	—	—
Developing countries	0.660	0.490	25.7	—	25.5	19.9	0.614	32.3	0.374	24.5	0.514	—	—	—
Regions														
Arab States	0.686	0.512	25.4	—	24.7	17.4	0.643	38.9	0.334	17.7	0.626	—	—	—
East Asia and the Pacific	0.710	0.572	19.4	—	19.2	11.7	0.734	18.4	0.491	27.4	0.520	—	—	—
Europe and Central Asia	0.748	0.651	13.0	—	12.9	14.3	0.690	7.9	0.655	16.6	0.611	—	—	—
Latin America and the Caribbean	0.748	0.570	23.7	—	23.2	13.3	0.734	21.0	0.522	35.2	0.485	—	—	—
South Asia	0.607	0.433	28.7	—	27.9	24.4	0.563	41.5	0.288	17.9	0.499	—	—	—
Sub-Saharan Africa	0.518	0.345	33.3	—	33.1	36.6	0.375	35.3	0.285	27.5	0.385	—	—	—
Least developed countries	0.502	0.347	30.9	—	30.7	32.3	0.451	36.4	0.253	23.4	0.367	—	—	—
Small island developing states	0.660	0.493	25.3	—	24.9	18.6	0.628	21.3	0.457	34.9	0.418	—	—	—
Organisation for Economic Co-operation and Development	0.880	0.763	13.3	—	12.9	5.6	0.873	9.5	0.744	23.6	0.685	—	—	—
World	**0.711**	**0.548**	**22.8**	**—**	**22.7**	**17.4**	**0.654**	**26.8**	**0.442**	**24.0**	**0.570**	**—**	**—**	**—**

NOTES

a See http://hdr.undp.org for the list of surveys used to estimate inequalities.

b Based on countries for which the Inequality-adjusted Human Development Index is calculated.

c Calculated from the 2010–2015 period life tables from UNDESA (2013a).

d Data refer to 2014 or the most recent year available.

e Data refer to the most recent year available during the period specified.

DEFINITIONS

Human Development Index (HDI): A composite index measuring average achievement in three basic dimensions of human development—a long and healthy life, knowledge and a decent standard of living. See *Technical note 1* at http://hdr.undp.org for details on how the HDI is calculated.

Inequality-adjusted HDI (IHDI): HDI value adjusted for inequalities in the three basic dimensions of human development. See *Technical note 2* at http://hdr.undp.org for details on how the IHDI is calculated.

Overall loss: Percentage difference between the IHDI and the HDI.

Difference from HDI rank: Difference in ranks on the IHDI and the HDI, calculated only for countries for which the IHDI is calculated.

Coefficient of human inequality: Average inequality in three basic dimensions of human development. See *Technical note 2* at http://hdr.undp.org.

Inequality in life expectancy: Inequality in distribution of expected length of life based on data from life tables estimated using the Atkinson inequality index.

Inequality-adjusted life expectancy index: The HDI life expectancy index adjusted for inequality in distribution of expected length of life based on data from life tables listed in *Main data sources*.

Inequality in education: Inequality in distribution of years of schooling based on data from household surveys estimated using the Atkinson inequality index.

Inequality-adjusted education index: The HDI education index adjusted for inequality in distribution of years of schooling based on data from household surveys listed in *Main data sources*.

Inequality in income: Inequality in income distribution based on data from household surveys estimated using the Atkinson inequality index.

Inequality-adjusted income index: The HDI income index adjusted for inequality in income distribution based on data from household surveys listed in *Main data sources*.

Quintile ratio: Ratio of the average income of the richest 20% of the population to the average income of the poorest 20% of the population.

Palma ratio: Ratio of the richest 10% of the population's share of gross national income (GNI) divided by the poorest 40%'s share. It is based on the work of Palma (2011), who found that middle class incomes almost always account for about half of GNI and that the other half is split between the richest 10% and poorest 40%, though their shares vary considerably across countries.

Gini coefficient: Measure of the deviation of the distribution of income among individuals or households within a country from a perfectly equal distribution. A value of 0 represents absolute equality, a value of 100 absolute inequality.

MAIN DATA SOURCES

Column 1: HDRO calculations based on data from UNDESA (2015), UNESCO Institute for Statistics (2015), United Nations Statistics Division (2015), World Bank (2015a), Barro and Lee (2014) and IMF (2015).

Column 2: Calculated as the geometric mean of the values in columns 7, 9 and 11 using the methodology described in *Technical note 2* (available at http://hdr.undp.org).

Column 3: Calculated based on data in columns 1 and 2.

Column 4: Calculated based on data in column 2 and recalculated HDI ranks for countries for which the IHDI is calculated.

Column 5: Calculated as the arithmetic mean of the values in columns 6, 8 and 10 using the methodology described in *Technical note 2* (available at http://hdr.undp.org).

Column 6: Calculated based on abridged life tables from UNDESA (2013a).

Column 7: Calculated based on data in column 6 and the life expectancy index.

Columns 8 and 10: Calculated based on data from the Luxembourg Income Study database, Eurostat's European Union Statistics on Income and Living Conditions, the World Bank's International Income Distribution Database, United Nations Children's Fund Multiple Indicator Cluster Surveys and ICF Macro Demographic and Health Surveys using the methodology in *Technical note 2* (available at http://hdr.undp.org).

Column 9: Calculated based on data in column 8 and the unadjusted education index.

Column 11: Calculated based on data in column 10 and the unadjusted income index.

Columns 12 and 13: HDRO calculations based on data from World Bank (2015b).

Column 14: World Bank (2015b).

TABLE 3   Inequality-adjusted Human Development Index | 219

TABLE 4

Gender Development Index

TABLE 4

HDI rank	Gender Development Index Value 2014	Group[b] 2014	Human Development Index (HDI) Value — Female 2014	Male 2014	Life expectancy at birth (years) — Female 2014	Male 2014	Expected years of schooling (years) — Female 2014[c]	Male 2014[c]	Mean years of schooling (years) — Female 2014[c]	Male 2014[c]	Estimated gross national income per capita[a] (2011 PPP $) — Female 2014	Male 2014
VERY HIGH HUMAN DEVELOPMENT												
1 Norway	0.996	1	0.940	0.944	83.6	79.5	18.2	16.8	12.7	12.5	57,140	72,825
2 Australia	0.976	1	0.922	0.945	84.5	80.3	20.7	19.7	13.1	12.9	33,688	50,914
3 Switzerland	0.950	2	0.898	0.945	85.0	80.8	15.7	15.9	11.5	13.1	44,132	69,077
4 Denmark	0.977	1	0.912	0.934	82.2	78.3	19.3	18.1	12.8	12.7	36,439	51,727
5 Netherlands	0.947	3	0.893	0.943	83.3	79.7	18.0	17.9	11.6	12.2	29,500	61,641
6 Germany	0.963	2	0.901	0.936	83.3	78.5	16.3	16.6	12.9	13.8	34,886	53,290
6 Ireland	0.973	2	0.901	0.926	83.0	78.8	18.5	18.7	12.3	12.0	30,104	49,166
8 United States	0.995	1	0.911	0.916	81.4	76.7	17.2	15.7	13.0	12.9	43,054	63,158
9 Canada	0.982	1	0.904	0.921	84.0	80.0	16.3	15.5	13.1	13.0	33,587	50,853
9 New Zealand	0.961	2	0.894	0.930	83.6	80.0	20.0	18.3	12.5	12.6	24,309	41,372
11 Singapore	0.985	1	0.898	0.912	86.0	79.9	15.5[d]	15.3[d]	10.1[e]	10.9[e]	59,994	93,699[f]
12 Hong Kong, China (SAR)	0.958	2	0.892	0.931	86.8	81.2	15.7	15.5	10.9	11.9	38,060	72,052
13 Liechtenstein	..	..	..	..	..	..	13.8	16.1	..	..	..	..
14 Sweden	0.999	1	0.906	0.906	83.9	80.4	16.6	15.1	12.2	12.0	40,222	51,084
14 United Kingdom	0.965	2	0.888	0.920	82.6	78.7	16.6	15.8	12.9	13.2	27,259	51,628
16 Iceland	0.975	1	0.886	0.909	84.0	81.0	20.1	17.9	10.8	10.4	28,792	41,486
17 Korea (Republic of)	0.930	3	0.861	0.926	85.0	78.5	16.0	17.7	11.2	12.7	21,896	46,018
18 Israel	0.971	2	0.879	0.905	84.1	80.5	16.5	15.5	12.5	12.6	22,451	39,064
19 Luxembourg	0.971	2	0.877	0.903	83.9	79.3	14.0	13.7	11.3	12.1	47,723	69,800
20 Japan	0.961	2	0.870	0.905	86.7	80.2	15.2	15.5	11.3	11.7	24,975	49,541
21 Belgium	0.975	1	0.872	0.895	83.2	78.3	16.7	16.0	10.6[e]	11.1[e]	31,879	50,845
22 France	0.987	1	0.881	0.893	85.1	79.2	16.4	15.6	11.0	11.3	31,073	45,497
23 Austria	0.943	3	0.856	0.909	83.8	78.8	16.1	15.4	9.9	11.9	29,598	58,826
24 Finland	0.996	1	0.879	0.882	83.6	78.0	17.7	16.5	10.2	10.2	31,644	45,994
25 Slovenia	0.996	1	0.877	0.881	83.4	77.3	17.7	16.0	11.8	12.0	22,180	33,593
26 Spain	0.975	1	0.863	0.885	85.3	79.8	17.7	17.0	9.4	9.8	24,059	40,221
27 Italy	0.964	2	0.851	0.883	85.5	80.6	16.5	15.6	9.5	10.2	22,526	44,148
28 Czech Republic	0.980	1	0.859	0.877	81.5	75.7	16.9	15.8	12.1	12.5	19,929	33,604
29 Greece	0.961	2	0.844	0.879	83.8	78.0	17.7	17.6	9.8	10.5	17,288	31,952
30 Estonia	1.030	2	0.872	0.847	81.3	72.0	17.4	15.6	12.7	12.2	20,854	30,254
31 Brunei Darussalam	0.977	1	0.840	0.860	80.7	77.0	14.9	14.2	8.5	9.0	54,228	90,437
32 Cyprus	0.971	2	0.836	0.861	82.4	78.0	14.2	13.8	11.4	11.8	22,613	34,400
32 Qatar	0.998	1	0.853	0.854	79.9	77.3	14.0	13.9	10.4	8.8	55,123	143,979[f]
34 Andorra	..	..	..	..	..	..	..	..	..	..	..	..
35 Slovakia	0.999	1	0.841	0.842	79.9	72.5	15.7	14.5	12.3	12.2	19,903	32,122
36 Poland	1.007	1	0.844	0.839	81.4	73.4	16.3	14.7	11.7	11.9	18,423	28,271
37 Lithuania	1.030	2	0.851	0.827	78.9	67.7	16.9	15.9	12.3	12.5	20,955	28,656
37 Malta	0.937	3	0.806	0.861	82.2	78.8	14.8	14.0	9.9	10.8	16,435	39,432
39 Saudi Arabia	0.901	4	0.778	0.864	75.7	73.0	16.6	16.1	7.8	9.3	20,094	77,044[f]
40 Argentina	0.982	1	0.819	0.834	80.1	72.4	19.1	16.8	9.8	9.8	14,202	30,237
41 United Arab Emirates	0.954	2	0.796	0.835	78.5	76.3	13.9[g]	12.9[g]	9.9	8.5	22,391	77,300[f]
42 Chile	0.967	2	0.815	0.843	84.5	78.6	15.5	15.0	9.7	9.9	14,732	27,992
43 Portugal	0.985	1	0.823	0.836	83.8	77.9	16.5	16.1	8.1	8.4	21,259	30,543
44 Hungary	0.976	1	0.818	0.838	78.7	71.5	15.7	15.1	11.3[e]	12.2[e]	17,443	28,960
45 Bahrain	0.940	3	0.789	0.839	77.6	75.8	15.1[h]	13.7[h]	9.1	9.6	20,038	49,890
46 Latvia	1.029	2	0.829	0.805	78.9	69.1	15.9	14.6	11.7[e]	11.2[e]	18,437	26,845
47 Croatia	0.987	1	0.812	0.823	80.6	74.0	15.4	14.2	10.5	11.6	16,200	22,853
48 Kuwait	0.972	2	0.793	0.816	75.8	73.5	15.2	14.2	7.3	7.1	42,292	111,988[f]
49 Montenegro	0.954	2	0.782	0.819	78.4	74.1	15.5	14.8	10.5	11.8	11,106	18,094
HIGH HUMAN DEVELOPMENT												
50 Belarus	1.021	1	0.806	0.789	77.2	65.5	16.2	15.1	11.9[i]	12.1[i]	12,922	21,010
50 Russian Federation	1.019	1	0.804	0.789	75.8	64.4	15.1	14.3	11.9	12.0	17,269	28,287
52 Oman	0.909	4	0.741	0.815	79.2	75.1	13.9	13.5	7.0	8.5	14,709	46,400
52 Romania	0.989	1	0.787	0.796	78.3	71.2	14.6	13.8	10.3	11.1	15,250	21,117
52 Uruguay	1.018	1	0.797	0.783	80.6	73.5	16.6	14.4	8.7	8.2	14,721	24,166
55 Bahamas	..	..	..	..	78.3	72.3	..	..	11.1	10.7	17,868	24,957
56 Kazakhstan	1.002	1	0.787	0.785	74.1	64.6	15.4	14.7	11.3	11.5	15,408	26,746
57 Barbados	1.018	1	0.791	0.777	78.0	73.2	17.2	13.8	10.6	10.2	10,245	14,739
58 Antigua and Barbuda	..	..	..	..	78.5	73.5	14.6	13.3	..	..	..	..[f]
59 Bulgaria	0.991	1	0.777	0.784	77.7	70.8	14.6	14.1	10.6[e]	10.4[e]	12,448	18,926
60 Palau	..	..	..	..	..	..	13.9	13.5	..	..	..	..

		Gender Development Index		Human Development Index (HDI)		Life expectancy at birth		Expected years of schooling		Mean years of schooling		Estimated gross national income per capita[a]	
				Value		(years)		(years)		(years)		(2011 PPP $)	
		Value	Group[b]	Female	Male	Female	Male	Female	Male	Female	Male	Female	Male
HDI rank		2014	2014	2014	2014	2014	2014	2014[c]	2014[c]	2014[c]	2014[c]	2014	2014
60	Panama	0.996	1	0.776	0.779	80.7	74.6	13.8	12.8	9.6	9.1	13,699	22,597
62	Malaysia	0.947	3	0.753	0.795	77.1	72.4	12.7	12.7	9.4	10.1	15,635	30,320
63	Mauritius	0.950	2	0.752	0.792	78.0	70.9	15.9	15.2	8.0	9.1	10,541	24,581
64	Seychelles	..	..	..	..	78.1	68.9	13.3	13.4	..	..	..	..
64	Trinidad and Tobago	0.985	1	0.763	0.774	74.0	67.0	12.5	12.1	10.7[e]	10.9[e]	19,669	32,656
66	Serbia	0.966	2	0.757	0.784	77.7	72.1	14.9	13.9	9.8	11.2	9,697	14,799
67	Cuba	0.954	2	0.747	0.783	81.5	77.4	14.4	13.3	11.4[i]	11.6[i]	4,912	9,665
67	Lebanon	0.899	5	0.718	0.800	81.3	77.6	13.6	13.9	7.6[i]	8.2[i]	7,334	25,391
69	Costa Rica	0.974	2	0.753	0.774	81.9	77.0	14.3	13.4	8.4	8.3	9,680	17,033
69	Iran (Islamic Republic of)	0.858	5	0.689	0.804	76.5	74.3	15.0	15.2	7.7	8.6	4,828	25,924
71	Venezuela (Bolivarian Republic of)	1.030	2	0.772	0.749	78.5	70.2	15.3	13.1	9.2	8.6	12,458	19,840
72	Turkey	0.902	4	0.716	0.793	78.5	72.0	14.0	15.1	6.7	8.5	10,024	27,645
73	Sri Lanka	0.948	3	0.730	0.769	78.2	71.5	14.2	13.3	10.7	10.9	5,452	14,307
74	Mexico	0.943	3	0.731	0.775	79.2	74.4	13.2	12.9	8.2	8.8	10,233	22,252
75	Brazil	0.997	1	0.752	0.754	78.3	70.7	15.6	14.8	7.8	7.5	11,393	19,084
76	Georgia	0.962	2	0.736	0.765	78.4	71.2	14.0	13.6	12.0[i]	12.3[i]	4,887	9,718
77	Saint Kitts and Nevis	..	..	..	..	..	..	13.4	12.4	..	..	..	..
78	Azerbaijan	0.942	3	0.721	0.766	74.0	67.7	11.8	12.0	10.5[i]	11.2[i]	10,120	22,814
79	Grenada	..	..	..	..	75.9	71.0	16.3	15.3	..	..	..	..
80	Jordan	0.860	5	0.674	0.784	75.8	72.4	13.7	13.3	9.3	10.5	3,587	18,831
81	The former Yugoslav Republic of Macedonia	0.949	3	0.726	0.765	77.7	73.1	13.5	13.2	8.7[i]	9.8[i]	8,796	14,754
81	Ukraine	1.003	1	0.747	0.745	75.9	66.0	15.4	14.9	11.3	11.4	6,518	10,120
83	Algeria	0.837	5	0.637	0.761	77.2	72.5	14.2	13.8	4.8	7.8	3,898	22,009
84	Peru	0.947	3	0.712	0.752	77.2	71.9	13.1	13.0	8.5	9.6	8,040	13,977
85	Albania	0.948	3	0.711	0.750	80.4	75.4	11.9	11.8	8.9	9.6	7,217	12,655
85	Armenia	1.008	1	0.734	0.728	78.6	70.9	13.6	11.2	10.8	10.9	6,042	10,089
85	Bosnia and Herzegovina	..	..	..	..	79.0	74.0	..	..	6.6[i]	8.6[i]	6,514	12,912
88	Ecuador	0.980	1	0.722	0.737	78.7	73.2	14.5	13.9	7.4	7.6	8,487	12,723
89	Saint Lucia	0.991	1	0.725	0.731	77.8	72.4	13.0	12.1	9.4	9.3	8,018	11,576
90	China	0.943	3	0.705	0.747	77.3	74.3	13.2	12.9	6.9	8.2	10,128	14,795
90	Fiji	0.941	3	0.699	0.743	73.2	67.2	16.0	15.5	9.8	10.0	4,274	10,592
90	Mongolia	1.028	2	0.737	0.716	73.9	65.3	15.3	13.9	9.5	9.0	9,029	12,462
93	Thailand	1.000	1	0.726	0.726	77.9	71.1	13.9	13.1	7.1	7.5	11,820	14,888
94	Dominica	..	..	..	..	..	..	..	..	..	..	..	..
94	Libya	0.950	2	0.699	0.736	74.6	68.9	14.3	13.8	7.7	7.0	7,427	22,392
96	Tunisia	0.894	5	0.671	0.751	77.3	72.5	15.0	14.0	5.9	7.8	4,748	16,159
97	Colombia	0.997	1	0.719	0.721	77.7	70.5	13.9	13.2	7.4	7.3	9,785	14,372
97	Saint Vincent and the Grenadines	..	..	..	..	75.1	70.9	13.4	13.1	..	..	7,283	12,541
99	Jamaica	0.995	1	0.715	0.719	78.1	73.3	13.1	11.6	10.0	9.4	5,820	9,059
100	Tonga	0.967	2	0.704	0.727	75.8	69.9	15.0	14.4	10.7	10.8	3,796	6,336
101	Belize	0.958	2	0.696	0.727	72.9	67.4	13.9	13.2	10.5	10.5	5,034	10,198
101	Dominican Republic	0.995	1	0.710	0.713	76.7	70.4	13.6	12.6	7.7	7.2	8,860	14,903
103	Suriname	0.975	1	0.702	0.720	74.4	68.0	13.6	11.8	7.3	8.0	10,241	20,970
104	Maldives	0.937	3	0.678	0.723	77.8	75.8	12.8	12.5	5.7[e]	6.0[e]	8,531	16,073
105	Samoa	0.956	2	0.681	0.713	76.8	70.4	13.2[k]	12.5[k]	10.3[k]	10.3[k]	3,416	7,124
MEDIUM HUMAN DEVELOPMENT													
106	Botswana	0.982	1	0.691	0.704	66.8	62.1	12.6	12.4	8.7	9.1	15,179	18,096
107	Moldova (Republic of)	1.003	1	0.694	0.692	75.7	67.4	12.2	11.6	11.1	11.3	4,599	5,915
108	Egypt	0.868	5	0.633	0.729	73.4	69.0	13.3	13.8	5.4	7.7	4,928	16,049
109	Turkmenistan	..	..	..	..	69.9	61.5	10.6	11.0	..	..	8,725	17,552
110	Gabon	..	..	..	..	65.0	63.8	..	..	8.9	6.6	13,527	19,177
110	Indonesia	0.927	3	0.655	0.706	71.0	66.9	13.1	12.9	7.0	8.2	6,485	13,052
112	Paraguay	0.956	2	0.662	0.692	75.1	70.8	12.2	11.7	7.5	7.9	5,576	9,678
113	Palestine, State of	0.860	5	0.607	0.706	74.9	71.0	13.8	12.2	8.4	9.3	1,580	7,726
114	Uzbekistan	0.945	3	0.640	0.678	71.8	65.0	11.3	11.7	9.5[i]	9.9[i]	3,811	7,342
115	Philippines	0.977	1	0.649	0.664	71.8	64.9	11.5	11.1	8.4[e]	7.9[e]	5,382	10,439
116	El Salvador	0.965	2	0.652	0.676	77.4	68.3	12.1	12.4	6.2	6.9	5,497	9,406
116	South Africa	0.948	3	0.646	0.681	59.3	55.2	13.7	13.4	9.7	10.2	8,713	15,737
116	Viet Nam	..	..	..	..	80.5	71.0	..	..	7.0	7.9	4,624	5,570
119	Bolivia (Plurinational State of)	0.931	3	0.637	0.684	70.9	65.9	12.9	13.4	7.5	8.9	4,383	7,140
120	Kyrgyzstan	0.961	2	0.638	0.664	74.6	66.6	12.7	12.3	10.5	10.6	2,122	3,992
121	Iraq	0.787	5	0.561	0.712	71.7	67.2	8.7	11.4	5.1	7.7	4,279	23,515

TABLE 4   Gender Development Index | 221

TABLE 4 GENDER DEVELOPMENT INDEX

HDI rank	Gender Development Index Value 2014	Group[b] 2014	Human Development Index (HDI) Value Female 2014	Male 2014	Life expectancy at birth (years) Female 2014	Male 2014	Expected years of schooling (years) Female 2014[c]	Male 2014[c]	Mean years of schooling (years) Female 2014[c]	Male 2014[c]	Estimated gross national income per capita[a] (2011 PPP $) Female 2014	Male 2014
122 Cabo Verde	..	..	..	..	75.0	71.4	13.9	13.1	..	..	4,098	8,105
123 Micronesia (Federated States of)	..	..	..	..	70.1	68.1	..	..	..	..	..	..
124 Guyana	0.984	1	0.626	0.636	68.8	64.2	11.2	9.4	8.9	8.0	4,164	8,804
125 Nicaragua	0.960	2	0.615	0.640	77.9	71.9	11.8	11.3	6.2	5.8	2,967	5,979
126 Morocco	0.828	5	0.555	0.670	75.1	73.0	10.6	11.6	3.2	5.3	3,222	10,573
126 Namibia	0.981	1	0.620	0.632	67.3	62.1	11.4	11.3	6.3	6.1	7,672	11,267
128 Guatemala	0.949	3	0.608	0.641	75.3	68.3	10.2	11.0	5.5	5.7	5,021	8,934
129 Tajikistan	0.926	3	0.600	0.648	73.2	66.2	10.5	12.0	9.6[i]	11.2[i]	2,014	3,017
130 India	0.795	5	0.525	0.660	69.5	66.6	11.3	11.8	3.6	7.2	2,116	8,656
131 Honduras	0.944	3	0.583	0.618	75.7	70.7	11.6	10.6	5.5	5.4	2,365	5,508
132 Bhutan	0.897	5	0.572	0.638	69.7	69.2	12.8	12.6	2.0	4.1	5,733	8,418
133 Timor-Leste	0.868	5	0.548	0.631	70.1	66.5	11.3	12.0	3.6[i]	5.3[i]	3,122	7,530
134 Syrian Arab Republic	0.834	5	0.520	0.623	76.5	64.0	12.2	12.3	5.4	7.4	864	4,523
134 Vanuatu	0.903	4	0.587	0.650	74.0	69.9	10.2	10.9	8.0	10.0	2,141	3,445
136 Congo	0.922	4	0.561	0.609	63.9	60.8	10.9	11.3	4.9	6.5	5,165	6,859
137 Kiribati	..	..	..	..	69.2	62.8	12.7	11.9	..	..	..	..
138 Equatorial Guinea	..	..	..	..	59.0	56.3	..	..	4.0	7.2	17,073	24,850
139 Zambia	0.917	4	0.558	0.609	62.0	58.2	13.0	13.9	5.8	7.3	3,019	4,452
140 Ghana	0.885	5	0.540	0.610	62.3	60.4	11.0	12.0	5.6	7.9	3,200	4,515
141 Lao People's Democratic Republic	0.896	5	0.543	0.606	67.5	64.8	10.1	11.0	3.9	6.1	4,086	5,279
142 Bangladesh	0.917	4	0.541	0.590	72.9	70.4	10.3	9.7	4.5	5.5	2,278	4,083
143 Cambodia	0.890	5	0.519	0.584	70.3	66.2	10.3	11.5	3.2[i]	5.4[i]	2,526	3,393
143 Sao Tome and Principe	0.891	5	0.520	0.584	68.4	64.4	11.4	11.2	4.0[i]	5.5[i]	1,886	3,976
LOW HUMAN DEVELOPMENT												
145 Kenya	0.913	4	0.527	0.577	63.4	59.9	10.7	11.3	5.9	7.3	2,255	3,270
145 Nepal	0.908	4	0.521	0.574	71.1	68.2	12.5	12.2	2.3	4.5	1,956	2,690
147 Pakistan	0.726	5	0.436	0.601	67.2	65.3	7.0	8.5	3.1	6.2	1,450	8,100
148 Myanmar	..	..	..	..	68.0	63.9	..	..	4.3	3.8	3,873	5,386
149 Angola	..	..	..	..	53.8	50.8	8.7	14.0	..	..	5,497	8,169
150 Swaziland	0.879	5	0.494	0.561	48.2	49.6	10.9	11.8	7.4	6.8	3,894	7,235
151 Tanzania (United Republic of)	0.938	3	0.504	0.538	66.4	63.5	9.0	9.3	4.5	5.8	2,320	2,502
152 Nigeria	0.841	5	0.468	0.556	53.1	52.4	8.2	9.8	4.9[i]	7.1[i]	4,052	6,585
153 Cameroon	0.879	5	0.478	0.544	56.7	54.4	9.5	11.2	5.3	6.7	2,266	3,341
154 Madagascar	0.945	3	0.497	0.526	66.6	63.6	10.2	10.5	6.6[i]	6.1[i]	1,098	1,560
155 Zimbabwe	0.922	4	0.487	0.529	58.9	56.2	10.7	11.0	6.7	7.7	1,387	1,850
156 Mauritania	0.816	5	0.446	0.546	64.5	61.6	8.5	8.5	2.7	4.8	1,625	5,468
156 Solomon Islands	..	..	..	..	69.4	66.5	8.8	9.7	..	..	1,046	2,019
158 Papua New Guinea	..	..	..	..	64.8	60.5	..	..	3.2	4.8	2,145	2,768
159 Comoros	0.813	5	0.443	0.545	65.0	61.6	11.0	11.9	3.7	5.6	778	2,123
160 Yemen	0.739	5	0.414	0.560	65.2	62.5	7.7	10.6	1.3	3.8	1,595	5,412
161 Lesotho	0.953	2	0.482	0.505	49.8	49.6	11.7	10.6	6.5[e]	5.2[e]	2,613	4,017
162 Togo	0.831	5	0.439	0.527	60.4	58.9	10.9	13.4	3.0	6.3	1,084	1,376
163 Haiti	..	..	..	..	64.9	60.7	..	..	4.2	5.6	1,379	1,966
163 Rwanda	0.957	2	0.472	0.493	67.0	61.1	10.4	10.2	3.2	4.3	1,312	1,612
163 Uganda	0.886	5	0.452	0.510	60.3	56.7	9.7	9.9	4.5	6.3	1,226	1,997
166 Benin	0.823	5	0.431	0.524	61.0	58.1	9.4	12.7	2.1	4.6	1,493	2,043
167 Sudan	0.830	5	0.428	0.516	65.0	62.0	6.7	7.3	2.5	3.8	1,882	5,722
168 Djibouti	..	..	..	..	63.7	60.4	5.9	6.9	..	..	2,019	4,522
169 South Sudan	..	..	..	..	56.7	54.7	..	..	4.5	5.8	..	..
170 Senegal	0.883	5	0.436	0.494	68.3	64.5	7.8	8.1	1.8	3.2	1,657	2,739
171 Afghanistan	0.600	5	0.328	0.546	61.6	59.2	7.2	11.3	1.2	5.1	506	3,227
172 Côte d'Ivoire	0.810	5	0.410	0.507	52.4	50.7	7.9	10.0	3.2	5.3	2,146	4,157
173 Malawi	0.907	4	0.423	0.467	63.7	61.7	10.8	10.7	3.4	5.2	679	815
174 Ethiopia	0.840	5	0.403	0.479	66.0	62.2	8.0	9.0	1.4	3.6	1,090	1,765
175 Gambia	0.889	5	0.414	0.466	61.6	58.9	8.7	8.9	2.0	3.7	1,267	1,753
176 Congo (Democratic Republic of the)	0.833	5	0.393	0.472	60.1	57.2	8.6	10.8	4.5	7.7	597	765
177 Liberia	0.789	5	0.387	0.491	61.8	59.9	8.9	12.4	2.6	5.8	678	930
178 Guinea-Bissau	..	..	..	..	57.0	53.4	..	..	..	..	1,135	1,593
179 Mali	0.776	5	0.363	0.468	57.8	58.2	7.5	9.3	1.5	2.7	961	2,195
180 Mozambique	0.881	5	0.390	0.443	56.5	53.6	8.8	9.8	2.4[i]	4.3[i]	1,040	1,210
181 Sierra Leone	0.814	5	0.370	0.454	51.4	50.4	7.2	10.0	2.2	4.0	1,582	1,981
182 Guinea	0.778	5	0.358	0.460	59.2	58.3	7.3	10.0	1.4[i]	3.7[i]	877	1,314

HDI rank	Gender Development Index		Human Development Index (HDI)		Life expectancy at birth		Expected years of schooling		Mean years of schooling		Estimated gross national income per capita[a]	
	Value	Group[b]	Value		(years)		(years)		(years)		(2011 PPP $)	
			Female	Male	Female	Male	Female	Male	Female	Male	Female	Male
	2014	2014	2014	2014	2014	2014	2014[c]	2014[c]	2014[c]	2014[c]	2014	2014
183 Burkina Faso	0.881	5	0.376	0.427	59.9	57.3	7.4	8.1	1.0[l]	1.9[l]	1,325	1,859
184 Burundi	0.911	4	0.381	0.418	58.7	54.8	9.6	10.7	2.2	3.1	693	825
185 Chad	0.768	5	0.338	0.440	52.7	50.5	5.9	8.9	1.0	2.9	1,657	2,513
186 Eritrea	..	..	..	..	65.9	61.6	..	..	..	..	971	1,290
187 Central African Republic	0.773	5	0.303	0.392	52.6	48.8	5.9	8.6	2.8	5.7	476	689
188 Niger	0.729	5	0.287	0.394	62.4	60.6	4.8	6.1	0.8	2.0	491	1,319
OTHER COUNTRIES OR TERRITORIES												
Korea (Democratic People's Rep. of)	..	..	..	..	73.7	66.7	..	..	..	..	..	..
Marshall Islands	..	..	..	..	..	..	..	..	..	..	..	..
Monaco	..	..	..	..	..	..	..	..	..	..	..	..
Nauru	..	..	..	..	..	..	9.9	8.9	..	..	..	..
San Marino	..	..	..	..	..	..	15.9	14.7	..	..	..	..
Somalia	..	..	..	..	57.0	53.8	..	..	..	..	..	..
Tuvalu	..	..	..	..	..	..	..	..	..	..	..	..
Human development groups												
Very high human development	0.978	—	0.884	0.903	83.1	77.8	16.8	16.0	11.7	12.0	30,991	52,315
High human development	0.954	—	0.724	0.758	77.4	72.8	13.8	13.4	7.7	8.5	10,407	17,443
Medium human development	0.861	—	0.574	0.667	70.6	66.8	11.5	11.8	4.9	7.3	3,333	9,257
Low human development	0.830	—	0.456	0.549	61.8	59.3	8.3	9.8	3.4	5.5	1,983	4,201
Developing countries	0.899	—	0.617	0.686	71.7	68.0	11.6	11.9	5.4	7.3	5,926	12,178
Regions												
Arab States	0.849	—	0.611	0.719	72.7	68.8	11.6	12.3	4.9	6.9	5,686	24,985
East Asia and the Pacific	0.948	—	0.692	0.730	76.0	72.2	13.0	12.8	6.9	8.0	9,017	13,780
Europe and Central Asia	0.945	—	0.719	0.760	76.1	68.5	13.5	13.8	9.0	10.0	8,238	17,607
Latin America and the Caribbean	0.976	—	0.736	0.754	78.2	71.7	14.4	13.7	8.0	8.1	10,194	18,435
South Asia	0.801	—	0.525	0.655	69.9	67.1	10.8	11.3	3.7	6.9	2,198	8,827
Sub-Saharan Africa	0.872	—	0.480	0.550	59.7	57.1	9.1	10.3	4.2	6.0	2,626	4,148
Least developed countries	0.866	—	0.465	0.537	64.8	61.9	8.9	10.0	3.2	4.9	1,783	3,005
Small island developing states	..	—	..	..	72.6	67.8	13.4	12.6	..	..	5,045	8,849
Organisation for Economic Co-operation and Development	0.973	—	0.862	0.887	82.7	77.5	16.0	15.5	11.0	11.5	28,430	47,269
World	**0.924**	**—**	**0.670**	**0.725**	**73.7**	**69.5**	**12.2**	**12.4**	**6.2**	**7.9**	**10,296**	**18,373**

TABLE
4

NOTES

a Because disaggregated income data are not available, data are crudely estimated. See *Definitions* and *Technical note 3* at http://hdr.undp.org for details on how the Gender Development Index is calculated.

b Countries are divided into five groups by absolute deviation from gender parity in HDI values.

c Data refer to 2014 or the most recent year available.

d Calculated by the Singapore Ministry of Education.

e HDRO updates based on data on educational attainment from UNESCO Institute for Statistics (2015) and methodology from Barro and Lee (2014).

f For the purpose of calculating the male HDI value, estimated gross national income per capita is capped at $75,000.

g Based on data from UNESCO Institute for Statistics (2011).

h Based on data on school life expectancy from the UNESCO Institute for Statistics (2013).

i HDRO calculations based on data from recent United Nations Children's Fund (UNICEF) Multiple Indicator Cluster Surveys.

j Based on the estimate of educational attainment distribution from UNESCO Institute for Statistics (2015).

k HDRO calculations based on data from the 2011 population census from Samoa Bureau of Statistics (2013).

l HDRO estimate based on data from the country's most recent ICF Macro Demographic and Health Survey.

DEFINITIONS

Gender Development Index: Ratio of female to male HDI values. See *Technical note 3* at http://hdr.undp.org for details on how the Gender Development Index is calculated.

Gender Development Index groups: Countries are divided into five groups by absolute deviation from gender parity in HDI values. Group 1 comprises countries with high equality in HDI achievements between women and men (absolute deviation of less than 2.5 percent); group 2 comprises countries with medium to high equality in HDI achievements between women and men (absolute deviation of 2.5–5 percent); group 3 comprises countries with medium equality in HDI achievements between women and men (absolute deviation of 5–7.5

percent); group 4 comprises countries with medium to low equality in HDI achievements between women and men (absolute deviation of 7.5–10 percent); and group 5 comprises countries with low equality in HDI achievements between women and men (absolute deviation from gender parity of more than 10 percent).

Human Development Index (HDI): A composite index measuring average achievement in three basic dimensions of human development—a long and healthy life, knowledge and a decent standard of living. See *Technical note 1* at http://hdr.undp.org for details on how the HDI is calculated.

Life expectancy at birth: Number of years a newborn infant could expect to live if prevailing patterns of age-specific mortality rates at the time of birth stay the same throughout the infant's life.

Expected years of schooling: Number of years of schooling that a child of school entrance age can expect to receive if prevailing patterns of age-specific enrolment rates persist throughout the child's life.

Mean years of schooling: Average number of years of education received by people ages 25 and older, converted from educational attainment levels using official durations of each level.

Estimated gross national income per capita: derived from the ratio of female to male wages, female and male shares of economically active population and GNI (in 2011 purchasing power parity terms). See *Technical note 3* at http://hdr.undp.org for details.

MAIN DATA SOURCES

Columns 1: Calculated based on data in columns 3 and 4.

Columns 2: Calculated based on data in column 1.

Columns 3 and 4: HDRO calculations based on data from UNDESA (2015), UNESCO Institute for Statistics (2015), Barro and Lee (2014), World Bank (2015a), ILO (2015a) and IMF (2015).

Columns 5 and 6: UNDESA (2015).

Columns 7 and 8: UNESCO Institute for Statistics (2015).

Columns 9 and 10: UNESCO Institute for Statistics (2015), Barro and Lee (2014), UNICEF Multiple Indicator Cluster Surveys and ICF Macro Demographic and Health Surveys.

Columns 11 and 12: HDRO calculations based on ILO (2015a), UNDESA (2013a), World Bank (2015a) and IMF (2015).

TABLE 4   Gender Development Index | 223

TABLE 5

Gender Inequality Index

TABLE 5

	Gender Inequality Index		Maternal mortality ratio	Adolescent birth rate	Share of seats in parliament	Population with at least some secondary education		Labour force participation rate[a]	
						(% ages 25 and older)		(% ages 15 and older)	
	Value	Rank	(deaths per 100,000 live births)	(births per 1,000 women ages 15–19)	(% held by women)	Female	Male	Female	Male
HDI rank	2014	2014	2013	2010/2015[b]	2014	2005–2014[c]	2005–2014[c]	2013	2013
VERY HIGH HUMAN DEVELOPMENT									
1 Norway	0.067	9	4	7.8	39.6	97.4	96.7	61.2	68.7
2 Australia	0.110	19	6	12.1	30.5	94.3[d]	94.6[d]	58.8	71.8
3 Switzerland	0.028	2	6	1.9	28.5	95.0	96.6	61.8	74.9
4 Denmark	0.048	4	5	5.1	38.0	95.5[e]	96.6[e]	58.7	66.4
5 Netherlands	0.062	7	6	6.2	36.9	87.7	90.5	58.5	70.6
6 Germany	0.041	3	7	3.8	36.9	96.3	97.0	53.6	66.4
6 Ireland	0.113	21	9	8.2	19.9	80.5	78.6	53.1	68.1
8 United States	0.280	55	28	31.0	19.4	95.1	94.8	56.3	68.9
9 Canada	0.129	25	11	14.5	28.2	100.0	100.0	61.6	71.0
9 New Zealand	0.157	32	8	25.3	31.4	95.0	95.3	62.0	73.8
11 Singapore	0.088	13	6	6.0	25.3	74.1	81.0	58.8	77.2
12 Hong Kong, China (SAR)	..	..	..	3.3	..	72.2	79.2	51.3	67.8
13 Liechtenstein	..	..	..	..	20.0	..	..	..	..
14 Sweden	0.055	6	4	6.5	43.6	86.5	87.3	60.3	67.9
14 United Kingdom	0.177	39	8	25.8	23.5	99.8	99.9	55.7	68.7
16 Iceland	0.087	12	4	11.5	41.3	91.0	91.6	70.5	77.4
17 Korea (Republic of)	0.125	23	27	2.2	16.3	77.0[f]	89.1[f]	50.1	72.1
18 Israel	0.101	18	2	7.8	22.5	84.4	87.3	57.9	69.1
19 Luxembourg	0.100	17	11	8.3	28.3	100.0[e]	100.0[e]	50.7	64.6
20 Japan	0.133	26	6	5.4	11.6	87.0	85.8	48.8	70.4
21 Belgium	0.063	8	6	6.7	42.4	77.5	82.9	47.5	59.3
22 France	0.088	13	12	5.7	25.7	78.0	83.2	50.7	61.6
23 Austria	0.053	5	4	4.1	30.3	100.0	100.0	54.6	67.7
24 Finland	0.075	11	4	9.2	42.5	100.0	100.0	55.7	64.0
25 Slovenia	0.016	1	7	0.6	27.7	95.8	98.0	52.3	63.2
26 Spain	0.095	16	4	10.6	38.0	66.8	73.1	52.5	65.8
27 Italy	0.068	10	4	4.0	30.1	71.2	80.5	39.6	59.5
28 Czech Republic	0.091	15	5	4.9	18.9	99.9	99.7	51.1	68.3
29 Greece	0.146	29	5	11.9	21.0	59.5	67.0	44.2	62.5
30 Estonia	0.164	33	11	16.8	19.8	100.0[e]	100.0[e]	56.2	68.9
31 Brunei Darussalam	..	..	27	23.0	..	63.9[f]	67.8[f]	52.6	75.3
32 Cyprus	0.124	22	10	5.5	12.5	76.0	81.7	56.0	71.1
32 Qatar	0.524	116	6	9.5	0.0[g]	66.7	59.0	50.8	95.5
34 Andorra	..	..	..	..	50.0	49.5	49.3	..	..
35 Slovakia	0.164	33	7	15.9	18.7	99.1	99.5	51.1	68.6
36 Poland	0.138	28	3	12.2	22.1	79.4	85.5	48.9	64.9
37 Lithuania	0.125	23	11	10.6	23.4	89.1	94.3	55.8	67.3
37 Malta	0.227	46	9	18.2	13.0	68.6	78.2	37.9	66.3
39 Saudi Arabia	0.284	56	16	10.2	19.9	60.5	70.3	20.2	78.3
40 Argentina	0.376	75	69	54.4	36.8	56.3[f]	57.6[f]	47.5	75.0
41 United Arab Emirates	0.232	47	8	27.6	17.5	73.1	61.2	46.5	92.0
42 Chile	0.338	65	22	55.3	15.8	73.3	76.4	49.2	74.8
43 Portugal	0.111	20	8	12.6	31.3	47.7	48.2	54.9	66.2
44 Hungary	0.209	42	14	12.1	10.1	97.9[e]	98.7[e]	44.8	60.0
45 Bahrain	0.265	51	22	13.8	15.0	56.7[f]	51.4[f]	39.2	86.9
46 Latvia	0.167	36	13	13.5	18.0	98.9	99.0	54.9	67.6
47 Croatia	0.149	30	13	12.7	25.8	85.0	93.6	44.7	58.4
48 Kuwait	0.387	79	14	14.5	1.5	55.6	56.3	43.6	83.1
49 Montenegro	0.171	37	7	15.2	17.3	84.2	94.7	43.0	57.3
HIGH HUMAN DEVELOPMENT									
50 Belarus	0.151	31	1	20.6	30.1	87.0	92.2	50.1	63.1
50 Russian Federation	0.276	54	24	25.7	14.5	89.6	92.5	57.1	71.7
52 Oman	0.275	53	11	10.6	9.6	47.2	57.1	29.0	82.6
52 Romania	0.333	64	33	31.0	12.0	86.1	92.0	48.7	64.9
52 Uruguay	0.313	61	14	58.3	11.5	54.4[f]	50.3	55.6	76.8
55 Bahamas	0.298	58	37	28.5	16.7	91.2	87.6[f]	69.3	79.3
56 Kazakhstan	0.267	52	26	29.9	20.1	95.3	98.8	67.7	77.9
57 Barbados	0.357	69	52	48.4	19.6	89.5[f]	87.7[f]	65.9	76.6
58 Antigua and Barbuda	..	..	..	49.3	25.7	..	..	..	..
59 Bulgaria	0.212	44	5	35.9	20.4	93.0	95.7	47.9	59.0
60 Palau	..	..	..	..	10.3	..	..	..	..

		Gender Inequality Index		Maternal mortality ratio	Adolescent birth rate	Share of seats in parliament	Population with at least some secondary education		Labour force participation rate[a]	
							(% ages 25 and older)		(% ages 15 and older)	
		Value	Rank	(deaths per 100,000 live births)	(births per 1,000 women ages 15–19)	(% held by women)	Female	Male	Female	Male
HDI rank		2014	2014	2013	2010/2015[b]	2014	2005–2014[c]	2005–2014[c]	2013	2013
60	Panama	0.454	96	85	78.5	19.3	54.0[f]	49.9[f]	49.0	81.8
62	Malaysia	0.209	42	29	5.7	14.2	65.1[f]	71.9[f]	44.4	75.5
63	Mauritius	0.419	88	73	30.9	11.6	49.4	58.0	43.6	74.2
64	Seychelles	..	..	..	56.3	43.8	66.9	66.6	..	..
64	Trinidad and Tobago	0.371	73	84	34.8	24.7	59.7	60.9	53.0	75.5
66	Serbia	0.176	38	16	16.9	34.0	58.4	73.6	44.5	60.9
67	Cuba	0.356	68	80	43.1	48.9	74.3[f]	78.8[f]	43.4	70.0
67	Lebanon	0.385	78	16	12.0	3.1	53.0	55.4	23.3	70.9
69	Costa Rica	0.349	66	38	60.8	33.3	50.7[f]	50.5[f]	46.6	79.0
69	Iran (Islamic Republic of)	0.515	114	23	31.6	3.1	62.2	67.6	16.6	73.6
71	Venezuela (Bolivarian Republic of)	0.476	103	110	83.2	17.0	56.6	50.8	51.1	79.2
72	Turkey	0.359	71	20	30.9	14.4	39.0	60.0	29.4	70.8
73	Sri Lanka	0.370	72	29	16.9	5.8	72.7	76.4	35.1	76.3
74	Mexico	0.373	74	49	63.4	37.1	55.7	60.6	45.1	79.9
75	Brazil	0.457	97	69	70.8	9.6	54.6	52.4	59.4	80.8
76	Georgia	0.382	77	41	46.8	11.3	89.7	92.7	56.5	75.1
77	Saint Kitts and Nevis	..	..	..	..	6.7	..	..	..	..
78	Azerbaijan	0.303	59	26	40.0	15.6	93.7	97.4	62.9	69.6
79	Grenada	..	..	23	35.4	25.0	..	..	..	..
80	Jordan	0.473	102	50	26.5	11.6	69.5	78.5	15.6	66.6
81	The former Yugoslav Republic of Macedonia	0.164	33	7	18.3	33.3	40.2	55.6	43.1	67.5
81	Ukraine	0.286	57	23	25.7	11.8	91.7[f]	95.9[f]	53.2	66.9
83	Algeria	0.413	85	89	10.0	25.7	26.7	31.0	15.2	72.2
84	Peru	0.406	82	89	50.7	22.3	56.3	66.1	68.2	84.4
85	Albania	0.217	45	21	15.3	20.7	81.8	87.9	44.9	65.5
85	Armenia	0.318	62	29	27.1	10.7	94.0[f]	95.0[f]	54.2	72.6
85	Bosnia and Herzegovina	0.201	41	8	15.1	19.3	44.9	69.8	34.1	57.3
88	Ecuador	0.407	83	87	77.0	41.6	40.1	39.4	54.7	82.7
89	Saint Lucia	..	..	34	56.3	20.7	..	..	62.7	76.2
90	China	0.191	40	32	8.6	23.6	58.7	71.9	63.9	78.3
90	Fiji	0.418	87	59	42.8	14.0	64.2	64.5	37.5	72.0
90	Mongolia	0.325	63	68	18.7	14.9	85.3[f]	84.1[f]	56.6	69.3
93	Thailand	0.380	76	26	41.0	6.1	35.7	40.8	64.3	80.7
94	Dominica	..	..	..	..	21.9	29.7	23.2	..	..
94	Libya	0.134	27	15	2.5	16.0	55.5[f]	41.9[f]	30.0	76.4
96	Tunisia	0.240	48	46	4.6	31.3	32.8	46.1	25.1	70.9
97	Colombia	0.429	92	83	68.5	20.9	56.9	55.6	55.8	79.7
97	Saint Vincent and the Grenadines	..	..	45	54.5	13.0	..	..	55.7	78.0
99	Jamaica	0.430	93	80	70.1	16.7	74.0[f]	70.2[f]	56.1	70.9
100	Tonga	0.666	148	120	18.1	0.0[g]	87.5	88.3	53.5	74.6
101	Belize	0.426	90	45	71.4	13.3	76.4[f]	75.8[f]	49.2	82.3
101	Dominican Republic	0.477	104	100	99.6	19.1	55.6	53.1	51.3	78.6
103	Suriname	0.463	100	130	35.2	11.8	44.6	47.1	40.5	68.8
104	Maldives	0.243	49	31	4.2	5.9	27.3	32.7	56.2	77.5
105	Samoa	0.457	97	58	28.3	6.1	64.3	60.0	23.5	58.4
MEDIUM HUMAN DEVELOPMENT										
106	Botswana	0.480	106	170	44.2	9.5	73.6[f]	77.9[f]	71.9	81.6
107	Moldova (Republic of)	0.248	50	21	29.3	20.8	93.6	96.6	37.6	44.2
108	Egypt	0.573	131	45	43.0	2.2[h]	43.9[f]	60.6[f]	23.7	74.8
109	Turkmenistan	..	..	61	18.0	25.8	..	..	46.9	76.9
110	Gabon	0.514	113	240	103.0	16.2	53.9[f]	36.1[f]	56.2	65.4
110	Indonesia	0.494	110	190	48.3	17.1	39.9	49.2	51.4	84.2
112	Paraguay	0.472	101	110	67.0	16.8	36.8	43.0	55.7	84.8
113	Palestine, State of	..	..	..	45.8	..	53.9	59.4	15.4	66.4
114	Uzbekistan	..	..	36	38.8	16.4	..	..	48.1	75.6
115	Philippines	0.420	89	120	46.8	27.1	65.9	63.7	51.1	79.7
116	El Salvador	0.427	91	69	76.0	27.4	36.8	43.6	47.8	79.0
116	South Africa	0.407	83	140	50.9	40.7[i]	72.7	75.9	44.5	60.5
116	Viet Nam	0.308	60	49	29.0	24.3	59.4	71.2	73.0	82.2
119	Bolivia (Plurinational State of)	0.444	94	200	71.9	51.8	47.6	59.1	64.2	80.9
120	Kyrgyzstan	0.353	67	75	29.3	23.3	94.5	96.8	56.0	79.5
121	Iraq	0.539	123	67	68.7	26.5	27.8	50.2	14.9	69.8

TABLE 5   Gender Inequality Index | 225

TABLE 5

TABLE 5 GENDER INEQUALITY INDEX

HDI rank	Gender Inequality Index		Maternal mortality ratio (deaths per 100,000 live births)	Adolescent birth rate (births per 1,000 women ages 15–19)	Share of seats in parliament (% held by women)	Population with at least some secondary education (% ages 25 and older)		Labour force participation rate[a] (% ages 15 and older)	
	Value	Rank				Female	Male	Female	Male
	2014	2014	2013	2010/2015[b]	2014	2005–2014[c]	2005–2014[c]	2013	2013
122 Cabo Verde	..	..	53	70.6	20.8	..	..	51.5	83.7
123 Micronesia (Federated States of)	..	..	96	18.6	0.0[g]	..	..	..	..
124 Guyana	0.515	114	250	88.5	31.3	60.3[f]	47.8[f]	42.6	80.5
125 Nicaragua	0.449	95	100	100.8	39.1	39.4[f]	38.3[f]	47.4	80.3
126 Morocco	0.525	117	120	35.8	11.0	20.7[f]	30.2[f]	26.5	75.8
126 Namibia	0.401	81	130	54.9	37.7	33.3[f]	34.4[f]	54.7	63.7
128 Guatemala	0.533	119	140	97.2	13.3	21.9	23.2	49.3	88.2
129 Tajikistan	0.357	69	44	42.8	15.2	95.1	91.2	58.9	77.1
130 India	0.563	130	190	32.8	12.2	27.0	56.6	27.0	79.9
131 Honduras	0.480	106	120	84.0	25.8	28.0	25.8	42.8	82.9
132 Bhutan	0.457	97	120	40.9	8.3	34.0	34.5	66.7	77.2
133 Timor-Leste	..	..	270	52.2	38.5	..	..	24.6	50.8
134 Syrian Arab Republic	0.533	119	49	41.6	12.4	29.5	40.5	13.5	72.7
134 Vanuatu	..	..	86	44.8	0.0[g]	..	..	61.5	80.0
136 Congo	0.593	137	410	126.7	11.5	39.7[f]	47.0[f]	68.5	73.0
137 Kiribati	..	..	130	16.6	8.7	..	..	..	..
138 Equatorial Guinea	..	..	290	112.6	19.7	..	..	80.7	92.2
139 Zambia	0.587	132	280	125.4	12.7	25.8[f]	44.0[f]	73.1	85.6
140 Ghana	0.554	127	380	58.4	10.9	45.2	64.7	67.3	71.4
141 Lao People's Democratic Republic	..	..	..	65.0	25.0	22.9[f]	37.0[f]	76.3	79.1
142 Bangladesh	0.503	111	170	80.6	20.0	34.1[f]	41.3[f]	57.4	84.1
143 Cambodia	0.477	104	170	44.3	19.0	9.9	22.9	78.8	86.5
143 Sao Tome and Principe	..	..	210	65.1	18.2	..	..	45.3	77.8
LOW HUMAN DEVELOPMENT									
145 Kenya	0.552	126	400	93.6	20.8	25.3	31.4	62.2	72.4
145 Nepal	0.489	108	190	73.7	29.5	17.7[f]	38.2[f]	79.9	87.1
147 Pakistan	0.536	121	170	27.3	19.7	19.3	46.1	24.6	82.9
148 Myanmar	0.413	85	200	12.1	4.7	22.9[f]	15.3[f]	75.2	82.3
149 Angola	..	..	460	170.2	36.8	..	..	63.3	76.9
150 Swaziland	0.557	128	310	72.0	14.7	21.9[f]	26.0[f]	43.9	71.6
151 Tanzania (United Republic of)	0.547	125	410	122.7	36.0	5.6[f]	9.5[f]	88.1	90.2
152 Nigeria	..	..	560	119.6	6.6	..	..	48.2	63.7
153 Cameroon	0.587	132	590	115.8	27.1	21.3[f]	34.9[f]	63.8	76.8
154 Madagascar	..	..	440	122.8	20.5	..	..	86.6	90.5
155 Zimbabwe	0.504	112	470	60.3	35.1	48.7	62.0	83.2[i]	89.7[i]
156 Mauritania	0.610	139	320	73.3	22.2	8.3[f]	20.9[f]	28.7	79.1
156 Solomon Islands	..	..	130	64.9	2.0	..	..	53.4	79.0
158 Papua New Guinea	0.611	140	220	62.1	2.7	7.6[f]	14.5[f]	70.5	74.0
159 Comoros	..	..	350	51.1	3.0	..	..	35.2	80.1
160 Yemen	0.744	155	270	47.0	0.7	8.6[f]	26.7[f]	25.4	72.2
161 Lesotho	0.541	124	490	89.4	26.8	21.9	19.0	59.0	73.5
162 Togo	0.588	134	450	91.5	17.6	16.1[f]	40.3[f]	80.6	81.3
163 Haiti	0.603	138	380	42.0	3.5	22.4[f]	35.2[f]	60.9	71.0
163 Rwanda	0.400	80	320	33.6	57.5	8.0[f]	8.8[f]	86.4	85.3
163 Uganda	0.538	122	360	126.6	35.0	22.9	33.5	75.8	79.2
166 Benin	0.614	142	340	90.2	8.4	11.3[f]	27.0[f]	67.6	78.3
167 Sudan	0.591	135	360	84.0	23.8	12.1[f]	18.2[f]	31.3	76.0
168 Djibouti	..	..	230	18.6	12.7	..	..	36.3	67.7
169 South Sudan	..	..	730	75.3	24.3	..	..	..	..
170 Senegal	0.528	118	320	94.4	42.7	7.2	15.4	66.0	88.0
171 Afghanistan	0.693	152	400	86.8	27.6	5.9[f]	29.8[f]	15.8	79.5
172 Côte d'Ivoire	0.679	151	720	130.3	9.2	14.0[f]	30.1[f]	52.4	81.4
173 Malawi	0.611	140	510	144.8	16.7	11.1	21.6	84.6	81.5
174 Ethiopia	0.558	129	420	78.4	25.5	7.8	18.2	78.2	89.3
175 Gambia	0.622	143	430	115.8	9.4	17.4[f]	31.5[f]	72.2	82.9
176 Congo (Democratic Republic of the)	0.673	149	730	135.3	8.2	12.8[f]	32.4[f]	70.7	73.2
177 Liberia	0.651	146	640	117.4	10.7	15.4[f]	39.3[f]	58.2	64.8
178 Guinea-Bissau	..	..	560	99.3	13.7	..	..	68.2	78.5
179 Mali	0.677	150	550	175.6	9.5	7.7	15.1	50.8	81.4
180 Mozambique	0.591	135	480	137.8	39.6	1.4[f]	6.2[f]	85.5	82.8

HDI rank	Gender Inequality Index Value	Gender Inequality Index Rank	Maternal mortality ratio (deaths per 100,000 live births)	Adolescent birth rate (births per 1,000 women ages 15–19)	Share of seats in parliament (% held by women)	Population with at least some secondary education (% ages 25 and older) Female	Population with at least some secondary education (% ages 25 and older) Male	Labour force participation rate[a] (% ages 15 and older) Female	Labour force participation rate[a] (% ages 15 and older) Male
	2014	2014	2013	2010/2015[b]	2014	2005–2014[c]	2005–2014[c]	2013	2013
181 Sierra Leone	0.650	145	1,100	100.7	12.4	10.0[f]	21.7[f]	65.7	69.0
182 Guinea	..	..	650	131.0	21.9	..	..	65.6	78.3
183 Burkina Faso	0.631	144	400	115.4	13.3	0.9	3.2	77.1	90.0
184 Burundi	0.492	109	740	30.3	34.9	5.3[f]	8.3[f]	83.3	82.0
185 Chad	0.706	153	980	152.0	14.9	1.7	9.9	64.0	79.2
186 Eritrea	..	..	380	65.3	22.0	..	..	80.0	89.8
187 Central African Republic	0.655	147	880	98.3	12.5[h]	10.1[f]	26.7[f]	72.6	85.1
188 Niger	0.713	154	630	204.8	13.3	2.4[f]	7.8[f]	40.0	89.7
OTHER COUNTRIES OR TERRITORIES									
Korea (Democratic People's Rep. of)	..	..	87	0.6	16.3	..	..	72.2	84.2
Marshall Islands	..	..	..	..	3.0	..	..	..	..
Monaco	..	..	..	..	20.8	..	..	..	..
Nauru	..	..	..	..	5.3	..	..	..	..
San Marino	..	..	..	..	16.7	..	..	..	..
Somalia	..	..	850	110.4	13.8	..	..	37.2	75.5
Tuvalu	..	..	..	..	6.7	..	..	..	..
Human Development Index groups									
Very high human development	0.199	—	18	19.0	26.5	86.2	87.9	52.1	68.7
High human development	0.310	—	41	28.8	20.6	60.6	69.5	57.0	77.2
Medium human development	0.506	—	168	43.4	18.8	34.8	55.3	37.5	79.8
Low human development	0.583	—	461	92.1	20.5	14.8	28.3	57.2	79.1
Developing countries	0.478	—	225	51.5	20.2	44.2	58.4	49.5	78.7
Regions									
Arab States	0.537	—	155	45.4	14.0	34.7	47.6	23.2	75.3
East Asia and the Pacific	0.328	—	72	21.2	18.7	54.7	66.3	62.6	79.4
Europe and Central Asia	0.300	—	28	30.8	19.0	70.8	80.6	45.6	70.0
Latin America and the Caribbean	0.415	—	85	68.3	27.0	54.3	55.2	53.7	79.8
South Asia	0.536	—	183	38.7	17.5	29.1	54.6	29.8	80.3
Sub-Saharan Africa	0.575	—	506	109.7	22.5	22.1	31.5	65.4	76.6
Least developed countries	0.566	—	439	97.0	20.4	17.2	26.4	65.7	82.6
Small island developing states	0.474	—	220	61.5	22.8	51.1	55.1	53.0	73.3
Organisation for Economic Co-operation and Development	0.231	—	21	25.4	26.9	82.9	86.3	50.9	68.9
World	**0.449**	—	**210**[T]	**47.4**	**21.8**	**54.5**	**65.4**	**50.3**	**76.7**

NOTES

a Modeled International Labour Organization estimates.

b Data are annual average of projected values for 2010–2015.

c Data refer to the most recent year available during the period specified.

d Refers to population ages 25–64.

e Refers to population ages 25–74.

f Barro and Lee (2014) estimate for 2010 based on data from the United Nations Educational, Scientific and Cultural Organization Institute for Statistics.

g For the purpose of calculating the Gender Inequality Index value, 0.1% was used.

h Refers to 2013.

i Excludes the 36 special rotating delegates appointed on an ad hoc basis.

j Refers to 2012.

T From original data source.

DEFINITIONS

Gender Inequality Index: A composite measure reflecting inequality in achievement between women and men in three dimensions: reproductive health, empowerment and the labour market. See *Technical note 4* at http://hdr.undp.org/en for details on how the Gender Inequality Index is calculated.

Maternal mortality ratio: Number of deaths due to pregnancy-related causes per 100,000 live births.

Adolescent birth rate: Number of births to women ages 15–19 per 1,000 women ages 15–19.

Share of seats in national parliament: Proportion of seats held by women in the national parliament expressed as percentage of total seats. For countries with bicameral legislative systems, the share of seats is calculated based on both houses.

Population with at least some secondary education: Percentage of the population ages 25 and older who have reached (but not necessarily completed) a secondary level of education.

Labour force participation rate: Proportion of a country's working-age population (ages 15 and older) that engages in the labour market, either by working or actively looking for work, expressed as a percentage of the working-age population.

MAIN DATA SOURCES

Column 1: HDRO calculations based on data in columns 3–9.

Column 2: Calculated based on data in column 1.

Column 3: UN Maternal Mortality Estimation Group (2014).

Column 4: UNDESA (2013a).

Column 5: IPU (2015).

Columns 6 and 7: UNESCO Institute for Statistics (2015).

Columns 8 and 9: ILO (2015a).

TABLE 5

TABLE 5 Gender Inequality Index | 227

TABLE 6

Multidimensional Poverty Index: developing countries

Country	Year and survey[a]	Multidimensional Poverty Index[b] HDRO specifications[c] Index Value	HDRO Headcount (%)	2010 specifications[d] Index Value	2010 Headcount (%)	Population in multidimensional poverty[c] Headcount (thousands)	Intensity of deprivation (%)	Population near multidimensional poverty[c] (%)	Population in severe multidimensional poverty[c] (%)	Contribution of deprivation to overall poverty (%) Education	Health	Living standards	Population living below income poverty line (%) National poverty line 2004–2014[e]	PPP $1.25 a day 2002–2012[e]
Afghanistan	2010/2011 M	0.293 [f]	58.8 [f]	0.353 [f]	66.2 [f]	17,116 [f]	49.9 [f]	16.0 [f]	29.8 [f]	45.6 [f]	19.2 [f]	35.2 [f]	35.8	..
Albania	2008/2009 D	0.005	1.2	0.005	1.4	38	38.3	7.2	0.1	22.4	47.1	30.5	14.3	0.5
Argentina	2005 N	0.015 [g]	3.7 [g]	0.011 [g]	2.9 [g]	1,438 [g]	39.1 [g]	5.2 [g]	0.5 [g]	38.2 [g]	27.8 [g]	34.0 [g]	..	1.4
Armenia	2010 D	0.002	0.6	0.001	0.3	18	37.0	3.0	0.1	3.4	87.8	8.7	32.0	1.8
Azerbaijan	2006 D	0.009	2.4	0.021	5.3	210	38.2	11.5	0.2	20.0	50.7	29.3	5.3	0.3
Bangladesh	2011 D	0.237	49.5	0.253	51.3	75,610	47.8	18.8	21.0	28.4	26.6	44.9	31.5	43.3
Barbados	2012 M	0.004 [h]	1.2 [h]	0.003 [h]	0.9 [h]	3 [h]	33.7 [h]	0.3 [h]	0.0 [h]	1.5 [h]	95.9 [h]	2.6 [h]	..	..
Belarus	2005 M	0.001	0.4	0.000	0.0	41	34.5	1.1	0.0	2.6	89.7	7.7	5.5	0.0
Belize	2011 M	0.030	7.4	0.018	4.6	23	41.2	6.4	1.5	36.2	34.8	29.0	..	..
Benin	2011/2012 D	0.343	64.2	0.307	62.2	6,455	53.3	16.9	37.7	33.1	24.8	42.1	36.2	51.6
Bhutan	2010 M	0.128	29.4	0.119	27.2	211	43.5	18.0	8.8	40.3	26.3	33.4	12.0	2.4
Bolivia (Plurinational State of)	2008 D	0.097	20.6	0.089	20.5	2,022	47.0	17.3	7.8	21.9	27.9	50.2	45.0	8.0
Bosnia and Herzegovina	2011/2012 M	0.006 [h]	1.7 [h]	0.002 [h]	0.5 [h]	65 [h]	37.3 [h]	3.2 [h]	0.0 [h]	7.8 [h]	79.5 [h]	12.7 [h]	17.9	0.0
Brazil	2013 N	0.011 [f,i]	2.9 [f,i]	..	..	5,738 [f,i]	40.2 [f,i]	7.2 [f,i]	0.4 [f,i]	27.6 [f,i]	40.7 [f,i]	31.7 [f,i]	8.9	3.8
Burkina Faso	2010 D	0.508	82.8	0.535	84.0	12,875	61.3	7.6	63.8	39.0	22.5	38.5	46.7	44.5
Burundi	2010 D	0.442	81.8	0.454	80.8	7,553	54.0	12.0	48.2	25.0	26.3	48.8	66.9	81.3
Cambodia	2010 D	0.211	46.8	0.212	45.9	6,721	45.1	20.4	16.4	25.9	27.7	46.4	17.7	10.1
Cameroon	2011 D	0.260	48.2	0.248	46.0	10,187	54.1	17.8	27.1	24.5	31.3	44.2	39.9	27.6
Central African Republic	2010 M	0.424	76.3	0.430	77.6	3,320	55.6	15.7	48.5	23.8	26.2	50.0	62.0	62.8
Chad	2010 M	0.545	86.9	0.554	87.2	10,186	62.7	8.8	67.6	32.3	22.5	45.2	46.7	36.5
China	2012 N	0.023 [i]	5.2 [i]	0.023 [i]	5.2 [i]	71,939 [i]	43.3 [i]	22.7 [i]	1.0 [i]	30.0 [i]	36.6 [i]	33.4 [i]	..	6.3
Colombia	2010 D	0.032	7.6	0.022	5.4	3,534	42.2	10.2	1.8	34.3	24.7	41.0	30.6	5.6
Comoros	2012 D/M	0.165	34.3	0.173	36.0	247	48.1	23.1	14.9	29.1	25.9	45.0	44.8	46.1
Congo	2011/2012 D	0.192	43.0	0.181	39.7	1,866	44.7	26.2	12.2	10.6	32.8	56.6	46.5	32.8
Congo (Democratic Republic of the)	2013/2014 D	0.369	72.5	0.401	75.1	50,312	50.8	18.5	36.7	15.6	31.0	53.4	63.6	87.7
Côte d'Ivoire	2011/2012 D	0.307	59.3	0.310	58.7	11,772	51.7	17.9	32.4	36.5	25.8	37.7	42.7	35.0
Djibouti	2006 M	0.127	26.9	0.139	29.3	212	47.3	16.0	11.1	36.1	22.7	41.2	..	18.8
Dominican Republic	2013 D	0.025	6.0	0.020	5.1	620	41.6	20.6	1.0	28.4	39.6	32.0	41.1	2.3
Ecuador	2013/2014 N	0.015	3.7	0.013	3.5	588	39.6	8.4	0.5	23.6	42.4	34.0	22.5	4.0
Egypt	2014 D	0.016 [j]	4.2 [j]	0.014 [j]	3.6 [j]	3,491 [j]	37.4 [j]	5.6 [j]	0.4 [j]	45.6 [j]	46.7 [j]	7.8 [j]	25.2	1.7
Ethiopia	2011 D	0.537	88.2	0.564	87.3	78,887	60.9	6.7	67.0	27.4	25.2	47.4	29.6	36.8
Gabon	2012 D	0.073	16.7	0.070	16.5	273	43.4	19.9	4.4	15.2	43.8	40.9	32.7	6.1
Gambia	2013 D	0.289	57.2	0.323	60.4	1,058	50.5	21.3	31.7	32.9	30.9	36.2	48.4	33.6
Georgia	2005 M	0.008	2.2	0.003	0.8	99	37.6	4.1	0.1	7.4	67.4	25.2	14.8	14.1
Ghana	2011 M	0.144	30.5	0.139	30.4	7,559	47.3	18.7	12.1	27.7	27.1	45.2	24.2	28.6
Guinea	2012 D/M	0.425	73.8	0.459	75.1	8,456	57.6	12.7	49.8	36.6	22.8	40.6	55.2	40.9
Guinea-Bissau	2006 M	0.495	80.4	0.462	77.5	1,168	61.6	10.5	58.4	30.5	27.9	41.6	69.3	48.9
Guyana	2009 D	0.031	7.8	0.030	7.7	61	40.0	18.8	1.2	16.8	51.2	32.0	..	..
Haiti	2012 D	0.242	50.2	0.248	49.4	5,104	48.1	22.2	20.1	24.8	23.4	51.8	58.5	..
Honduras	2011/2012 D	0.098 [k]	20.7 [k]	0.072 [k]	15.8 [k]	1,642 [k]	47.4 [k]	28.6 [k]	7.2 [k]	36.6 [k]	23.1 [k]	40.3 [k]	64.5	16.5
India	2005/2006 D	0.282	55.3	0.283	53.7	631,999	51.1	18.2	27.8	22.7	32.5	44.8	21.9	23.6
Indonesia	2012 D	0.024 [f]	5.9 [f]	0.066 [f]	15.5 [f]	14,574 [f]	41.3 [f]	8.1 [f]	1.1 [f]	24.7 [f]	35.1 [f]	40.2 [f]	11.3	16.2
Iraq	2011 M	0.052	13.3	0.045	11.6	4,236	39.4	7.4	2.5	50.1	38.6	11.3	18.9	3.9
Jamaica	2010 N	0.014 [h,i]	3.7 [h,i]	0.008 [h,i]	2.0 [h,i]	102 [h,i]	38.8 [h,i]	9.1 [h,i]	0.5 [h,i]	7.7 [h,i]	59.3 [h,i]	33.0 [h,i]	19.9	0.2
Jordan	2012 D	0.004	1.2	0.006	1.7	85	35.3	1.0	0.1	31.5	65.0	3.5	14.4	0.1
Kazakhstan	2010/2011 M	0.004	1.1	0.001	0.2	173	36.4	2.3	0.0	4.3	83.9	11.8	2.9	0.1
Kenya	2008/2009 D	0.226	48.2	0.229	47.8	19,190	47.0	29.1	15.7	11.2	32.4	56.4	45.9	43.4
Kyrgyzstan	2012 D	0.006	1.8	0.007	2.0	96	36.9	10.7	0.1	6.6	70.5	22.9	37.0	5.1
Lao People's Democratic Republic	2011/2012 M	0.186	36.8	0.174	34.1	2,447	50.5	18.5	18.8	37.7	25.4	36.9	23.2	30.3
Lesotho	2009 D	0.227	49.5	0.156	35.3	984	45.9	20.4	18.2	14.8	33.8	51.4	57.1	56.2
Liberia	2013 D	0.356	70.1	0.374	71.2	3,010	50.8	21.5	35.4	23.0	25.6	51.4	63.8	83.8
Libya	2007 N	0.005	1.4	0.006	1.5	79	37.5	6.3	0.1	31.9	47.9	20.2	..	..
Madagascar	2008/2009 D	0.420	77.0	0.357	66.9	15,774	54.6	11.7	48.0	31.6	24.5	43.9	75.3	87.7
Malawi	2010 D	0.332	66.7	0.334	66.7	10,012	49.8	24.5	29.8	18.9	27.7	53.4	50.7	72.2
Maldives	2009 D	0.008	2.0	0.018	5.2	6	37.5	8.5	0.1	27.8	60.2	11.9	..	1.5
Mali	2012/2013 D	0.456	78.4	0.457	77.7	11,998	58.2	10.8	55.9	37.9	22.4	39.7	43.6	50.6
Mauritania	2011 M	0.291	55.6	0.285	52.2	2,060	52.4	16.8	29.9	34.5	20.3	45.3	42.0	23.4
Mexico	2012 N	0.024	6.0	0.011	2.8	7,272	39.9	10.1	1.1	31.4	25.6	43.0	52.3	1.0
Moldova (Republic of)	2012 M	0.004	1.1	0.003	0.8	38	38.4	2.2	0.1	11.0	66.9	22.1	12.7	0.2
Mongolia	2010 M	0.047	11.1	0.037	9.2	302	42.5	19.3	2.3	18.1	27.7	54.2	27.4	..

Country	Year and survey[a]	Multidimensional Poverty Index[b]				Population in multidimensional poverty[c]		Population near multidimensional poverty[c]	Population in severe multidimensional poverty[c]	Contribution of deprivation to overall poverty			Population living below income poverty line	
		HDRO specifications[c]		2010 specifications[d]						(%)			(%)	
		Index	Headcount	Index	Headcount	Headcount	Intensity of deprivation						National poverty line	PPP $1.25 a day
	2005–2014	Value	(%)	Value	(%)	(thousands)	(%)	(%)	(%)	Education	Health	Living standards	2004–2014[e]	2002–2012[e]
Montenegro	2013 M	0.002	0.5	0.001	0.3	3	38.9	2.0	0.0	22.0	59.9	18.1	11.3	0.2
Morocco	2011 N	0.069	15.6	0.067	15.4	5,016	44.3	12.6	4.9	44.8	21.8	33.4	8.9	2.57
Mozambique	2011 D	0.390	70.2	0.389	69.6	17,246	55.6	14.8	44.1	30.4	22.3	47.3	54.7	60.7
Namibia	2013 D	0.205	44.9	0.193	42.0	1,034	45.5	19.3	13.4	11.0	39.2	49.8	28.7	23.5
Nepal	2011 D	0.197	41.4	0.217	44.2	11,255	47.4	18.1	18.6	27.3	28.2	44.5	25.2	23.7
Nicaragua	2011/2012 D	0.088	19.4	0.072	16.1	1,146	45.6	14.8	6.9	37.8	12.6	49.6	42.5	8.5
Niger	2012 D	0.584	89.8	0.605	89.3	15,408	65.0	5.9	73.5	35.9	24.0	40.0	48.9	40.8
Nigeria	2013 D	0.279	50.9	0.303	53.2	88,425	54.8	18.4	30.0	29.8	29.8	40.4	46.0	62.0
Pakistan	2012/2013 D	0.237	45.6	0.230	44.2	83,045	52.0	14.9	26.5	36.2	32.3	31.6	22.3	12.7
Palestine, State of	2010 M	0.007	1.9	0.006	1.5	75	37.4	6.2	0.1	13.9	68.8	17.3	25.8	0.1
Peru	2012 D	0.043	10.4	0.043	10.5	3,132	41.4	12.3	2.1	19.4	29.8	50.8	23.9	2.9
Philippines	2013 D	0.033 [f,l]	6.3 [f,l]	0.052 [f,l]	11.0 [f,l]	6,221 [f,l]	51.9 [f,l]	8.4 [f,l]	4.2 [f,l]	35.3 [f,l]	30.2 [f,l]	34.5 [f,l]	25.2	19.0
Rwanda	2010 D	0.352	70.8	0.350	69.0	7,669	49.7	17.9	34.6	23.8	27.2	49.0	44.9	63.0
Saint Lucia	2012 M	0.003	0.8	0.003	1.0	2	34.5	0.9	0.0	15.8	65.2	19.0	..	..
Sao Tome and Principe	2008/2009 D	0.217	47.5	0.154	34.5	82	45.5	21.5	16.4	29.1	26.5	44.4	61.7	43.5
Senegal	2014 D	0.278	51.9	0.309	56.9	7,556	53.5	18.1	30.8	43.6	23.1	33.4	46.7	34.1
Serbia	2014 M	0.002	0.4	0.001	0.2	41	40.6	2.7	0.1	30.7	40.7	28.7	24.6	0.1
Sierra Leone	2013 D	0.411	77.5	0.464	81.0	4,724	53.0	14.6	43.9	25.7	28.5	45.9	52.9	56.6
Somalia	2006 M	0.500	81.8	0.514	81.2	7,104	61.1	8.3	63.6	33.7	18.8	47.5	..	..
South Africa	2012 N	0.041	10.3	0.044	11.1	5,400	39.6	17.1	1.3	8.4	61.4	30.2	53.8	9.4
South Sudan	2010 M	0.551	89.3	0.557	91.1	8,877	61.7	8.5	69.6	39.3	14.3	46.3	50.6	..
Sudan	2010 M	0.290	53.1	0.321	57.8	18,916	54.6	17.9	31.9	30.4	20.7	48.9	46.5	19.8
Suriname	2010 M	0.033 [h]	7.6 [h]	0.024 [h]	5.9 [h]	40 [h]	43.1 [h]	4.7 [h]	2.0 [h]	31.0 [h]	37.2 [h]	31.8 [h]	..	..
Swaziland	2010 M	0.113	25.9	0.086	20.4	309	43.5	20.5	7.4	13.7	41.0	45.3	63.0	39.3
Syrian Arab Republic	2009 N	0.028	7.2	0.016	4.4	1,519	39.1	7.4	1.3	54.7	34.0	11.3	35.2	1.7
Tajikistan	2012 D	0.031	7.9	0.054	13.2	629	39.0	23.4	1.2	13.4	52.6	34.0	47.2	6.5
Tanzania (United Republic of)	2010 D	0.335	66.4	0.332	65.6	29,842	50.4	21.5	32.1	16.9	28.2	54.9	28.2	43.5
Thailand	2005/2006 M	0.004	1.0	0.006	1.6	664	38.8	4.4	0.1	19.4	51.3	29.4	12.6	0.3
The former Yugoslav Republic of Macedonia	2011 M	0.007 [h]	1.7 [h]	0.002 [h]	0.7 [h]	36 [h]	38.4 [h]	2.4 [h]	0.1 [h]	18.5 [h]	57.2 [h]	24.3 [h]	27.1	0.3
Timor-Leste	2009/2010 D	0.322	64.3	0.360	68.1	694	50.1	21.4	31.5	20.0	30.4	49.6	49.9	34.9
Togo	2013/2014 D	0.242	48.5	0.252	50.1	3,394	49.9	19.9	23.2	26.4	28.8	44.9	58.7	52.5
Trinidad and Tobago	2006 M	0.007 [f]	1.7 [f]	0.020 [f]	5.6 [f]	23 [f]	38.0 [f]	0.5 [f]	0.2 [f]	2.2 [f]	86.1 [f]	11.7 [f]	..	..
Tunisia	2011/2012 M	0.006	1.5	0.004	1.2	161	39.3	3.2	0.2	33.7	48.2	18.1	15.5	0.7
Uganda	2011 D	0.359	70.3	0.367	69.9	24,712	51.1	20.6	33.3	18.0	30.2	51.9	19.5	37.8
Ukraine	2012 M	0.001 [f]	0.4 [f]	0.004 [f]	1.2 [f]	162 [f]	34.5 [f]	0.0 [f]	0.0 [f]	19.0 [f]	77.5 [f]	3.5 [f]	8.4	0.0
Uzbekistan	2006 M	0.013	3.5	0.008	2.3	935	36.6	6.2	0.1	3.7	83.4	12.8	16.0	..
Vanuatu	2007 M	0.135	31.2	0.129	30.1	69	43.1	32.6	7.3	24.4	24.1	51.6	..	..
Viet Nam	2010/2011 M	0.026	6.4	0.017	4.2	5,796	40.7	8.7	1.3	35.9	25.7	38.4	17.2	2.4
Yemen	2013 D	0.200	40.0	..	..	9,754	50.1	22.4	19.4	29.5	32.2	38.2	34.8	9.8
Zambia	2013/2014 D	0.264	54.4	0.281	56.6	8,173	48.6	23.1	22.5	17.9	29.8	52.3	60.5	74.3
Zimbabwe	2014 M	0.128	28.9	0.127	29.7	4,222	44.1	29.3	7.8	10.8	34.5	54.8	72.3	..

TABLE
6

NOTES

a *D* indicates data from Demographic and Health Surveys, *M* indicates data from Multiple Indicator Cluster Surveys and *N* indicates data from national surveys (see http://hdr.undp.org for the list of national surveys).

b Not all indicators were available for all countries, so caution should be used in cross-country comparisons. Where an indicator is missing, weights of available indicators are adjusted to total 100%. See *Technical note 5* at http://hdr.undp.org for details.

c The HDRO specifications refer to somewhat modified definitions of deprivations in some indicators compared to the 2010 specifications. See *Technical note 5* for details.

d Based on a methodology from Alkire and Santos (2010).

e Data refer to the most recent year available during the period specified.

f Missing indicators on nutrition.

g Refers to urban areas only.

h Missing indicator on child mortality.

i Missing indicator on type of floor.

j Missing indicator on cooking fuel.

k Missing indicator on electricity.

l Missing indicator on school attendance.

DEFINITIONS

Multidimensional Poverty Index: Percentage of the population that is multidimensionally poor adjusted by the intensity of the deprivations. See *Technical note 5* for details on how the Multidimensional Poverty Index is calculated.

Multidimensional poverty headcount: Percentage of the population with a weighted deprivation score of at least 33 percent. It is also expressed in thousands of the population in the survey year.

Intensity of deprivation of multidimensional poverty: Average percentage of deprivation experienced by people in multidimensional poverty.

Population near multidimensional poverty: Percentage of the population at risk of suffering multiple deprivations—that is, those with a deprivation score of 20–33 percent.

Population in severe multidimensional poverty: Percentage of the population in severe multidimensional poverty—that is, those with a deprivation score of 50 percent or more.

Contribution of deprivation to overall poverty: Percentage of the Multidimensional Poverty Index attributed to deprivations in each dimension.

Population below national poverty line: Percentage of the population living below the national poverty line, which is the poverty line deemed appropriate for a country by its authorities. National estimates are based on population-weighted subgroup estimates from household surveys.

Population below PPP $1.25 a day: Percentage of the population living below the international poverty line $1.25 (in purchasing power parity terms) a day.

MAIN DATA SOURCES

Column 1: Refers to the year and the survey whose data were used to calculate the values in columns 2–10.

Columns 2–3 and 6–12: HDRO calculations based on data on household deprivations in education, health and living standards from various household surveys listed in column 1 using revised methodology described in *Technical note 5*.

Columns 4 and 5: Alkire and Robles (2015).

Columns 13 and 14: World Bank (2015a).

TABLE 6   Multidimensional Poverty Index: developing countries | 229

TABLE 7

Multidimensional Poverty Index: changes over time

Country	Year and survey[a]	Multidimensional Poverty Index[b] HDRO specifications[c] Value	Population in multidimensional poverty[c] Headcount (%)	Headcount (thousands)	Intensity of deprivation (%)	Population near multidimensional poverty[c] (%)	Population in severe multidimensional poverty[c] (%)	Contribution of deprivation to overall poverty (%) Education	Health	Living standards
Bangladesh	2011 D	0.237	49.5	75,610	47.8	18.8	21.0	28.4	26.6	44.9
Bangladesh	2007 D	0.294	59.5	87,185	49.3	18.7	27.2	26.0	26.5	47.5
Belize	2011 M	0.030	7.4	23	41.2	6.4	1.5	36.2	34.8	29.0
Belize	2006 M	0.028	6.9	19	40.8	6.5	1.2	13.8	52.6	33.6
Benin	2011/2012 D	0.343	64.2	6,455	53.3	16.9	37.7	33.1	24.8	42.1
Benin	2006 D	0.401	69.8	5,897	57.4	18.8	45.7	35.0	24.9	40.1
Bosnia and Herzegovina	2011/2012 M	0.006[d]	1.7[d]	65[d]	37.3[d]	3.2[d]	0.0[d]	7.8[d]	79.5[d]	12.7[d]
Bosnia and Herzegovina	2006 M	0.013[d]	3.5[d]	134[d]	38.1[d]	5.3[d]	0.1[d]	7.9[d]	76.3[d]	15.8[d]
Brazil	2013 N[e]	0.011[f,g]	2.9[f,g]	5,738[f,g]	40.2[f,g]	7.2[f,g]	0.4[f,g]	27.6[f,g]	40.7[f,g]	31.7[f,g]
Brazil	2012 N[e]	0.012[f,g]	3.1[f,g]	6,083[f,g]	40.8[f,g]	7.4[f,g]	0.5[f,g]	27.7[f,g]	38.4[f,g]	33.9[f,g]
Brazil	2006 N[e]	0.017[h]	4.0[h]	7,578[h]	41.4[h]	11.2[h]	0.7[h]	41.4[h]	20.4[h]	38.2[h]
Burkina Faso	2010 D	0.508	82.8	12,875	61.3	7.6	63.8	39.0	22.5	38.5
Burkina Faso	2006 M	0.538	85.2	11,775	63.2	6.9	67.1	38.0	22.3	39.6
Burundi	2010 D	0.442	81.8	7,553	54.0	12.0	48.2	25.0	26.3	48.8
Burundi	2005 M	0.485[f]	87.9[f]	6,833[f]	55.2[f]	8.5[f]	53.5[f]	37.8[f]	11.1[f]	51.1[f]
Cambodia	2010 D	0.211	46.8	6,721	45.1	20.4	16.4	25.9	27.7	46.4
Cambodia	2005 D	0.282	58.0	7,746	48.7	17.5	26.4	29.0	26.3	44.7
Cameroon	2011 D	0.260	48.2	10,187	54.1	17.8	27.1	24.5	31.3	44.2
Cameroon	2006 M	0.304[d]	51.8[d]	9,644[d]	58.7[d]	14.0[d]	35.9[d]	24.8[d]	31.7[d]	43.5[d]
Central African Republic	2010 M	0.424	76.3	3,320	55.6	15.7	48.5	23.8	26.2	50.0
Central African Republic	2006 M	0.464	80.5	3,245	57.7	12.1	54.5	30.2	24.3	45.6
China	2012 N[e]	0.023[g]	5.2[g]	71,939[g]	43.3[g]	22.7[g]	1.0[g]	30.0[g]	36.6[g]	33.4[g]
China	2009 N[e]	0.026[g,i]	6.0[g,i]	80,784[g,i]	43.4[g,i]	19.0[g,i]	1.3[g,i]	21.0[g,i]	44.4[g,i]	34.6[g,i]
Congo	2011/2012 D	0.192	43.0	1,866	44.7	26.2	12.2	10.6	32.8	56.6
Congo	2009 D	0.154[f]	32.7[f]	1,308[f]	47.1[f]	29.9[f]	15.1[f]	16.2[f]	25.6[f]	58.2[f]
Congo (Democratic Republic of the)	2013/2014 D	0.369	72.5	50,312	50.8	18.5	36.7	15.6	31.0	53.4
Congo (Democratic Republic of the)	2010 M	0.399	74.4	46,278	53.7	15.5	46.2	18.5	25.5	55.9
Côte d'Ivoire	2011/2012 D	0.307	59.3	11,772	51.7	17.9	32.4	36.5	25.8	37.7
Côte d'Ivoire	2005 D	0.269[f,h]	50.0[f,h]	8,693[f,h]	53.9[f,h]	22.7[f,h]	26.7[f,h]	42.8[f,h]	20.8[f,h]	36.5[f,h]
Dominican Republic	2013 D	0.025	6.0	620	41.6	20.6	1.0	28.4	39.6	32.0
Dominican Republic	2007 D	0.026	6.2	599	41.9	10.8	1.4	36.2	30.4	33.3
Ecuador	2013/2014 N	0.015	3.7	588	39.6	8.4	0.5	23.6	42.4	34.0
Ecuador	2006 N	0.043	10.6	1,486	40.9	9.4	2.1	22.2	44.3	33.5
Egypt	2014 D	0.016[h]	4.2[h]	3,491[h]	37.4[h]	5.6[h]	0.4[h]	45.6[h]	46.7[h]	7.8[h]
Egypt	2008 D	0.036[h]	8.9[h]	6,740[h]	40.3[h]	8.6[h]	1.5[h]	41.8[h]	45.6[h]	12.6[h]
Gambia	2013 D	0.289	57.2	1,058	50.5	21.3	31.7	32.9	30.9	36.2
Gambia	2005/2006 M	0.329	60.8	901	54.1	15.7	35.9	34.0	30.5	35.5
Ghana	2011 M	0.144	30.5	7,559	47.3	18.7	12.1	27.7	27.1	45.2
Ghana	2008 D	0.186	39.2	9,057	47.4	20.3	15.4	26.5	28.5	45.0
Guinea	2012 D/M	0.425	73.8	8,456	57.6	12.7	49.8	36.6	22.8	40.6
Guinea	2005 D	0.548	86.5	8,283	63.4	7.7	68.6	34.4	22.3	43.3
Guyana	2009 D	0.031	7.8	61	40.0	18.8	1.2	16.8	51.2	32.0
Guyana	2007 M	0.032	7.9	61	40.1	10.7	1.5	16.9	44.8	38.3
Haiti	2012 D	0.242	50.2	5,104	48.1	22.2	20.1	24.8	23.4	51.8
Haiti	2005/2006 D	0.315	59.3	5,566	53.2	18.1	32.8	28.8	22.8	48.5
Honduras	2011/2012 D	0.098[i]	20.7[i]	1,642[i]	47.4[i]	28.6[i]	7.2[i]	36.6[i]	23.1[i]	40.3[i]
Honduras	2005/2006 D	0.156[i]	31.5[i]	2,214[i]	49.6[i]	26.6[i]	13.3[i]	38.4[i]	22.6[i]	39.0[i]
Indonesia	2012 D	0.024[f]	5.9[f]	14,574[f]	41.3[f]	8.1[f]	1.1[f]	24.7[f]	35.1[f]	40.2[f]
Indonesia	2007 D	0.043[f]	10.1[f]	23,432[f]	42.4[f]	15.4[f]	2.3[f]	30.4[f]	21.0[f]	48.7[f]
Iraq	2011 M	0.052	13.3	4,236	39.4	7.4	2.5	50.1	38.6	11.3
Iraq	2006 M	0.077	18.5	5,182	41.8	15.0	4.3	45.7	33.9	20.4
Jordan	2012 D	0.004	1.2	85	35.3	1.0	0.1	31.5	65.0	3.5
Jordan	2009 D	0.004	1.0	64	36.8	4.1	0.1	33.7	56.3	10.0
Kazakhstan	2010/2011 M	0.004	1.1	173	36.4	2.3	0.0	4.3	83.9	11.8
Kazakhstan	2006 M	0.007	1.8	277	38.5	4.7	0.2	5.5	73.4	21.2
Kyrgyzstan	2012 D	0.006	1.8	96	36.9	10.7	0.1	6.6	70.5	22.9
Kyrgyzstan	2005/2006 M	0.013	3.4	173	37.9	10.1	0.3	5.0	63.9	31.2
Lao People's Democratic Republic	2011/2012 M	0.186	36.8	2,447	50.5	18.5	18.8	37.7	25.4	36.9
Lao People's Democratic Republic	2006 M	0.320[d]	55.0[d]	3,242[d]	58.3[d]	11.1[d]	35.2[d]	32.3[d]	32.6[d]	35.2[d]
Liberia	2013 D	0.356	70.1	3,010	50.8	21.5	35.4	23.0	25.6	51.4
Liberia	2007 D	0.459	81.9	2,883	56.1	12.9	52.8	30.4	21.8	47.8
Mali	2012/2013 D	0.456	78.4	11,998	58.2	10.8	55.9	37.9	22.4	39.7
Mali	2006 D	0.533	85.6	10,545	62.4	7.8	66.8	37.4	22.6	40.1

Country	Year and survey[a]	Multidimensional Poverty Index[b] HDRO specifications[c] Value	Population in multidimensional poverty[c] Headcount (%)	Population in multidimensional poverty[c] Headcount (thousands)	Population in multidimensional poverty[c] Intensity of deprivation (%)	Population near multidimensional poverty[e] (%)	Population in severe multidimensional poverty[c] (%)	Contribution of deprivation to overall poverty (%) Education	Contribution of deprivation to overall poverty (%) Health	Contribution of deprivation to overall poverty (%) Living standards
	2005–2014	Value	(%)	(thousands)	(%)	(%)	(%)	Education	Health	Living standards
Mauritania	2011 M	0.291	55.6	2,060	52.4	16.8	29.9	34.5	20.3	45.3
Mauritania	2007 M	0.362	66.0	2,197	54.9	12.8	42.3	33.5	18.2	48.3
Mexico	2012 N	0.024	6.0	7,272	39.9	10.1	1.1	31.4	25.6	43.0
Mexico	2006 N	0.028	6.9	7,779	40.9	10.7	1.6	32.0	29.0	39.0
Moldova (Republic of)	2012 M	0.004	1.1	38	38.4	2.2	0.1	11.0	66.9	22.1
Moldova (Republic of)	2005 D	0.005	1.3	49	38.8	5.2	0.2	17.7	46.6	35.6
Mongolia	2010 M	0.047	11.1	302	42.5	19.3	2.3	18.1	27.7	54.2
Mongolia	2005 M	0.077	18.3	462	42.0	19.0	4.2	13.5	35.7	50.8
Montenegro	2013 M	0.002	0.5	3	38.9	2.0	0.0	22.0	59.9	18.1
Montenegro	2005/2006 M	0.012[d]	3.0[d]	19[d]	40.1[d]	1.3[d]	0.5[d]	21.0[d]	63.8[d]	15.3[d]
Mozambique	2011 D	0.390	70.2	17,246	55.6	14.8	44.1	30.4	22.3	47.3
Mozambique	2009 D	0.395[f]	70.0[f]	16,343[f]	56.5[f]	14.7[f]	43.2[f]	31.3[f]	20.3[f]	48.4[f]
Namibia	2013 D	0.205	44.9	1,034	45.5	19.3	13.4	11.0	39.2	49.8
Namibia	2006/2007 D	0.200	42.1	876	47.5	22.6	15.7	14.8	33.4	51.8
Nepal	2011 D	0.197	41.4	11,255	47.4	18.1	18.6	27.3	28.2	44.5
Nepal	2006 D	0.314	62.1	15,910	50.6	15.5	31.6	26.0	28.0	46.0
Nicaragua	2011/2012 D	0.088	19.4	1,146	45.6	14.8	6.9	37.8	12.6	49.6
Nicaragua	2006/2007 D	0.137	27.9	1,561	49.2	15.3	12.9	38.1	12.3	49.7
Niger	2012 D	0.584	89.8	15,408	65.0	5.9	73.5	35.9	24.0	40.0
Niger	2006 D	0.677	93.4	12,774	72.5	3.4	86.1	35.2	24.5	40.3
Nigeria	2013 D	0.279	50.9	88,425	54.8	18.4	30.0	29.8	29.8	40.4
Nigeria	2011 M	0.239	43.3	71,014	55.2	17.0	25.7	26.9	32.6	40.4
Nigeria	2008 D	0.294	53.8	81,357	54.7	18.2	31.4	27.2	30.8	42.0
Pakistan	2012/2013 D	0.237	45.6	83,045	52.0	14.9	26.5	36.2	32.3	31.6
Pakistan	2006/2007 D	0.218[f]	43.5[f]	71,378[f]	50.0[f]	13.2[f]	21.7[f]	43.0[f]	19.7[f]	37.3[f]
Palestine (State of)	2010 M	0.007	1.9	75	37.4	6.2	0.1	13.9	68.8	17.3
Palestine (State of)	2006/2007 N	0.007	2.0	74	36.9	7.4	0.1	16.6	72.3	11.1
Peru	2012 D	0.043	10.4	3,132	41.4	12.3	2.1	19.4	29.8	50.8
Peru	2011 D	0.051	12.2	3,607	42.2	12.3	2.8	20.2	29.0	50.8
Peru	2010 D	0.056	13.2	3,859	42.1	14.3	3.1	18.3	30.3	51.4
Peru	2008 D	0.069	16.1	4,605	42.7	15.1	3.9	17.9	29.1	53.0
Philippines	2013 D	0.033[f,k]	6.3[f,k]	6,221[f,k]	51.9[f,k]	8.4[f,k]	4.2[f,k]	35.3[f,k]	30.2[f,k]	34.5[f,k]
Philippines	2008 D	0.038[f,k]	7.3[f,k]	6,559[f,k]	51.9[f,k]	12.2[f,k]	5.0[f,k]	37.1[f,k]	25.7[f,k]	37.2[f,k]
Rwanda	2010 D	0.352	70.8	7,669	49.7	17.9	34.6	23.8	27.2	49.0
Rwanda	2005 D	0.481	86.5	8,155	55.6	9.7	60.4	23.3	22.3	54.4
Senegal	2014 D	0.278	51.9	7,556	53.5	18.1	30.8	43.6	23.1	33.4
Senegal	2012/2013 D	0.296	54.8	7,744	54.0	17.8	32.6	41.8	23.6	34.5
Senegal	2010/2011 D	0.390	69.4	9,247	56.2	14.4	45.1	36.7	33.1	30.2
Senegal	2005 D	0.436	71.1	8,018	61.3	11.7	51.6	38.4	26.1	35.5
Serbia	2014 M	0.002	0.4	41	40.6	2.7	0.1	30.7	40.7	28.7
Serbia	2010 M	0.001	0.3	25	39.9	3.1	0.0	24.7	48.6	26.7
Serbia	2005/2006 M	0.011[d]	3.0[d]	296[d]	38.3[d]	3.8[d]	0.3[d]	18.1[d]	60.1[d]	21.8[d]
Sierra Leone	2013 D	0.411	77.5	4,724	53.0	14.6	43.9	25.7	28.5	45.9
Sierra Leone	2010 M	0.405	72.7	4,180	55.8	16.7	46.4	24.2	28.3	47.4
Sierra Leone	2008 D	0.451	79.7	4,409	56.6	12.5	51.7	32.0	22.7	45.3
South Africa	2012 N	0.041	10.3	5,400	39.6	17.1	1.3	8.4	61.4	30.2
South Africa	2008 N	0.039[g]	9.4[g]	4,701[g]	41.5[g]	21.4[g]	1.4[g]	13.4[g]	45.6[g]	41.1[g]
Suriname	2010 M	0.033[d]	7.6[d]	40[d]	43.1[d]	4.7[d]	2.0[d]	31.0[d]	37.2[d]	31.8[d]
Suriname	2006 M	0.044	9.2	46	47.4	6.3	3.6	36.7	21.1	42.2
Syrian Arab Republic	2009 N	0.028	7.2	1,519	39.1	7.4	1.3	54.7	34.0	11.3
Syrian Arab Republic	2006 M	0.024	6.4	1,197	38.0	7.7	0.9	44.4	43.1	12.5
Tajikistan	2012 D	0.031	7.9	629	39.0	23.4	1.2	13.4	52.6	34.0
Tajikistan	2005 M	0.059	14.7	1,002	39.8	18.6	2.3	11.0	57.3	31.7
The former Yugoslav Republic of Macedonia	2011 M	0.007[d]	1.7[d]	36[d]	38.4[d]	2.4[d]	0.1[d]	18.5[d]	57.2[d]	24.3[d]
The former Yugoslav Republic of Macedonia	2005 M	0.013	3.0	64	42.2	7.1	0.7	50.7	22.3	27.0
Togo	2013/2014 D	0.242	48.5	3,394	49.9	19.9	23.2	26.4	28.8	44.9
Togo	2010 M	0.260	50.9	3,207	51.2	20.3	26.4	28.9	25.0	46.1
Togo	2006 M	0.277	53.1	3,021	52.2	20.3	28.8	31.4	23.2	45.4
Uganda	2011 D	0.359	70.3	24,712	51.1	20.6	33.3	18.0	30.2	51.9
Uganda	2006 D	0.399	74.5	22,131	53.6	18.2	41.5	17.1	30.4	52.5
Ukraine	2012 M	0.001[f]	0.4[f]	162[f]	34.5[f]	0.0[f]	0.0[f]	19.0[f]	77.5[f]	3.5[f]
Ukraine	2007 D	0.002[f]	0.6[f]	264[f]	34.3[f]	0.2[f]	0.0[f]	1.0[f]	95.1[f]	3.8[f]
Yemen	2013 D	0.200	40.0	9,754	50.1	22.4	19.4	29.5	32.2	38.2

TABLE 7   Multidimensional Poverty Index : changes over time | 231

TABLE
7

TABLE 7 MULTIDIMENSIONAL POVERTY INDEX: CHANGES OVER TIME

Country	Year and survey[a] 2005–2014	Multidimensional Poverty Index[b] HDRO specifications[c] Value	Population in multidimensional poverty[c] Headcount (%)	Population in multidimensional poverty[c] Headcount (thousands)	Intensity of deprivation (%)	Population near multidimensional poverty[c] (%)	Population in severe multidimensional poverty[c] (%)	Contribution of deprivation to overall poverty (%) Education	Contribution of deprivation to overall poverty (%) Health	Contribution of deprivation to overall poverty (%) Living standards
Yemen	2006 M	0.191[f]	37.5[f]	7,741[f]	50.9[f]	16.7[f]	18.4[f]	33.4[f]	21.3[f]	45.3[f]
Zambia	2013/2014 D	0.264	54.4	8,173	48.6	23.1	22.5	17.9	29.8	52.3
Zambia	2007 D	0.318	62.8	7,600	50.7	18.7	31.3	16.3	29.4	54.3
Zimbabwe	2014 M	0.128	28.9	4,222	44.1	29.3	7.8	10.8	34.5	54.8
Zimbabwe	2010/2011 D	0.181	41.0	5,482	44.1	24.9	12.2	7.8	37.9	54.3
Zimbabwe	2006 D	0.193	42.4	5,399	45.4	22.8	15.7	11.5	29.6	58.9

NOTES

a *D* indicates data from Demographic and Health Surveys, *M* indicates data from Multiple Indicator Cluster Surveys and *N* indicates data from national surveys (see http://hdr.undp.org for the list of national surveys).

b Not all indicators were available for all countries, so caution should be used in cross-country comparisons. Where an indicator is missing, weights of available indicators are adjusted to total 100%. See *Technical note 5* at http://hdr.undp.org for details.

c The HDRO specifications refer to somewhat modified definitions of deprivations in some indicators compared to the 2010 specifications. See *Technical note 5* for details.

d Missing indicator on child mortality.

e The estimates based on national household surveys that refer to different years are not necessarily comparable. Caution should be used in comparing estimates over time.

f Missing indicators on nutrition.

g Missing indicator on type of floor.

h Missing indicator on cooking fuel.

i Refers to only part of the country (nine provinces).

j Missing indicator on electricity.

k Missing indicator on school attendance.

DEFINITIONS

Multidimensional Poverty Index: Percentage of the population that is multidimensionally poor adjusted by the intensity of the deprivations. See *Technical note 5* for details on how the Multidimensional Poverty Index is calculated.

Multidimensional poverty headcount: Population with a weighted deprivation score of at least 33 percent.

Intensity of deprivation of multidimensional poverty: Average percentage of deprivation experienced by people in multidimensional poverty.

Population near multidimensional poverty: Percentage of the population at risk of suffering multiple deprivations—that is, those with a deprivation score of 20–33 percent.

Population in severe multidimensional poverty: Percentage of the population in severe multidimensional poverty—that is, those with a deprivation score of 50 percent or more.

Contribution of deprivation in dimension to overall poverty: Percentage of the Multidimensional Poverty Index attributed to deprivation in each dimension.

MAIN DATA SOURCES

Column 1: Refers to the year and the survey whose data were used to calculate the values in columns 2–10.

Columns 2–10: HDRO calculations based on data on household deprivations in education, health and living standards from various household surveys listed in column 1 using the revised methodology described in *Technical note 5*.

TABLE 7

Human development indicators

TABLE 8

Population trends

		Population									Dependency ratio				Sex ratio at birth[b]
		Total		Average annual growth		Urban[a]	Under age 5	Ages 15–64	Ages 65 and older	Median age	(per 100 people ages 15–64)		Total fertility rate		
		(millions)		(%)		(%)	(millions)			(years)	Young age (0–14)	Old age (65 and older)	(births per woman)		(male to female births)
HDI rank		2014[c]	2030[c]	2000/2005	2010/2015[c]	2014[c]	2014[c]	2014[c]	2014[c]	2015[c]	2015[c]	2015[c]	2000/2005[c]	2010/2015[c]	2010/2015[c]
VERY HIGH HUMAN DEVELOPMENT															
1	Norway[d]	5.1	5.8	0.6	1.0	80.2	0.3	3.3	0.8	39.2	28.6	25.2	1.8	1.9	1.06
2	Australia[e]	23.6	28.3	1.3	1.3	89.6	1.6	15.6	3.5	37.4	29.1	22.7	1.8	1.9	1.06
3	Switzerland	8.2	9.5	0.7	1.0	73.9	0.4	5.5	1.5	42.3	21.9	27.1	1.4	1.5	1.05
4	Denmark	5.6	6.0	0.3	0.4	87.3	0.3	3.6	1.0	41.5	27.0	29.1	1.8	1.9	1.06
5	Netherlands	16.8	17.3	0.6	0.3	84.3	0.9	11.0	3.0	42.4	25.8	27.8	1.7	1.8	1.06
6	Germany	82.7	79.6	0.1	−0.1	74.3	3.5	54.3	17.6	46.3	19.7	32.7	1.4	1.4	1.06
6	Ireland	4.7	5.3	1.8	1.1	63.1	0.4	3.1	0.6	35.9	32.9	19.2	2.0	2.0	1.07
8	United States	322.6	362.6	0.9	0.8	83.1	21.0	213.6	46.2	37.7	29.4	22.2	2.0	2.0	1.05
9	Canada	35.5	40.6	1.0	1.0	81.0	2.0	24.1	5.5	40.5	24.4	23.7	1.5	1.7	1.06
9	New Zealand	4.6	5.2	1.4	1.0	86.4	0.3	3.0	0.7	37.3	30.8	22.5	1.9	2.1	1.06
11	Singapore	5.5	6.6	2.7	2.0	100.0	0.3	4.1	0.6	38.7	20.8	15.2	1.3	1.3	1.07
12	Hong Kong, China (SAR)	7.3	7.9	0.2	0.7	100.0	0.3	5.4	1.1	43.2	16.0	20.5	1.0	1.1	1.07
13	Liechtenstein	0.0	0.0	1.0	0.7	..	..	..	..	..	..	..	..	..	..
14	Sweden	9.6	10.7	0.4	0.7	85.7	0.6	6.1	1.9	41.2	27.6	31.8	1.7	1.9	1.06
14	United Kingdom	63.5	68.6	0.5	0.6	80.0	3.9	41.0	11.3	40.5	27.4	28.1	1.7	1.9	1.05
16	Iceland	0.3	0.4	1.1	1.1	94.1	0.0	0.2	0.0	35.9	31.2	20.3	2.0	2.1	1.05
17	Korea (Republic of)	49.5	52.2	0.5	0.5	84.0	2.4	36.1	6.2	40.5	19.5	17.9	1.2	1.3	1.07
18	Israel	7.8	9.6	1.9	1.3	92.1	0.8	4.8	0.8	30.1	45.8	17.8	2.9	2.9	1.05
19	Luxembourg	0.5	0.6	1.0	1.3	86.1	0.0	0.4	0.1	39.1	25.4	21.2	1.7	1.7	1.05
20	Japan	127.0	120.6	0.2	−0.1	93.0	5.4	77.8	32.8	46.5	21.2	43.6	1.3	1.4	1.06
21	Belgium	11.1	11.7	0.5	0.4	97.6	0.7	7.2	2.0	41.9	26.7	29.0	1.7	1.9	1.05
22	France	64.6	69.3	0.7	0.5	87.4	4.0	41.1	11.8	41.0	28.6	29.6	1.9	2.0	1.05
23	Austria	8.5	9.0	0.5	0.4	68.3	0.4	5.7	1.6	43.3	21.6	27.9	1.4	1.5	1.06
24	Finland[f]	5.4	5.6	0.3	0.3	84.1	0.3	3.5	1.1	42.6	26.1	32.3	1.8	1.9	1.04
25	Slovenia	2.1	2.1	0.1	0.2	49.8	0.1	1.4	0.4	43.0	21.4	26.4	1.2	1.5	1.05
26	Spain[g]	47.1	48.2	1.5	0.4	77.9	2.5	31.3	8.5	42.2	23.4	27.6	1.3	1.5	1.06
27	Italy	61.1	61.2	0.6	0.2	68.9	2.9	39.4	13.1	45.0	21.8	33.8	1.3	1.5	1.06
28	Czech Republic	10.7	11.1	0.0	0.4	73.4	0.6	7.3	1.8	40.9	23.0	26.3	1.2	1.6	1.06
29	Greece	11.1	11.0	0.1	0.0	62.2	0.6	7.3	2.2	43.5	22.6	31.1	1.3	1.5	1.07
30	Estonia	1.3	1.2	−0.6	−0.3	69.7	0.1	0.8	0.2	41.3	24.7	28.2	1.4	1.6	1.06
31	Brunei Darussalam	0.4	0.5	2.1	1.4	77.1	0.0	0.3	0.0	31.1	34.6	6.9	2.3	2.0	1.06
32	Cyprus[h]	1.2	1.3	1.8	1.1	71.1	0.1	0.8	0.1	35.9	23.5	18.1	1.6	1.5	1.07
32	Qatar	2.3	2.8	6.5	5.9	99.2	0.1	1.9	0.0	31.7	15.9	1.1	3.0	2.1	1.05
34	Andorra	0.1	0.1	4.3	0.8	..	..	..	..	..	..	..	..	..	..
35	Slovakia	5.5	5.4	0.0	0.1	54.6	0.3	3.9	0.7	38.9	21.4	19.1	1.2	1.4	1.05
36	Poland	38.2	37.4	−0.1	0.0	60.7	2.0	26.8	5.7	39.4	21.7	22.0	1.3	1.4	1.06
37	Lithuania	3.0	2.8	−1.2	−0.5	67.4	0.2	2.1	0.5	39.7	22.4	22.8	1.3	1.5	1.05
37	Malta	0.4	0.4	0.4	0.3	95.3	0.0	0.3	0.1	41.4	20.8	26.0	1.4	1.4	1.06
39	Saudi Arabia	29.4	35.6	4.1	1.8	82.9	2.9	20.1	0.9	28.4	41.2	4.4	3.5	2.7	1.03
40	Argentina	41.8	46.9	0.9	0.9	93.0	3.4	27.1	4.6	31.6	36.7	17.3	2.4	2.2	1.04
41	United Arab Emirates	9.4	12.3	6.3	2.5	85.2	0.7	7.9	0.0	31.4	19.4	0.6	2.4	1.8	1.05
42	Chile	17.8	19.8	1.1	0.9	89.8	1.2	12.3	1.8	33.7	29.9	15.3	2.0	1.8	1.04
43	Portugal	10.6	10.4	0.4	0.0	62.7	0.5	7.0	2.0	43.0	21.8	29.3	1.5	1.3	1.06
44	Hungary	9.9	9.5	−0.3	−0.2	70.9	0.5	6.7	1.7	41.0	21.9	26.1	1.3	1.4	1.06
45	Bahrain	1.3	1.6	5.5	1.7	88.9	0.1	1.0	0.0	30.2	28.3	3.0	2.7	2.1	1.04
46	Latvia	2.0	1.9	−1.3	−0.6	67.7	0.1	1.4	0.4	41.7	23.5	28.2	1.3	1.6	1.05
47	Croatia	4.3	4.0	−0.4	−0.4	58.7	0.2	2.8	0.8	43.1	22.0	28.6	1.4	1.5	1.06
48	Kuwait	3.5	4.8	3.7	3.6	98.3	0.3	2.5	0.1	29.7	33.6	3.3	2.6	2.6	1.04
49	Montenegro	0.6	0.6	0.2	0.0	63.9	0.0	0.4	0.1	37.6	26.9	20.2	1.8	1.7	1.07
HIGH HUMAN DEVELOPMENT															
50	Belarus	9.3	8.5	−0.6	−0.5	76.3	0.5	6.6	1.3	39.5	22.4	19.7	1.2	1.5	1.06
50	Russian Federation	142.5	133.6	−0.4	−0.2	74.3	8.4	100.8	18.6	38.5	23.4	18.8	1.3	1.5	1.06
52	Oman	3.9	4.9	2.8	7.9	74.2	0.4	2.9	0.1	27.1	29.2	4.0	3.2	2.9	1.05
52	Romania	21.6	20.2	−0.2	−0.3	52.9	1.1	15.1	3.3	40.0	21.8	22.3	1.3	1.4	1.06
52	Uruguay	3.4	3.6	0.0	0.3	92.8	0.2	2.2	0.5	34.8	33.4	22.3	2.2	2.1	1.06
55	Bahamas	0.4	0.4	2.0	1.4	84.8	0.0	0.3	0.0	32.5	29.4	11.7	1.9	1.9	1.06
56	Kazakhstan	16.6	18.6	0.7	1.0	53.3	1.6	11.2	1.1	29.7	39.4	10.1	2.0	2.4	1.07
57	Barbados	0.3	0.3	0.5	0.5	46.0	0.0	0.2	0.0	37.4	26.7	16.2	1.8	1.9	1.04
58	Antigua and Barbuda	0.1	0.1	1.2	1.0	29.8	0.0	0.1	0.0	30.9	35.2	10.4	2.3	2.1	1.03

HDI rank	Total (millions) 2014	Total (millions) 2030	Avg annual growth (%) 2000/2005	Avg annual growth (%) 2010/2015	Urban (%) 2014	Under age 5 (millions) 2014	Ages 15–64 (millions) 2014	Ages 65 and older (millions) 2014	Median age (years) 2015	Dependency ratio Young age (0–14) 2015	Dependency ratio Old age (65 and older) 2015	Total fertility rate 2000/2005	Total fertility rate 2010/2015	Sex ratio at birth 2010/2015
59 Bulgaria	7.2	6.2	−0.8	−0.8	74.8	0.3	4.8	1.4	43.4	21.2	30.1	1.2	1.5	1.06
60 Palau	0.0	0.0	0.8	0.8	..	..	..	..	..	..	..	..	..	..
60 Panama	3.9	4.9	1.9	1.6	77.0	0.4	2.5	0.3	28.5	42.5	11.7	2.8	2.5	1.05
62 Malaysia[i]	30.2	36.8	2.0	1.6	74.8	2.6	20.8	1.7	28.2	36.6	8.3	2.5	2.0	1.06
63 Mauritius[j]	1.2	1.3	0.5	0.4	41.8	0.1	0.9	0.1	35.5	26.4	13.3	1.9	1.5	1.04
64 Seychelles	0.1	0.1	1.8	0.6	54.8	0.0	0.1	0.0	33.2	31.7	11.2	2.2	2.2	1.06
64 Trinidad and Tobago	1.3	1.3	0.5	0.3	14.5	0.1	0.9	0.1	34.2	29.9	13.8	1.8	1.8	1.04
66 Serbia[k]	9.5	8.6	−0.6	−0.5	57.4	0.5	6.6	1.4	39.3	22.9	21.7	1.6	1.4	1.05
67 Cuba	11.3	10.8	0.3	−0.1	75.1	0.5	7.9	1.5	41.3	22.1	19.9	1.6	1.5	1.06
67 Lebanon	5.0	5.2	4.2	3.0	87.6	0.3	3.5	0.4	30.7	27.1	12.3	2.0	1.5	1.05
69 Costa Rica	4.9	5.8	1.9	1.4	66.0	0.4	3.4	0.4	30.6	32.5	10.8	2.3	1.8	1.05
69 Iran (Islamic Republic of)	78.5	91.3	1.2	1.3	69.5	7.1	55.5	4.2	29.5	34.2	7.8	2.0	1.9	1.05
71 Venezuela (Bolivarian Republic of)	30.9	37.2	1.8	1.5	94.1	3.0	20.2	2.0	27.7	42.6	10.1	2.7	2.4	1.05
72 Turkey	75.8	86.8	1.4	1.2	74.3	6.3	50.9	5.7	30.1	37.0	11.4	2.3	2.1	1.05
73 Sri Lanka	21.4	23.3	1.1	0.8	15.3	1.9	14.2	1.9	32.0	38.1	13.7	2.3	2.4	1.04
74 Mexico	123.8	143.7	1.3	1.2	79.0	11.2	81.1	8.1	27.7	41.7	10.3	2.5	2.2	1.05
75 Brazil	202.0	222.7	1.3	0.8	85.4	14.7	138.6	15.7	31.2	33.6	11.6	2.3	1.8	1.05
76 Georgia[l]	4.3	4.0	−1.2	−0.4	53.2	0.3	2.9	0.6	38.1	27.6	22.0	1.6	1.8	1.11
77 Saint Kitts and Nevis	0.1	0.1	1.5	1.1	..	..	..	..	..	..	..	..	..	..
78 Azerbaijan[m]	9.5	10.5	1.1	1.1	54.4	0.8	6.9	0.5	30.4	30.8	7.8	2.0	1.9	1.15
79 Grenada	0.1	0.1	0.3	0.4	40.2	0.0	0.1	0.0	27.2	40.0	10.7	2.4	2.2	1.05
80 Jordan	7.5	9.4	1.9	3.5	83.4	1.0	4.7	0.3	24.0	53.0	5.8	3.9	3.3	1.05
81 The former Yugoslav Republic of Macedonia	2.1	2.1	0.4	0.1	59.7	0.1	1.5	0.3	37.8	23.2	18.3	1.6	1.4	1.05
81 Ukraine	44.9	39.8	−0.8	−0.6	69.5	2.5	31.6	6.7	39.9	21.4	21.2	1.2	1.5	1.06
83 Algeria	39.9	48.6	1.4	1.8	75.5	4.6	26.9	1.8	27.5	42.4	7.0	2.4	2.8	1.05
84 Peru	30.8	36.5	1.3	1.3	78.3	2.9	20.0	2.0	27.1	42.9	10.3	2.8	2.4	1.05
85 Albania	3.2	3.3	−0.7	0.3	56.6	0.2	2.2	0.3	33.5	28.1	16.3	2.2	1.8	1.08
85 Armenia	3.0	3.0	−0.4	0.2	64.2	0.2	2.1	0.3	33.4	29.2	15.0	1.7	1.7	1.14
85 Bosnia and Herzegovina	3.8	3.7	0.2	−0.1	49.9	0.2	2.6	0.6	40.1	21.2	22.9	1.2	1.3	1.07
88 Ecuador	16.0	19.6	1.9	1.6	69.1	1.6	10.2	1.1	26.7	45.8	10.7	3.0	2.6	1.05
89 Saint Lucia	0.2	0.2	1.1	0.8	15.5	0.0	0.1	0.0	31.2	34.1	13.2	2.1	1.9	1.03
90 China	1,393.8	1,453.3	0.6	0.6	54.4	91.0	1,014.3	127.2	36.0	25.1	13.1	1.6	1.7	1.16
90 Fiji	0.9	0.9	0.3	0.7	53.4	0.1	0.6	0.0	27.5	43.9	8.9	3.0	2.6	1.06
90 Mongolia	2.9	3.4	1.0	1.5	71.2	0.3	2.0	0.1	27.5	40.4	5.6	2.1	2.4	1.05
93 Thailand	67.2	67.6	1.0	0.3	35.2	3.6	48.5	6.8	38.0	24.2	14.5	1.6	1.4	1.06
94 Dominica	0.1	0.1	0.2	0.4	..	..	..	..	..	..	..	..	..	..
94 Libya	6.3	7.5	1.6	0.9	78.2	0.6	4.1	0.3	27.2	44.7	7.6	2.9	2.4	1.06
96 Tunisia	11.1	12.6	1.0	1.1	67.0	0.9	7.7	0.8	31.2	33.4	10.8	2.0	2.0	1.05
97 Colombia	48.9	57.2	1.6	1.3	76.1	4.5	32.4	3.1	28.3	40.7	10.0	2.6	2.3	1.05
97 Saint Vincent and the Grenadines	0.1	0.1	0.2	0.0	50.5	0.0	0.1	0.0	29.8	36.0	10.7	2.2	2.0	1.03
99 Jamaica	2.8	2.9	0.8	0.5	52.3	0.2	1.8	0.2	28.2	39.5	12.3	2.5	2.3	1.05
100 Tonga	0.1	0.1	0.6	0.4	23.7	0.0	0.1	0.0	21.3	64.3	10.2	4.2	3.8	1.05
101 Belize	0.3	0.5	2.6	2.4	44.2	0.0	0.2	0.0	23.7	52.1	6.5	3.4	2.7	1.03
101 Dominican Republic	10.5	12.2	1.5	1.2	71.4	1.1	6.7	0.7	26.4	46.4	10.3	2.8	2.5	1.05
103 Suriname	0.5	0.6	1.4	0.9	70.9	0.0	0.4	0.0	29.1	39.6	10.2	2.6	2.3	1.08
104 Maldives	0.4	0.4	1.7	1.9	44.5	0.0	0.2	0.0	26.0	42.2	7.3	2.8	2.3	1.06
105 Samoa	0.2	0.2	0.6	0.8	19.3	0.0	0.1	0.0	21.2	64.9	9.1	4.4	4.2	1.08
MEDIUM HUMAN DEVELOPMENT														
106 Botswana	2.0	2.3	1.3	0.9	63.6	0.2	1.3	0.1	22.8	52.3	6.0	3.2	2.6	1.03
107 Moldova (Republic of)[n]	3.5	3.1	−1.7	−0.8	49.8	0.2	2.5	0.4	36.3	23.6	16.4	1.5	1.5	1.06
108 Egypt	83.4	102.6	1.6	1.6	44.0	9.3	52.7	4.9	25.8	48.8	9.4	3.2	2.8	1.05
109 Turkmenistan	5.3	6.2	1.1	1.3	49.7	0.5	3.6	0.2	26.4	41.7	6.1	2.8	2.3	1.05
110 Gabon	1.7	2.4	2.4	2.4	87.1	0.2	1.0	0.1	20.9	67.6	8.9	4.5	4.1	1.03
110 Indonesia	252.8	293.5	1.4	1.2	53.0	23.3	167.4	13.4	28.4	42.2	8.2	2.5	2.4	1.05
112 Paraguay	6.9	8.7	2.0	1.7	63.5	0.8	4.3	0.4	24.4	50.8	9.1	3.5	2.9	1.05
113 Palestine, State of[o]	4.4	6.4	2.1	2.5	75.0	0.6	2.5	0.1	19.7	67.3	5.3	5.0	4.1	1.05
114 Uzbekistan	29.3	34.1	1.0	1.4	36.3	3.0	19.8	1.3	26.0	41.5	6.4	2.6	2.3	1.05
115 Philippines	100.1	127.8	2.0	1.7	49.6	11.5	62.3	4.0	23.4	53.4	6.5	3.7	3.1	1.06
116 El Salvador	6.4	6.9	0.4	0.7	66.2	0.6	4.1	0.5	24.7	45.2	11.5	2.6	2.2	1.05
116 South Africa	53.1	58.1	1.5	0.8	63.3	5.3	34.5	3.0	26.5	45.1	8.8	2.8	2.4	1.03

TABLE 8 Population trends | 235

TABLE 8

TABLE 8 POPULATION TRENDS

HDI rank		Population Total (millions)		Average annual growth (%)		Urban[a] (%)	Under age 5 (millions)	Ages 15–64 (millions)	Ages 65 and older (millions)	Median age (years)	Dependency ratio (per 100 people ages 15–64) Young age (0–14)	Old age (65 and older)	Total fertility rate (births per woman)		Sex ratio at birth[b] (male to female births)
		2014[c]	2030[c]	2000/2005	2010/2015[c]	2014[c]	2014[c]	2014[c]	2014[c]	2015[c]	2015[c]	2015[c]	2000/2005	2010/2015[c]	2010/2015[c]
116	Viet Nam	92.5	101.8	1.0	1.0	33.0	7.1	65.5	6.2	30.7	31.7	9.6	1.9	1.8	1.10
119	Bolivia (Plurinational State of)	10.8	13.7	1.9	1.6	68.1	1.3	6.6	0.5	22.8	56.1	8.3	4.0	3.3	1.05
120	Kyrgyzstan	5.6	6.9	0.4	1.4	35.6	0.7	3.7	0.2	25.1	47.6	6.3	2.5	3.1	1.06
121	Iraq	34.8	51.0	2.8	2.9	66.4	5.0	19.9	1.1	20.0	68.1	5.5	4.8	4.1	1.07
122	Cabo Verde	0.5	0.6	1.6	0.8	64.9	0.0	0.3	0.0	25.2	42.4	7.9	3.3	2.3	1.03
123	Micronesia (Federated States of)	0.1	0.1	−0.2	0.2	22.9	0.0	0.1	0.0	21.5	55.3	7.1	4.1	3.3	1.07
124	Guyana	0.8	0.9	0.4	0.5	28.6	0.1	0.5	0.0	23.0	55.7	5.7	2.7	2.6	1.05
125	Nicaragua	6.2	7.4	1.3	1.4	58.5	0.7	3.9	0.3	23.8	50.4	7.6	3.0	2.5	1.05
126	Morocco	33.5	39.2	1.0	1.4	58.1	3.6	22.5	1.7	27.5	41.7	7.6	2.5	2.8	1.06
126	Namibia	2.3	3.0	1.3	1.9	40.1	0.3	1.4	0.1	21.8	57.0	5.9	3.8	3.1	1.03
128	Guatemala	15.9	22.6	2.5	2.5	51.1	2.3	8.8	0.7	19.7	71.3	8.4	4.6	3.8	1.05
129	Tajikistan	8.4	11.4	1.9	2.4	26.7	1.2	5.1	0.3	22.0	59.4	5.2	3.7	3.9	1.05
130	India	1,267.4	1,476.4	1.6	1.2	32.4	122.0	835.2	67.9 ·	26.9	42.9	8.3	3.0	2.5	1.11
131	Honduras	8.3	10.8	2.0	2.0	53.9	1.0	5.0	0.4	22.5	56.1	7.5	3.7	3.0	1.05
132	Bhutan	0.8	0.9	2.8	1.6	37.9	0.1	0.5	0.0	26.7	39.9	7.3	3.1	2.3	1.04
133	Timor-Leste	1.2	1.6	3.1	1.7	29.5	0.2	0.6	0.0	16.9	86.5	6.6	7.0	5.9	1.05
134	Syrian Arab Republic	22.0	29.9	2.1	0.7	57.3	2.6	13.4	0.9	22.7	56.4	7.1	3.7	3.0	1.05
134	Vanuatu	0.3	0.4	2.5	2.2	25.8	0.0	0.2	0.0	22.1	60.3	6.7	4.1	3.4	1.07
136	Congo	4.6	6.8	2.5	2.6	64.9	0.8	2.5	0.2	18.7	78.5	6.3	5.1	5.0	1.03
137	Kiribati	0.1	0.1	1.8	1.5	44.3	0.0	0.1	0.0	24.1	47.8	6.7	3.6	3.0	1.07
138	Equatorial Guinea	0.8	1.1	3.1	2.8	40.0	0.1	0.5	0.0	20.9	65.6	4.8	5.6	4.9	1.03
139	Zambia	15.0	25.0	2.5	3.2	40.5	2.7	7.6	0.4	16.7	90.6	5.0	6.0	5.7	1.02
140	Ghana	26.4	35.3	2.5	2.1	53.9	3.7	15.4	0.9	20.9	65.0	5.9	4.6	3.9	1.05
141	Lao People's Democratic Republic	6.9	8.8	1.4	1.9	37.6	0.9	4.2	0.3	22.0	55.6	6.2	3.7	3.1	1.05
142	Bangladesh	158.5	185.1	1.6	1.2	29.9	15.2	104.1	7.6	25.8	43.8	7.3	2.9	2.2	1.05
143	Cambodia	15.4	19.1	1.8	1.7	20.5	1.8	9.8	0.8	25.0	49.0	8.9	3.5	2.9	1.05
143	Sao Tome and Principe	0.2	0.3	2.1	2.6	64.7	0.0	0.1	0.0	19.4	74.8	5.8	4.6	4.1	1.03
LOW HUMAN DEVELOPMENT															
145	Kenya	45.5 ·	66.3	2.7 ·	2.7	25.2	7.1	25.2	1.2	19.0	75.4	5.0	5.0	4.4	1.03
145	Nepal	28.1	32.9	1.7	1.2	18.0	2.9	17.1	1.5	23.1	53.4	8.6	3.7	2.3	1.07
147	Pakistan	185.1	231.7	1.9	1.7	37.2	21.5	115.4	8.1	23.2	52.3	7.0	4.0	3.2	1.09
148	Myanmar	53.7	58.7	0.7	0.8	34.4	4.4	37.7	2.8	29.8	34.4	7.7	2.2	2.0	1.03
149	Angola	22.1	34.8	3.4	3.1	61.5	4.1	11.1	0.5	16.4	92.9	4.8	6.8	5.9	1.03
150	Swaziland	1.3	1.5	0.8	1.5	21.1	0.2	0.7	0.0	20.5	63.1	6.1	4.0	3.4	1.03
151	Tanzania (United Republic of)[p]	50.8	79.4	2.6	3.0	28.1	8.8	26.4	1.6	17.6	85.9	6.2	5.7	5.2	1.03
152	Nigeria	178.5	273.1	2.6	2.8	51.5	31.4	94.4	4.9	17.7	83.9	5.1	6.1	6.0	1.06
153	Cameroon	22.8	33.1	2.6	2.5	53.8	3.7	12.3	0.7	18.5	78.4	5.9	5.5	4.8	1.03
154	Madagascar	23.6	36.0	3.0	2.8	34.5	3.7	13.0	0.7	18.7	75.2	5.1	5.3	4.5	1.03
155	Zimbabwe	14.6	20.3	0.3	2.8	40.1	2.1	8.4	0.6	20.1	66.9	6.7	4.0	3.5	1.02
156	Mauritania	4.0	5.6	3.0	2.5	42.3	0.6	2.3	0.1	20.0	69.4	5.6	5.2	4.7	1.05
156	Solomon Islands	0.6	0.8	2.6	2.1	21.8	0.1	0.3	0.0	19.9	69.4	5.9	4.6	4.1	1.07
158	Papua New Guinea	7.5	10.0	2.5	2.1	12.7	1.0	4.4	0.2	21.2	62.2	5.0	4.4	3.8	1.08
159	Comoros	0.8	1.1	2.6	2.4	28.3	0.1	0.4	0.0	19.1	75.1	5.1	5.3	4.7	1.05
160	Yemen	25.0	34.0	2.8	2.3	34.1	3.5	14.3	0.7	19.7	67.5	5.1	5.9	4.2	1.05
161	Lesotho	2.1	2.4	0.7	1.1	29.8	0.3	1.3	0.1	21.2	59.2	6.9	3.8	3.1	1.03
162	Togo	7.0	10.0	2.6	2.6	39.5	1.1	3.9	0.2	19.0	74.6	4.9	5.1	4.7	1.02
163	Haiti	10.5	12.5	1.5	1.4	57.4	1.3	6.4	0.5	22.7	55.8	7.5	4.0	3.2	1.05
163	Rwanda	12.1	17.8	2.3	2.7	20.0	1.9	6.7	0.3	18.4	74.1	4.5	5.6	4.6	1.02
163	Uganda	38.8	63.4	3.4	3.3	16.8	7.3	19.2	0.9	15.9	96.6	4.9	6.7	5.9	1.03
166	Benin	10.6	15.5	3.3	2.7	46.9	1.7	5.8	0.3	18.6	76.7	5.3	5.8	4.9	1.04
167	Sudan	38.8	55.1	2.6	2.1	33.7	5.8	21.7	1.3	19.4	72.1	5.9	5.3	4.5	1.04
168	Djibouti	0.9	1.1	1.4	1.5	77.3	0.1	0.6	0.0	23.4	53.9	6.6	4.2	3.4	1.04
169	South Sudan	11.7	17.3	3.8	4.0	18.6	1.8	6.4	0.4	18.9	75.3	6.4	5.9	5.0	1.04
170	Senegal	14.5	21.9	2.7	2.9	43.5	2.4	7.8	0.4	18.2	80.5	5.4	5.4	5.0	1.04
171	Afghanistan	31.3	43.5	3.8	2.4	24.5	4.8	16.2	0.8	17.0	85.4	4.7	7.4	5.0	1.06
172	Côte d'Ivoire	20.8	29.2	1.5	2.3	53.5	3.3	11.6	0.7	19.1	73.4	5.7	5.2	4.9	1.03
173	Malawi	16.8	26.0	2.6	2.8	16.1	2.9	8.7	0.5	17.3	86.3	6.3	6.1	5.4	1.03
174	Ethiopia	96.5	137.7	2.9	2.6	17.8	14.4	52.6	3.3	18.6	75.2	6.3	6.1	4.6	1.04
175	Gambia	1.9	3.1	3.1	3.2	58.9	0.3	1.0	0.0	17.0	87.9	4.5	5.9	5.8	1.03
176	Congo (Democratic Republic of the)	69.4	103.7	2.8	2.7	35.9	12.1	36.3	2.0	17.5	84.7	5.4	6.9	6.0	1.03

HDI rank	Total (millions) 2014	Total (millions) 2030	Avg annual growth (%) 2000/2005	Avg annual growth (%) 2010/2015	Urban (%) 2014	Under age 5 (millions) 2014	Ages 15–64 (millions) 2014	Ages 65 and older (millions) 2014	Median age (years) 2015	Dependency ratio Young age (0–14) 2015	Dependency ratio Old age (65 and older) 2015	Total fertility rate 2000/2005	Total fertility rate 2010/2015	Sex ratio at birth 2010/2015
177 Liberia	4.4	6.4	2.5	2.6	49.3	0.7	2.4	0.1	18.6	77.4	5.5	5.7	4.8	1.05
178 Guinea-Bissau	1.7	2.5	2.2	2.4	46.0	0.3	1.0	0.1	19.3	73.3	5.3	5.7	5.0	1.03
179 Mali	15.8	26.0	3.0	3.0	36.9	3.0	7.8	0.4	16.2	95.5	5.4	6.8	6.9	1.05
180 Mozambique	26.5	38.9	2.8	2.5	32.0	4.5	13.6	0.9	17.3	87.4	6.4	5.7	5.2	1.03
181 Sierra Leone	6.2	8.1	4.3	1.9	40.4	0.9	3.5	0.2	19.3	72.4	4.7	5.7	4.8	1.02
182 Guinea	12.0	17.3	1.8	2.5	36.9	1.9	6.6	0.4	18.8	75.9	5.6	5.8	5.0	1.02
183 Burkina Faso	17.4	26.6	2.9	2.8	29.0	3.0	9.1	0.4	17.3	85.6	4.6	6.4	5.7	1.05
184 Burundi	10.5	16.4	3.0	3.2	11.8	2.0	5.5	0.2	17.6	85.3	4.5	6.9	6.1	1.03
185 Chad	13.2	20.9	3.8	3.0	22.1	2.5	6.5	0.3	15.9	96.3	4.8	7.2	6.3	1.03
186 Eritrea	6.5	9.8	4.2	3.2	22.7	1.1	3.6	0.2	18.5	78.8	4.3	5.7	4.7	1.05
187 Central African Republic	4.7	6.3	1.7	2.0	39.8	0.7	2.7	0.2	20.0	68.7	6.7	5.3	4.4	1.03
188 Niger	18.5	34.5	3.6	3.9	18.6	3.8	8.8	0.5	15.0	106.0	5.5	7.7	7.6	1.05
OTHER COUNTRIES OR TERRITORIES														
Korea (Democratic People's Rep. of)	25.0	26.7	0.8	0.5	60.7	1.7	17.3	2.4	33.9	30.5	13.8	2.0	2.0	1.05
Marshall Islands	0.1	0.1	0.0	0.2	..	..	..	..	..	..	..	..	..	..
Monaco	0.0	0.0	1.0	0.8	..	..	..	..	..	..	..	..	..	..
Nauru	0.0	0.0	0.1	0.2	..	..	..	..	..	..	..	..	..	..
San Marino	0.0	0.0	2.0	0.6	..	..	..	..	..	..	..	..	..	..
Somalia	10.8	16.9	2.7	2.9	39.2	2.0	5.4	0.3	16.5	92.6	5.6	7.4	6.6	1.03
Tuvalu	0.0	0.0	0.6	0.2	..	..	..	..	..	..	..	..	..	..
Human development groups														
Very high human development	1,185.3	1,266.3	0.7	0.6	81.9	69.1	783.5	197.4	40.2	26.1	25.8	1.7	1.8	1.05
High human development	2,516.7	2,676.6	0.7	0.7	62.3	178.5	1,782.2	223.9	34.2	28.7	13.0	1.8	1.8	1.06
Medium human development	2,288.2	2,712.0	1.6	1.3	38.7	228.9	1,493.3	119.0	26.5	44.6	8.1	3.0	2.6	1.05
Low human development	1,185.2	1,692.9	2.5	2.4	34.8	181.1	660.6	39.5	19.5	72.6	6.0	5.3	4.6	1.04
Developing countries	5,962.5	7,091.5	1.4	1.3	47.9	591.3	3,912.5	369.4	28.1	42.7	9.6	2.8	2.7	1.04
Regions														
Arab States	373.1	481.3	2.2	2.0	58.1	44.4	236.4	15.9	24.6	50.8	6.8	3.6	3.2	1.05
East Asia and the Pacific	2,051.5	2,211.9	0.8	0.8	51.8	149.5	1,456.1	166.1	33.7	29.5	11.8	1.8	1.9	1.05
Europe and Central Asia	234.9	251.0	0.4	0.7	60.9	18.8	160.1	21.3	32.2	33.4	13.4	2.0	2.0	1.07
Latin America and the Caribbean	618.0	711.1	1.3	1.1	79.8	53.6	408.3	45.6	29.0	39.4	11.4	2.5	2.2	1.05
South Asia	1,771.5	2,085.5	1.6	1.3	33.7	175.5	1,158.5	92.0	26.4	44.2	8.1	3.1	2.6	1.06
Sub-Saharan Africa	911.9	1,348.9	2.6	2.7	37.8	149.4	492.1	28.4	18.5	78.9	5.8	5.7	5.1	1.03
Least developed countries	919.1	1,287.0	2.4	2.3	29.8	133.9	521.4	32.6	20.2	69.1	6.2	5.0	4.2	1.04
Small island developing states	54.9	63.4	1.3	1.1	53.3	5.4	34.9	3.8	27.9	45.4	11.0	3.1	2.7	1.06
Organisation for Economic Co-operation and Development	1,272.4	1,366.6	0.7	0.6	80.6	77.8	837.3	202.0	39.0	27.8	24.7	1.8	1.8	1.06
World	7,243.8 T	8,424.9 T	1.2 T	1.1 T	53.5 T	663.0 T	4,765.8 T	586.3 T	30.2 T	39.6 T	12.5 T	2.6 T	2.5 T	1.05 T

NOTES

a Because data are based on national definitions of what constitutes a city or metropolitan area, cross-country comparison should be made with caution.

b The natural sex ratio at birth is commonly assumed and empirically confirmed to be 1.05 male births to 1 female birth.

c Projections based on medium-fertility variant.

d Includes Svalbard and Jan Mayen Islands.

e Includes Christmas Island, Cocos (Keeling) Islands and Norfolk Island.

f Includes Åland Islands.

g Includes Canary Islands, Ceuta and Melilla.

h Includes Northern Cyprus.

i Includes Sabah and Sarawak.

j Includes Agalega, Rodrigues and Saint Brandon

k Includes Kosovo.

l Includes Abkhazia and South Ossetia.

m Includes Nagorno-Karabakh.

n Includes Transnistria.

o Includes East Jerusalem.

p Includes Zanzibar.

T From original data source.

DEFINITIONS

Total population: De facto population in a country, area or region as of 1 July.

Population average annual growth: Average annual exponential growth rate for the period specified.

Urban population: De facto population living in areas classified as urban according to the criteria used by each country or area as of 1 July.

Population under age 5: De facto population in a country, area or region under age 5 as of 1 July.

Population ages 15–64: De facto population in a country, area or region ages 15–64 as of 1 July.

Population ages 65 and older: De facto population in a country, area or region ages 65 and older as of 1 July.

Median age: Age that divides the population distribution into two equal parts—that is, 50 percent of the population is above that age and 50 percent is below it.

Young age dependency ratio: Ratio of the population ages 0–14 to the population ages 15–64, expressed as the number of dependants per 100 people of working age (ages 15–64).

Old age dependency ratio: Ratio of the population ages 65 and older to the population ages 15–64, expressed as the number of dependants per 100 people of working age (ages 15–64).

Total fertility rate: Number of children that would be born to a woman if she were to live to the end of her child-bearing years and bear children at each age in accordance with prevailing age-specific fertility rates.

Sex ratio at birth: Number of male births per female birth.

MAIN DATA SOURCES

Columns 1–4 and 6–14: UNDESA (2013a).

Column 5: UNDESA (2014).

TABLE 8

TABLE 8   Population trends | 237

TABLE 9

Health outcomes

HDI rank		Infants exclusively breastfed (% ages 0–5 months) 2008–2013ᵃ	Infants lacking immunization (% of one-year-olds) DTP 2013	Measles 2013	Mortality rates (per 1,000 live births) Infant 2013	Under-five 2013	Child malnutrition (% under age 5) Stunting (moderate or severe) 2008–2013ᵃ	Adult mortality rate (per 1,000 people) Female 2013	Male 2013	Deaths due to (per 100,000 people) Malaria 2012	Tuberculosis 2012	HIV prevalence, adult (% ages 15–49) 2013	Life expectancy at age 60 (years) 2010/2015ᵇ	Physicians (per 10,000 people) 2001–2013ᵃ	Public health expenditure (% of GDP) 2013
VERY HIGH HUMAN DEVELOPMENT															
1	Norway	..	1	7	2.3	2.8	..	47	73	..	0.1	..	24.0	37.4	9.6
2	Australia	..	8	6	3.4	4.0	..	45	78	..	0.2	0.2	25.1	32.7	9.0
3	Switzerland	..	2	7	3.6	4.2	..	40	66	..	0.2	0.3	25.0	39.4	11.5
4	Denmark	..	3	11	2.9	3.5	..	60	100	..	0.4	0.2	22.4	34.2	10.6
5	Netherlands	..	1	4	3.3	4.0	..	54	69	..	0.2	..	23.5	31.5	12.9
6	Germany	..	2	3	3.2	3.9	1.3ᶜ	50	92	..	0.4	..	23.5	38.1	11.3
6	Ireland	..	2	7	3.2	3.8	..	49	82	..	0.4	..	23.4	27.2	8.9
8	United States	..	2	9	5.9	6.9	2.1	76	128	..	0.1	..	23.2	24.5	17.1
9	Canada	..	2	5	4.6	5.2	..	52	81	..	0.2	..	24.4	20.7	10.9
9	New Zealand	..	7	8	5.2	6.3	..	52	80	..	0.1	..	24.1	27.4	9.7
11	Singapore	..	2	5	2.2	2.8	4.4ᶜ	38	69	..	1.7	..	24.5	19.2	4.6
12	Hong Kong, China (SAR)	..	..	..	..	..	..	..	..	..	2.6	..	25.4	..	..
13	Liechtenstein	..	..	..	..	..	..	..	..	..	..	..	..	..	..
14	Sweden	..	1	3	2.4	3.0	..	43	69	..	0.1	..	24.1	32.7	9.7
14	United Kingdom	..	2	5	3.9	4.6	..	55	88	..	0.5	0.3	23.5	27.9	9.1
16	Iceland	..	3	9	1.6	2.1	..	37	61	..	0.3	..	24.3	34.8	9.1
17	Korea (Republic of)	..	1	1	3.2	3.7	2.5	38	93	0.0	5.4	..	24.0	21.4	7.2
18	Israel	..	5	3	3.2	4.0	..	41	72	..	0.2	..	24.3	33.5	7.2
19	Luxembourg	..	1	5	1.6	2.0	..	50	79	..	0.4	..	23.4	28.2	7.1
20	Japan	..	1	5	2.1	2.9	..	42	81	..	1.7	..	26.1	23.0	10.3
21	Belgium	..	1	8	3.5	4.4	..	57	98	..	0.4	..	23.6	29.9	11.2
22	France	..	1	11	3.5	4.2	..	52	109	..	0.5	..	25.1	31.8	11.7
23	Austria	..	7	24	3.2	3.9	..	46	91	..	0.4	..	23.9	48.3	11.0
24	Finland	..	1	3	2.1	2.6	..	51	114	..	0.3	..	23.8	29.1	9.4
25	Slovenia	..	2	6	2.3	2.9	..	49	112	..	1.0	..	22.8	25.2	9.2
26	Spain	..	2	5	3.6	4.2	..	40	86	..	0.6	0.4	24.8	37.0	8.9
27	Italy	..	1	10	3.0	3.6	..	38	69	..	0.4	0.3	24.7	40.9	9.1
28	Czech Republic	..	1	1	2.9	3.6	2.6ᶜ	57	127	..	0.4	0.1	21.1	36.2	7.2
29	Greece	..	1	1	3.7	4.4	..	41	98	..	0.7	..	23.5	43.8	9.8
30	Estonia	..	4	6	2.7	3.4	..	64	195	..	2.8	1.3	20.2	32.6	5.7
31	Brunei Darussalam	..	4	1	8.4	9.9	..	69	101	..	3.0	..	21.4	15.0	2.5
32	Cyprus	..	1	14	2.8	3.6	..	36	75	..	0.2	0.1	22.0	22.9	7.4
32	Qatar	..	1	1	7.0	8.2	11.6ᶜ	50	72	..	0.2	..	21.2	77.4	2.2
34	Andorra	..	2	5	2.2	3.0	..	43	90	..	0.9	..	..	37.0	8.1
35	Slovakia	..	1	2	6.0	7.2	..	67	168	..	0.6	..	19.8	30.0	8.2
36	Poland	..	1	2	4.5	5.2	..	70	186	..	1.8	..	21.1	22.0	6.7
37	Lithuania	..	3	7	4.0	4.9	..	88	254	..	3.0	..	19.1	41.2	6.2
37	Malta	..	1	1	5.3	6.1	..	41	75	..	0.4	..	22.3	35.0	8.7
39	Saudi Arabia	..	2	2	13.4	15.5	9.3ᶜ	67	89	0.0	3.9	..	19.2	7.7	3.2
40	Argentina	32.7	7	9	11.9	13.3	8.2ᶜ	83	151	..	1.3	..	21.4	31.6	7.3
41	United Arab Emirates	..	6	6	7.0	8.2	..	59	84	..	0.1	..	19.8	19.3	3.2
42	Chile	..	8	10	7.1	8.2	1.8	55	107	..	1.2	0.3	23.6	10.2	7.7
43	Portugal	..	1	2	3.1	3.8	..	48	111	..	1.3	..	23.2	34.2	9.7
44	Hungary	..	1	1	5.2	6.1	..	91	201	..	0.7	..	19.9	29.6	8.0
45	Bahrain	..	1	1	5.2	6.1	13.6ᶜ	54	70	..	0.3	..	19.5	9.1	4.9
46	Latvia	..	4	4	7.4	8.4	..	85	224	..	2.6	..	19.1	28.8	5.7
47	Croatia	..	2	6	3.8	4.5	..	58	135	..	1.4	..	20.6	28.4	7.3
48	Kuwait	..	1	1	8.1	9.5	4.3	42	59	..	0.9	..	17.6	17.9	2.9
49	Montenegro	19.3ᶜ	2	12	4.9	5.3	9.4	79	152	..	0.2	..	19.2	19.8	6.5
HIGH HUMAN DEVELOPMENT															
50	Belarus	19.0	1	1	3.7	4.9	4.5ᶜ	100	299	..	6.0	0.5	17.1	37.6	6.1
50	Russian Federation	..	3	2	8.6	10.1	..	126	339	..	13.0	..	17.5	43.1	6.5
52	Oman	..	1	1	9.8	11.4	9.8	73	116	..	0.9	..	20.5	22.2	2.6
52	Romania	15.8ᶜ	4	8	10.5	12.0	12.8ᶜ	81	205	..	5.6	0.1	19.4	23.9	5.3
52	Uruguay	..	2	4	9.5	11.1	10.7	79	148	..	1.5	0.7	21.8	37.4	8.8
55	Bahamas	..	1	8	10.4	12.9	..	88	141	..	0.4	3.2	22.3	28.2	7.3
56	Kazakhstan	31.8	1	1	14.6	16.3	13.1	146	322	..	7.8	..	16.5	35.8	4.3
57	Barbados	19.7	7	10	13.3	14.4	..	65	116	..	0.7	0.9	19.5	18.1	6.8

	Infants exclusively breastfed	Infants lacking immunization (% of one-year-olds)		Mortality rates (per 1,000 live births)		Child malnutrition (% under age 5)	Adult mortality rate (per 1,000 people)		Deaths due to (per 100,000 people)		HIV prevalence, adult	Life expectancy at age 60	Physicians	Public health expenditure
	(% ages 0–5 months)	DTP	Measles	Infant	Under-five	Stunting (moderate or severe)	Female	Male	Malaria	Tuberculosis	(% ages 15–49)	(years)	(per 10,000 people)	(% of GDP)
HDI rank	2008–2013[a]	2013	2013	2013	2013	2008–2013[a]	2013	2013	2012	2012	2013	2010/2015[b]	2001–2013[a]	2013
58 Antigua and Barbuda	..	1	2	7.7	9.3	..	145	201	..	1.4	..	21.5	..	4.9
59 Bulgaria	..	4	6	10.1	11.6	8.8[c]	83	189	..	2.0	..	18.8	38.1	7.6
60 Palau	..	1	1	15.1	17.5	..	106	156	..	4.4	..	..	13.8	9.9
60 Panama	..	7	8	15.4	17.9	19.1	81	149	0.0	4.9	0.6	23.9	15.5	7.2
62 Malaysia	..	1	5	7.2	8.5	17.2[c]	86	169	1.0	5.4	0.4	19.0	12.0	4.0
63 Mauritius	21.0[c]	1	1	12.5	14.3	13.6[c]	95	202	..	1.0	1.1	19.3	10.6	4.8
64 Seychelles	..	1	3	12.2	14.2	7.7[c]	99	214	..	1.8	..	19.4	15.1	4.0
64 Trinidad and Tobago	12.8[c]	6	9	19.0	21.3	5.3[c]	129	229	..	2.1	1.7	17.8	11.8	5.3
66 Serbia	13.7	2	8	5.8	6.6	6.6	84	172	..	1.5	0.1	18.7	21.1	10.6
67 Cuba	48.6	2	1	5.0	6.2	7.0[c]	73	115	..	0.3	0.2	22.9	67.2	8.8
67 Lebanon	14.8	16	21	7.8	9.1	16.5[c]	46	70	..	1.5	..	22.7	32.0	7.2
69 Costa Rica	32.5	2	9	8.4	9.6	5.6	64	111	0.0	0.8	0.2	23.8	11.1	9.9
69 Iran (Islamic Republic of)	53.1	2	2	14.4	16.8	6.8	83	153	0.0	2.9	0.1	19.9	8.9	6.7
71 Venezuela (Bolivarian Republic of)	..	10	15	12.9	14.9	13.4	88	198	2.2	2.4	0.6	21.1	19.4	3.4
72 Turkey	41.6[c]	1	2	16.5	19.2	12.3	73	147	0.0	0.5	..	20.9	17.1	5.6
73 Sri Lanka	75.8[c]	1	1	8.2	9.6	14.7	75	184	0.0	1.1	0.1[d]	19.6	6.8	3.2
74 Mexico	14.4	10	11	12.5	14.5	13.6	93	174	0.0	1.8	0.2	22.7	21.0	6.2
75 Brazil	38.6[c]	1	1	12.3	13.7	7.1[c]	97	197	0.6	2.5	0.5	21.8	18.9	9.7
76 Georgia	54.8	1	4	11.7	13.1	11.3	66	174	0.0	4.5	0.3	19.8	42.4	9.4
77 Saint Kitts and Nevis	..	1	1	7.8	10.2	..	79	165	..	2.5	..	..	11.7	6.3
78 Azerbaijan	11.8[c]	5	2	29.9	34.2	26.8[c]	83	167	0.1	4.2	0.2	18.3	34.3	5.6
79 Grenada	..	1	6	10.7	11.8	..	120	194	..	1.0	..	18.5	6.6	6.3
80 Jordan	22.7	1	3	16.0	18.7	7.8	96	131	..	0.5	..	19.0	25.6	7.2
81 The former Yugoslav Republic of Macedonia	23.0	1	4	5.8	6.6	4.9	71	134	..	0.8	0.1[d]	19.1	26.2	6.4
81 Ukraine	19.7	10	21	8.6	10.0	3.7[c]	114	295	..	13.0	0.8	17.4	35.3	7.8
83 Algeria	6.9[c]	1	5	21.6	25.2	15.9[c]	121	164	0.0	15.0	0.1	17.9	12.1	6.6
84 Peru	72.3	3	15	12.9	16.7	18.4	90	116	0.7	5.1	0.3	21.5	11.3	5.3
85 Albania	38.6	1	1	13.3	14.9	23.1	85	118	..	0.3	0.1[d]	21.1	11.5	5.9
85 Armenia	34.6	3	3	14.0	15.6	20.8	95	227	..	6.3	0.2	20.0	26.9	4.5
85 Bosnia and Herzegovina	18.5	5	6	5.7	6.6	8.9	64	138	..	5.2	..	20.2	16.9	9.6
88 Ecuador	40.0[c]	1	3	19.1	22.5	25.3	85	157	0.0	2.7	0.4	23.6	16.9	6.4[e]
89 Saint Lucia	..	1	1	12.7	14.5	2.5	85	177	..	1.2	..	21.0	4.7	8.5
90 China	27.6[c]	1	1	10.9	12.7	9.4	76	103	0.0	3.2	..	19.5	14.6	5.6
90 Fiji	39.8[c]	1	6	20.0	23.6	7.5[c]	143	239	..	1.7	0.1	17.0	4.3	4.1
90 Mongolia	65.7	2	3	26.4	31.8	15.9	148	309	..	7.2	0.1[d]	16.3	27.6	6.0
93 Thailand	12.3	1	1	11.3	13.1	16.3	90	177	0.9	14.0	1.1	21.4	3.9	4.6
94 Dominica	..	2	7	10.2	11.4	..	116	219	..	2.0	..	..	15.9	6.0
94 Libya	..	1	2	12.4	14.5	21.0[c]	80	117	..	6.8	..	19.7	19.0	4.3
96 Tunisia	8.5	1	6	13.1	15.2	10.1	69	130	..	2.9	0.1	20.2	12.2	7.1
97 Colombia	42.8	3	8	14.5	16.9	12.7	73	148	0.9	1.6	0.5	21.3	14.7	6.8
97 Saint Vincent and the Grenadines	..	4	1	17.2	19.0	..	110	169	..	2.6	..	19.7	5.3	5.2
99 Jamaica	23.8	3	6	14.3	16.6	4.8	105	173	..	0.2	1.8	21.3	4.1	5.8
100 Tonga	52.2	1	1	10.4	12.1	2.2[c]	245	115	..	2.5	..	18.6	5.6	4.7
101 Belize	14.7	4	1	14.3	16.7	19.3	78	145	0.0	4.3	1.5	21.5	8.3	5.4
101 Dominican Republic	6.7	9	21	23.6	28.1	10.1[c]	146	160	0.1	4.4	0.7	21.9	14.9	5.4
103 Suriname	2.8	5	7	20.3	22.8	8.8	94	171	1.2	2.6	0.9	18.5	9.1	4.8
104 Maldives	47.8	1	1	8.4	9.9	20.3	55	86	..	2.0	0.1[d]	21.0	14.2	10.8
105 Samoa	51.3	1	1	15.5	18.1	6.4[c]	97	166	..	3.2	..	18.9	4.5	7.5
MEDIUM HUMAN DEVELOPMENT														
106 Botswana	20.3[c]	2	6	36.3	46.6	31.4[c]	254	321	0.4	21.0	21.9	16.4	3.4	5.4
107 Moldova (Republic of)	36.4	4	9	13.3	15.4	11.3[c]	106	277	..	18.0	0.6	16.2	28.6	11.8
108 Egypt	53.2[c]	3	4	18.6	21.8	30.7	117	193	..	0.5	0.1[d]	17.5	28.3	5.1
109 Turkmenistan	13.0[c]	1	1	46.6	55.2	28.1[c]	200	376	..	8.4	..	17.0	41.8	2.0
110 Gabon	6.0	20	30	39.1	56.1	17.5	235	296	67.4	44.0	3.9	18.2	2.9	3.8
110 Indonesia	41.5	2	16	24.5	29.3	36.4	121	176	9.8	27.0	0.5	17.8	2.0	3.1
112 Paraguay	24.4[c]	7	8	18.7	21.9	17.5[c]	96	178	0.0	3.0	0.4	20.8	11.1	9.0
113 Palestine, State of	28.8	1	1	18.6	21.8	10.9	..	..	..	0.2	..	18.7	..	..
114 Uzbekistan	26.4[c]	1	3	36.7	42.5	19.6[c]	130	210	..	2.1	0.2	18.3	23.8	6.1
115 Philippines	34.0[c]	2	10	23.5	29.9	30.3	136	255	0.1	24.0	..	17.0	11.5	4.4

TABLE 9
TABLE 9   Health outcomes | 239

TABLE 9 HEALTH OUTCOMES

HDI rank	Infants exclusively breastfed (% ages 0–5 months) 2008–2013[a]	Infants lacking immunization (% of one-year-olds) DTP 2013	Infants lacking immunization (% of one-year-olds) Measles 2013	Mortality rates (per 1,000 live births) Infant 2013	Mortality rates (per 1,000 live births) Under-five 2013	Child malnutrition (% under age 5) Stunting (moderate or severe) 2008–2013[a]	Adult mortality rate (per 1,000 people) Female 2013	Adult mortality rate (per 1,000 people) Male 2013	Deaths due to (per 100,000 people) Malaria 2012	Deaths due to (per 100,000 people) Tuberculosis 2012	HIV prevalence, adult (% ages 15–49) 2013	Life expectancy at age 60 (years) 2010/2015[b]	Physicians (per 10,000 people) 2001–2013[a]	Public health expenditure (% of GDP) 2013
116 El Salvador	31.4[c]	3	6	13.5	15.7	20.6	136	290	0.0	1.0	0.5	22.0	16.0	6.9
116 South Africa	8.3[c]	31	34	32.8	43.9	23.9	320	441	2.2	59.0	19.1	16.0	7.8	8.9
116 Viet Nam	17.0	17	2	19.0	23.8	23.3	69	189	0.2	20.0	0.4	22.4	11.6	6.0
119 Bolivia (Plurinational State of)	60.4[c]	2	5	31.2	39.1	27.2	172	247	0.1	21.0	0.2	18.6	4.7	6.1
120 Kyrgyzstan	56.1	2	1	21.6	24.2	17.8	130	272	0.0	9.5	0.2	16.8	19.6	6.7
121 Iraq	19.6	18	37	28.0	34.0	22.6	104	203	..	2.9	..	17.5	6.1	5.2
122 Cabo Verde	59.6[c]	7	9	21.9	26.0	21.4[c]	68	144	0.0	23.0	0.5	19.9	3.0	4.4
123 Micronesia (Federated States of)	..	2	9	29.8	36.4	..	154	181	..	24.0	..	17.3	1.8	12.6
124 Guyana	33.2	2	1	29.9	36.6	19.5	256	377	23.6	15.0	1.4	16.6	2.1	6.5
125 Nicaragua	30.6[c]	1	1	20.0	23.5	23.0[c]	116	200	0.1	3.1	0.2	22.2	3.7	8.3
126 Morocco	31.0[c]	1	1	26.1	30.4	14.9	121	170	..	9.2	0.2	17.9	6.2	6.0
126 Namibia	23.9[c]	6	18	35.2	49.8	29.6[c]	177	255	0.1	14.0	14.3	17.3	3.7	7.7
128 Guatemala	49.6	3	15	25.8	31.0	48.0	126	236	0.0	2.1	0.6	21.5	9.3	6.5
129 Tajikistan	34.3	2	8	40.9	47.7	26.8	153	176	0.0	7.6	0.3	18.2	19.0	6.8
130 India	46.4[c]	12	26	41.4	52.7	47.9[c]	158	239	4.1	22.0	0.3	17.0	7.0	4.0
131 Honduras	31.2	12	11	18.9	22.2	22.7	120	173	0.1	2.9	0.5	22.1	3.7	8.6
132 Bhutan	48.7	3	6	29.7	36.2	33.6	212	219	0.0	14.0	0.1	19.5	2.6	3.6
133 Timor-Leste	51.5	14	30	46.2	54.6	57.7	164	208	16.2	74.0	..	16.9	0.7	1.3
134 Syrian Arab Republic	42.6	45	39	11.9	14.6	27.5	73	116	..	2.1	..	19.9	15.0	3.3
134 Vanuatu	40.0[c]	22	48	14.6	16.9	25.9[c]	113	161	3.7	7.9	..	18.0	1.2	3.9
136 Congo	20.5	10	35	35.6	49.1	25.0	280	323	103.8	42.0	2.5	17.1	1.0	4.1
137 Kiribati	69.0	5	9	45.1	58.2	..	134	206	..	17.0	..	17.4	3.8	10.1
138 Equatorial Guinea	7.4	45	58	69.3	95.8	26.2	319	368	69.3	0.0	..	15.9	3.0	3.5
139 Zambia	60.9[c]	14	20	55.8	87.4	45.8[c]	303	356	79.2	28.0	12.5	17.0	0.7	5.0
140 Ghana	45.7	6	11	52.3	78.4	22.7	222	261	67.0	6.9	1.3	15.5	1.0	5.4
141 Lao People's Democratic Republic	40.4	11	18	53.8	71.4	43.8	158	197	9.5	11.0	0.1	17.1	1.8	2.0
142 Bangladesh	64.1	1	7	33.2	41.1	41.4	126	156	13.9	45.0	0.1[d]	18.4	3.6	3.7
143 Cambodia	73.5	5	10	32.5	37.9	40.9	157	210	3.7	63.0	0.7	23.8	2.3	7.5
143 Sao Tome and Principe	51.4	1	9	36.7	51.0	31.6	168	221	42.5	16.0	0.6	18.2	4.9	6.9
LOW HUMAN DEVELOPMENT														
145 Kenya	31.9	18	7	47.5	70.7	35.2	250	299	49.6	22.0	6.0	17.8	1.8	4.5
145 Nepal	69.6	6	12	32.2	39.7	40.5	159	192	0.2	20.0	0.2	17.1	2.1	6.0
147 Pakistan	37.7	21	39	69.0	85.5	45.0	155	189	1.8	34.0	0.1	17.4	8.3	2.8
148 Myanmar	23.6	10	14	39.8	50.5	35.1	183	240	11.3	48.0	0.6	16.6	6.1	1.8
149 Angola	..	3	9	101.6	167.4	29.2[c]	322	372	100.9	42.0	2.4	15.7	1.7	3.8
150 Swaziland	44.1	1	15	55.9	80.0	31.0	496	515	1.2	63.0	27.4	16.3	1.7	8.4
151 Tanzania (United Republic of)	49.8	1	1	36.4	51.8	34.8	244	314	50.5	13.0	5.0	17.9	0.1	7.3
152 Nigeria	17.4	37	41	74.3	117.4	36.4	325	357	106.6	16.0	3.2	13.7	4.1	3.9
153 Cameroon	20.4	5	17	60.8	94.5	32.6	341	370	64.7	29.0	4.3	16.4	0.8	5.1
154 Madagascar	41.9	20	37	39.6	56.0	49.2	208	257	41.4	46.0	0.4	16.9	1.6	4.2
155 Zimbabwe	31.4	2	7	55.0	88.5	32.3	288	385	18.4	33.0	15.0	18.8	0.6	..
156 Mauritania	26.9	5	20	67.1	90.1	22.0	187	234	67.2	93.0	..	16.4	1.3	3.8
156 Solomon Islands	73.7[c]	12	24	25.1	30.1	32.8[c]	162	203	5.5	15.0	..	16.9	2.2	5.1
158 Papua New Guinea	56.1[c]	12	30	47.3	61.4	49.5	243	319	40.3	54.0	0.6	14.9	0.5	4.5
159 Comoros	12.1	15	18	57.9	77.9	32.1	234	281	70.4	6.3	..	15.9	1.5	5.8
160 Yemen	10.3	6	22	40.4	51.3	46.6	211	255	10.0	5.6	0.1[d]	16.2	2.0	5.4
161 Lesotho	53.5	3	8	73.0	98.0	39.0	492	577	..	17.0	22.9	15.5	0.5	11.5
162 Togo	62.4	6	28	55.8	84.7	29.8	279	323	82.8	8.7	2.3	14.5	0.5	8.6
163 Haiti	39.7	14	35	54.7	72.8	21.9	221	263	5.1	25.0	2.0	17.2	2.4	9.4
163 Rwanda	84.9	1	3	37.1	52.0	44.3	196	246	33.2	10.0	2.9	17.8	0.6	11.1
163 Uganda	63.2	11	18	43.8	66.1	33.7	307	380	57.9	13.0	7.4	17.5	1.2	9.8
166 Benin	32.5	15	37	56.2	85.3	44.7[c]	238	284	79.6	9.4	1.1	15.6	0.6	4.6
167 Sudan	41.0	1	15	51.2	76.6	35.0	212	274	16.5	22.0	0.2	17.4	2.8	6.5
168 Djibouti	1.3[c]	13	20	57.4	69.6	33.5	245	286	27.9	76.0	0.9	17.5	2.3	8.9
169 South Sudan	45.1	43	70	64.1	99.2	31.1	323	353	55.4	30.0	2.2	16.4	..	2.2
170 Senegal	39.0	4	16	43.9	55.3	19.2	192	244	59.5	20.0	0.5	16.2	0.6	4.2
171 Afghanistan	..	14	25	70.2	97.3	59.3[c]	232	252	0.2	37.0	0.1[d]	15.9	2.3	8.1
172 Côte d'Ivoire	12.1	7	26	71.3	100.0	29.6	398	410	70.6	22.0	2.7	13.9	1.4	5.7
173 Malawi	71.4	4	12	44.2	67.9	47.8	290	362	62.9	9.0	10.3	17.0	0.2	8.3

HDI rank	Infants exclusively breastfed (% ages 0–5 months) 2008–2013[a]	Infants lacking immunization (% of one-year-olds) DTP 2013	Infants lacking immunization (% of one-year-olds) Measles 2013	Mortality rates (per 1,000 live births) Infant 2013	Mortality rates (per 1,000 live births) Under-five 2013	Child malnutrition (% under age 5) Stunting (moderate or severe) 2008–2013[a]	Adult mortality rate (per 1,000 people) Female 2013	Adult mortality rate (per 1,000 people) Male 2013	Deaths due to (per 100,000 people) Malaria 2012	Deaths due to (per 100,000 people) Tuberculosis 2012	HIV prevalence, adult (% ages 15–49) 2013	Life expectancy at age 60 (years) 2010/2015[b]	Physicians (per 10,000 people) 2001–2013[a]	Public health expenditure (% of GDP) 2013
174 Ethiopia	52.0	16	38	44.4	64.4	44.2	198	239	48.1	18.0	1.2	17.8	0.3	5.1
175 Gambia	33.5	1	4	49.4	73.8	23.4	240	295	83.7	51.0	1.2	15.2	1.1	6.0
176 Congo (Democratic Republic of the)	37.0	14	27	86.1	118.5	43.5	320	379	106.6	54.0	1.1	15.2	1.1	3.5
177 Liberia	55.2	3	26	53.6	71.1	41.8	240	279	69.2	46.0	1.1	15.4	0.1	10.0
178 Guinea-Bissau	38.3	8	31	77.9	123.9	32.2	325	393	96.2	29.0	3.7	14.9	0.7	5.5
179 Mali	20.4	18	28	77.6	122.7	38.5 c	275	277	92.1	9.0	0.9	15.4	0.8	7.1
180 Mozambique	42.8	7	15	61.5	87.2	43.1	432	438	71.4	53.0	10.8	16.8	0.4	6.8
181 Sierra Leone	31.6	2	17	107.2	160.6	44.9	423	444	108.7	143.0	1.6	12.5	0.2	11.8
182 Guinea	20.5	24	38	64.9	100.7	31.3	267	301	104.8	23.0	1.7	14.8	1.0	4.7
183 Burkina Faso	38.2	6	18	64.1	97.6	32.9	256	298	103.3	8.5	0.9	15.1	0.5	6.4
184 Burundi	69.3	2	2	54.8	82.9	57.5	300	359	63.7	18.0	1.0	16.0	0.3	8.0
185 Chad	3.4	45	41	88.5	147.5	38.7	377	410	152.6	18.0	2.5	15.6	0.4	3.6
186 Eritrea	68.7	3	4	36.1	49.9	50.3	229	301	3.6	4.6	0.6	15.1	0.5	3.0
187 Central African Republic	34.3	65	75	96.1	139.2	40.7	412	433	114.9	50.0	3.8	15.9	0.5	3.9
188 Niger	23.3	15	33	59.9	104.2	43.0	241	252	131.1	16.0	0.4	15.5	0.2	6.5
OTHER COUNTRIES OR TERRITORIES														
Korea (Democratic People's Rep. of)	68.9	6	1	21.7	27.4	27.9	111	183	0.0	9.0	..	16.8	32.9	..
Marshall Islands	31.3 c	27	30	30.6	37.5	..	104	153	..	111.0	..	..	4.4	16.5
Monaco	..	1	1	3.0	3.7	..	48	105	..	0.1	..	..	71.7	4.0
Nauru	67.2 c	2	4	29.9	36.6	24.0 c	44	88	..	9.5	..	..	7.1	..
San Marino	..	28	26	2.8	3.1	..	46	54	..	0.0	..	..	51.3	6.5
Somalia	9.1 c	48	54	89.8	145.6	42.1 c	289	339	33.5	64.0	0.5	16.1	0.4	..
Tuvalu	34.7 c	1	4	24.4	29.2	10.0 c	182	236	..	37.0	..	..	10.9	19.7
Human development groups														
Very high human development	..	2	7	5.1	6.0	..	57	106	..	0.9	0.3	23.0	27.9	12.2
High human development	30.0	2	3	12.0	13.9	10.6	85	143	..	4.2	0.4	20.0	17.6	6.2
Medium human development	43.8	10	20	35.2	44.5	40.2	148	225	6.6	23.0	0.9	18.5	7.7	4.6
Low human development	36.0	17	27	61.5	89.4	39.5	249	291	54.5	27.3	2.4	16.2	2.9	4.5
Developing countries	37.5	10	17	37	49.3	31.0	134	192	14.0	15.8	1.2	19.0	10.3	5.6
Regions														
Arab States	30.1	9	15	28.6	37.6	25.7	124	172	..	8.2	0.1	19.0	13.8	4.1
East Asia and the Pacific	30.7	3	5	16.1	19.5	18.1	89	130	1.8	10.2	0.6	18.5	12.2	5.3
Europe and Central Asia	32.1	3	5	20.9	23.8	14.3	102	216	..	5.2	0.4	18.7	25.9	5.9
Latin America and the Caribbean	35.0	5	8	15.2	17.9	13.9	96	177	0.6	3.0	0.5	21.2	18.9	7.6
South Asia	47.9	12	24	43.2	54.9	45.1	151	222	4.4	24.4	0.2	18.6	6.8	4.3
Sub-Saharan Africa	35.4	18	26	60.8	91.2	37.2	288	337	72.9	26.4	4.7	16.6	1.9	5.6
Least developed countries	45.9	11	21	54.6	78.8	40.5	223	266	47.3	32.1	1.9	16.8	1.8	5.2
Small island developing states	35.9	9	22	36.8	47.8	23.7	153	202	..	16.5	1.1	19.3	22.6	6.5
Organisation for Economic Co-operation and Development	..	3	7	6.5	7.6	..	60	113	..	0.9	..	23.4	27.0	12.3
World	**37.4**	**9**	**16**	**33.6** T	**45.6** T	**29.7**	**120**	**181**	**..**	**13.4**	**1.1**	**20.7**	**13.8**	**9.9**

NOTES

a Data refer to the most recent year available during the period specified.

b Data are annual average of projected values for 2010–2015.

c Refers to a year earlier than that specified.

d 0.1 or less.

e Refers to 2012.

T From original data source.

DEFINITIONS

Infants exclusively breastfed: Percentage of children ages 0–5 months who are fed exclusively with breast milk in the 24 hours prior to the survey.

Infants lacking immunization against DPT: Percentage of surviving infants who have not received their first dose of diphtheria, pertussis and tetanus vaccine.

Infants lacking immunization against measles: Percentage of surviving infants who have not received the first dose of measles vaccine.

Infant mortality rate: Probability of dying between birth and exactly age 1, expressed per 1,000 live births.

Under-five mortality rate: Probability of dying between birth and exactly age 5, expressed per 1,000 live births.

Stunted children: Percentage of children ages 0–59 months who are more than two standard deeviations below the median height-for-age of the World Health Organization (WHO) Child Growth Standards.

Adult mortality rate: Probability that a 15-year-old will die before reaching age 60, expressed per 1,000 people.

Deaths due to malaria: Number of deaths due to malaria from confirmed and probable cases, expressed per 100,000 people.

Deaths due to tuberculosis: Number of deaths due to tuberculosis from confirmed and probable cases, expressed per 100,000 people.

HIV prevalence, adult: Percentage of the population ages 15–49 who are living with HIV.

Life expectancy at age 60: Additional number of years that a 60-year-old could expect to live if prevailing patterns of age-specific mortality rates stay the same throughout the rest of his or her life.

Physicians: Number of medical doctors (physicians), both generalists and specialists, expressed per 10,000 people.

Public health expenditure: Current and capital spending on health from government (central and local) budgets, external borrowing and grants (including donations from international agencies and nongovernmental organizations) and social (or compulsory) health insurance funds, expressed as a percentage of GDP.

MAIN DATA SOURCES

Columns 1–6: UNICEF (2015).

Columns 7–11 and 13: WHO (2015).

Column 12: UNDESA (2013a).

Column 14: World Bank (2015b).

TABLE 9

9

TABLE 9   Health outcomes | 241

TABLE 10

Education achievements

TABLE 10

		Literacy rate		Population with at least some secondary education	Gross enrolment ratio				Primary school dropout rate	Education quality			Primary school teachers trained to teach	Pupil–teacher ratio, primary school	Public expenditure on education
	Adult (% ages 15 and older)	Youth (% ages 15–24)			Pre-primary	Primary	Secondary	Tertiary		Performance of 15-year-old students					
		Female	Male	(% ages 25 and older)	(% of preschool-age children)	(% of primary school-age population)	(% of secondary school-age population)	(% of tertiary school-age population)	(% of primary school cohort)	Reading[a]	Mathematics[b]	Science[c]	(%)	(number of pupils per teacher)	(% of GDP)
HDI rank	2005–2013[d]	2005–2013[d]	2005–2013[d]	2005–2013[d]	2008–2014[d]	2008–2014[d]	2008–2014[d]	2008–2014[d]	2008–2014[d]	2012	2012	2012	2008–2014[d]	2008–2014[d]	2005–2014[d]
VERY HIGH HUMAN DEVELOPMENT															
1 Norway	..	..	..	97.1	99	99	111	74	1.5	504	489	495	..	..	6.6
2 Australia	..	..	..	94.4[e]	108	105	136	86	..	512	504	521	..	..	5.1
3 Switzerland	..	..	..	95.7	100	103	96	56	..	509	531	515	..	11	5.3
4 Denmark	..	..	..	96.1[f]	102	101	125	80	1.1	496	500	498	..	..	8.7
5 Netherlands	..	..	..	89.0	91	106	130	77	..	511	523	522	..	12	5.9
6 Germany	..	..	..	96.6	113	100	101	62	3.8	508	514	524	..	12	5.0
6 Ireland	..	..	..	79.6	52	104	119	71	..	523	501	522	..	16	6.2
8 United States	..	..	..	95.0	74	98	94	94	..	498	481	497	..	14	5.2
9 Canada	..	..	..	100.0	72	98	103	..	..	523	518	525	..	..	5.3
9 New Zealand	..	..	..	95.2	92	99	120	80	..	512	500	516	..	15	7.4
11 Singapore	96.4	99.8	99.8	77.4	..	..	..	..	1.3	542	573	551	94	17	2.9
12 Hong Kong, China (SAR)	..	..	..	75.4	101	105	99	67	1.0	545	561	555	96	14	3.8
13 Liechtenstein	..	..	..	..	99	104	110	42	20.6	516	535	525	..	7	2.1
14 Sweden	..	..	..	86.9	95	102	98	70	4.4	483	478	485	..	10	6.8
14 United Kingdom	..	..	..	99.9	84	109	95	62	..	499	494	514	..	18	6.0
16 Iceland	..	..	..	91.3	97	98	112	81	2.1	483	493	478	..	10	7.4
17 Korea (Republic of)	..	..	..	82.9	118	103	97	98	0.8	536	554	538	..	18	4.9
18 Israel	97.8	99.4	99.7	85.8	101	106	101	68	1.1	486	466	470	..	13	5.6
19 Luxembourg	..	..	..	100.0[f]	92	97	100	20	..	488	490	491	..	8	..
20 Japan	..	..	..	86.4	88	102	102	61	0.2	538	536	547	..	17	3.8
21 Belgium	..	..	..	80.1	118	103	107	71	13.5	509	515	505	..	11	6.5
22 France	..	..	..	80.5	110	107	110	58	..	505	495	499	..	18	5.7
23 Austria	..	..	..	100.0	103	101	98	72	0.3	490	506	506	..	11	5.8
24 Finland	..	..	..	100.0	70	100	108	94	0.4	524	519	545	..	14	6.8
25 Slovenia	99.7	99.9	99.8	96.9	94	99	98	86	1.2	481	501	514	..	17	5.7
26 Spain	97.9	99.7	99.6	69.8	127	103	131	85	2.9	488	484	496	..	13	5.0
27 Italy	99.0	99.9	99.9	75.7	99	99	99	62	0.7	490	485	494	..	10 [g]	4.3
28 Czech Republic	..	..	..	99.8	103	100	97	64	0.7	493	499	508	..	19	4.5
29 Greece	97.4	99.3	99.4	63.2	76	102	109	117	5.6	477	453	467	..	9	4.1
30 Estonia	99.9	100.0	99.9	100.0[f]	93	98	107	77	3.0	516	521	541	..	12	5.2
31 Brunei Darussalam	95.4	99.7	99.8	65.9[h]	64	94	106	24	3.6	..	..	..	85	10	3.8
32 Cyprus	98.7	99.8	99.8	78.7	78	100	95	46	4.7 [g]	449	440	438	..	14	7.2
32 Qatar	96.7	99.8	98.7	60.5	58	103 [i]	112	14	2.3	388	376	384	58	10	2.4
34 Andorra	..	..	..	49.4	..	..	..	..	..	..	..	..	100	9	3.1
35 Slovakia	..	..	..	99.3	91	102	94	55	2.0	463	482	471	..	15	4.1
36 Poland	99.7	100.0	100.0	82.3	78	101	98	73	1.5	518	518	526	..	10	4.9
37 Lithuania	99.8	99.9	99.8	91.4	76	99	106	74	2.5	477	479	496	..	12	5.2
37 Malta	92.4	99.1	97.5	73.3	116	96	86	41	5.5	..	..	..	..	11	8.0
39 Saudi Arabia	94.4	99.1	99.3	66.5	13	106	116	58	1.3	..	..	..	91 [g]	10	5.1
40 Argentina	97.9	99.4	99.1	56.9	76	124	107	80	2.9	396	388	406	..	16	5.1
41 United Arab Emirates	90.0	97.0	93.6	64.3	79	108	..	..	8.0	442	434	448	100	16	..
42 Chile	98.6	98.9	98.9	74.8	114	101	89	74	1.3	441	423	445	..	21	4.6
43 Portugal	94.5	99.5	99.4	48.0	86	106	113	69	..	488	487	489	..	12	5.3
44 Hungary	99.4	99.5	99.3	98.3[f]	87	100	102	60	1.9	488	477	494	..	10	4.7
45 Bahrain	94.6	97.6	98.6	54.9[h]	50	..	96	33	2.2	..	..	..	82	12	2.7
46 Latvia	99.9	99.9	99.8	98.9	92	103	98	65	9.7	489	491	502	..	11	4.9
47 Croatia	99.1	99.7	99.7	89.1	63	97	98	62	0.6	485	471	491	100 [i]	14	4.2
48 Kuwait	95.5	98.8	98.7	56.0	81 [g]	106 [g]	100 [g]	28	5.9	..	..	..	77	9	3.8
49 Montenegro	98.4	99.1	99.4	89.2	61	101	91	56	19.5	422	410	410	..	..	..
HIGH HUMAN DEVELOPMENT															
50 Belarus	99.6	99.8	99.8	89.3	104	98	105	93	1.3	..	..	..	100	15	5.1
50 Russian Federation	99.7	99.8	99.7	90.9	91	101	95	76	3.4	475	482	486	..	20	4.1
52 Oman	86.9	98.2	97.4	53.9	52	113	94	28	6.4	..	..	..	..	20 [i]	4.2
52 Romania	98.6	99.0	99.0	88.9	77	94	95	52	6.0	438	445	439	..	18	3.1
52 Uruguay	98.4	99.3	98.6	52.5	89	112	90	63	5.3	411	409	416	..	14	4.4
55 Bahamas	..	..	..	..	..	108	93	..	10.5	..	..	..	92	14	..
56 Kazakhstan	99.7	99.9	99.8	99.3	58	106	98	45	0.7	393	432	425	..	16	3.1
57 Barbados	..	..	..	88.7[h]	79	105	105	61	6.6	..	..	..	55	13	5.6

HDI rank		Literacy rate			Population with at least some secondary education	Gross enrolment ratio				Primary school dropout rate	Education quality			Primary school teachers trained to teach	Pupil–teacher ratio, primary school	Public expenditure on education
		Adult (% ages 15 and older)	Youth (% ages 15–24)			Pre-primary	Primary	Secondary	Tertiary		Performance of 15-year-old students					
			Female	Male	(% ages 25 and older)	(% of preschool-age children)	(% of primary school–age population)	(% of secondary school–age population)	(% of tertiary school–age population)	(% of primary school cohort)	Readingª	Mathematicsᵇ	Scienceᶜ	(%)	(number of pupils per teacher)	(% of GDP)
		2005–2013ᵈ	2005–2013ᵈ	2005–2013ᵈ	2005–2013ᵈ	2008–2014ᵈ	2008–2014ᵈ	2008–2014ᵈ	2008–2014ᵈ	2008–2014ᵈ	2012	2012	2012	2008–2014ᵈ	2008–2014ᵈ	2005–2014ᵈ
58	Antigua and Barbuda	99.0	..	..	..	89	98	105	23	8.7 ᵍ	..	..	..	60	14	2.6
59	Bulgaria	98.4	97.7	98.1	94.3	86	100	93	63	3.1	436	439	446	..	17	3.8
60	Palau	99.5	99.8	99.8	..	65	103	89	46	..	..	..	..	..	..	..
60	Panama	94.1	97.3	97.9	52.0	70	99	73	43	8.4	..	..	..	90	25	3.3
62	Malaysia	93.1	98.5	98.4	68.2 ʰ	84	101 ⁱ	71	37	0.9	398	421	420	..	12	5.9
63	Mauritius	89.2	98.6	97.7	53.6	113	108	96	41	4.2	..	..	..	100	20	3.7
64	Seychelles	91.8	99.4	98.8	..	113	108	80	1	15.1 ʲ	..	..	..	87	13	3.6
64	Trinidad and Tobago	98.8	99.6	99.6	59.3	83 ᵍ	106	86 ᵏ	12 ᵏ	10.6	..	..	..	88	18	..
66	Serbia	98.2	99.2	99.3	65.6	58	101	94	56	0.8	446	449	445	56	15	0.1
67	Cuba	99.8	100.0	100.0	76.5 ʰ	98	98	92	48	4.2	..	..	..	100	9	12.8
67	Lebanon	89.6	99.1	98.4	54.2	102	113	75	48	6.7	..	..	..	91	12	2.6
69	Costa Rica	97.4	99.3	99.0	50.6	77	103	109	48	12.3	441	407	429	97	16	6.9
69	Iran (Islamic Republic of)	84.3	97.7	98.3	65.1	38	119	86	55	3.8	..	..	..	100	26	3.7
71	Venezuela (Bolivarian Republic of)	95.5	98.8	98.3	53.7	76	102	93	78	4.9	..	..	..	..	..	6.9
72	Turkey	94.9	98.4	99.6	49.4	31	100	86	69	10.0	475	448	463	..	20	2.9
73	Sri Lanka	91.2	98.6	97.7	74.0	89	98	99	17	3.4	..	..	..	82	24	1.7
74	Mexico	94.2	99.0	98.7	58.0	101	105	86	29	4.2	424	413	415	96	28	5.1
75	Brazil	91.3	99.0	98.2	53.6	69 ⁱ	136 ⁱ	105 ⁱ	26 ⁱ	19.4 ⁱ	410	391	405	..	21	5.8
76	Georgia	99.7	99.9	99.7	..	58	103	101	33	0.2	..	..	..	95	9	2.0
77	Saint Kitts and Nevis	..	..	..	..	82	85	101	18	7.2	..	..	..	65	15	4.2
78	Azerbaijan	99.8	99.9	100.0	95.5	25	98	100	20	1.8	..	..	..	100	12	2.4
79	Grenada	..	..	..	..	99	103	108	53	..	..	..	..	65	16	..
80	Jordan	97.9	99.2	99.0	74.1	34	98	88	47	2.1	399	386	409	..	20 ʲ	..
81	The former Yugoslav Republic of Macedonia	97.5	98.5	98.8	..	29	89	83	38	2.5 ˡ	..	..	..	..	15	..
81	Ukraine	99.7	99.8	99.7	93.6 ʰ	83	105	99	79	0.6	..	..	..	100	16	6.7
83	Algeria	72.6	89.1	94.4	28.9	79	117	98	31	7.2	..	..	..	99 ᵍ	23	4.3
84	Peru	93.8	98.7	98.7	61.1	86	102	94	41	26.1	384	368	373	..	18	3.3
85	Albania	96.8	98.9	98.7	84.8	71	100 ʲ	82	56	1.3	394	394	397	32	19	3.3
85	Armenia	99.6	99.8	99.7	94.5 ʰ	46	102	97	46	4.4	..	..	..	77 ⁱ	19 ᵍ	2.3
85	Bosnia and Herzegovina	98.2	99.7	99.7	56.8	..	..	..	..	17.2	..	..	..	..	17	..
88	Ecuador	93.3	98.6	98.6	39.8	167	112	103	41	11.1	..	..	..	85	19	4.4
89	Saint Lucia	..	..	..	..	60	100 ᵍ	88	14	10.4	..	..	..	89	17	4.7
90	China	95.1	99.6	99.7	65.3 ʰ	70	128	89	27	..	570 ᵐ	613 ᵐ	580 ᵐ	..	18	..
90	Fiji	..	..	..	64.3	18	105	88	16 ⁱ	3.5	..	..	..	100	28	4.2
90	Mongolia	98.3	98.9	98.0	84.8 ʰ	86	109	92	62	9.1 ʲ	..	..	..	100	28	5.5
93	Thailand	96.4	96.6	96.6	38.1	119	93	87	51	..	441	427	444	..	16	7.6
94	Dominica	..	..	..	..	99	118	97	..	13.6	..	..	..	65	15	..
94	Libya	89.9	99.9	99.9	48.5 ʰ	10 ᵍ	114 ˡ	104 ˡ	61 ʲ	..	..	..	..	..	..	..
96	Tunisia	79.7	96.3	98.2	39.3	40	110	91	35	5.9	404	388	398	100	17	6.2
97	Colombia	93.6	98.7	97.8	56.3	49	115	93	48	15.3	403	376	399	97	25	4.9
97	Saint Vincent and the Grenadines	..	..	..	..	78	105	103	..	31.4	..	..	..	83	15	5.1
99	Jamaica	87.5	98.6	93.3	72.2 ʰ	92	92	78	29	13.9	..	..	..	70	21	6.3
100	Tonga	99.4	99.5	99.4	87.9	35	110	91	6 ʲ	9.6 ʲ	..	..	..	..	21	..
101	Belize	..	..	..	76.1	49	118	86	26	9.1	..	..	..	49	23	6.6
101	Dominican Republic	90.9	98.3	96.8	54.4	42	103	76	46	8.9	..	..	..	85	24	3.8
103	Suriname	94.7	98.8	98.0	45.9	96	113	76	..	9.7	..	..	..	94	13	..
104	Maldives	98.4	99.4	99.2	30.1	82 ᵍ	98	72 ᵏ	13	17.2	..	..	..	77	11	6.2
105	Samoa	98.9	99.6	99.4	..	34	105	86	..	10.0	..	..	..	..	30	5.8
MEDIUM HUMAN DEVELOPMENT																
106	Botswana	86.7	97.9	94.2	75.7 ʰ	18	106	82	18	7.0	..	..	..	100	25	9.5
107	Moldova (Republic of)	99.1	100.0	100.0	95.0	82	94	88	41	5.4	..	..	..	..	16	8.3
108	Egypt	73.9	86.1	92.4	52.1 ʰ	27	113	86	30	3.9	..	..	..	..	28	3.8
109	Turkmenistan	99.6	99.9	99.8	..	47	89	85	8	..	..	..	..	..	..	3.0
110	Gabon	82.3	89.4	87.4	45.2 ʰ	35	165	..	9 ʲ	..	..	..	..	100 ʲ	25	..
110	Indonesia	92.8	98.8	98.8	44.5	48	109	83	32	11.0	396	375	382	..	19	3.6
112	Paraguay	93.9	98.7	98.5	38.8	35	95	70	35	19.9	..	..	..	..	22	5.0
113	Palestine, State of	95.9	99.2	99.3	56.7	48	95	82	46	3.5	..	..	..	100	24	..
114	Uzbekistan	99.5	100.0	99.9	..	25	93	105	9	1.9	..	..	..	100	16	..
115	Philippines	95.4	98.5	97.0	64.8	51	106	85	28	24.2	..	..	..	..	31	3.4

TABLE 10   Education achievements | 243

TABLE 10

TABLE 10 EDUCATION ACHIEVEMENTS

		Literacy rate		Population with at least some secondary education	Gross enrolment ratio				Primary school dropout rate	Education quality			Primary school teachers trained to teach	Pupil–teacher ratio, primary school	Public expenditure on education
	Adult (% ages 15 and older)	Youth (% ages 15–24)			Pre-primary	Primary	Secondary	Tertiary		Performance of 15-year-old students					
		Female	Male	(% ages 25 and older)	(% of preschool-age children)	(% of primary school-age population)	(% of secondary school-age population)	(% of tertiary school-age population)	(% of primary school cohort)	Reading[a]	Mathematics[b]	Science[c]	(%)	(number of pupils per teacher)	(% of GDP)
HDI rank	2005–2013[d]	2005–2013[d]	2005–2013[d]	2005–2013[d]	2008–2014[d]	2008–2014[d]	2008–2014[d]	2008–2014[d]	2008–2014[d]	2012	2012	2012	2008–2014[d]	2008–2014[d]	2005–2014[d]
116 El Salvador	85.5	96.9	96.2	39.8	62	110	70	25	16.2	..	..	..	96	24	3.4
116 South Africa	93.7	99.3	98.5	74.3	76	101	111	20	23.0[j]	..	..	..	87	29	6.2
116 Viet Nam	93.5	96.8	97.4	65.0	82	105	..	25	5.5	508	511	528	100	19	6.3
119 Bolivia (Plurinational State of)	94.5	98.8	99.2	53.1	60	91	80	38[g]	3.3	..	..	..	..	24[g]	6.4
120 Kyrgyzstan	99.2	99.8	99.7	95.6	25	106	88	48	2.9	..	..	..	72	24	6.8
121 Iraq	79.0	80.6	83.7	39.0[h]	7[g]	107[g]	53[g]	16[i]	..	..	..	..	100[k]	17[g]	..
122 Cabo Verde	85.3	98.4	97.9	..	76	112	93	23	8.6	..	..	..	96	23	5.0
123 Micronesia (Federated States of)	..	..	..	..	29	98	83[i]	..	..	..	..	..	..	..	..
124 Guyana	85.0	93.7	92.4	54.5[h]	66	75	101	13	7.8	..	..	..	70	23	3.2
125 Nicaragua	78.0	88.8	85.2	38.9[h]	55	117	69	..	51.6[g]	..	..	..	75	30	4.4
126 Morocco	67.1	74.0	88.8	25.3	62	118	69	16	10.7	..	..	..	100	26	6.6
126 Namibia	76.5	90.6	83.2	33.9[h]	16	109	65[g]	9	15.5	..	..	..	96	30	8.5
128 Guatemala	78.3	91.9	95.5	22.6	59	108	65	19	33.3	..	..	..	..	26	2.8
129 Tajikistan	99.7	99.9	99.9	93.2	9	96	87	22	2.0	..	..	..	100	22	4.0
130 India	62.8	74.4	88.4	42.1[h]	58	113	69	25	..	..	..	..	..	35	3.8
131 Honduras	85.4	96.0	94.0	27.0	41	105	71	21	30.3	..	..	..	87[k]	34	..
132 Bhutan	52.8	68.0	80.0	34.4	14	107	78	9	21.1	..	..	..	91	24	5.5
133 Timor-Leste	58.3	78.6	80.5	..	10[i]	125	57	18	16.4	..	..	..	..	31	9.4
134 Syrian Arab Republic	85.1	94.5	96.6	34.1	6	74	48	28	6.8	..	..	..	..	..	4.9
134 Vanuatu	83.4	95.1	94.7	..	61	122	60	5[k]	28.5	..	..	..	100[g]	22	5.0
136 Congo	79.3	76.9	85.7	43.2	14	109	54	10	29.7[g]	..	..	..	80	44	6.2
137 Kiribati	..	..	..	..	71[i]	116	86	..	21.1[i]	..	..	..	85	25	..
138 Equatorial Guinea	94.5	98.5	97.7	..	73	91	28[i]	..	27.9	..	..	..	49	26	..
139 Zambia	61.4	58.5	70.3	35.0[h]	..	108	..	..	44.5	..	..	..	93	48	1.3
140 Ghana	71.5	83.2	88.3	54.3[h]	117	107	67	12	16.3	..	..	..	52	30	8.1
141 Lao People's Democratic Republic	72.7	78.7	89.2	29.8[h]	26	121	50	18	26.7	..	..	..	98	26	2.8
142 Bangladesh	58.8	81.9	78.0	37.8[h]	26	114	54	13	33.8	..	..	..	58	40	2.2
143 Cambodia	73.9	85.9	88.4	15.5	15	125	45	16	35.8	..	..	..	100	47	2.6
143 Sao Tome and Principe	69.5	77.3	83.1	..	45	117	80	8	33.9	..	..	..	48	31	9.5
LOW HUMAN DEVELOPMENT															
145 Kenya	72.2	81.6	83.2	28.6	60	114	67	4	22.4[k]	..	..	..	97	57	6.6
145 Nepal	57.4	77.5	89.2	27.4[h]	87	133	67	14	39.6	..	..	..	94	24	4.7
147 Pakistan	54.7	63.1	78.0	33.2	82	92	38	10	37.8	..	..	..	85	43	2.5
148 Myanmar	92.6	95.8	96.2	19.2[h]	9	114	50	13	25.2	..	..	..	100	28	0.8
149 Angola	70.6	66.4	79.8	..	87	140	32	7	68.1	..	..	..	47	43	3.5
150 Swaziland	83.1	94.7	92.2	23.8[h]	25	114	61	5	32.7	..	..	..	68	29	7.8
151 Tanzania (United Republic of)	67.8	72.8	76.5	7.5[h]	33	90	33	4	33.3	..	..	..	99	43	6.2
152 Nigeria	51.1	58.0	75.6	..	13	85	44	10[i]	20.7	..	..	..	66	38	..
153 Cameroon	71.3	76.4	85.4	28.0[h]	30	111	50	12	30.2	..	..	..	79	46	3.0
154 Madagascar	64.5	64.0	65.9	..	12	145	38	4	62.0	..	..	..	19	40	2.7
155 Zimbabwe	83.6	92.1	89.6	55.3[h]	34	109	47	6	..	..	..	..	88	36	2.0
156 Mauritania	45.5	47.7	66.4	14.4[h]	2[k]	97	30	5	35.9	..	..	..	100	35	3.8
156 Solomon Islands	..	..	..	..	43	114	48	..	25.1	..	..	..	63	21	9.9
158 Papua New Guinea	62.9	75.8	66.8	11.1[h]	100	114	40	..	..	..	..	..	..	..	..
159 Comoros	75.9	86.5	86.3	..	22	103	64	10	44.6[j]	..	..	..	75	28	7.6
160 Yemen	66.4	77.8	96.7	17.6[h]	1	101	49	10	40.5[k]	..	..	..	..	30	4.6
161 Lesotho	75.8	92.1	74.2	20.9	37	108	53	11	43.2	..	..	..	72	33	13.0
162 Togo	60.4	72.7	86.9	26.8	14	134	55	10	36.1	..	..	..	82	41	4.0
163 Haiti	48.7	70.5	74.4	28.5[h]	..	..	..	..	..	..	..	..	..	..	..
163 Rwanda	65.9	78.0	76.7	8.4[h]	14	134	33	7	65.3	..	..	..	96	60	5.1
163 Uganda	73.2	85.5	89.6	28.8	11	107	27	4	75.2	..	..	..	95	46	3.3
166 Benin	28.7	30.8	54.9	19.2[h]	19	124	54	12	46.8	..	..	..	47	44	5.3
167 Sudan	73.4	85.5	90.3	15.2[h]	38	70	41	17	20.1	..	..	..	68	46	2.2
168 Djibouti	..	..	..	..	4	68	48	5	15.6	..	..	..	96	33	4.5
169 South Sudan	..	..	..	..	6	86	..	..	..	..	..	..	44	50	0.7
170 Senegal	52.1	59.0	74.0	10.8	15	84	41	8	38.6	..	..	..	48	32	5.6
171 Afghanistan	31.7	32.1	61.9	18.2[h]	1[i]	106	54	4	..	..	..	..	..	45	..
172 Côte d'Ivoire	41.0	38.8	58.3	22.4[h]	6	96	39	9	26.9	..	..	..	83	41	4.6
173 Malawi	61.3	70.0	74.3	16.3[h]	..	141	37	1	40.3	..	..	..	91	69	5.4

	Literacy rate			Population with at least some secondary education	Gross enrolment ratio				Primary school dropout rate	Education quality			Primary school teachers trained to teach	Pupil–teacher ratio, primary school	Public expenditure on education
	Adult (% ages 15 and older)	Youth (% ages 15–24)			Pre-primary	Primary	Secondary	Tertiary		Performance of 15-year-old students					
		Female	Male	(% ages 25 and older)	(% of preschool-age children)	(% of primary school-age population)	(% of secondary school-age population)	(% of tertiary school-age population)	(% of primary school cohort)	Reading[a]	Mathematics[b]	Science[c]	(%)	(number of pupils per teacher)	(% of GDP)
HDI rank	2005–2013[d]	2005–2013[d]	2005–2013[d]	2005–2013[d]	2008–2014[d]	2008–2014[d]	2008–2014[d]	2008–2014[d]	2008–2014[d]	2012	2012	2012	2008–2014[d]	2008–2014[d]	2005–2014[d]
174 Ethiopia	39.0	47.0	63.0	12.5	3[g]	87[l]	29[l]	3[i]	63.4	..	..	..	57	54	4.7
175 Gambia	52.0	65.5	73.4	24.3[h]	32	87	57	..	27.0	..	..	..	82	36	4.1
176 Congo (Democratic Republic of the)	61.2	53.3	78.9	22.4[h]	4	111	43	8	29.3	..	..	..	94	35	1.6
177 Liberia	42.9	37.2	63.5	26.7[h]	..	96	38	12	32.2	..	..	..	56	26	2.8
178 Guinea-Bissau	56.7	68.9	79.7	..	7	116	34[l]	3[l]	..	..	..	..	39	52	..
179 Mali	33.6	39.0	56.3	10.9	4	84	45	7	38.4	..	..	..	52	41	4.8
180 Mozambique	50.6	56.5	79.8	3.6[h]	..	105	26	5	68.4	..	..	..	87	55	5.0
181 Sierra Leone	44.5	53.8	71.6	15.7[h]	10	134	45	..	52.2	..	..	..	57	35	2.9
182 Guinea	25.3	21.8	37.6	..	16	91	38	10	41.4	..	..	..	75	44	2.5
183 Burkina Faso	28.7	33.1	46.7	2.0	4	87	28	5	30.9	..	..	..	86	46	3.4
184 Burundi	86.9	88.1	89.6	6.7[h]	9	134	33	3	52.5	..	..	..	95	45	5.8
185 Chad	37.3	44.0	53.8	5.5	1	103	23	2	49.0	..	..	..	65	62	2.3
186 Eritrea	70.5	88.7	93.2	..	..	..	..	2	31.0	..	..	..	90	41	2.1
187 Central African Republic	36.8	27.0	48.9	18.1[h]	6	95	18	3	53.4	..	..	..	58	80	1.2
188 Niger	15.5	15.1	34.5	5.2[h]	6	71	16	2	30.7	..	..	..	97	39	4.4
OTHER COUNTRIES OR TERRITORIES															
Korea (Democratic People's Rep. of)	100.0	100.0	100.0	..	..	..	..	..	..	..	..	..	..	..	..
Marshall Islands	..	..	..	..	48	105	103	43	16.5	..	..	..	..	..	..
Monaco	..	..	..	..	..	..	..	..	..	..	..	..	..	..	1.6
Nauru	..	..	..	..	79	94	72	..	..	..	..	..	74	22	..
San Marino	..	..	..	..	108	92	95	64	3.8	..	..	..	..	6	..
Somalia	..	..	..	..	..	29[g]	7[g]	..	..	..	..	..	..	36[g]	..
Tuvalu	..	..	..	..	84	88	84	..	..	..	..	..	..	19[k]	..
Human development groups															
Very high human development	..	..	..	87.0	86	103	102	77	2.0	—	—	—	..	14	5.1
High human development	94.5	99.0	99.1	64.9	72	118	91	35	9.0	—	—	—	95	19	4.9
Medium human development	71.8	82.2	90.1	45.0	52	110	70	24	18.1	—	—	—	83	30	4.1
Low human development	57.1	62.7	75.7	21.6	27	101	41	8	39.4	—	—	—	78	41	3.6
Developing countries	79.9	84.1	90.4	51.2	50	110	70	25	25.3	—	—	—	84	27	4.7
Regions															
Arab States	78.0	86.9	93.1	41.5	33	104	74	29	8.8	—	—	—	91	23	4.3
East Asia and the Pacific	94.5	98.7	98.8	60.5	64	118	85	28	17.3	—	—	—	..	19	4.9
Europe and Central Asia	98.0	99.3	99.7	75.5	42	100	93	51	3.7	—	—	—	94	17	3.4
Latin America and the Caribbean	92.3	98.0	97.6	54.6	78	114	94	38	13.8	—	—	—	93	22	5.5
South Asia	62.5	74.3	86.3	42.0	57	111	64	23	22.8	—	—	—	80	35	3.5
Sub-Saharan Africa	58.4	62.7	74.7	26.5	22	101	43	8	37.9	—	—	—	75	42	5.1
Least developed countries	58.4	65.9	75.5	21.6	18	106	42	8	39.6	—	—	—	76	41	3.4
Small island developing states	80.1	86.7	86.1	..	66	106	71	..	15.1	—	—	—	88	18	7.6
Organisation for Economic Co-operation and Development	..	..	..	84.5	87	102	98	71	2.8	—	—	—	..	16	5.1
World	**81.2**	**84.7**	**90.8**	**59.7**	**54**	**109**	**74**	**32**	**17.6**	—	—	—	**..**	**25**	**5.0**

TABLE 10

NOTES

a Average score for Organisation for Economic Co-operation and Development (OECD) countries is 496.

b Average score for OECD countries is 494.

c Average score for OECD countries is 501.

d Data refer to the most recent year available during the period specified.

e Refers to population ages 25–64.

f Refers to population ages 25–74.

g Refers to 2007.

h Barro and Lee (2014) estimate for 2010 based on data from the United Nations Educational, Scientific and Cultural Organization Institute for Statistics.

i Refers to 2005.

j Refers to 2003.

k Refers to 2004.

l Refers to 2006.

m Refers to Shanghai only.

DEFINITIONS

Adult literacy rate: Percentage of the population ages 15 and older who can, with understanding, both read and write a short simple statement on their everyday life.

Youth literacy rate: Percentage of the population ages 15–24 who can, with understanding, both read and write a short simple statement on their everyday life.

Population with at least some secondary education: Percentage of the population ages 25 and older that has reached at least a secondary level of education.

Gross enrolment ratio: Total enrolment in a given level of education (pre-primary, primary, secondary or tertiary), regardless of age, expressed as a percentage of the official school-age population for the same level of education.

Primary school dropout rate: Percentage of students from a given cohort that have enrolled in primary school but that drop out before reaching the last grade of primary education. It is calculated as 100 minus the survival rate to the last grade of primary education and assumes that observed flow rates remain unchanged throughout the cohort life and that dropouts do not re-enter school.

Performance of 15-year-old students in reading, mathematics and science: Score obtained in testing of skills and knowledge of 15-year-old students in these subjects essential for participation in society.

Primary school teachers trained to teach: Percentage of primary school teachers that have received the minimum organized teacher training (pre-service or in-service) required for teaching at the primary level.

Pupil–teacher ratio, primary school: Average number of pupils per teacher in primary education in a given school year.

Public expenditure on education: Current and capital spending on education, expressed as a percentage of GDP.

MAIN DATA SOURCES

Columns 1–9, 13 and 14: UNESCO Institute for Statistics (2015).

Columns 10–12: OECD (2014).

Column 15: World Bank (2015b).

TABLE 10 Education achievements | 245

TABLE 11

National income and composition of resources

		Gross domestic product (GDP)		Gross fixed capital formation	General government final consumption expenditure		Total tax revenue	Taxes on income, profit and capital gain	Research and development expenditure	Domestic credit provided by financial sector	External debt stock	Total debt service	Consumer price index	Domestic food price level	
		Total (2011 PPP $ billions)	Per capita (2011 PPP $)	(% of GDP)	Total (% of GDP)	Average annual growth (%)	(% of GDP)	(% of total tax revenue)	(% of GDP)	(% of GDP)	(% of GNI)	(% of GNI)	(2010=100)	Index	Volatility index
HDI rank		2013	2013	2005–2013ᵃ	2005–2013ᵃ	2005–2013ᵃ	2005–2013ᵃ	2005–2013ᵃ	2005–2012ᵃ	2013	2005–2013ᵃ	2013	2013	2009–2014ᵃ	2009–2014ᵃ
VERY HIGH HUMAN DEVELOPMENT															
1	Norway	317.5	62,448	22.6	21.9	1.8	27.3	31.8	1.7	87.0ᵇ	..	..	104	1.5	11.3
2	Australia	990.7	42,831	28.3	17.7	0.0	21.4	65.3	2.4	158.8	..	..	108	1.4	..
3	Switzerland	442.0	54,697	23.4	11.0	1.4	9.8	22.7	2.9	173.4	..	..	99	1.4	6.6
4	Denmark	235.7	41,991	18.3	26.7	−0.5	33.4	39.7	3.0	199.6	..	..	106	1.3	6.0
5	Netherlands	755.3	44,945	18.2	26.3	−0.3	19.7	23.5	2.2	193.0	..	..	107	1.4	5.6
6	Germany	3,483.4	43,207	19.8	19.3	0.7	11.5	16.3	2.9	113.5	..	..	106	1.5	5.6
6	Ireland	206.5	44,931	15.2	17.5	0.0	22.0	37.8	1.7	186.1	..	..	105	1.2	3.3
8	United States	16,230.2	51,340	19.3	15.2	−1.3	10.6	53.2	2.8	240.5	..	..	107	1.0	0.0
9	Canada	1,472.9	41,894	23.7	21.7	0.6	11.5	51.6	1.7	173.1ᶜ	..	..	105	1.3	7.1
9	New Zealand	146.7	32,808	20.6	18.7	1.9	29.3	36.3	1.3	154.0ᵈ	..	..	107	2.0	..
11	Singapore	411.6	76,237	25.9	10.2	9.9	14.0	34.7	2.1	112.6	..	..	113	1.0	4.0
12	Hong Kong, China (SAR)	370.2	51,509	23.9	9.3	2.3	12.5	36.2	0.7	224.0	..	..	114	..	..
13	Liechtenstein	..	..	..	..	..	..	..	..	..	..	..	..	..	..
14	Sweden	419.6	43,741	22.1	26.2	1.6	20.7	9.8	3.4	138.1	..	..	104	1.5	6.7
14	United Kingdom	2,372.7	37,017	16.4	20.2	0.7	25.3	31.8	1.7	184.1	..	..	110	1.2	5.0
16	Iceland	13.3	41,250	15.1	24.3	0.8	22.3	27.4	2.6	130.9	..	..	114	1.8	5.4
17	Korea (Republic of)	1,642.6	32,708	29.7	14.9	2.7	14.4	30.3	4.0	155.9	..	..	108	1.9	9.1
18	Israel	249.3	30,927	19.5	22.5	3.5	22.1	27.9	3.9	81.4ᵉ	..	..	107	2.2	5.9
19	Luxembourg	47.7	87,737	17.1	17.3	5.0	25.5	30.0	1.4	163.9	..	..	108	1.3	8.9
20	Japan	4,535.1	35,614	21.7	20.6	1.9	10.1	46.0	3.4	366.5	..	..	100	1.9	5.6
21	Belgium	454.6	40,607	22.3	24.4	1.1	24.9	35.2	2.2	111.2	..	..	108	1.7	6.0
22	France	2,453.3	37,154	22.1	24.1	2.0	21.4	24.6	2.3	130.8	..	..	105	1.7	4.8
23	Austria	376.0	44,376	22.2	19.8	0.7	18.3	23.7	2.8	127.9	..	..	108	1.4	5.9
24	Finland	211.3	38,846	21.2	24.9	1.5	20.0	14.8	3.5	104.9	..	..	108	1.6	6.2
25	Slovenia	56.8	27,576	19.7	20.4	−1.1	17.5	10.4	2.8	82.8	..	..	106	2.2	9.4
26	Spain	1,473.9	31,596	18.5	19.5	−2.9	7.1	20.1	1.3	205.1	..	..	107	2.0	8.4
27	Italy	2,044.3	34,167	17.8	19.4	−0.7	22.4	32.8	1.3	161.8	..	..	107	2.0	5.0
28	Czech Republic	294.2	27,959	24.9	19.6	2.3	13.4	14.5	1.9	67.0	..	..	107	2.3	10.7
29	Greece	270.7	24,540	11.2	20.0	−6.5	22.4	19.0	0.7	134.3	..	..	104	2.6	11.2
30	Estonia	33.3	25,132	27.3	19.1	2.8	16.3	8.9	2.2	71.6	..	..	112	2.8	7.4
31	Brunei Darussalam	29.0	69,474	15.3	18.3	1.1	..	..	..	20.8	..	..	103	3.0	4.7
32	Cyprus	23.8	27,394	18.4	19.7	0.5	25.5	26.5	0.5	335.8	..	..	105	2.0	12.7
32	Qatar	276.6	127,562	..	12.9	10.1	14.7	40.2	..	73.9	..	..	107	1.8	6.3
34	Andorra	..	..	..	..	..	..	..	..	..	..	..	..	..	..
35	Slovakia	142.2	26,263	20.4	18.1	2.4	12.2	10.3	0.8	53.1ᶜ	..	..	109	2.6	9.2
36	Poland	881.5	22,877	18.8	18.1	2.1	16.0	13.4	0.9	65.8	..	..	109	2.7	7.0
37	Lithuania	72.4	24,483	17.4	18.7	−3.0	13.4	9.0	0.9	51.0	..	..	109	3.5	5.5
37	Malta	12.2	28,828	14.3	20.2	0.9	27.0	33.1	0.8	146.7	..	..	107	2.6	8.6
39	Saudi Arabia	1,501.1	52,068	23.2	22.1	9.2	..	..	0.1	−7.9	..	..	113	2.9	3.8
40	Argentina	..	..	17.0	15.5	5.1	..	..	0.6	33.3	22.7	2.3	..	..	..
41	United Arab Emirates	525.1ᶠ	57,045ᶠ	22.0	6.8	4.2	0.4	..	0.5	76.5ᶠ	..	..	103	..	..
42	Chile	382.6	21,714	23.6	12.4	4.2	19.0	30.4	0.4	115.5	..	..	108	2.6	7.4
43	Portugal	267.7	25,596	15.1	19.0	−1.9	20.3	22.2	1.5	183.3	..	..	107	2.5	9.0
44	Hungary	226.8	22,914	19.9	19.9	3.2	22.9	15.9	1.3	64.7	170.8	99.0	112	2.4	5.8
45	Bahrain	56.5	42,444	19.2	14.4	10.7	1.1	0.7	..	78.6	..	..	106	2.2	18.5
46	Latvia	43.9	21,825	22.2	15.5	−11.0	13.8	8.7	0.7	58.6	..	..	107	2.9	7.9
47	Croatia	85.3	20,063	19.3	20.0	0.5	19.6	7.9	0.8	94.1	..	..	108	3.2	2.7
48	Kuwait	267.7ᶠ	82,358ᶠ	18.0	16.7	15.0	0.7	0.6	0.1	47.9ᶠ	..	..	111	2.6	3.7
49	Montenegro	8.8	14,152	19.2	19.8	1.4	..	..	0.4	61.0	65.5	8.1	109	5.6	9.1
HIGH HUMAN DEVELOPMENT															
50	Belarus	161.4	17,055	36.9	14.2	−2.6	15.1	3.5	0.7	39.9	56.7	6.7	289	5.3	6.0
50	Russian Federation	3,381.5	23,564	21.5	19.5	0.5	15.1	1.9	1.1	48.3	..	..	122	4.3	5.2
52	Oman	147.6ᶠ	44,532ᶠ	22.3	19.2	10.0	2.6	2.6	0.1	35.7	..	..	108	3.3	9.2
52	Romania	363.3	18,200	23.5	6.2	2.3	18.8	18.4	0.5	52.0	72.9	16.6	114	3.7	4.3
52	Uruguay	64.6	18,966	22.9	13.8	4.2	19.3	18.4	0.4	36.3	..	..	127	3.1	6.4
55	Bahamas	8.5	22,518	26.3	16.1	3.7	15.5	..	..	104.9	..	..	106	1.6	5.4
56	Kazakhstan	382.8	22,467	21.2	10.3	1.7	..	..	0.2	39.1	74.6	15.0	121	..	..
57	Barbados	4.3ᶠ	15,299ᶠ	18.4	16.1	..	25.2	31.6	..	106.7ᵉ	..	..	116	2.4	5.4

		Gross domestic product (GDP)		Gross fixed capital formation	General government final consumption expenditure		Total tax revenue	Taxes on income, profit and capital gain	Research and development expenditure	Debts			Prices		
										Domestic credit provided by financial sector	External debt stock	Total debt service	Consumer price index	Domestic food price level	
		Total (2011 PPP $ billions)	Per capita (2011 PPP $)	(% of GDP)	Total (% of GDP)	Average annual growth (%)	(% of GDP)	(% of total tax revenue)	(% of GDP)	(% of GDP)	(% of GNI)	(% of GNI)	(2010=100)	Index	Volatility index
HDI rank		2013	2013	2005–2013[a]	2005–2013[a]	2005–2013[a]	2005–2013[a]	2005–2013[a]	2005–2012[a]	2013	2005–2013[a]	2013	2013	2009–2014[a]	2009–2014[a]
58	Antigua and Barbuda	1.8	20,353	23.2	18.5	..	18.6	12.0	..	90.0	..	..	108	2.6	..
59	Bulgaria	114.0	15,695	21.3	16.5	2.8	19.0	14.6	0.6	71.1	104.9	9.5	108	3.2	5.9
60	Palau	0.3	14,612	..	..	..	..	..	..	..	..	..	..	..	..
60	Panama	72.6	18,793	26.5	10.0	−1.7	..	..	0.2	67.6	38.9	3.9	116	3.0	2.1
62	Malaysia	671.3	22,589	26.9	13.6	6.3	16.1	52.0	1.1	142.6	70.7	3.2	107	2.9	4.3
63	Mauritius	21.6	16,648	21.2	14.4	0.7	19.0	17.9	0.4	122.4	91.4	28.4	115	4.9	11.7
64	Seychelles	2.1	23,799	33.6	25.6	6.9	31.2	27.9	0.3	35.2	222.4	4.5	115	6.7	7.2
64	Trinidad and Tobago	39.5	29,469	9.7	9.5	−0.6	28.3	49.6	0.0	33.7	..	..	121	4.0	16.5
66	Serbia	92.4	12,893	21.2	19.6	2.1	19.7	7.6	1.0	49.5	88.1	19.4	128	4.0	8.5
67	Cuba	211.9[g]	18,796[g]	12.2	37.9	2.7	..	..	0.4	..	..	..	..	..	..
67	Lebanon	74.3	16,623	27.9	14.7	23.2	15.5	19.0	..	187.6	68.9	7.8	112[f]	..	..
69	Costa Rica	65.4	13,431	21.0	17.9	3.9	13.6	15.1	0.5	56.5	35.9	6.2	115	3.2	7.6
69	Iran (Islamic Republic of)	1,168.7	15,090	25.8	11.2	−4.3	8.4	19.3	0.7	18.0[d]	2.1	0.1	214	4.5	13.0
71	Venezuela (Bolivarian Republic of)	535.6	17,615	20.3	12.2	3.3	15.5	21.5	..	52.5	27.5	4.6	215	4.5	12.8
72	Turkey	1,398.3	18,660	20.3	15.1	6.2	20.4	17.6	0.9	84.3	47.9	7.6	125	3.8	12.9
73	Sri Lanka	193.1	9,426	29.2	13.1	5.5	12.0	16.2	0.2	48.4[f]	38.5	2.8	123	6.9	8.3
74	Mexico	1,992.9	16,291	21.0	11.9	1.1	..	..	0.4	49.5	35.9	3.4	112	3.7	4.7
75	Brazil	2,916.3	14,555	18.2	22.0	2.0	15.4	26.5	1.2	110.1	21.9	3.8	119	2.6	4.4
76	Georgia	31.1	6,946	21.9	16.7	..	24.1	35.2	0.2	42.9	86.4	11.2	107	..	..
77	Saint Kitts and Nevis	1.1	20,709	28.7	19.2	..	20.2	9.8	..	65.9	..	..	109	2.9	..
78	Azerbaijan	156.3	16,594	24.6	11.2	2.2	13.0	13.6	0.2	25.5	13.3	3.6	112	..	..
79	Grenada	1.2	11,272	20.1	16.8	..	18.7	16.9	..	80.0	72.6	4.4	105	3.4	..
80	Jordan	73.7	11,407	27.2	19.7	−11.4	15.3	13.6	0.4	111.9	71.9	3.0	115	4.5	6.1
81	The former Yugoslav Republic of Macedonia	24.5	11,609	24.8	17.5	−3.6	16.7	10.7	0.2	52.4	69.5	9.1	110	5.1	7.9
81	Ukraine	387.0	8,508	18.2	19.4	2.2	18.2	11.8	0.7	95.7	81.6	20.9	108	5.2	3.9
83	Algeria	505.5	12,893	33.8	18.9	80.4	37.4	60.2	0.1	3.0	2.5	0.3	118	5.1	5.5
84	Peru	346.2	11,396	26.6	11.2	6.7	16.5	34.0	..	22.0	29.0	3.6	110	3.9	3.4
85	Albania	28.9	10,405	26.2	10.7	0.1	..	..	0.2	66.9	60.1	3.2	108	6.4	10.3
85	Armenia	22.4	7,527	20.9	14.5	16.3	18.7	21.3	0.3	46.0	79.4	18.5	117	8.9	11.9
85	Bosnia and Herzegovina	35.9	9,387	17.9	22.8	−0.5	20.9	6.7	0.0	67.7	60.9	6.2	106	4.8	6.3
88	Ecuador	165.9	10,541	27.9	14.1	7.7	..	..	0.2	29.6	22.9	3.4	113	3.4	5.7
89	Saint Lucia	1.9	10,152	23.3	14.8	..	23.0	27.1	..	123.1	37.2	2.9	109	3.4	12.3
90	China	15,643.2	11,525	47.3	14.1	8.2	10.6	24.9	2.0	163.0	9.5	0.4	111	3.3	8.1
90	Fiji	6.6	7,502	21.1	15.0	..	23.2	32.5	..	121.8	20.7	1.2	116	5.1	8.3
90	Mongolia	25.9	9,132	44.2	11.3	−25.0	18.2	11.6	0.3	63.6	176.0	13.0	137	4.8	16.7
93	Thailand	933.6	13,932	26.7	13.8	4.9	16.5	36.2	0.3	173.3	37.2	3.6	109	4.5	2.8
94	Dominica	0.7	10,011	12.0	19.1	..	21.8	16.2	..	61.9	59.4	3.7	104	..	..
94	Libya	126.3	20,371	27.9	9.3	..	..	..	..	−51.1	..	..	126	..	..
96	Tunisia	117.2	10,768	20.2	19.0	2.8	21.0	26.7	1.1	83.4	55.5	5.9	116	3.9	4.7
97	Colombia	581.1	12,025	24.5	16.7	5.8	13.2	19.7	0.2	70.1	25.3	2.8	109	2.7	4.5
97	Saint Vincent and the Grenadines	1.1	10,154	24.9	18.5	..	23.0	24.5	..	58.4	40.6	4.0	107	3.4	4.8
99	Jamaica	23.4	8,607	19.7	16.3	0.3	27.1	33.0	..	51.4	100.6	8.8	126	5.0	7.0
100	Tonga	0.5	5,134	33.1	18.9	..	..	..	..	27.1	41.6	1.4	108	..	..
101	Belize	2.7	8,215	17.8	15.1	0.7	22.6	28.7	..	58.3	80.5	9.0	98	3.0	27.9
101	Dominican Republic	122.7	11,795	21.5	10.1	−6.4	12.2	20.5	..	47.8	41.2	4.9	118	4.1	5.2
103	Suriname	8.4	15,556	24.9	23.3	..	19.4	31.9	..	31.5	..	..	126	6.2	9.7
104	Maldives	3.9	11,283	40.4	16.8	..	15.5	2.8	..	86.9	42.0	3.7	129	3.5	14.2
105	Samoa	1.1	5,584	..	..	..	0.0	20.3	..	40.8	67.2	1.9	108	..	..
MEDIUM HUMAN DEVELOPMENT															
106	Botswana	30.8	15,247	33.9	19.7	4.3	27.1	23.9	0.5	13.6	16.6	1.3	123	2.9	3.6
107	Moldova (Republic of)	16.1	4,521	22.9	20.3	−2.0	18.6	2.7	0.4	44.0	75.0	7.6	118	4.8	5.7
108	Egypt	880.8	10,733	13.8	11.7	3.4	13.2	26.2	0.4	86.2	16.7	1.3	129	7.5	9.8
109	Turkmenistan	71.0	13,555	47.2	8.9	..	..	..	..	..	1.3	0.1	..	..	..
110	Gabon	31.2	18,646	33.3	8.9	18.2	..	..	0.6	11.7	25.0	6.5	104	5.2	21.0
110	Indonesia	2,312.4	9,254	31.7	9.1	4.9	11.4	36.8	0.1	45.6	30.8	4.8	117	6.7	10.7
112	Paraguay	53.3	7,833	15.1	12.1	5.3	12.8	11.6	0.1	38.3	47.2	6.8	115	4.3	11.2
113	Palestine, State of	18.7	4,484	22.2	27.7	−9.0	5.1	2.1	..	..	..	..	119	..	..
114	Uzbekistan	151.3	5,002	23.2	22.4	..	..	..	..	..	18.1	1.2	..	..	..
115	Philippines	622.5	6,326	20.5	11.1	7.7	12.9	42.1	0.1	51.9	18.6	1.8	111	6.8	2.6

TABLE 11 National income and composition of resources | 247

TABLE
11

TABLE 11 NATIONAL INCOME AND COMPOSITION OF RESOURCES

HDI rank		Gross domestic product (GDP)		Gross fixed capital formation	General government final consumption expenditure		Total tax revenue	Taxes on income, profit and capital gain	Research and development expenditure	Debts Domestic credit provided by financial sector	External debt stock	Total debt service	Prices Consumer price index	Domestic food price level	
		Total (2011 PPP $ billions)	Per capita (2011 PPP $)	(% of GDP)	Total (% of GDP)	Average annual growth (%)	(% of GDP)	(% of total tax revenue)	(% of GDP)	(% of GDP)	(% of GNI)	(% of GNI)	(2010=100)	Index	Volatility index
		2013	2013	2005–2013ᵃ	2005–2013ᵃ	2005–2013ᵃ	2005–2013ᵃ	2005–2013ᵃ	2005–2012ᵃ	2013	2005–2013ᵃ	2013	2013	2009–2014ᵃ	2009–2014ᵃ
116	El Salvador	47.6	7,515	15.1	12.0	6.0	14.5	23.0	0.0	72.1	57.1	4.7	108	4.3	3.0
116	South Africa	641.4	12,106	19.3	22.2	2.4	26.5	49.4	0.8	190.2	40.7	2.8	117	3.0	6.2
116	Viet Nam	459.7	5,125	23.8	6.2	7.3	..	..	..	108.2	40.2	3.1	138	..	..
119	Bolivia (Plurinational State of)	63.3	5,934	19.1	13.9	9.3	17.0	9.6	0.2	50.4	27.5	1.9	121	5.9	12.2
120	Kyrgyzstan	17.8	3,110	30.5	18.1	0.2	18.1	19.1	0.2	14.0ʰ	98.4	5.6	128	..	..
121	Iraq	483.6	14,471	16.7	21.3	..	..	..	0.0	−1.4	..	..	114	5.1	16.4
122	Cabo Verde	3.1	6,210	35.9	18.5	..	17.8	23.3	0.1	82.8	80.9	2.2	109	5.7	5.4
123	Micronesia (Federated States of)	0.3	3,286	..	..	..	..	..	..	−27.2	..	..	..	..	..
124	Guyana	5.1	6,336	24.9	14.2	..	..	..	..	53.7	74.9	2.6	109	..	..
125	Nicaragua	27.3	4,494	22.9	5.7	1.8	14.8	30.0	..	44.8	87.7	5.8	124	4.5	6.4
126	Morocco	233.9	6,967	30.2	19.0	3.7	24.5	26.2	0.7	115.5	38.7	5.0	104	5.7	4.9
126	Namibia	21.4	9,276	25.7	27.6	8.9	23.1	32.6	0.1	49.7	..	..	118	3.5	7.2
128	Guatemala	109.2	7,063	14.3	10.5	4.8	10.8	29.4	0.0	40.6	32.0	2.4	115	7.1	5.5
129	Tajikistan	20.0	2,432	14.1	11.7	1.1	..	..	0.1	19.0	41.8	5.0	125	..	..
130	India	6,558.7	5,238	28.3	11.8	3.8	10.7	44.8	0.8	77.1	23.0	2.2	132	4.7	8.4
131	Honduras	36.0	4,445	24.5	16.6	4.1	14.7	21.6	..	57.3	39.6	5.4	118	4.8	4.8
132	Bhutan	5.4	7,167	47.3	17.5	−6.6	9.2	15.9	..	50.2	83.6	4.5	129	5.1	6.4
133	Timor-Leste	2.3ᶠ	2,040ᶠ	55.7	85.7	3.6	..	..	..	−53.6ᶠ	..	..	141	..	..
134	Syrian Arab Republic	..	..	20.4	12.3	23.6	14.2	30.2	..	36.2ʰ	14.3	1.7ʰ	143ᶠ	..	..
134	Vanuatu	0.7	2,895	25.2	16.3	2.2	16.0	..	..	68.7	16.7	1.0	104	..	..
136	Congo	25.3	5,680	30.7	13.6	−11.8	5.9	4.1	..	−7.2	30.4	2.7	112	6.3	18.8
137	Kiribati	0.2	1,796	..	..	..	16.1	8.7	..	..	..	..	..	..	..
138	Equatorial Guinea	24.7	32,685	58.4	6.7	9.8	20.5	35.7	..	−3.5	..	..	121	..	..
139	Zambia	55.2	3,800	25.9	2.8	..	16.0	48.0	0.3	27.5	25.9	1.2	121	10.1	3.2
140	Ghana	100.1	3,864	22.7	16.6	−7.9	14.9	24.7	0.4	34.8	33.8	2.0	132	5.4	18.3
141	Lao People's Democratic Republic	31.6	4,667	29.2	14.5	33.9	14.8	17.6	..	26.5ᵈ	81.4	2.9	119	8.6	3.6
142	Bangladesh	446.8	2,853	28.4	5.1	5.8	8.7	22.4	..	57.9	19.5	1.0	126	8.0	4.5
143	Cambodia	44.6	2,944	18.1	5.2	21.3	11.6	13.8	..	40.3	44.4	1.1	112	7.8	4.7
143	Sao Tome and Principe	0.6	2,876	41.7	13.5	..	14.0	12.8	..	28.8	69.6	1.9	132	9.1	50.5
LOW HUMAN DEVELOPMENT															
145	Kenya	120.0	2,705	20.4	14.0	1.1	15.9	40.9	1.0	42.8	30.8	1.1	132	5.8	6.0
145	Nepal	60.4	2,173	22.6	9.9	−6.9	15.3	18.7	0.3	69.1	19.7	1.1	130	9.5	10.2
147	Pakistan	811.3	4,454	13.0	11.0	10.2	11.1	27.9	0.3	49.0	22.8	3.3	132	7.1	13.2
148	Myanmar	..	..	..	..	..	..	25.2	..	24.6ⁱ	..	..	112	8.5	8.1
149	Angola	160.8	7,488	14.7	19.9	−2.7	18.8	31.9	..	18.9	22.0	4.3	136	7.2	13.7
150	Swaziland	8.1	6,471	9.6	19.9	5.5	..	..	..	18.4	13.1	0.9	122	..	..
151	Tanzania (United Republic of)	82.2	1,718	32.1	19.0	36.2	16.1	21.9	0.5	24.3	39.7	0.5	141	11.5	4.8
152	Nigeria	941.5	5,423	14.5	8.1	1.4	1.6	28.3	0.2	22.3	2.8	0.1	135	6.3	4.0
153	Cameroon	61.0	2,739	19.4	11.6	6.5	..	..	..	15.5	17.1	0.7	108	7.8	10.0
154	Madagascar	31.4	1,369	15.7	8.6	−14.3	10.1	21.2	0.1	15.6	27.3	0.7	123	7.1	3.5
155	Zimbabwe	25.1	1,773	13.0	20.0	2.7	..	..	..	75.5ᵇ	69.5	21.2	109	..	..
156	Mauritania	11.5	2,945	38.0	17.0	2.8	..	..	..	39.1ᶠ	91.7	4.2	115	10.1	3.1
156	Solomon Islands	1.1	2,003	13.4	39.2	..	..	..	..	20.3	21.2	4.6	120	..	..
158	Papua New Guinea	18.0	2,458	..	..	..	..	..	..	49.2	148.4	30.3	115	..	..
159	Comoros	1.0	1,400	17.8	16.2	2.2	..	..	..	26.9	22.3	0.1	106	..	..
160	Yemen	93.5	3,832	16.4	13.8	..	..	..	..	33.9	22.1	0.8	146	7.6	11.0
161	Lesotho	5.2	2,494	36.7	37.7	12.3	58.7	17.4	0.0	1.7	30.9	1.4	117	4.4	6.4
162	Togo	9.2	1,346	18.3	9.6	15.8	16.4	10.1	0.3	36.0	24.4	1.5	108	6.8	15.5
163	Haiti	17.0	1,648	..	9.1	..	..	..	..	20.4	14.9	0.1	122	9.7	3.4
163	Rwanda	16.8	1,426	25.5	14.2	1.0	13.4	25.7	..	8.0ⁱ	23.0	0.6	117	8.6	10.5
163	Uganda	51.4	1,368	23.8	8.3	7.5	13.0	30.6	0.6	16.3	21.0	0.4	143	5.2	21.8
166	Benin	17.9	1,733	26.3	11.2	4.2	15.6	16.8	..	21.5	28.7	1.4	111	8.1	21.8
167	Sudan	123.9	3,265	20.0	7.5	20.9	..	..	..	24.0	47.9	0.5	218	..	..
168	Djibouti	2.5	2,903	37.5	25.1	8.0	..	..	..	33.9	62.5	2.3ⁱ	112	..	..
169	South Sudan	22.2	1,965	11.9	23.0	10.9	..	..	..	..	..	..	149ᵍ	..	..
170	Senegal	30.7	2,170	25.1	15.5	−1.1	19.2	23.1	0.5	35.1	34.9	2.7	106	8.4	8.7
171	Afghanistan	57.6	1,884	17.7	12.3	..	7.5	3.7	..	−3.9	12.3	0.1	127	..	..
172	Côte d'Ivoire	63.1	3,107	17.0	8.4	8.7	14.2	21.4	..	26.9	37.9	4.2	109	6.7	8.8
173	Malawi	12.4	755	19.7	22.1	11.6	..	..	..	31.2	43.6	1.2	166	7.6	23.6

		Gross domestic product (GDP)		Gross fixed capital formation	General government final consumption expenditure		Total tax revenue	Taxes on income, profit and capital gain	Research and development expenditure	Debts			Prices		
										Domestic credit provided by financial sector	External debt stock	Total debt service	Consumer price index	Domestic food price level	
		Total (2011 PPP $ billions)	Per capita (2011 PPP $)	(% of GDP)	Total (% of GDP)	Average annual growth (%)	(% of GDP)	(% of total tax revenue)	(% of GDP)	(% of GDP)	(% of GNI)	(% of GNI)	(2010=100)	Index	Volatility index
HDI rank		2013	2013	2005–2013[a]	2005–2013[a]	2005–2013[a]	2005–2013[a]	2005–2013[a]	2005–2012[a]	2013	2005–2013[a]	2013	2013	2009–2014[a]	2009–2014[a]
174	Ethiopia	125.7	1,336	35.8	8.3	..	9.2	16.0	0.2	36.9[c]	26.8	1.4	177	6.3	9.0
175	Gambia	3.0	1,608	20.7	7.5	3.0	15.1	18.0	0.1	50.1	59.0	3.1	115	7.3	2.7
176	Congo (Democratic Republic of the)	52.9	783	20.6	12.4	14.2	8.4	11.9	0.1	7.3	21.9	1.1	129	..	..
177	Liberia	3.6	850	25.4	15.5	6.2	20.9	27.7	..	38.7	30.9	0.3	125	..	..
178	Guinea-Bissau	2.3	1,362	..	..	..	..	..	..	18.6	32.3	0.2	108	..	..
179	Mali	24.3	1,589	16.3	11.3	–4.2	15.8	21.9	0.7	20.9	33.3	0.9	108	7.7	9.4
180	Mozambique	27.6	1,070	17.4	20.4	16.5	20.8	29.5	0.5	29.3	45.0	0.9	117	8.6	6.7
181	Sierra Leone	9.1	1,495	16.6	10.4	6.5	11.7	31.0	..	14.5	31.1	0.6	145	6.8	3.3
182	Guinea	14.2	1,213	15.0	9.7	5.8	..	..	..	32.2[g]	20.8	1.1	156	9.9	7.3
183	Burkina Faso	26.8	1,582	17.6	18.2	8.4	16.3	20.1	0.2	25.4	23.2	0.7	107	8.4	11.8
184	Burundi	7.6	747	28.7	22.1	2.0	..	..	0.1	23.9	23.5	1.2	140	7.0	8.3
185	Chad	25.9	2,022	27.4	7.6	10.3	..	..	..	7.0	17.2	0.8	110	8.0	11.7
186	Eritrea	7.3	1,157	10.0	21.1	–9.5	..	..	..	104.0[g]	27.7	2.6	..	..	..
187	Central African Republic	2.7	584	14.8	7.7	1.0	9.5	6.9	..	36.7	37.4	0.4	109	..	..
188	Niger	15.8	887	34.4	14.6	7.9	11.3	11.6	..	11.8	36.3	0.6	106	7.2	9.4
OTHER COUNTRIES OR TERRITORIES															
	Korea (Democratic People's Rep. of)	..	..	..	..	..	..	..	..	..	..	..	..	..	..
	Marshall Islands	0.2	3,776	..	..	..	..	..	..	..	..	..	..	..	..
	Monaco	..	..	..	..	..	..	..	0.0	..	..	..	..	..	..
	Nauru	..	..	..	..	..	..	..	..	..	..	..	..	..	..
	San Marino	..	..	..	..	..	22.3	14.6	..	..	..	..	107	..	..
	Somalia	..	..	..	..	..	..	..	..	..	..	..	..	..	..
	Tuvalu	0.0	3,528	..	..	..	..	..	..	..	..	..	..	..	..
Human development groups															
	Very high human development	46,814.6	41,395	20.5	18.2	3.0	14.3	36.8	2.4	196.7	..	..	—	—	—
	High human development	33,466.1	13,549	33.8	15.5	4.1	13.3	22.9	1.4	114.6	21.8	2.5	—	—	—
	Medium human development	13,654.0	6,106	25.9	12.5	6.5	13.1	40.0	0.5	73.0	27.2	2.7	—	—	—
	Low human development	3,205.5	2,904	17.3	11.4	5.3	8.7	25.5	..	28.7	20.2	1.8	—	—	—
Developing countries		49,538.3	8,696	31.3	14.4	6.1	12.5	29.5	1.1	101.3	22.2	2.4	—	—	—
Regions															
	Arab States	5,508.7	16,697	22.2	16.2	11.8	12.2	41.3	..	37.8	25.5	1.9	—	—	—
	East Asia and the Pacific	20,776.2	10,779	43.8	13.5	..	11.1	28.4	..	149.6	15.0	1.2	—	—	—
	Europe and Central Asia	3,005.8	12,929	22.0	15.0	1.8	19.3	..	0.7	70.5	54.5	9.8	—	—	—
	Latin America and the Caribbean	7,911.4	13,877	20.0	16.7	2.7	15.6	26.0	..	74.1	27.4	3.5	—	—	—
	South Asia	9,305.8	5,324	26.6	11.3	–0.1	10.3	38.4	0.7	64.3	20.3	2.0	—	—	—
	Sub-Saharan Africa	2,977.6	3,339	19.5	14.2	6.1	13.5	42.8	0.4	62.7	23.7	1.8	—	—	—
Least developed countries		1,770.8	2,122	23.8	11.9	7.5	..	..	..	30.4	28.1	1.3	—	—	—
Small island developing states		513.1	9,391	18.0	20.7	..	..	..	..	54.7	65.9	8.7	—	—	—
Organisation for Economic Co-operation and Development		46,521.4	36,923	20.5	18.2	1.1	14.6	36.4	2.5	199.1	..	..	—	—	—
World		**97,140.4**	**13,964**	**24.3**	**17.0**	**3.6**	**13.9**	**33.7**	**2.0**	**164.0**	**23.6**	**2.9**	**—**	**—**	**—**

NOTES

a Data refer to the most recent year available during the period specified.

b Refers to 2006.

c Refers to 2008.

d Refers to 2010.

e Refers to 2009.

f Refers to 2012.

g Refers to 2011.

h Refers to 2007.

i Refers to 2004.

j Refers to 2005.

DEFINITIONS

Gross domestic product (GDP): Sum of gross value added by all resident producers in the economy plus any product taxes and minus any subsidies not included in the value of the products, expressed in 2011 international dollars using purchasing power parity (PPP) rates.

GDP per capita: GDP in a particular period divided by the total population for the same period.

Gross fixed capital formation: Value of acquisitions of new or existing fixed assets by the business sector, governments and households (excluding their unincorporated enterprises) less disposals of fixed assets, expressed as a percentage of GDP. No adjustment is made for depreciation of fixed assets.

General government final consumption expenditure: All government current expenditures for purchases of goods and services (including compensation of employees and most expenditures on national defense and security but excluding government military expenditures that are part of government capital formation), expressed as a percentage of GDP.

Tax revenue: Compulsory transfers to the central government for public purposes, expressed as a percentage of GDP.

Taxes on income, profit and capital gain: Taxes levied on the actual or presumptive net income of individuals, on the profits of corporations and enterprises and on capital gains, whether realized or not, on land, securities and other assets.

Research and development expenditure: Current and capital expenditures (both public and private) on creative work undertaken systematically to increase knowledge and the use of knowledge for new applications, expressed as a percentage of GDP. It covers basic research, applied research and experimental development.

Domestic credit provided by financial sector: Credit to various sectors on a gross basis (except credit to the central government, which is net), expressed as a percentage of GDP.

External debt stock: Debt owed to nonresidents repayable in foreign currency, goods or services, expressed as a percentage of gross national income (GNI).

Total debt service: Sum of principal repayments and interest actually paid in foreign currency, goods or services on long-term debt; interest paid on short-term debt; and repayments (repurchases and charges) to the International Monetary Fund, expressed as a percentage of GNI.

Consumer price index: Index that reflects changes in the cost to the average consumer of acquiring a basket of goods and services that may be fixed or changed at specified intervals, such as yearly.

Domestic food price level index: Food purchasing power parity rate divided by the general PPP rate. The index shows the price of food in a country relative to the price of the generic consumption basket in the country.

Domestic food price level volatility index: A measure of variation of the domestic food price level index, computed as the standard deviation of the deviations from the trend over the previous five years.

MAIN DATA SOURCES

Columns 1–12: World Bank (2015b).

Columns 13 and 14: FAO (2015a).

TABLE **11**

TABLE 11 National income and composition of resources | 249

TABLE 12

Environmental sustainability

HDI rank	Primary energy supply — Fossil fuels (% of total) 2012[a]	Primary energy supply — Renewable sources (% of total) 2012[a]	Electrification rate — Total (% of population) 2012	Electrification rate — Rural (% of rural population) 2012	Carbon dioxide emissions per capita (tonnes) 2011	Carbon dioxide emissions per capita — Average annual growth (%) 1970/2011	Natural resource depletion (% of GNI) 2008–2013[b]	Forest area (% of total land area) 2012	Forest area (% change) 1990/2012	Fresh water withdrawals (% of total renewable water resources) 2005–2014[b]	Deaths of children under age 5 due to — Outdoor air pollution 2008	Deaths of children under age 5 due to — Indoor air pollution 2004	Deaths of children under age 5 due to — Poor water, sanitation or hygiene 2004	Population living on degraded land (%) 2010	Natural disasters — Population affected (average annual per million people) 2005/2012
VERY HIGH HUMAN DEVELOPMENT															
1 Norway	57.3	47.8	100.0	100.0	9.2	0.8	9.0	28.0	11.9	0.8	0	0	0	0.2	12
2 Australia	95.4	4.6	100.0	100.0	16.5	1.1	3.8	19.2	−4.6	3.9	0	0	0	9.0	1,337
3 Switzerland	51.1	49.7 c	100.0	100.0	4.6	−1.1	0.0	31.6	8.6	..	0	0	0	0.5	73
4 Denmark	70.6	26.8	100.0	100.0	7.2	−1.8	1.5	12.9	23.0	10.8	0	0	1	8.5	0
5 Netherlands	91.4	6.7 c	100.0	100.0	10.1	−0.3	0.9	10.8	5.9	11.7	0	0	0	5.4	0 d
6 Germany	80.2	20.4 c	100.0	100.0	8.9	..	0.1	31.8	3.3	21.0	0	0	0	8.1	10
6 Ireland	84.7	6.4	100.0	100.0	7.9	0.6	0.1	11.0	62.7	..	0	0	..	0.5	14
8 United States	83.6	16.3 c	100.0	100.0	17.0	−0.7	1.0	33.3	3.0	15.5	0	0	0	1.1	5,074
9 Canada	73.7	27.9 c	100.0	100.0	14.1	−0.4	2.4	34.1	0.0	..	0	0	0	2.7	364
9 New Zealand	61.4	38.4	100.0	100.0	7.1	1.1	1.6	31.3	6.9	..	0	0	..	5.3	14,226
11 Singapore	97.2	2.8	100.0	99.0	4.3	−2.3	..	3.3	−4.3	..	0	0	0	..	0
12 Hong Kong, China (SAR)	94.8	0.4	100.0	100.0	5.7	3.2	..	..	..	..	..	..	..	..	221
13 Liechtenstein	..	..	100.0	100.0	1.4	..	..	43.1	6.2	..	..	..	..	..	..
14 Sweden	31.7	70.5 c	100.0	100.0	5.5	−2.3	0.3	69.2	4.1	1.5	0	0	0	0.3	0
14 United Kingdom	85.1	14.4 c	100.0	100.0	7.1	−1.6	0.9	12.0	10.9	8.8	0	0	..	2.7	665
16 Iceland	15.3	84.7	100.0	100.0	5.9	−0.5	0.0	0.3	264.4	0.1	0	0	0	..	0
17 Korea (Republic of)	82.8	17.2 c	100.0	100.0	11.8	6.5	0.0	63.8	−3.4	..	0	0	..	2.9	206
18 Israel	96.7	4.8	100.0	100.0	9.0	1.5	0.2	7.1	16.4	..	0	0	..	12.9	27,775
19 Luxembourg	87.4	4.0	100.0	100.0	20.9	−2.1	0.0	33.5	..	..	0	0	2	..	0
20 Japan	94.8	5.2 c	100.0	100.0	9.3	0.8	0.0	68.6	0.2	..	0	0	0	0.3	921
21 Belgium	70.1	28.3 c	100.0	100.0	8.8	−1.2	0.0	22.5	..	34.0	0	0	0	10.5	9
22 France	49.1	52.4 c	100.0	100.0	5.2	−1.6	0.0	29.3	10.4	14.8	0	0	..	3.9	819
23 Austria	67.1	32.2	100.0	100.0	7.8	0.4	0.2	47.3	3.3	..	0	0	0	2.7	19
24 Finland	43.0	47.5 c	100.0	100.0	10.2	0.5	0.2	72.9	1.5	1.5	0	0	0	0.0	7
25 Slovenia	66.6	34.5 c	100.0	100.0	7.5	..	0.3	62.4	5.8	3.0	0	0	..	8.4	3,114
26 Spain	75.9	24.9 c	100.0	100.0	5.8	1.7	0.0	37.1	34.2	28.6	0	0	0	1.4	51
27 Italy	83.7	13.9	100.0	100.0	6.7	0.6	0.1	31.6	22.6	..	0	0	..	2.2	153
28 Czech Republic	76.9	26.5 c	100.0	100.0	10.4	..	0.2	34.5	1.3	12.9	0	0	1	4.2	12,572
29 Greece	90.6	8.8	100.0	100.0	7.6	3.3	0.1	30.7	20.1	13.8	0	0	..	1.1	827
30 Estonia	88.1	14.6	100.0	100.0	14.0	..	1.1	51.8	5.1	14.0	0	0	0	5.0	8
31 Brunei Darussalam	100.0	0.0	76.2	67.1	24.0	−3.1	29.8	71.4	−8.9	..	0	0	..	..	0
32 Cyprus	94.9	5.1	100.0	100.0	6.7	2.9	0.0	18.8	7.6	17.6	0	0	13	11.4	0
32 Qatar	100.0	0.0	97.7	92.9	43.9	−1.5	17.4	0.0	—	374.1	1	0	6	0.1	..
34 Andorra	..	..	100.0	100.0	6.3	..	..	34.0	0.0	..	0	0	0	..	..
35 Slovakia	67.5	32.3 c	100.0	100.0	6.4	..	0.5	40.2	0.6	1.4	0	0	0	9.1	19
36 Poland	90.7	9.6	100.0	100.0	8.3	−0.4	1.0	30.7	5.8	19.4	0	0	..	13.2	279
37 Lithuania	74.0	14.5	100.0	100.0	4.5	..	0.4	34.7	11.9	9.5	0	0	..	4.8	..
37 Malta	94.5	5.5	100.0	100.0	6.0	3.3	..	0.9	0.0	..	0	0	..	..	..
39 Saudi Arabia	100.0	0.0	97.7	92.9	18.7	2.9	20.6	0.5	0.0	943.3	2	0	..	4.3	41
40 Argentina	89.7	9.3 c	99.8	95.8	4.7	1.0	3.1	10.6	−16.9	4.3	0	0	3	1.7	1,667
41 United Arab Emirates	..	0.1	97.7	92.9	20.0	−3.8	11.7	3.8	30.4	1,867.0	1	0	10	1.9	..
42 Chile	75.6	24.2	99.6	97.8	4.6	1.9	8.3	21.9	6.8	3.8	0	0	1	1.1	24,051
43 Portugal	74.9	22.0	100.0	100.0	4.7	3.2	0.1	37.8	4.0	..	0	0	..	2.3	48
44 Hungary	71.1	26.0 c	100.0	100.0	4.9	−1.1	0.4	22.6	12.9	5.4	0	0	0	17.1	1,055
45 Bahrain	99.9	0.0	97.7	92.9	18.1	1.3	9.7	0.7	154.2	..	0	0	..	..	..
46 Latvia	63.7	33.8	100.0	100.0	3.8	..	1.0	54.3	6.4	..	0	0	0	1.8	0
47 Croatia	81.6	10.6	100.0	100.0	4.8	..	1.6	34.4	4.1	0.6	0	0	0	17.5	277
48 Kuwait	100.0	0.0	97.7	92.9	29.1	−0.4	25.1	0.4	86.6	..	1	0	..	0.6	0
49 Montenegro	60.2	28.4	100.0	100.0	4.1	..	..	40.4	0.0	..	..	..	..	8.0	2,000
HIGH HUMAN DEVELOPMENT															
50 Belarus	90.4	5.9	100.0	100.0	6.7	..	1.5	42.9	11.9	..	0	0	1	4.7	472
50 Russian Federation	91.0	9.2 c	100.0	100.0	12.6	..	11.8	49.4	0.1	..	0	0	5	3.1	161
52 Oman	100.0	0.0	97.7	92.9	21.4	14.5	25.7	0.0	0.0	..	1	0	..	5.8	682
52 Romania	77.7	22.8 c	100.0	100.0	4.2	−1.1	1.5	28.9	4.0	3.3	1	6	..	13.5	562
52 Uruguay	57.0	42.1	99.5	95.1	2.3	0.4	1.6	10.5	99.3	..	0	0	3	5.7	3,682
55 Bahamas	..	..	100.0	100.0	5.2	−3.4	0.0	51.4	0.0	..	0	0	2	..	6,305
56 Kazakhstan	98.9	1.0	100.0	100.0	15.8	..	17.2	1.2	−3.6	18.4	5	3	249	23.5	634
57 Barbados	..	..	90.9	79.8	5.6	3.7	0.8	19.4	0.0	87.5	0	0	0	..	894
58 Antigua and Barbuda	..	..	90.9	79.8	5.8	−0.6	..	22.3	−4.9	8.5	0	1	0	..	35,508

HDI rank		Primary energy supply		Electrification rate		Carbon dioxide emissions per capita		Natural resource depletion	Forest area		Fresh water withdrawals	Deaths of children under 5 due to (per 100,000 children under age 5)			Population living on degraded land	Natural disasters
		Fossil fuels	Renewable sources	Total	Rural							Outdoor air pollution	Indoor air pollution	Poor water, sanitation or hygiene		
		(% of total)		(% of population)	(% of rural population)	(tonnes)	Average annual growth (%)	(% of GNI)	(% of total land area)	(% change)	(% of total renewable water resources)				(%)	Population affected (average annual per million people)
		2012[a]	2012[a]	2012	2012	2011	1970/2011	2008–2013[b]	2012	1990/2012	2005–2014[b]	2008	2004	2004	2010	2005/2012
59	Bulgaria	75.0	29.4[c]	100.0	100.0	6.7	−0.2	0.9	37.2	23.7	28.7	1	2	2	7.8	808
60	Palau	..	..	59.3	45.5	10.9	−0.3	..	87.6	..	..	0	0	40	..	0
60	Panama	79.7	20.2	90.9	79.8	2.6	1.9	0.2	43.4	−14.9	0.7	0	16	55	4.1	2,457
62	Malaysia	94.5	5.5	100.0	100.0	7.8	5.9	8.1	61.7	−9.4	1.9	0	0	33	1.2	10,160
63	Mauritius	..	..	100.0	100.0	3.1	5.4	0.0	17.3	−9.7	..	0	0	7	..	214
64	Seychelles	..	..	100.0	17.3	6.8	8.5	0.1	88.5	0.0	..	0	0	..	..	14,228
64	Trinidad and Tobago	99.9	0.1	99.8	99.0	37.2	4.5	29.3	43.9	−6.5	8.8	0	1	5	..	0
66	Serbia	89.1	11.1	100.0	100.0	6.8	..	..	32.1	21.4	2.5	..	..	..	18.5	18,081
67	Cuba	86.7	13.3	100.0	95.4	3.2	1.3	3.6	27.6	44.1	18.3	0	1	1	17.0	30,624
67	Lebanon	95.5	3.3	100.0	100.0	4.7	3.3	0.0	13.4	4.6	24.3	1	0	40	1.2	0[d]
69	Costa Rica	48.3	51.8	99.5	98.7	1.7	2.9	0.8	51.9	3.4	2.1	0	2	4	1.3	9,470
69	Iran (Islamic Republic of)	99.5	0.7[c]	100.0	97.3	7.8	2.9	17.6	6.8	0.0	..	6	3	..	25.1	1,467
71	Venezuela (Bolivarian Republic of)	88.9	11.2	100.0	100.0	6.4	−0.3	10.0	51.8	−12.2	1.7	0	1	30	1.9	545
72	Turkey	89.5	10.3	100.0	100.0	4.4	4.2	0.3	15.0	19.5	..	2	11	85	5.5	217
73	Sri Lanka	48.7	51.3	88.7	86.0	0.7	3.1	0.4	29.2	−22.1	24.5	0	8	42	21.1	46,648
74	Mexico	90.1	9.9[c]	99.1	97.2	3.9	1.9	6.4	33.2	−8.3	17.2	1	8	23	3.8	9,882
75	Brazil	54.6	44.2[c]	99.5	97.0	2.2	2.7	3.8	61.6	−10.4	0.9	0	18	123	7.9	4,506
76	Georgia	72.8	28.3	100.0	100.0	1.8	..	0.7	39.4	−1.5	2.9	2	70	169	1.9	3,301
77	Saint Kitts and Nevis	..	..	90.9	79.8	5.1	7.3	..	42.3	0.0	51.3	0	0	28	..	0
78	Azerbaijan	97.9	2.6	100.0	100.0	3.6	..	26.0	11.3	0.7	34.5	2	132	269	3.8	1,079
79	Grenada	..	..	90.9	79.8	2.4	5.4	..	50.0	0.0	7.1	0	12	5	..	1,578
80	Jordan	96.0	2.0	99.5	99.4	3.6	4.1	0.9	1.1	−0.6	92.4	3	0	59	22.0	0
81	The former Yugoslav Republic of Macedonia	82.1	10.4	100.0	100.0	4.4	..	2.7	39.9	11.4	16.1	0	1	..	7.1	48,256
81	Ukraine	79.6	20.7[c]	100.0	100.0	6.3	..	4.6	16.8	5.2	13.8	0	0	3	6.2	944
83	Algeria	99.9	0.1	100.0	100.0	3.2	3.8	18.0	0.6	−11.6	..	1	5	101	28.8	343
84	Peru	76.0	24.0	91.2	72.9	1.8	0.9	5.7	52.9	−3.5	0.7	2	21	69	−0.7	13,408
85	Albania	60.5	26.6	100.0	100.0	1.6	−0.3	3.5	28.2	−1.9	4.3	0	5	50	5.7	20,568
85	Armenia	71.5	32.7[c]	100.0	100.0	1.7	..	2.2	8.9	−26.9	37.9	2	17	65	9.6	2,549
85	Bosnia and Herzegovina	93.9	7.9	100.0	100.0	6.2	..	..	42.8	−1.1	0.9	1	1	2	6.1	27,578
88	Ecuador	86.3	12.9	97.2	92.3	2.3	3.9	8.6	38.1	−23.6	2.2	1	2	63	1.6	6,002
89	Saint Lucia	..	..	90.9	79.8	2.3	4.2	0.1	77.0	7.3	14.3	0	3	2	..	115,690
90	China	88.3	11.7[c]	100.0	100.0	6.7	6.5	4.2	22.6	35.2	19.5	2	10	55	8.6	73,314
90	Fiji	..	..	59.3	45.5	1.4	1.1	1.7	55.9	7.1	..	1	18	11	..	9,681
90	Mongolia	95.4	4.1	89.8	69.9	6.9	3.7	19.8	6.9	−14.4	1.6	19	78	195	31.5	29,190
93	Thailand	80.4	18.9	100.0	99.8	4.6	8.0	4.7	37.2	−2.8	13.1	0	21	59	17.0	70,701
94	Dominica	..	..	92.7	79.8	1.7	5.2	0.1	58.8	−11.8	10.0	0	1	0	..	10,905
94	Libya	98.7	1.3	100.0	100.0	6.4	−2.8	23.5	0.1	0.0	..	3	2	..	8.5	34
96	Tunisia	85.3	14.8	100.0	100.0	2.4	3.9	4.6	6.7	61.6	69.7	1	3	64	36.7	62
97	Colombia	75.6	24.8	97.0	87.9	1.5	0.5	9.2	54.3	−3.6	0.5	1	6	33	2.0	18,001
97	Saint Vincent and the Grenadines	..	..	75.9	31.8	2.2	6.3	0.1	68.9	6.2	7.9	0	2	..	..	22,280
99	Jamaica	82.1	17.9	92.6	86.7	2.9	0.2	0.8	31.0	−2.4	7.5	1	15	47	3.3	13,248
100	Tonga	..	..	95.9	82.8	1.0	3.9	0.1	12.5	0.0	..	0	16	55	..	4,364
101	Belize	..	..	100.0	100.0	1.7	1.8	5.9	60.2	−13.4	..	0	21	27	1.1	22,279
101	Dominican Republic	89.3	10.7	98.0	96.7	2.2	3.8	0.2	40.8	0.0	30.4	2	12	73	7.0	4,057
103	Suriname	..	..	100.0	100.0	3.6	−0.6	20.5	94.6	−0.2	0.6	0	0	43	..	6,040
104	Maldives	..	..	100.0	100.0	3.3	..	0.0	3.0	0.0	15.7	1	41	167	..	908
105	Samoa	..	..	100.0	92.8	1.3	6.0	0.8	60.4	31.5	..	0	26	63	..	9,854
MEDIUM HUMAN DEVELOPMENT																
106	Botswana	65.4	22.3	53.2	23.9	2.4	..	1.6	19.6	−19.0	..	4	210	341	22.0	1,610
107	Moldova (Republic of)	94.9	3.4	100.0	100.0	1.4	..	0.2	12.0	23.9	9.1	1	13	15	21.8	6,840
108	Egypt	96.5	3.7	100.0	100.0	2.8	5.1	8.1	0.1	61.8	..	2	2	86	25.3	5
109	Turkmenistan	..	0.0	100.0	100.0	12.2	..	37.0	8.8	0.0	..	2	2	449	11.1	0
110	Gabon	38.9	61.1	89.3	44.9	1.4	−2.9	29.1	85.4	0.0	0.1	9	33	102	..	6,531
110	Indonesia	66.4	33.6	96.0	92.9	2.3	6.7	4.8	51.4	−21.5	..	2	41	130	3.1	4,292
112	Paraguay	33.8	145.2	98.2	96.3	0.8	3.2	4.6	43.4	−18.6	0.6	1	21	56	1.3	39,146
113	Palestine, State of	..	..	97.7	92.9	0.6	..	..	1.5	1.0	48.8	..	..	..	..	1,624
114	Uzbekistan	98.2	1.8	100.0	100.0	3.9	..	13.8	7.7	7.3	100.6	1	192	325	27.0	6
115	Philippines	59.7	40.3	87.5	81.5	0.9	0.7	2.2	26.1	18.3	17.0	1	37	96	2.2	105,941
116	El Salvador	47.9	51.9	93.7	85.7	1.1	3.4	1.7	13.4	−26.2	8.1	1	24	82	6.3	9,378
116	South Africa	87.2	12.9[c]	85.4	66.9	9.3	1.0	4.8	7.6	0.0	..	2	23	104	17.5	860

TABLE 12 Environmental sustainability | 251

TABLE 12 ENVIRONMENTAL SUSTAINABILITY

HDI rank		Primary energy supply — Fossil fuels (% of total) 2012ᵃ	Primary energy supply — Renewable sources (% of total) 2012ᵃ	Electrification rate — Total (% of population) 2012	Electrification rate — Rural (% of rural population) 2012	Carbon dioxide emissions per capita (tonnes) 2011	Carbon dioxide emissions per capita — Average annual growth (%) 1970/2011	Natural resource depletion (% of GNI) 2008–2013ᵇ	Forest area (% of total land area) 2012	Forest area (% change) 1990/2012	Fresh water withdrawals (% of total renewable water resources) 2005–2014ᵇ	Deaths of children under age 5 due to — Outdoor air pollution 2008	Deaths of children under age 5 due to — Indoor air pollution 2004	Deaths of children under age 5 due to — Poor water, sanitation or hygiene 2004	Population living on degraded land (%) 2010	Natural disasters — Population affected (average annual per million people) 2005/2012
116	Viet Nam	71.0	28.2	99.0	97.7	2.0	3.6	6.7	45.4	57.9	9.3	1	27	65	8.0	20,060
119	Bolivia (Plurinational State of)	72.7	27.3	90.5	72.5	1.6	3.2	13.0	52.2	−9.9	0.4	0	93	245	2.0	25,572
120	Kyrgyzstan	68.4	39.4	100.0	100.0	1.2	..	7.9	5.1	18.1	32.6	1	115	245	9.7	38,560
121	Iraq	97.5	1.0	100.0	96.9	4.2	1.8	19.5	1.9	3.3	..	12	12	383	4.5	231
122	Cabo Verde	..	..	70.6	46.8	0.9	6.2	0.5	21.3	48.3	..	0	26	93	..	4,351
123	Micronesia (Federated States of)	..	..	59.3	45.5	1.2	..	0.1	91.7	..	..	0	30	83	..	0
124	Guyana	..	..	79.5	75.1	2.3	0.1	8.8	77.2	0.0	0.5	0	38	132	..	52,340
125	Nicaragua	49.8	50.3	77.9	42.7	0.8	1.1	4.7	24.7	−34.1	0.9	1	49	102	13.9	10,726
126	Morocco	93.6	4.1	100.0	100.0	1.8	4.4	1.5	11.5	2.0	35.7	6	8	114	39.1	737
126	Namibia	66.0	21.0	47.3	17.4	1.3	..	1.0	8.7	−18.5	..	1	11	21	28.5	63,965
128	Guatemala	33.5	66.2	78.5	72.1	0.8	1.9	3.6	33.1	−25.3	2.6	2	57	126	9.1	50,204
129	Tajikistan	42.9	57.5	100.0	100.0	0.4	..	1.3	2.9	0.5	51.1	1	343	551	10.5	38,572
130	India	72.3	27.6ᶜ	78.7	69.7	1.7	5.2	3.6	23.1	7.5	33.9	5	131	316	9.6	11,986
131	Honduras	51.6	48.8	82.2	65.8	1.1	2.4	2.8	44.3	−39.1	..	1	49	106	15.0	23,856
132	Bhutan	..	..	75.6	52.8	0.8	14.2	16.6	85.8	32.1	0.4	0	124	324	0.1	2,821
133	Timor-Leste	..	..	41.6	26.8	0.2	..	..	48.4	−25.5	..	0	0	149	..	951
134	Syrian Arab Republic	98.7	1.4	96.3	81.1	2.6	3.0	..	2.7	35.3	84.2	2	12	54	33.3	6,280
134	Vanuatu	..	..	27.1	17.8	0.6	0.7	0.0	36.1	0.0	..	0	9	41	..	28,826
136	Congo	48.9	51.0	41.6	11.7	0.5	0.7	54.4	65.6	−1.5	..	19	149	220	0.1	1,463
137	Kiribati	..	..	59.3	45.5	0.6	0.7	0.1	15.0	0.0	..	0	0	206	..	314
138	Equatorial Guinea	..	..	66.0	43.0	9.3	14.9	67.6	57.1	−13.8	..	10	0	505	..	0
139	Zambia	8.8	91.8	22.1	5.8	0.2	−4.4	10.5	66.1	−6.9	..	12	378	503	4.6	26,183
140	Ghana	37.4	63.1	64.1	41.0	0.4	0.9	12.6	20.7	−36.8	..	3	152	226	1.4	3,055
141	Lao People's Democratic Republic	..	..	70.0	54.8	0.2	−0.5	8.3	67.6	−9.9	1.1	1	157	242	4.1	22,280
142	Bangladesh	71.5	28.5	59.6	49.3	0.4	..	2.8	11.0	−3.8	2.9	2	142	334	11.3	28,112
143	Cambodia	26.2	71.1	31.1	18.8	0.3	2.0	2.5	55.7	−24.0	0.5	3	346	595	39.3	28,828
143	Sao Tome and Principe	..	..	60.5	47.0	0.6	3.4	2.3	28.1	0.0	..	9	225	428	..	0
LOW HUMAN DEVELOPMENT																
145	Kenya	19.7	80.3	23.0	6.7	0.3	0.5	3.3	6.1	−7.1	..	4	217	362	31.0	46,271
145	Nepal	12.5	86.9	76.3	71.6	0.2	7.0	4.2	25.4	−24.7	4.5	1	139	337	2.3	8,366
147	Pakistan	60.9	39.1ᶜ	93.6	90.5	0.9	2.7	3.1	2.1	−36.6	74.4	22	132	205	4.5	29,014
148	Myanmar	21.3	78.7	52.4	31.2	0.2	0.5	..	47.7	−20.5	..	3	181	378	19.2	6,406
149	Angola	39.3	60.7	37.0	6.0	1.5	2.9	31.0	46.7	−4.5	0.5	11	1,073	1,266	3.3	13,473
150	Swaziland	..	..	42.0	24.5	0.9	0.2	1.8	33.2	21.1	..	2	148	252	..	35,652
151	Tanzania (United Republic of)	10.7	89.3	15.3	3.6	0.2	0.6	3.2	36.8	−21.4	..	4	239	322	25.0	11,026
152	Nigeria	17.4	82.6	55.6	34.4	0.5	1.1	8.1	9.0	−52.3	4.6	14	370	559	11.5	5,667
153	Cameroon	26.8	73.2	53.7	18.5	0.3	3.4	5.1	41.2	−19.9	..	14	361	497	15.3	607
154	Madagascar	..	..	15.4	8.1	0.1	−0.9	3.7	21.4	−9.2	..	2	390	540	0.0	22,638
155	Zimbabwe	28.3	70.3	40.5	16.1	0.7	−2.4	5.6	38.7	−32.5	..	5	168	256	29.4	46,023
156	Mauritania	..	..	21.8	4.4	0.6	1.7	32.6	0.2	−44.1	11.8	16	220	390	23.8	45,968
156	Solomon Islands	..	..	22.8	12.6	0.4	1.2	31.3	78.7	−5.3	..	0	54	84	..	19,098
158	Papua New Guinea	..	..	18.1	10.4	0.7	3.2	18.7	62.8	−9.8	0.1	1	108	288	..	7,920
159	Comoros	..	..	69.3	61.4	0.2	1.9	3.5	1.2	−81.7	..	2	108	177	..	55,515
160	Yemen	98.5	1.5	48.4	33.5	1.0	2.8	7.5	1.0	0.0	168.6	5	174	377	32.4	360
161	Lesotho	..	..	20.6	10.2	1.1	..	4.5	1.5	11.0	..	2	19	44	63.6	60,491
162	Togo	15.2	82.4	31.5	8.9	0.3	3.0	9.2	4.9	−61.0	..	5	302	419	5.1	4,818
163	Haiti	22.0	78.0	37.9	15.0	0.2	3.3	1.9	3.6	−14.3	10.3	5	297	428	15.2	53,388
163	Rwanda	..	..	18.0	7.7	0.1	4.4	6.1	18.4	43.1	..	2	803	970	10.1	369
163	Uganda	..	..	18.2	8.1	0.1	−1.1	13.2	14.1	−40.8	1.1	2	327	427	23.5	10,376
166	Benin	41.7	56.2	38.4	14.5	0.5	5.5	1.9	39.6	−22.6	..	8	394	518	1.6	13,001
167	Sudan	29.5	70.5	32.6	17.8	0.4	0.0	4.8	23.2ᵉ	−27.9	71.2	11	181	255	39.9	27,986
168	Djibouti	..	..	53.3	13.0	0.6	−1.4	..	0.2	0.0	..	31	41	454	7.5	88,442
169	South Sudan	..	..	5.1	3.5	..	..	..	..	..	1.3	..	..	..	..	7,598
170	Senegal	53.2	46.4	56.5	26.6	0.6	2.1	1.5	43.6	−10.2	..	14	292	530	16.2	12,059
171	Afghanistan	..	..	43.0	32.0	0.4	3.3	1.2	2.1	0.0	..	21	1,183	1,405	11.0	17,311
172	Côte d'Ivoire	21.5	79.0	55.8	29.0	0.3	−1.1	5.4	32.7	1.8	1.8	9	370	561	1.3	104
173	Malawi	..	..	9.8	2.0	0.1	−0.8	12.5	33.6	−18.6	7.9	3	498	617	19.4	54,758
174	Ethiopia	5.7	94.3	26.6	7.6	0.1	1.2	14.0	12.0	−21.0	..	2	538	705	72.3	25,871
175	Gambia	..	..	34.5	25.7	0.2	2.7	6.8	47.8	9.4	..	7	197	286	17.9	29,355

TABLE 12

	Primary energy supply		Electrification rate		Carbon dioxide emissions per capita		Natural resource depletion	Forest area		Fresh water withdrawals	Deaths of children under age 5 due to (per 100,000 children under age 5)			Population living on degraded land	Natural disasters
	Fossil fuels	Renewable sources	Total	Rural							Outdoor air pollution	Indoor air pollution	Poor water, sanitation or hygiene		Population affected (average annual per million people)
	(% of total)		(% of population)	(% of rural population)	(tonnes)	Average annual growth (%)	(% of GNI)	(% of total land area)	(% change)	(% of total renewable water resources)				(%)	
HDI rank	2012ᵃ	2012ᵃ	2012	2012	2011	1970/ 2011	2008–2013ᵇ	2012	1990/ 2012	2005–2014ᵇ	2008	2004	2004	2010	2005/ 2012
176 Congo (Democratic Republic of the)	4.2	95.8	16.4	5.8	0.1	−3.0	31.0	67.7	−4.3	0.1	16	644	786	0.1	471
177 Liberia	..	..	9.8	1.2	0.2	−4.9	36.4	44.3	−13.4	..	6	676	885	..	14,150
178 Guinea-Bissau	..	..	60.6	21.5	0.2	1.3	15.0	71.2	−9.7	..	12	648	873	1.0	8,168
179 Mali	..	..	25.6	11.9	0.1	2.9	8.1	10.1	−12.4	4.3	9	703	880	59.5	38,351
180 Mozambique	9.5	93.3	20.2	5.4	0.1	−2.7	4.0	49.1	−11.0	..	11	270	388	1.9	18,424
181 Sierra Leone	..	..	14.2	1.2	0.2	−2.6	8.0	37.2	−13.8	0.1	11	1,207	1,473	..	1,028
182 Guinea	..	..	26.2	2.9	0.2	0.7	21.6	26.3	−10.9	..	11	324	480	0.8	1,473
183 Burkina Faso	..	..	13.1	1.4	0.1	5.2	13.9	20.2	−19.2	6.1	9	632	786	73.2	48,243
184 Burundi	..	..	6.5	1.2	0.0	0.6	26.9	6.6	−41.7	..	4	897	1,088	18.5	27,356
185 Chad	..	..	6.4	3.1	0.0	0.9	11.0	9.0	−13.3	1.9	14	488	618	45.4	43,132
186 Eritrea	21.7	78.3	36.1	12.0	0.1	..	17.3	15.1	−6.0	..	3	237	379	58.8	29,975
187 Central African Republic	..	..	10.8	8.2	0.1	−1.8	0.2	36.2	−2.8	0.1	10	411	511	..	2,515
188 Niger	..	..	14.4	5.2	0.1	1.8	16.1	0.9	−39.4	2.9	6	1,023	1,229	25.0	97,330
OTHER COUNTRIES OR TERRITORIES															
Korea (Democratic People's Rep. of)	88.4	11.6	29.6	12.8	3.0	..	..	45.0	−34.0	11.2	3	0	245	2.9	22,195
Marshall Islands	..	..	59.3	45.5	2.0	..	..	70.2	..	..	..	45	201	..	14,022
Monaco	..	..	100.0	100.0	..	..	..	0.0ᶠ	—	..	0	0	2	..	..
Nauru	..	..	..	..	..	..	..	..	..	..	0	1	..	..	..
San Marino	..	..	100.0	100.0	..	..	..	0.0	—	..	0	0	..	..	..
Somalia	..	..	32.7	17.3	0.1	−0.2	..	10.5	−20.4	..	19	710	885	26.3	120,989
Tuvalu	..	..	44.6	31.8	..	..	..	33.3	0.0	..	0	18	148	..	0
Human development groups															
Very high human development	81.8	17.9	99.9	99.7	11.1	−1.2	1.9	27.5	1.5	..	0	0	..	3.2	—
High human development	87.2	12.8	99.6	98.9	6.1	0.3	6.1	35.8	−1.1	4.8	2	10	60	8.9	—
Medium human development	74.5	25.3	81.9	72.8	1.9	0.9	5.3	29.5	−8.9	..	4	106	260	10.2	—
Low human development	..	..	43.6	27.7	0.4	−0.2	7.8	26.2	−14.4	..	10	405	552	20.2	—
Developing countries	80.6	19.2	81.2	68.5	3.3	−1.1	6.3	27.2	−7.2	..	5	161	287	11.6	—
Regions															
Arab States	96.3	3.2	86.9	74.2	4.9	−1.6	15.3	5.9	−22.6	..	6	76	219	24.3	—
East Asia and the Pacific	..	..	95.7	92.5	5.3	1.5	4.4	29.7	2.8	..	2	29	93	..	—
Europe and Central Asia	88.8	10.5	100.0	100.0	5.5	..	6.2	9.1	8.2	33.5	2	63	169	10.7	—
Latin America and the Caribbean	74.2	25.8	96.4	87.0	2.9	0.6	5.5	46.8	−9.6	1.7	1	22	79	5.3	—
South Asia	76.3	23.7	78.9	69.9	1.7	3.1	5.2	14.6	3.4	25.2	7	156	333	10.0	—
Sub-Saharan Africa	..	..	35.4	14.9	0.8	−0.1	10.0	28.2	−11.3	..	8	428	578	22.3	—
Least developed countries	..	..	34.2	20.2	0.3	−0.1	9.6	28.7	−12.4	..	7	440	599	23.5	—
Small island developing states	..	..	70.5	48.0	2.6	0.3	5.6	62.9	−3.8	..	2	121	216	..	—
Organisation for Economic Co-operation and Development	81.2	18.7	99.9	99.7	9.9	−0.5	1.0	30.5	1.2	..	0	2	14	3.4	—
World	**81.2**	**18.6**	**84.5**	**70.9**	**4.6**	**−0.9**	**4.0**	**30.9**	**−3.7**	**..**	**5**	**144**	**263**	**10.2**	**—**

TABLE 12

NOTES

a Data refer to 2012 or the most recent year available.

b Data refer to the most recent year available during the period specified.

c Includes nuclear energy.

d Less than 0.5.

e Refers to 2010.

f Refers to 2011.

DEFINITIONS

Fossil fuels: Percentage of total energy supply that comes from natural resources formed from biomass in the geological past (such as coal, oil and natural gas).

Renewable energy sources: Percentage of total energy supply that comes from constantly replenished natural processes, including solar, wind, biomass, geothermal, hydropower and ocean resources, and some waste. Excludes nuclear energy, unless otherwise noted.

Electrification rate: People with access to electricity, expressed as a percentage of the total population. It includes electricity sold commercially (both on grid and off grid) and self-generated electricity but excludes unauthorized connections.

Carbon dioxide emissions per capita: Human-originated carbon dioxide emissions stemming from the burning of fossil fuels, gas flaring and the production of cement, divided by midyear population. Includes carbon dioxide emitted by forest biomass through depletion of forest areas.

Natural resource depletion: Monetary expression of energy, mineral and forest depletion, expressed as a percentage of gross national income (GNI).

Forest area: Land spanning more than 0.5 hectare with trees taller than 5 metres and a canopy cover of more than 10 percent or trees able to reach these thresholds in situ. Excludes land predominantly under agricultural or urban land use, tree stands in agricultural production systems (for example, in fruit plantations and agroforestry systems) and trees in urban parks and gardens. Areas under reforestation that have not yet reached but are expected to reach a canopy cover of 10 percent and a tree height of 5 meters are included, as are temporarily unstocked areas resulting from human intervention or natural causes that are expected to regenerate.

Fresh water withdrawals: Total fresh water withdrawn, expressed as a percentage of total renewable water resources.

Deaths of children under age 5 due to outdoor air pollution: Deaths of children under age 5 due to respiratory infections and diseases, lung cancer and selected cardiovascular diseases attributable to outdoor air pollution.

Deaths of children under age 5 due to indoor air pollution: Deaths of children of age under 5 due to acute respiratory infections attributable to indoor smoke from solid fuels.

Deaths of children under age 5 due to poor water, sanitation or hygiene: Deaths of children under age 5 due to diarrhoea attributable to unsafe water, unimproved sanitation or poor hygiene.

Population living on degraded land: Percentage of the population living on severely or very severely degraded land. Land degradation estimates consider biomass, soil health, water quantity and biodiversity.

Population affected by natural disasters: Average annual number of people requiring immediate assistance during a period of emergency as a result of a natural disaster, including displaced, evacuated, homeless and injured people, expressed per million people.

MAIN DATA SOURCES

Columns 1, 3–5, 7 and 8: World Bank (2015a).

Column 2: HDRO calculations based on data on total primary energy supply from World Bank (2015a).

Column 6: HDRO calculations based on data on carbon dioxide emissions per capita from World Bank (2015a).

Column 9: HDRO calculations based on data on forest and total land area from World Bank (2015a).

Column 10: FAO (2015b). .

Columns 11–13: WHO (2015).

Column 14: FAO (2011).

Column 15: CRED EM-DAT (2015) and UNDESA (2013a).

TABLE 12 Environmental sustainability | 253

TABLE 13

Work and employment

		Employment								Unemployment				Labour productivity	
		Employment to population ratio[a]	Labour force participation rate[a]	Employment in agriculture		Employment in services		Labour force with tertiary education	Vulnerable employment	Total	Long term	Youth	Youth not in school or employment	Output per worker	Hours worked per week
		(% ages 15 and older)		(% of total employment)				(%)	(% of total employment)	(% of labour force)		(% ages 15–24)		(2005 PPP $)	(per employed person)
HDI rank		2013	2013	1990[b]	2012[c]	1990[b]	2012[c]	2007–2012[d]	2008–2013[d]	2008–2013[d]	2008–2013[d]	2008–2014[d]	2008–2013[d]	2005–2012[d]	2003–2012[d]
VERY HIGH HUMAN DEVELOPMENT															
1	Norway	62.6	64.9	6.4	2.2	69.2	77.4	41.9	5.1	3.5	0.7	9.2	5.6	92,694	27.3
2	Australia	61.5	65.2	5.6	3.3	69.3	75.5	37.3	9.0	5.2	1.1	12.2	4.7	69,987	33.2
3	Switzerland	65.2	68.2	4.2	3.5	63.6	72.5	38.8	9.1	4.4	1.3	8.5	7.1	70,738	31.5
4	Denmark	58.1	62.5	5.5	2.6	66.6	77.5	36.6	5.6	7.0	1.8	12.6	6.0	67,033	29.7
5	Netherlands	60.1	64.4	4.5	2.5	68.6	71.5	36.1	12.3	6.7	2.4	11.0	5.1	72,312	26.6
6	Germany	56.7	59.9	..	1.5	..	70.2	30.8	6.5	5.3	2.3	7.9	6.3	70,030	26.9
6	Ireland	52.6	60.5	12.7	4.7	59.6	76.9	43.7	12.7	13.0	7.8	26.8	16.1	91,507	29.4
8	United States	57.8	62.5	2.9	1.6	70.7	81.2	61.9	..	8.1	1.9	13.4	16.5	91,710	34.4
9	Canada	61.5	66.2	4.1	2.4	71.9	76.5	50.8	..	7.2	0.9	13.5	13.4	69,930	32.9
9	New Zealand	63.6	67.8	10.6	6.6	64.5	72.5	38.6	12.1	6.9	0.7	15.8	11.9	50,713	33.4
11	Singapore	65.9	67.8	..	1.1	63.4	77.1	..	8.7	2.8	0.6	7.0	..	96,573	..
12	Hong Kong, China (SAR)	57.0	58.9	0.9	0.2	62.4	87.7	25.3	6.9	3.4	..	9.4	6.6	88,809	44.0
13	Liechtenstein	..	..	..	..	..	..	..	..	..	..	..	..	..	..
14	Sweden	58.9	64.1	3.4	2.0	67.2	77.9	37.6	6.8	8.0	1.4	23.6	7.4	71,577	31.2
14	United Kingdom	57.4	62.1	2.1	1.2	64.9	78.9	41.2	12.1	7.7	2.7	20.9	13.3	69,955	31.8
16	Iceland	69.8	73.9	..	5.5	..	75.8	37.0	8.3	5.4	1.0	10.7	5.5	60,672	32.8
17	Korea (Republic of)	59.1	61.0	17.9	6.6	46.7	76.4	31.0	24.8	3.2	0.0	9.3	..	57,271	40.2
18	Israel	59.4	63.4	4.1	1.7	67.5	77.1	51.0	7.2	6.9	0.8	10.5	15.7	65,705	36.7
19	Luxembourg	54.2	57.6	3.3	1.3	66.4	84.1	43.3	6.2	5.8	1.8	15.5	5.0	149,978	30.9
20	Japan	56.8	59.2	7.2	3.7	58.2	69.7	37.5	10.5	4.3	1.6	6.9	3.9	64,383	33.6
21	Belgium	48.8	53.3	3.1	1.2	65.6	77.1	41.3	10.8	8.4	3.9	23.2	12.7	80,810	30.3
22	France	50.1	55.9	5.6	2.9	64.8	74.9	34.9	7.0	9.9	4.0	23.9	11.2	74,114	28.4
23	Austria	58.0	61.0	7.9	4.9	54.9	68.9	22.5	8.7	4.9	1.3	9.2	7.1	72,743	32.7
24	Finland	54.9	59.8	8.8	4.1	61.1	72.7	43.1	9.4	8.2	1.7	20.0	9.3	68,638	32.2
25	Slovenia	51.8	57.7	..	8.3	..	60.3	30.7	13.6	10.1	5.2	21.6	9.2	53,749	31.5
26	Spain	43.3	59.0	11.8	4.4	54.8	74.9	36.2	12.8	26.1	13.0	55.5	18.6	69,619	32.4
27	Italy	43.1	49.1	8.8	3.7	59.3	68.5	18.9	17.9	12.2	6.9	40.0	22.2	69,989	33.7
28	Czech Republic	55.4	59.5	..	3.1	..	58.8	21.0	14.5	7.0	3.0	19.0	9.1	50,197	34.6
29	Greece	38.7	53.2	23.9	13.0	48.3	70.3	29.9	30.3	27.3	18.0	58.3	20.4	61,648	39.1
30	Estonia	56.5	62.0	21.0	4.7	41.8	64.1	40.0	5.4	8.6	3.8	18.7	11.3	41,503	36.3
31	Brunei Darussalam	61.6	64.0	..	..	..	..	..	..	1.7	..	..	..	100,057	..
32	Cyprus	53.6	63.7	13.5	2.9	56.3	76.9	42.4	13.5	15.9	6.1	38.9	18.7	39,165	..
32	Qatar	86.2	86.7	..	1.4	..	46.8	..	0.2	0.3	0.1	1.1	9.4	96,237	..
34	Andorra	..	..	..	..	..	..	..	..	..	..	..	..	..	..
35	Slovakia	51.1	59.5	..	3.2	..	59.2	20.8	12.4	14.2	10.0	33.6	13.7	48,653	34.3
36	Poland	50.7	56.5	25.2	12.6	35.8	57.0	29.6	17.6	10.3	4.4	27.3	12.2	42,704	37.1
37	Lithuania	53.8	61.0	..	8.9	..	65.9	38.5	9.6	11.8	5.1	19.3	11.1	41,579	..
37	Malta	48.6	52.0	..	1.0	..	76.4	22.2	9.1	6.4	2.9	13.0	9.9	54,662	..
39	Saudi Arabia	51.8	54.9	..	4.7	..	70.7	..	..	5.6	1.1	29.5	18.6	78,918	..
40	Argentina	56.2	60.8	0.4	0.6	67.6	75.3	23.5	19.0	6.6	2.0	19.4	18.6	28,204	39.4
41	United Arab Emirates	76.9	79.9	..	3.8	..	73.1	19.5[e]	1.0	4.2	..	12.1	..	55,567	56.8
42	Chile	58.1	61.8	19.3	10.3	55.5	66.4	19.9	24.4	6.4	..	16.1	..	34,967	39.0
43	Portugal	50.4	60.3	17.9	10.5	47.6	63.8	20.3	16.7	16.2	9.1	38.1	14.1	47,474	32.5
44	Hungary	46.6	51.9	18.2	5.2	45.0	64.9	25.5	6.0	10.2	5.0	27.2	15.4	43,100	36.3
45	Bahrain	65.0	70.2	..	1.1	..	62.4	..	2.0	1.2	20.3[e]	5.3	..	41,315	..
46	Latvia	53.8	60.6	..	8.4	..	68.1	32.8	7.4	11.9	5.7	23.2	13.0	35,380	..
47	Croatia	42.2	51.3	..	13.7	..	58.7	23.1	13.7	17.3	12.0	50.0	19.6	43,551	..
48	Kuwait	66.3	68.4	..	2.7	73.2	76.0	18.4	2.2	3.6	..	14.6	..	80,172	..
49	Montenegro	40.1	50.0	..	5.7	..	76.2	..	..	19.6	15.8	41.1	..	32,875	..
HIGH HUMAN DEVELOPMENT															
50	Belarus	52.7	56.0	21.6	10.5	36.1	49.9	..	2.1	6.1	..	12.5	12.1	..	..
50	Russian Federation	60.1	63.7	13.9	9.7	45.6	62.3	54.9	5.7	5.5	1.7	13.8	12.0	29,974	38.1
52	Oman	59.9	65.1	..	5.2	..	57.9	..	..	..	..	..	..	..	..
52	Romania	52.4	56.5	29.1	29.0	27.4	42.4	18.4	30.9	7.3	3.0	23.6	17.2	24,556	..
52	Uruguay	61.3	65.7	0.0	10.9	67.0	68.0	26.3	22.2	6.4	..	19.2	..	28,774	33.1
55	Bahamas	64.1	74.1	..	3.7	79.6	83.0	..	..	16.2	7.1	30.8	..	54,282	..
56	Kazakhstan	68.7	72.5	..	25.5	..	55.1	..	28.6	5.2	..	3.9	..	24,289	..
57	Barbados	62.5	71.2	6.3	2.8	62.0	72.6	19.0[e]	14.0[e]	11.6	2.3	29.6	..	33,312	..
58	Antigua and Barbuda	..	..	..	2.8	..	81.6	..	..	8.4[e]	..	19.9	..	..	..

			Employment						Unemployment				Labour productivity	
		Employment to population ratio[a]	Labour force participation rate[a]	Employment in agriculture		Employment in services		Labour force with tertiary education	Vulnerable employment	Total	Long term	Youth	Youth not in school or employment	Output per worker	Hours worked per week
		(% ages 15 and older)		(% of total employment)				(%)	(% of total employment)	(% of labour force)		(% ages 15–24)		(2005 PPP $)	(per employed person)
HDI rank		2013	2013	1990[b]	2012[c]	1990[b]	2012[c]	2007–2012[d]	2008–2013[d]	2008–2013[d]	2008–2013[d]	2008–2014[d]	2008–2013[d]	2005–2012[d]	2003–2012[d]
59	Bulgaria	46.4	53.3	18.5	6.4	37.3	62.2	28.0	8.2	12.9	7.4	28.4	21.6	30,327	..
60	Palau	..	..	..	..	..	..	..	..	4.2[e]	..	..	..	..	..
60	Panama	62.8	65.5	..	16.7	52.8	65.2	38.7	29.2	4.1	..	10.8	17.6	32,080	33.3
62	Malaysia	57.5	59.4	26.0	12.6	46.5	59.0	18.9	22.2	3.0	..	10.4	..	35,036	43.2
63	Mauritius	53.7	58.6	16.7	7.8	40.0	64.7	9.8	17.1	7.6	2.0	23.2	..	32,602	39.7
64	Seychelles	..	..	..	..	..	..	..	..	4.1	..	11.0	1.2	..	..
64	Trinidad and Tobago	60.4	64.1	12.3	3.8	60.5	63.8	..	15.6[e]	3.6	..	9.2	52.5	48,012	..
66	Serbia	40.8	52.4	..	21.0	..	52.6	..	28.6	22.1	18.7	49.4	19.5	20,857	..
67	Cuba	54.9	56.7	..	19.7	..	63.2	15.9	..	3.2	..	6.1	..	..	..
67	Lebanon	44.4	47.6	..	..	..	..	22.5	27.8[e]	9.0[e]	..	22.1[e]	..	35,544	..
69	Costa Rica	58.2	63.0	25.9	13.4	47.5	66.9	24.7	20.2	8.5	0.7	21.8	19.2	25,232	36.0
69	Iran (Islamic Republic of)	39.2	45.1	..	21.2	..	46.5	..	42.0	10.4	..	28.7	34.4	35,432	..
71	Venezuela (Bolivarian Republic of)	60.2	65.1	13.4	7.7	61.2	70.7	39.7	30.3	7.8	..	17.1	19.2	27,251	32.2
72	Turkey	44.5	49.4	46.9	23.6	32.4	50.4	19.2	31.3	9.7	1.9	18.7	25.5	41,353	35.7
73	Sri Lanka	52.6	55.0	47.8	39.4	30.0	41.5	14.6[e]	43.1	4.4	1.9	20.1	0.5	13,234	44.0
74	Mexico	58.5	61.6	22.6	13.4	46.1	61.9	25.4	29.2	4.8	0.1	9.2	18.2	30,344	42.8
75	Brazil	65.6	69.8	22.8	15.3	54.5	62.7	18.3	25.1	6.7	9.8	15.0	19.6	20,921	32.5
76	Georgia	55.8	65.0	..	53.4	..	36.2	..	60.6	14.6	..	35.6	..	11,630	..
77	Saint Kitts and Nevis	..	..	..	..	43.0	..	..	..	..	..	11.0	..	..	..
78	Azerbaijan	62.5	66.1	30.9	37.7	31.1	48.0	14.5	56.4	5.0	..	13.8	..	18,958	..
79	Grenada	..	..	..	..	49.4	..	..	..	..	..	..	..	..	..
80	Jordan	36.3	41.6	..	2.0	76.3	80.5	..	9.7	12.2	4.3	29.3	..	20,007	..
81	The former Yugoslav Republic of Macedonia	39.2	55.2	..	17.3	..	52.8	22.0	23.4	29.0	23.9	51.9	24.2	29,528	..
81	Ukraine	54.7	59.4	19.8	17.2	15.4	62.1	..	18.1	7.5	2.1	17.4	..	13,670	..
83	Algeria	39.6	43.9	..	10.8	..	58.4	10.9[e]	26.9	9.8	7.1	24.8	21.5	25,678	..
84	Peru	73.3	76.2	1.2	25.8	71.5	56.8	15.7	46.3	4.0	..	8.8	16.5	18,191	40.2
85	Albania	46.3	55.1	..	41.5	..	37.7	..	58.1	13.4	11.4	30.2	30.5	21,813	..
85	Armenia	53.2	63.4	..	38.9	..	44.4	22.9	29.8	18.4	9.7	36.0	..	14,109	35.4
85	Bosnia and Herzegovina	32.5	45.3	..	20.5	..	49.0	..	25.3	27.5	25.4	62.8	..	27,060	..
88	Ecuador	65.7	68.6	7.5	27.8	67.3	54.4	21.5	51.2	4.1	..	10.9	3.8	18,547	37.0
89	Saint Lucia	..	69.2	..	14.8	..	59.4	..	..	22.2	..	27.5[e]	..	..	..
90	China	68.0	71.3	60.1	34.8	18.5	35.7	..	..	2.9	..	..	..	..	..
90	Fiji	50.5	55.0	..	..	..	..	..	38.8	6.9	2.6[e]	18.7[e]	..	11,894	37.5
90	Mongolia	59.8	62.9	..	32.6	..	49.6	26.8	51.4	7.9	3.4	16.5	1.5	10,921	..
93	Thailand	71.7	72.3	64.0	39.6	22.0	39.4	..	55.9	0.8	0.1	3.4	..	14,443	42.8
94	Dominica	..	..	..	..	38.8	..	..	..	..	..	26.0	..	..	..
94	Libya	42.6	53.0	..	..	..	50.2	..	..	19.0	..	48.7	..	..	..
96	Tunisia	41.3	47.6	..	16.2	39.1	49.6	..	28.8	17.6	..	37.6	..	26,335	..
97	Colombia	60.3	67.4	1.4	16.9	67.7	62.2	23.1	48.6	8.9	1.0	19.1	22.0	21,001	36.8
97	Saint Vincent and the Grenadines	..	67.0	..	..	..	..	..	8.0	18.8	16.9	33.8	..	..	..
99	Jamaica	53.8	63.3	..	18.1	..	66.5	..	37.5	13.7	..	34.0	..	17,128	..
100	Tonga	..	63.9	..	31.8	44.7	37.5	..	55.2[e]	1.1[e]	..	11.9[e]	..	..	..
101	Belize	56.0	65.6	..	19.5	..	61.9	12.3[e]	23.5[e]	11.7	..	25.0	19.0	17,017	..
101	Dominican Republic	55.2	64.9	..	14.5	..	67.8	21.1	37.1	15.0	..	31.4	..	22,415	..
103	Suriname	50.3	54.5	3.7	8.0	73.3	64.3	..	..	4.8	..	15.3	..	..	..
104	Maldives	59.1	66.8	25.2	11.5	48.5	60.0	..	29.6[e]	14.4[e]	..	25.4	56.4	18,670	..
105	Samoa	..	41.5	..	..	..	..	..	38.1	5.7	..	19.1	98.8	..	..
MEDIUM HUMAN DEVELOPMENT															
106	Botswana	62.6	76.7	..	29.9	31.4	54.9	..	12.9	17.9	10.4[e]	36.0	..	33,651	..
107	Moldova (Republic of)	38.6	40.7	33.8	26.4	33.9	54.3	25.3	30.5	5.6	1.7	12.2	28.6	11,587	..
108	Egypt	42.9	49.1	39.0	29.2	40.1	47.1	..	26.4	13.2	11.7	34.3	27.9	19,525	..
109	Turkmenistan	55.0	61.5	..	..	..	..	..	..	..	..	..	..	..	..
110	Gabon	48.9	60.8	..	24.2	..	64.0	..	52.9[e]	20.4	..	35.7	..	46,714	..
110	Indonesia	63.5	67.7	55.9	35.9	30.2	43.5	7.1	33.0	6.2	..	31.3	..	9,536	..
112	Paraguay	66.6	70.3	2.1	27.2	70.3	56.7	16.6	43.2	5.0	..	10.5	12.3	11,967	36.3
113	Palestine, State of	31.6	41.2	..	11.5	..	62.2	23.3	25.6	23.3	..	41.0	..	..	..
114	Uzbekistan	55.1	61.6	..	..	..	..	..	..	..	..	..	..	..	..
115	Philippines	60.6	65.2	45.2	32.2	39.7	52.5	25.3	39.8	7.1	0.1	15.7	24.8	9,571	..
116	El Salvador	58.2	62.1	7.4	21.0	63.4	57.9	14.1	37.6	5.9	..	12.4	..	15,306	35.6
116	South Africa	39.2	52.1	..	4.6	..	62.7	6.6	10.0	24.7	15.5	51.4	31.4	35,206	..

TABLE
13

TABLE 13   Work and employment | 255

TABLE 13 WORK AND EMPLOYMENT

		Employment								Unemployment				Labour productivity	
		Employment to population ratio[a]	Labour force participation rate[a]	Employment in agriculture		Employment in services		Labour force with tertiary education	Vulnerable employment	Total	Long term	Youth	Youth not in school or employment	Output per worker	Hours worked per week
		(% ages 15 and older)		(% of total employment)				(%)	(% of total employment)	(% of labour force)		(% ages 15–24)		(2005 PPP $)	(per employed person)
HDI rank		2013	2013	1990[b]	2012[c]	1990[b]	2012[c]	2007–2012[d]	2008–2013[d]	2008–2013[d]	2008–2013[d]	2008–2014[d]	2008–2013[d]	2005–2012[d]	2003–2012[d]
116	Viet Nam	75.9	77.5	..	47.4	..	31.5	..	62.6	2.0	0.3	6.0	9.3	5,250	..
119	Bolivia (Plurinational State of)	70.6	72.5	1.2	32.1	73.2	47.9	15.7	54.9	2.7	1.3	6.2	..	10,026	..
120	Kyrgyzstan	62.0	67.5	32.7	34.0	39.4	45.3	1.5[e]	47.3[e]	8.3	..	13.4	21.2	4,938	..
121	Iraq	35.5	42.3	..	23.4	..	58.3	..	..	15.3	..	..	..	17,067	..
122	Cabo Verde	62.8	67.5	..	..	..	..	..	..	10.7	..	..	..	..	..
123	Micronesia (Federated States of)	..	..	..	..	..	..	..	..	..	..	..	..	..	..
124	Guyana	54.5	61.4	..	..	..	..	..	..	..	..	24.0	..	9,652	..
125	Nicaragua	58.8	63.4	39.3	32.2	38.2	51.2	13.9	47.1	5.3	..	11.9	..	9,043	37.4
126	Morocco	45.9	50.5	3.9	39.2	59.5	39.3	9.5[e]	50.7	9.2	5.8	19.1	..	13,769	..
126	Namibia	49.0	59.0	..	27.4	..	58.7	10.8[e]	7.8	29.6	30.9	56.2	32.0	21,086	..
128	Guatemala	65.8	67.7	12.9	32.3	57.2	48.2	7.5	49.9[e]	3.0	..	6.3	29.8	11,461	..
129	Tajikistan	60.7	67.9	..	55.5	..	26.2	..	47.1	11.5	..	16.7	..	..	..
130	India	52.2	54.2	..	47.2	..	28.1	..	80.8	3.6	1.4	10.7	..	8,821	..
131	Honduras	60.0	62.6	50.1	35.3	33.2	44.9	8.0	53.3	3.9	..	8.0	41.4	9,564	34.9
132	Bhutan	70.9	72.5	..	62.2	..	29.1	..	53.1	2.1	..	9.6	..	11,438	..
133	Timor-Leste	36.2	37.9	..	50.6	..	39.8	2.4	69.6	3.9	0.4	14.8	..	8,156	46.3
134	Syrian Arab Republic	38.9	43.6	..	14.3	46.2	53.0	..	32.9	8.4	..	35.8	..	18,563	..
134	Vanuatu	..	70.8	..	60.5	..	31.1	..	70.0	4.6	..	10.6	..	..	..
136	Congo	66.1	70.7	..	35.4	..	42.2	..	75.1[e]	..	..	..	..	..	..
137	Kiribati	..	..	..	..	..	..	..	..	30.6	..	54.0	..	..	..
138	Equatorial Guinea	79.8	86.7	..	..	17.6	..	..	..	..	..	..	..	..	..
139	Zambia	68.8	79.3	49.8	72.2	20.8	20.6	..	81.3	7.8	..	15.2	..	4,015	..
140	Ghana	66.2	69.3	..	41.5	..	43.1	..	76.8	4.2	..	11.2	..	4,308	..
141	Lao People's Democratic Republic	76.6	77.7	..	..	..	..	..	88.0[e]	1.4[e]	..	..	..	5,114	..
142	Bangladesh	67.8	70.8	..	48.1	14.8	37.4	..	85.0[e]	4.5	..	8.7	..	3,457	44.3
143	Cambodia	82.3	82.5	..	51.0	..	30.4	..	64.1	0.3	..	0.5	79.2	3,849	..
143	Sao Tome and Principe	..	61.1	..	..	28.3	..	..	..	16.7[e]	..	..	..	..	..
LOW HUMAN DEVELOPMENT															
145	Kenya	61.1	67.3	..	61.1	..	32.2	..	..	..	..	..	..	..	..
145	Nepal	81.1	83.3	..	..	..	..	..	..	2.7	1.0	3.5	..	2,448	37.6
147	Pakistan	51.6	54.4	51.1	44.7	28.9	35.2	24.6	63.1	5.0	1.1	7.7	..	7,367	33.1
148	Myanmar	75.9	78.6	69.7	..	21.0	..	..	..	..	..	..	..	..	..
149	Angola	65.2	70.0	..	..	..	..	..	..	..	..	..	..	..	..
150	Swaziland	44.5	57.4	..	..	..	..	..	..	28.2[e]	..	..	..	..	..
151	Tanzania (United Republic of)	86.0	89.1	..	76.5	..	19.2	..	74.0	3.5	..	5.8	..	2,822	..
152	Nigeria	51.8	56.1	..	44.6	43.7	41.7	..	..	23.9	..	..	..	..	..
153	Cameroon	67.4	70.3	..	53.3	14.0	34.1	..	76.4	3.8	..	6.4	10.8	5,255	..
154	Madagascar	85.4	88.5	..	80.4	..	15.8	4.4[e]	88.3	3.8	..	2.6	..	1,722	..
155	Zimbabwe	81.9	86.5	..	64.8	..	15.3	..	..	11.1	..	8.7	..	..	..
156	Mauritania	37.2	53.9	..	..	..	..	..	..	31.2	..	..	..	10,112	..
156	Solomon Islands	63.7	66.3	..	..	..	..	..	..	..	..	..	..	..	..
158	Papua New Guinea	70.7	72.3	..	..	..	..	..	..	..	..	5.3	..	5,738	..
159	Comoros	53.9	57.6	..	..	..	..	..	..	..	..	..	..	..	..
160	Yemen	40.3	48.8	..	24.7	..	56.2	11.3	29.6	17.8	4.0	33.7	..	9,057	..
161	Lesotho	49.7	66.0	..	..	..	..	..	..	24.4	15.6	34.4	..	5,526	..
162	Togo	75.4	81.0	..	54.1	..	37.5	..	89.1[e]	..	..	..	..	..	..
163	Haiti	61.2	65.8	65.6	..	22.8	..	..	..	..	..	..	..	..	..
163	Rwanda	85.4	85.9	..	78.8	6.7	16.6	..	..	3.4	..	4.5	..	..	..
163	Uganda	74.5	77.5	..	65.6	..	28.4	..	80.6[e]	4.2	..	2.6	7.0	3,046	..
166	Benin	72.1	72.9	..	42.7	..	46.2	..	89.9[e]	1.0	..	2.4	3.5	3,317	..
167	Sudan	45.4	53.5	..	..	..	..	..	..	14.8	..	22.9	..	7,093	..
168	Djibouti	..	52.0	..	..	..	..	..	..	..	..	..	..	..	..
169	South Sudan	..	..	..	..	..	..	..	..	13.7	..	18.5	..	..	..
170	Senegal	68.7	76.5	..	33.7	..	36.1	..	58.0	10.4	..	12.7	..	4,308	..
171	Afghanistan	44.0	47.9	..	..	..	..	..	..	8.2	..	..	..	5,417	..
172	Côte d'Ivoire	64.6	67.3	..	..	..	..	..	..	9.4	..	..	..	..	..
173	Malawi	76.7	83.0	..	..	..	..	..	..	7.8[e]	..	8.6	..	1,857	..
174	Ethiopia	79.0	83.7	..	79.3	..	13.0	..	91.2[e]	4.5	1.3[e]	7.3	1.1	2,185	..
175	Gambia	72.0	77.4	..	..	..	..	..	..	..	..	..	..	..	..
176	Congo (Democratic Republic of the)	66.2	71.9	..	..	..	..	..	..	3.7[e]	..	..	..	..	..

TABLE 13

			Employment							Unemployment				Labour productivity	
	Employment to population ratio^a	Labour force participation rate^a	Employment in agriculture		Employment in services		Labour force with tertiary education	Vulnerable employment	Total	Long term	Youth	Youth not in school or employment	Output per worker	Hours worked per week
	(% ages 15 and older)		(% of total employment)				(%)	(% of total employment)	(% of labour force)		(% ages 15–24)		(2005 PPP $)	(per employed person)
HDI rank	2013	2013	1990^b	2012^c	1990^b	2012^c	2007–2012^d	2008–2013^d	2008–2013^d	2008–2013^d	2008–2014^d	2008–2013^d	2005–2012^d	2003–2012^d
177 Liberia	59.3	61.5	..	48.9	..	41.9	..	78.7	3.7	..	5.1	..	1,675	..
178 Guinea-Bissau	68.1	73.3	..	..	..	..	..	..	..	..	..	..	..	..
179 Mali	60.6	66.0	..	66.0	..	28.3	..	82.9^e	7.3	..	10.7	13.5	3,267	..
180 Mozambique	77.2	84.2	..	80.5	..	16.1	..	87.8^e	22.5	5.9^e	39.3	10.1	..	..
181 Sierra Leone	65.2	67.3	..	68.5	..	25.0	..	92.4^e	3.4^e	..	5.2^e	..	3,093	..
182 Guinea	70.7	72.0	..	..	..	..	..	..	1.7	..	1.0	..	..	..
183 Burkina Faso	80.8	83.4	..	84.8	..	12.2	..	89.6^e	3.3^e	..	3.8^e	..	2,973	..
184 Burundi	76.9	82.6	..	..	..	..	..	94.6^e	..	..	..	..	..	..
185 Chad	66.6	71.6	..	..	..	..	..	..	..	..	..	..	..	..
186 Eritrea	78.7	84.8	..	..	..	..	..	..	..	..	..	..	..	..
187 Central African Republic	72.7	78.7	..	..	..	..	..	..	..	..	..	..	..	..
188 Niger	61.4	64.7	..	56.9	..	31.1	..	84.8^e	..	..	3.2	..	..	..
OTHER COUNTRIES OR TERRITORIES														
Korea (Democratic People's Rep. of)	74.4	78.0	..	..	..	..	..	..	..	..	..	..	..	..
Marshall Islands	..	..	..	..	..	..	..	..	4.7	..	..	..	..	..
Monaco	..	..	..	..	..	..	..	..	..	..	6.9	..	..	..
Nauru	..	..	..	..	..	..	..	..	..	..	38.0	..	..	..
San Marino	..	..	2.5	0.3	53.7	62.7	12.2	..	6.0	..	..	..	..	..
Somalia	52.2	56.1	..	..	..	..	..	..	..	..	..	..	..	..
Tuvalu	..	..	..	..	..	..	..	..	6.5^e	..	..	..	..	..
Human development groups														
Very high human development	55.4	60.3	..	3.3	..	74.3	32.5	12.4	8.3	3.0	18.0	13.4	64,041	33.3
High human development	63.4	67.1	..	28.8	..	43.8	..	28.7	4.7	..	16.7	..	23,766	..
Medium human development	55.7	58.8	..	42.5	..	35.3	..	65.1	5.3	..	15.1	..	9,483	..
Low human development	63.9	68.1	..	..	..	..	..	..	9.7	..	9.8	..	..	..
Developing countries	60.7	64.3	..	36.9	..	39.1	..	54.0	5.6	..	14.6	..	..	..
Regions														
Arab States	44.7	50.3	..	20.8	..	53.9	..	29.5	11.8	..	29.0	..	26,331	..
East Asia and the Pacific	67.9	71.1	..	35.5	..	37.3	..	..	3.3	..	18.6	..	..	..
Europe and Central Asia	51.5	57.2	..	24.5	..	52.5	..	28.8	9.9	..	19.5	..	30,460	..
Latin America and the Caribbean	62.2	66.4	..	16.3	54.2	62.1	..	31.3	6.2	..	13.7	..	23,243	..
South Asia	53.2	55.6	..	45.9	..	30.6	..	77.3	4.2	..	10.9	..	8,117	..
Sub-Saharan Africa	65.7	70.9	..	59.0	..	30.0	..	..	11.9	..	13.5	..	..	..
Least developed countries	69.8	74.0	..	..	..	..	..	..	6.3	..	10.3	..	..	..
Small island developing states	58.3	63.1	..	..	..	..	..	..	8.7	..	18.0	..	..	..
Organisation for Economic Co-operation and Development	54.9	59.7	10.1	5.1	59.9	72.3	31.1	15.7	8.2	2.8	16.5	14.7	58,391	34.3
World	**59.7**	**63.5**	..	**30.3**	..	**46.0**	..	**47.6**^T	**6.1**	..	**15.1**	..	**24,280**	..

NOTES

a Modeled ILO estimates

b Data refer to 1990 or the most recent year available.

c Data refer to 2012 or the most recent year available.

d Data refer to the most recent year available during the period specified.

e Refers to a year between 2003 and the earliest year in the column heading.

T From original data source.

DEFINITIONS

Employment to population ratio: Percentage of the population ages 15 and older that is employed.

Labour force participation rate: Percentage of a country's working-age population that engages actively in the labour market, either by working or looking for work. It provides an indication of the relative size of the supply of labour available to engage in the production of goods and services.

Employment in agriculture: Share of total employment that is employed in agriculture.

Employment in services: Share of total employment that is employed in services.

Labour force with tertiary education: Percentage of the labour force that has attained the tertiary level of education, that is levels 5, 5A, 5B and 6 of the International Standard Classification of Education.

Vulnerable employment: Percentage of employed people engaged as unpaid family workers and own-account workers.

Unemployment rate: Percentage of the labour force population ages 15 and older that is not in paid employment or self-employed but is available for work and has taken steps to seek paid employment or self-employment.

Long-term unemployment rate: Percentage of the labour force ages 15 and older that has not worked for at least 12 months but is available for work and has taken specific steps to seek paid employment or self-employment.

Youth unemployment rate: Percentage of the labour force population ages 15–24 that is not in paid employment or self-employed but is available for work and has taken steps to seek paid employment or self-employment.

Youth not in school or employment: Percentage of young people ages 15–24 who are not in employment or in education or training.

Output per worker: Output per unit of labour input, expressed as GDP per person engaged, in 2005 international dollars using purchasing power parity rates.

Hours worked per week: The number of hours that employed people (wage and salaried workers as well as self-employed workers) work per week.

MAIN DATA SOURCES

Columns 1–7 and 9–13: ILO (2015a).

Columns 8 and 14: ILO (2015b).

TABLE
13

TABLE 13 Work and employment | 257

TABLE 14

Human security

HDI rank		Birth registration (% under age 5) 2005–2013[c]	Refugees by country of origin[a] (thousands) 2014	Internally displaced persons[b] (thousands) 2014	Homeless people due to natural disaster (average annual per million people) 2005/2014	Orphaned children (thousands) 2013	Prison population (per 100,000 people) 2002–2013[c]	Homicide rate (per 100,000 people) 2008–2012[c]	Suicide rate (per 100,000 people) Female 2012	Suicide rate (per 100,000 people) Male 2012	Depth of food deficit (kilocalories per person per day) 2012/2014	Violence against women — Intimate or nonintimate partner violence ever experienced (%) 2001–2011[c]
VERY HIGH HUMAN DEVELOPMENT												
1	Norway	100	0.0	..	0	..	72	2.2	5.2	13.0	..	26.8 [d]
2	Australia	100	0.0	..	33	..	130	1.1	5.2	16.1	..	39.9
3	Switzerland	100	0.0	..	0	..	82	0.6	5.1	13.6	..	39.0
4	Denmark	100	0.0	..	0	..	73	0.8	4.1	13.6	..	50.0
5	Netherlands	100	0.1	..	0	..	82	0.9	4.8	11.7	..	..
6	Germany	100	0.2	..	0	..	79	0.8	4.1	14.5	..	40.0
6	Ireland	100	0.0	..	0	..	88	1.2	5.2	16.9	..	14.5 [d,e]
8	United States	100	4.8	..	15	..	716	4.7	5.2	19.4	..	35.6 [d,e]
9	Canada	100	0.1	..	21	..	118 [f]	1.6	4.8	14.9	..	6.4 [d]
9	New Zealand	100	0.0	..	16	..	192	0.9	5.0	14.4	..	33.1 [d,g]
11	Singapore	..	0.1	..	0	..	230	0.2	5.3	9.8	..	9.2
12	Hong Kong, China (SAR)	..	0.0	..	0	..	128	0.4	..	..	..	21.0
13	Liechtenstein	100	..	..	..	..	24 [f]	0.0	..	..	..	..
14	Sweden	100	0.0	..	0	..	67	0.7	6.1	16.2	..	46.0
14	United Kingdom	100	0.1	..	48	..	147 [h]	1.0	2.6	9.8	..	28.4 [d]
16	Iceland	100	0.0	..	0	..	47	0.3	6.7	21.0	..	42.0
17	Korea (Republic of)	..	0.5	..	9	..	99	0.9	18.0	41.7	5	..
18	Israel	100	1.0	..	0	..	223	1.8	2.3	9.8	..	..
19	Luxembourg	100	0.0	..	0	..	122	0.8	4.4	13.0	..	..
20	Japan	100	0.3	..	32	..	51	0.3	10.1	26.9	..	18.5 [g]
21	Belgium	100	0.1	..	0	..	108	1.6	7.7	21.0	..	28.9
22	France	100	0.1	..	1	..	98 [f]	1.0	6.0	19.3	..	..
23	Austria	100	0.0	..	0	..	98	0.9	5.4	18.2	..	..
24	Finland	100	0.0	..	0	..	58	1.6	7.5	22.2	..	43.5
25	Slovenia	100	0.1	..	51	..	66	0.7	4.4	20.8	..	..
26	Spain	100	0.1	..	33	..	147	0.8	2.2	8.2	..	..
27	Italy	100	0.1	..	124	..	106	0.9	1.9	7.6	..	31.9
28	Czech Republic	100	1.3	..	0	..	154	1.0	3.9	21.5	..	58.0
29	Greece	100	0.1	..	37	..	111	1.7	1.3	6.3	..	..
30	Estonia	100	0.4	..	0	..	238	5.0	3.8	24.9	..	..
31	Brunei Darussalam	..	0.0	..	0	..	122	2.0	5.2	7.7	15	..
32	Cyprus	100	0.0	212.4	0	..	106 [f]	2.0	1.5	7.7	..	..
32	Qatar	..	0.0	..	..	..	60	1.1	1.2	5.7	..	..
34	Andorra	100	0.0	..	..	..	38	1.3	..	..	..	..
35	Slovakia	100	0.2	..	0	..	187	1.4	2.5	18.5	..	27.9 [d]
36	Poland	100	1.4	..	0	..	217	1.2	3.8	30.5	..	35.0
37	Lithuania	100	0.2	..	0	..	329	6.7	8.4	51.0	..	37.6 [d]
37	Malta	100	0.0	..	..	..	145	2.8	0.7	11.1	..	16.0 [d,e]
39	Saudi Arabia	..	0.6	..	37	..	162	0.8	0.2	0.6	11	..
40	Argentina	100 [i]	0.4	..	18	..	147	5.5	4.1	17.2	6	..
41	United Arab Emirates	100 [i]	0.1	..	..	..	238	0.7 [i]	1.7	3.9	26	..
42	Chile	99 [i]	0.6	..	4,832	..	266	3.1	5.8	19.0	23	35.7 [d]
43	Portugal	100	0.0	..	1	..	136	1.2	3.5	13.6	..	38.0
44	Hungary	100	1.2	..	0	..	186	1.3	7.4	32.4	..	..
45	Bahrain	..	0.3	..	..	..	275	0.5	2.9	11.6	..	..
46	Latvia	100	0.2	..	0	..	304	4.7	4.3	30.7	..	..
47	Croatia	..	40.2	..	0	..	108	1.2	4.5	19.8	..	..
48	Kuwait	..	1.0	..	0	..	137	0.4	0.8	1.0	17	..
49	Montenegro	99	0.5	..	0	..	208	2.7	6.4	24.7	..	..
HIGH HUMAN DEVELOPMENT												
50	Belarus	100 [i]	4.4	..	0	..	335	5.1	6.4	32.7	..	..
50	Russian Federation	100	75.0	25.4	10	..	475	9.2	6.2	35.1	..	..
52	Oman	..	0.0	..	0	..	61	1.1	0.6	1.2	28	..
52	Romania	..	2.3	..	27	..	155	1.7	2.9	18.4	..	28.5 [d,e]
52	Uruguay	100 [i]	0.1	..	297	..	281	7.9	5.2	20.0	25	..
55	Bahamas	..	0.2	..	0	..	444	29.8	1.3	3.6	..	..
56	Kazakhstan	100	2.2	..	55	..	295	7.8	9.3	40.6	20	..
57	Barbados	..	0.1	..	0	..	521	7.4	0.6	4.1	24	30.0 [d,k]
58	Antigua and Barbuda	..	0.0	..	0	..	403	11.2	..	..	..	..
59	Bulgaria	100	1.9	..	0	..	151	1.9	5.3	16.6	..	..

HDI rank		Birth registration (% under age 5) 2005–2013[c]	Refugees by country of origin[a] (thousands) 2014	Internally displaced persons[b] (thousands) 2014	Homeless people due to natural disaster (average annual per million people) 2005/2014	Orphaned children (thousands) 2013	Prison population (per 100,000 people) 2002–2013[c]	Homicide rate (per 100,000 people) 2008–2012[c]	Suicide rate (per 100,000 people) Female 2012	Suicide rate (per 100,000 people) Male 2012	Depth of food deficit (kilocalories per person per day) 2012/2014	Violence against women Intimate or nonintimate partner violence ever experienced (%) 2001–2011[c]
60	Palau	..	0.0	..	0	..	295	3.1	..	..	..	..
60	Panama	..	0.1	..	35	..	411	17.2	1.3	8.1	75	..
62	Malaysia	..	0.5	..	103	..	132	2.3	1.5	4.7	20	..
63	Mauritius	..	0.1	..	0	..	202	2.8	2.9	13.2	36	..
64	Seychelles	..	0.0	..	0	..	709	9.5	..	..	..	..
64	Trinidad and Tobago	97	0.3	..	0	..	281	28.3	6.2	20.4	64	..
66	Serbia	99	48.8	97.3	15	..	142	1.2	5.8	19.9	..	26.2[g]
67	Cuba	100[i]	6.5	..	666	..	510	4.2	4.5	18.5	10	..
67	Lebanon	100	4.2	19.7	0	..	118	2.2	0.6	1.2	34	..
69	Costa Rica	100	0.5	..	0	..	314	8.5	2.2	11.2	40	60.0
69	Iran (Islamic Republic of)	99[i]	76.4	..	9	..	284	4.1	3.6	6.7	41	..
71	Venezuela (Bolivarian Republic of)	81[i]	8.4	..	13	..	161	53.7	1.0	4.3	12	..
72	Turkey	94	65.9	953.7	45	..	179	2.6	4.2	11.8	1	19.7
73	Sri Lanka	97	123.0	90.0	5,767	..	132	3.4	12.8	46.4	209	..
74	Mexico	93[i]	9.4	281.4	319	..	210	21.5	1.7	7.1	30	46.7[d,e]
75	Brazil	93[i]	1.0	..	53	..	274	25.2	2.5	9.4	11	38.8[f]
76	Georgia	100	6.8	232.7	137	..	225[f]	4.3	1.0	5.7	68	9.1[d]
77	Saint Kitts and Nevis	..	0.0	..	0	..	714	33.6	..	..	..	..
78	Azerbaijan	94	10.9	568.9	77	..	413[f]	2.1	1.0	2.4	15	15.0
79	Grenada	..	0.3	..	0	..	424	13.3	..	..	..	..
80	Jordan	99	1.6	..	0	..	95	2.0	1.9	2.2	13	23.0[d]
81	The former Yugoslav Republic of Macedonia	100	1.8	0.2	0	..	122	1.4	3.2	7.3	..	..
81	Ukraine	100	6.3	646.5	7	..	305	4.3	5.3	30.3	..	18.3
83	Algeria	99	3.7	..	12	..	162	0.7	1.5	2.3	23	..
84	Peru	96[i]	4.8	150.0	196	..	202	9.6	2.1	4.4	58	38.9[d]
85	Albania	99	10.5	..	7	..	158	5.0	5.2	6.6	..	..
85	Armenia	100	12.2	8.4	0	..	164	1.8	0.9	5.0	48	9.5[d]
85	Bosnia and Herzegovina	100	22.3	100.4	0	..	78[h]	1.3	4.1	18.0	..	..
88	Ecuador	90	0.7	..	98	..	149	12.4	5.3	13.2	73	46.3[d,e]
89	Saint Lucia	92	0.7	..	0	..	317	21.6	..	..	..	..
90	China	..	205.0	..	212	..	121[f]	1.0	8.7	7.1	83	..
90	Fiji	..	1.1	..	0	..	174	4.0	4.1	10.6	30	..
90	Mongolia	99	2.1	..	0	..	287	9.7	3.7	16.3	174	..
93	Thailand	99[i]	0.2	35.0	24	..	398	5.0	4.5	19.1	60	43.8[f]
94	Dominica	..	0.0	..	135	..	391	21.1	..	..	..	..
94	Libya	..	3.4	400.0	0	..	81	1.7	1.4	2.2	..	..
96	Tunisia	99	1.4	..	0	..	199	2.2	1.4	3.4	4	..
97	Colombia	97	108.5	6,044.2	27	..	245	30.8	1.9	9.1	73	37.4[d]
97	Saint Vincent and the Grenadines	..	1.5	..	550	..	376	25.6	..	..	46	..
99	Jamaica	98	1.5	..	43	..	152	39.3	0.7	1.8	62	35.0[d,e]
100	Tonga	..	0.0	..	0	..	150	1.0	..	..	..	..
101	Belize	95	0.0	..	0	..	476	44.7	0.5	4.9	40	..
101	Dominican Republic	81	0.3	..	116	..	240	22.1	2.1	6.1	91	17.2[d]
103	Suriname	99	0.0	..	0	..	186	6.1	11.9	44.5	59	..
104	Maldives	93	0.0	..	0	..	307[f]	3.9	4.9	7.8	45	28.4
105	Samoa	48	0.0	..	0	..	228	3.6	..	..	23	75.8
MEDIUM HUMAN DEVELOPMENT												
106	Botswana	72	0.2	..	0	130	205	18.4	2.0	5.7	191	..
107	Moldova (Republic of)	100	2.2	..	0	..	188[f]	6.5	4.8	24.1	..	24.6[d]
108	Egypt	99[i]	13.1	..	1	..	80	3.4	1.2	2.4	12	33.7[d]
109	Turkmenistan	..	0.5	4.0	0	..	224	12.8	7.5	32.5	27	..
110	Gabon	90	0.2	..	52	66	196	9.1	4.5	12.1	20	..
110	Indonesia	67	9.8	84.0	327	..	59	0.6	4.9	3.7	55	3.1
112	Paraguay	76[i]	0.1	..	31	..	118	9.7	3.2	9.1	80	..
113	Palestine, State of	99	5,589.8[m]	275.0[n]	13	..	..	7.4	..	..	..	..
114	Uzbekistan	100	5.0	3.4	0	..	152	3.7	4.1	13.2	37	..
115	Philippines	90	0.7	77.7	98	..	111	8.8	1.2	4.8	97	23.6
116	El Salvador	99	9.7	288.9	287	..	422	41.2	5.7	23.5	88	26.3[d]
116	South Africa	85[i]	0.4	..	17	3,600	294	31.0	1.1	5.5	16	..
116	Viet Nam	95	314.1	..	948	..	145	3.3	2.4	8.0	95	38.5
119	Bolivia (Plurinational State of)	76[i]	0.6	..	1,490	..	140	12.1	8.5	16.2	120	67.6[e]

TABLE 14 Human security | 259

TABLE
14

TABLE 14   HUMAN SECURITY

HDI rank		Birth registration (% under age 5) 2005–2013[c]	Refugees by country of origin[a] (thousands) 2014	Internally displaced persons[b] (thousands) 2014	Homeless people due to natural disaster (average annual per million people) 2005/2014	Orphaned children (thousands) 2013	Prison population (per 100,000 people) 2002–2013[c]	Homicide rate (per 100,000 people) 2008–2012[c]	Suicide rate (per 100,000 people) Female 2012	Suicide rate (per 100,000 people) Male 2012	Depth of food deficit (kilocalories per person per day) 2012/2014	Violence against women Intimate or nonintimate partner violence ever experienced (%) 2001–2011[c]
120	Kyrgyzstan	98	2.3	..	59	..	181	9.1	4.5	14.2	42	..
121	Iraq	99	426.1	3,276.0	8	..	110	8.0	2.1	1.2	190	..
122	Cabo Verde	91	0.0	..	0	..	267	10.3	1.6	9.1	77	16.1 [d]
123	Micronesia (Federated States of)	..	..	..	0	..	80	4.6	..	..	..	..
124	Guyana	88	0.8	..	0	..	260	17.0	22.1	70.8	79	..
125	Nicaragua	85	1.5	..	99	..	153	11.3	4.9	15.4	130	29.3 [d]
126	Morocco	94 [i]	1.3	..	0	..	220	2.2	1.2	9.9	34	44.5
126	Namibia	78 [i]	1.1	..	0	150	191	17.2	1.4	4.4	315	42.5 [o]
128	Guatemala	97	6.6	248.5	433	..	105	39.9	4.3	13.7	99	27.6 [d]
129	Tajikistan	88	0.7	..	32	..	130	1.6	2.8	5.7	268	58.3 [d,p]
130	India	84 [i]	11.2	853.9	743	..	30	3.5	16.4	25.8	110	35.4
131	Honduras	94	3.3	29.4	141	..	153	90.4	2.8	8.3	89	..
132	Bhutan	100	26.7	..	0	..	135	1.7	11.2	23.1	..	..
133	Timor-Leste	55	0.0	0.9	0	..	25	3.6	5.8	10.2	198	39.2
134	Syrian Arab Republic	96	3,017.5	7,600.0	0	..	58	2.2	0.2	0.7	..	..
134	Vanuatu	43 [i]	0.0	..	0	..	76	2.9	..	..	40	48.0 [q]
136	Congo	91	11.7	7.8	519	220	31	12.5	4.6	14.7	188	..
137	Kiribati	94	0.0	..	88	..	114	8.2	..	..	24	73.0
138	Equatorial Guinea	54	0.2	..	0	..	95	19.3	8.6	24.1	..	..
139	Zambia	14	0.2	..	154	1,400	119	10.7	10.8	20.8	415	51.9
140	Ghana	63	22.5	..	0	990	54	6.1	2.2	4.2	25	44.5
141	Lao People's Democratic Republic	75	7.7	4.5	0	..	69	5.9	6.6	11.2	134	..
142	Bangladesh	31	10.0	431.0	76	..	42	2.7	8.7	6.8	122	53.3 [d]
143	Cambodia	62	13.6	..	0	..	106	6.5	6.5	12.6	108	22.3
143	Sao Tome and Principe	75	0.0	..	0	..	128	3.3	..	..	37	..
LOW HUMAN DEVELOPMENT												
145	Kenya	60	8.6	309.2	12	2,500	121	6.4	8.4	24.4	140	45.1
145	Nepal	42	8.1	50.0	316	..	48	2.9	20.0	30.1	50	26.0
147	Pakistan	34	32.8	1,900.0	2,920	..	39	7.7	9.6	9.1	170	..
148	Myanmar	72	223.7	645.3	14	..	120	15.2	10.3	16.5	113	..
149	Angola	36	10.3	..	520	1,100	105	10.0	7.3	20.7	104	..
150	Swaziland	50	0.1	..	22	100	284	33.8	4.1	8.6	186	..
151	Tanzania (United Republic of)	16	1.0	..	68	3,100	78	12.7	18.3	31.6	241	45.4
152	Nigeria	30	32.2	1,075.3	9	10,000	32	20.0	2.9	10.3	39	29.5
153	Cameroon	61	11.5	40.0	233	1,500	119	7.6	3.4	10.9	66	51.1 [d]
154	Madagascar	83	0.3	..	1,206	..	87	11.1	6.9	15.2	223	..
155	Zimbabwe	49	19.7	36.0	0	1,100	129	10.6	9.7	27.2	259	43.4
156	Mauritania	59	34.3	..	362	..	45	5.0	1.5	4.5	41	..
156	Solomon Islands	..	0.1	..	113	..	55	4.3	7.2	13.9	65	64.0
158	Papua New Guinea	..	0.2	7.5	327	300	48	10.4	9.1	15.9	..	..
159	Comoros	87	0.5	..	0	..	16	10.0	10.3	24.0	..	..
160	Yemen	17 [i]	2.5	334.1	17	..	55	4.8	3.0	4.3	177	..
161	Lesotho	45	0.0	..	130	220	121	38.0	3.4	9.2	75	..
162	Togo	78	10.3	10.0	214	360	64	10.3	2.8	8.5	96	..
163	Haiti	80	38.5	..	948	340	96	10.2	2.4	3.3	510	20.0 [d]
163	Rwanda	63	82.6	..	55	..	492 [f]	23.1	7.2	17.1	247	47.9
163	Uganda	30	6.7	29.8	853	2,400	97	10.7	12.3	26.9	167	62.2
166	Benin	80	0.3	..	1,638	450	75	8.4	3.1	8.8	60	..
167	Sudan	59	657.8 [f]	3,100.0	627	..	56	11.2	11.5	23.0	176 [s]	..
168	Djibouti	92	0.8	..	0	42	83	10.1	9.5	20.9	143	..
169	South Sudan	35	508.5 [t]	1,498.2	0	570	65	13.9	12.8	27.1	..	..
170	Senegal	73	21.8	24.0	0	..	64	2.8	2.8	8.6	134	..
171	Afghanistan	37	2,690.8	805.4	161	..	76	6.5	5.3	6.2	158	..
172	Côte d'Ivoire	65	70.9	300.9	53	1,300	34	13.6	4.1	10.6	96	..
173	Malawi	2 [i]	0.3	..	565	1,200	76	1.8	8.9	23.9	139	41.2
174	Ethiopia	7	74.5	397.2	1	4,000	136	12.0	6.7	16.5	250	55.9 [l]
175	Gambia	53	3.4	..	24	83	56	10.2	2.6	7.6	34	..
176	Congo (Democratic Republic of the)	28	493.3	2,756.6	36	4,000	33	28.3	4.8	15.8	..	64.1 [d]
177	Liberia	4 [i]	16.8	23.0	100	200	46	3.2	2.0	6.8	269	38.6 [d]
178	Guinea-Bissau	24	1.2	..	48	120	..	8.4	2.4	7.2	157	..
179	Mali	81	147.7	61.6	72	1,100	36	7.5	2.7	7.2	23	..

TABLE 14

HDI rank	Birth registration (% under age 5) 2005–2013[c]	Refugees by country of origin[a] (thousands) 2014	Internally displaced persons[b] (thousands) 2014	Homeless people due to natural disaster (average annual per million people) 2005/2014	Orphaned children (thousands) 2013	Prison population (per 100,000 people) 2002–2013[c]	Homicide rate (per 100,000 people) 2008–2012[c]	Suicide rate (per 100,000 people) Female 2012	Suicide rate (per 100,000 people) Male 2012	Depth of food deficit (kilocalories per person per day) 2012/2014	Violence against women — Intimate or nonintimate partner violence ever experienced (%) 2001–2011[c]
180 Mozambique	48	0.1	..	331	2,100	65	12.4	21.1	34.2	195	31.5 [d]
181 Sierra Leone	78	5.3	..	40	310	52	1.9	4.5	11.0	169	..
182 Guinea	58	15.2	..	37	670	25	8.9	2.4	7.1	120	..
183 Burkina Faso	77	1.6	..	412	990	28	8.0	2.8	7.3	167	15.4 [d]
184 Burundi	75	71.9	77.6	564	740	72	8.0	12.5	34.1	..	..
185 Chad	16	14.6	71.0	106	980	41	7.3	2.3	7.4	288	..
186 Eritrea	..	286.0	10.0	0	180	..	7.1	8.7	25.8	..	..
187 Central African Republic	61	377.1	438.5	1,044	320	19	11.8	5.3	14.1	302	..
188 Niger	64	0.7	11.0	517	..	42	4.7	1.9	5.3	59	..
OTHER COUNTRIES OR TERRITORIES											
Korea (Democratic People's Rep. of)	100	1.2	..	1,451	..	..	5.2	..	..	344	..
Marshall Islands	96	0.0	..	0	..	58	4.7	..	..	..	36.3
Monaco	100	0.0	..	..	..	73 [f]	0.0	..	..	..	..
Nauru	83	..	..	..	..	277	1.3	..	..	..	..
San Marino	100	0.0	..	..	..	6 [f]	0.7	..	..	..	..
Somalia	3	1,080.8	1,106.8	258	630	..	8.0	6.8	18.1	..	..
Tuvalu	50	0.0	..	0	..	120	4.2	..	..	..	46.6
Human development groups											
Very high human development	100	56.7	212.4	—	..	281	2.2	5.5	18.2	..	..
High human development	96	839.4	9,653.8	—	..	188	6.5	6.4	10.3	64	..
Medium human development	80	9,511.5	13,185.0	—	..	63	5.0	10.8	17.4	98	..
Low human development	39	5,984.7	14,012.2	—	4,125	71	12.1	7.7	15.2	147	..
Developing countries	67	17,341.9	38,144.8	—	..	109	6.9	8.3	13.4	94	..
Regions											
Arab States	81	10,806.0 [u]	16,111.6 [u]	—	..	116	4.0	2.6	5.5	68	..
East Asia and the Pacific	79	780.0 [u]	854.9 [u]	—	..	123	2.1	7.3	7.3	83	..
Europe and Central Asia	97	203.5 [u]	2,615.5 [u]	—	..	220	3.9	4.8	18.5	..	..
Latin America and the Caribbean	92	207.3 [u]	7,042.4 [u]	—	..	230	23.2	2.8	9.9	43	..
South Asia	72	2,979.0 [u]	4,130.3 [u]	—	..	46	3.9	14.2	21.5	115	..
Sub-Saharan Africa	41	2,366.1 [u]	7,177.7 [u]	—	3,987	90	14.5	6.3	15.6	133	..
Least developed countries	39	6,948.0 [u]	11,886.5 [u]	—	..	77	10.0	8.6	15.6	167	..
Small island developing states	..	54.4 [u]	8.4 [u]	—	..	230	13.0	..	..	..	..
Organisation for Economic Co-operation and Development	98	88.3 [u]	1,235.1 [u]	—	..	279	4.0	5.3	17.4	..	32.5
World	71	17,474.2 [u]	38,170.2 [u]	—	..	144	6.2	7.8	14.7	93	..

NOTES

a Data refer to people recognized as refugees under the 1951 UN Convention, the 1967 UN Protocol and the 1969 Organization of African Unity Convention. In the absence of government figures, the Office of the UN High Commissioner for Refugees (UNHCR) has estimated the refugee population in many industrialized countries based on 10 years of individual asylum-seeker recognition.

b For more detailed comments on the estimates, see www.internal-displacement.org.

c Data refer to the most recent year available during the period specified.

d Refers to intimate partner violence only.

e Includes forms of emotional violence.

f For more detailed country notes, see www. prisonstudies.org/highest-to-lowest/prison_population_rate.

g Refers to urban areas only.

h HDRO calculations based on data from ICPS (2014).

i Data differ from the standard definition or refer to only part of a country.

j Data were updated on 6 October 2014 and supersede data published in UNODC (2014).

k Refers to 1990.

l Refers to rural areas only.

m Includes Palestinian refugees under the responsibility of United Nations Relief and Works Agency for Palestine Refugees in the Near East.

n Based on secondary information and surveys that use different methodologies. It does not reflect the full scale of displacement. No organization or mechanism monitors the number of internally displaced persons systematically.

o Windhoek only.

p Khatlon region only.

q Refers to nonpartner violence only.

r May include citizens of South Sudan.

s Refers to average for 2009–2011 prior to South Sudan's independence.

t An unknown number of refugees and asylum-seekers from South Sudan may be included under data for Sudan.

u Unweighted sum of national estimates.

DEFINITIONS

Birth registration: Percentage of children under age 5 who were registered at the moment of the survey. It includes children whose birth certificate was seen by the interviewer and children whose mother or caretaker says the birth has been registered.

Refugees by country of origin: Number of people who have fled their country of origin because of a well founded fear of persecution due to their race, religion, nationality, political opinion or membership in a particular social group and who cannot or do not want to return to their country of origin.

Internally displaced persons: Number of people who have been forced to leave their homes or places of habitual residence—in particular, as a result of or to avoid the effects of armed conflict, situations of generalized violence, violations of human rights or natural or human-made disasters—and who have not crossed an internationally recognized state border.

Homeless people due to natural disaster: Average annual number of people who lack a shelter for living quarters as a result of natural disasters, who carry their few possessions with them and who sleep in the streets, in doorways or on piers, or in any other space, on a more or less random basis, expressed per million people.

Orphaned children: Number of children (ages 0–17) who have lost one or both parents due to any cause.

Prison population: Number of adult and juvenile prisoners—including pre-trial detainees, unless otherwise noted (see note f)—expressed per 100,000 people.

Homicide rate: Number of unlawful deaths purposefully inflicted on a person by another person, expressed per 100,000 people.

Suicide rate: Number of deaths from purposely self-inflicted injuries expressed per 100,000 people in the reference population.

Depth of food deficit: Number of kilocalories needed to lift the undernourished from their status, holding all other factors constant.

Intimate or nonintimate partner violence ever experienced by a woman: Percentage of the female population, ages 15 and older, that has ever experienced physical or sexual violence from an intimate or nonintimate partner.

MAIN DATA SOURCES

Columns 1 and 5: UNICEF (2015).

Column 2: UNHCR (2015).

Column 3: IDMC (2015).

Column 4: CRED EM-DAT (2015) and UNDESA (2013a).

Column 6: ICPS (2014).

Column 7: UNODC (2014).

Columns 8 and 9: WHO (2015).

Column 10: FAO (2015a).

Column 11: UN Women (2014).

TABLE 14

TABLE 14 Human security | 261

TABLE 15

International integration

HDI rank	Trade Exports and imports (% of GDP) 2013[b]	Financial flows Foreign direct investment, net inflows (% of GDP) 2013[b]	Private capital flows (% of GDP) 2013[b]	Net official development assistance received[a] (% of GNI) 2013[b]	Remittances, inflows (% of GDP) 2013[b]	Human mobility Net migration rate (per 1,000 people) 2010/2015[c]	Stock of immigrants (% of population) 2013	International student mobility (% of total tertiary enrolment) 2013[b]	International inbound tourists (thousands) 2013[b]	Communication Internet users (% of population) 2014	Mobile phone subscriptions (per 100 people) 2014	Mobile phone subscriptions (% change) 2009–2014
VERY HIGH HUMAN DEVELOPMENT												
1 Norway	67.0	0.5	11.5	..	0.15	6.0	13.8[d]	−3.4	4,963	96.3	116.5	5.2
2 Australia	41.0	3.3	−6.8	..	0.16	6.5	27.7[e]	17.1	6,382	84.6	131.2	30.3
3 Switzerland	132.2	−1.2	12.6	..	0.46	8.0	28.9	12.6	8,967	87.0	140.5	16.7
4 Denmark	102.8	0.5	−0.1	..	0.43	2.7	9.9	8.3	8,557	96.0	126.0	1.8
5 Netherlands	155.6	3.8	7.6	..	0.18	0.6	11.7	5.4	12,800	93.2	116.4	−4.3
6 Germany	85.3	1.4	6.7	..	0.42	1.3	11.9	2.8	31,500	86.2	120.4	−4.6
6 Ireland	189.8	21.5	19.3	..	0.31	2.2	15.9	−3.9	8,260	79.7	104.3	−2.3
8 United States	30.0	1.8	0.7	..	0.04	3.1	14.3	3.6	69,800	87.4	98.4	11.0
9 Canada	61.9	3.9	−2.0	..	0.07	6.3	20.7	..	16,600	87.1	83.0	17.6
9 New Zealand	57.5	−0.3	0.8	..	0.25	3.3	25.1	14.0	2,629	85.5	112.1	3.1
11 Singapore	358.0	21.4	3.4	..	..	15.0	42.9	10.3	11,900	82.0	158.1	14.0
12 Hong Kong, China (SAR)	458.3	28.0	21.1	..	0.13	4.2	38.9	−1.7	25,700	74.6	239.3	33.1
13 Liechtenstein	..	..	..	..	..	..	33.1	−16.8	52	95.2	104.3	6.9
14 Sweden	82.7	−0.9	−3.3	..	0.20	4.2	15.9	1.8	5,229	92.5	127.8	14.0
14 United Kingdom	61.6	1.8	−3.9	..	0.06	2.9	12.4	16.3	31,200	91.6	123.6	−0.3
16 Iceland	103.1	3.1	8.5	..	1.15	3.3	10.4	−8.2	800	98.2	111.1	2.6
17 Korea (Republic of)	102.8	0.9	1.9	..	0.49	1.2	2.5	−1.7	12,200	84.3	115.5	16.1
18 Israel	64.5	4.1	0.2	..	0.26	−2.0	26.5	−2.6	2,962	71.5	121.5	−2.1
19 Luxembourg	371.4	50.0	−168.9	..	3.02	9.7	43.3	−85.9	945	94.7	148.4	2.7
20 Japan	35.1	0.1	−2.7	..	0.05	0.6	1.9	3.0	10,400	90.6	120.2	31.7
21 Belgium	164.2	−0.6	−10.0	..	2.12	2.7	10.4	6.6	7,684	85.0	114.3	5.4
22 France	58.0	0.2	−3.6	..	0.83	2.0	11.6	6.6	84,700	83.8	100.4	9.0
23 Austria	103.4	3.6	0.0	..	0.66	3.5	15.7	13.1	24,800	81.0	151.9	11.2
24 Finland	77.3	−2.0	0.7	..	0.40	1.8	5.4	4.4	4,226	92.4	139.7	−3.1
25 Slovenia	143.4	−0.9	−9.6	..	1.43	2.1	11.3	−0.1	2,259	71.6	112.1	9.1
26 Spain	59.7	3.2	−5.8	..	0.69	2.6	13.8[f]	1.4	60,700	76.2	107.9	−3.3
27 Italy	54.8	0.6	−0.2	..	0.35	3.0	9.4	1.6	47,700	62.0	154.3	3.2
28 Czech Republic	148.6	2.4	−3.1	..	1.09	3.8	4.0	6.5	9,004	79.7	130.0	4.4
29 Greece	63.4	1.2	1.7	..	0.33	0.9	8.9	−0.8	17,900	63.2	115.0	−4.0
30 Estonia	171.3	3.9	0.3	..	1.72	0.0	16.3	−3.5	2,868	84.2	160.7	33.3
31 Brunei Darussalam	108.6	5.6	−1.0	..	..	0.8	49.3	−38.3	225	68.8	110.1	5.1
32 Cyprus	86.7	2.8	−71.9	..	0.38	6.2	18.2[g]	−61.0	2,405	69.3	96.3	7.5
32 Qatar	97.5	−0.4	11.4	..	0.28	48.8	73.8	16.6	2,611	91.5	145.8	17.0
34 Andorra	..	..	..	..	..	..	56.9	−182.3	2,335	95.9	82.6	0.7
35 Slovakia	181.4	2.2	−10.8	..	2.12	0.6	2.7	−10.9	6,235	80.0	116.9	15.4
36 Poland	90.3	−0.9	0.0	..	1.33	−0.2	1.7	0.2	15,800	66.6	156.5	33.4
37 Lithuania	155.7	1.6	3.1	..	4.48	−1.9	4.9	−5.0	2,012	72.1	147.0	−8.1
37 Malta	182.5	−19.4	52.7	..	0.35	2.1	8.0	−6.3	1,582	73.2	127.0	27.2
39 Saudi Arabia	82.4	1.2	0.3	0.0	0.04	2.1	31.4	−0.8	13,400	63.7	179.6	7.3
40 Argentina	29.3	1.9	−1.7	0.0	0.09	−0.5	4.5	..	5,571	64.7	158.7	21.1
41 United Arab Emirates	176.1	2.6	..	..	..	11.4	83.7	38.2	7,126	90.4	178.1	28.8
42 Chile	65.5	7.3	−5.2	0.0	0.00	0.3	2.3	−0.5	3,576	72.4	133.3	37.6
43 Portugal	77.5	3.5	−0.2	..	1.92	1.9	8.4	1.4	8,097	64.6	111.8	0.3
44 Hungary	169.9	−3.2	−2.4	..	3.24	1.5	4.7	3.4	10,700	76.1	118.1	0.4
45 Bahrain	122.2	3.0	−10.4	..	..	3.4	54.7	−4.7	9,163	91.0	173.3	47.3
46 Latvia	121.5	2.8	−0.6	..	2.46	−1.0	13.8	−2.9	1,536	75.8	124.2	13.9
47 Croatia	85.4	1.0	−5.7	0.3	2.59	−0.9	17.6	−5.2	11,000	68.6	104.4	−2.8
48 Kuwait	98.1	1.0	20.5	..	0.00	18.3	60.2	..	300	78.7	218.4	137.8
49 Montenegro	103.9	10.1	−11.0	2.8	9.59	−0.8	8.2	..	1,324	61.0	163.0	−22.0
HIGH HUMAN DEVELOPMENT												
50 Belarus	125.2	3.1	−2.7	0.2	1.69	−0.2	11.6	−4.0	137	59.0	122.5	20.5
50 Russian Federation	50.9	3.4	1.3	..	0.32	1.5	7.7	1.2	30,800	70.5	155.1	−3.1
52 Oman	98.6	2.0	−0.6	0.0	0.05	59.2	30.6	−11.8	1,551	70.2	157.8	5.8
52 Romania	84.5	2.2	−6.0	..	1.86	−0.4	0.9	−1.4	8,019	54.1	105.9	−7.5
52 Uruguay	51.3	5.0	−10.0	0.1	0.22	−1.8	2.2	..	2,683	61.5	160.8	31.4
55 Bahamas	97.7	4.5	−4.1	..	..	5.2	16.3	..	1,364	76.9	71.4	−29.4
56 Kazakhstan	64.9	4.2	−0.8	0.0	0.09	0.0	21.1	−5.1	4,926	54.9	168.6	55.6
57 Barbados	96.8	12.2	−9.6	0.4	1.85	1.4	11.3	3.9	509	76.7	106.8	−11.6
58 Antigua and Barbuda	103.5	11.2	−11.6	0.1	1.76	−0.1	31.9	−19.0	244	64.0	120.0	−23.2

	Trade	Financial flows				Human mobility				Communication		
	Exports and imports	Foreign direct investment, net inflows	Private capital flows	Net official development assistance received[a]	Remittances, inflows	Net migration rate	Stock of immigrants	International student mobility	International inbound tourists	Internet users	Mobile phone subscriptions	
	(% of GDP)	(% of GDP)	(% of GDP)	(% of GNI)	(% of GDP)	(per 1,000 people)	(% of population)	(% of total tertiary enrolment)	(thousands)	(% of population)	(per 100 people)	(% change)
HDI rank	2013[b]	2013[b]	2013[b]	2013[b]	2013[b]	2010/2015[c]	2013	2013[b]	2013[b]	2014	2014	2009–2014
59 Bulgaria	137.4	3.5	−2.0	..	3.06	−1.4	1.2	−4.6	6,898	55.5	137.7	−1.9
60 Palau	137.2	2.3	..	14.8	..	..	26.7	..	105	..	90.6	44.6
60 Panama	154.8	11.8	−11.1	0.0	1.06	1.5	4.1	..	1,658	44.9	158.1	−5.8
62 Malaysia	154.1	3.7	0.7	0.0	0.45	3.1	8.3[h]	−0.2	25,700	67.5	148.8	37.2
63 Mauritius	120.8	2.2	−34.1	1.2	0.00	0.0	3.6[i]	−11.6	993	41.4	132.3	49.3
64 Seychelles	164.0	12.3	−11.8	1.8	0.89	−3.4	13.0	−198.3	230	54.3	162.2	32.8
64 Trinidad and Tobago	103.2	7.0	..	0.0	0.53	−2.2	2.4	−27.8	434	65.1	147.3	5.5
66 Serbia	92.7	4.3	−9.2	1.8	8.84	−2.1	5.6[j]	−1.2	922	53.5	122.1	−1.8
67 Cuba	39.1	..	..	0.1	..	−2.5	0.1	4.1	2,829	30.0	22.5	308.5
67 Lebanon	138.7	6.8	−6.5	1.4	17.73	21.3	17.6	6.9	1,274	74.7	88.4	57.0
69 Costa Rica	73.9	6.5	−9.7	0.1	1.20	2.7	8.6	−0.1	2,428	49.4	143.8	239.3
69 Iran (Islamic Republic of)	53.7	0.8	..	0.0	0.25	−0.8	3.4	−1.0	4,769	39.4	87.8	22.9
71 Venezuela (Bolivarian Republic of)	50.4	1.6	−1.2	0.0	0.03	0.3	3.9	−0.5	986	57.0	99.0	0.6
72 Turkey	57.9	1.6	−4.1	0.3	0.14	0.9	2.5	0.2	37,800	51.0	94.8	7.6
73 Sri Lanka	54.5	1.4	−4.4	0.6	9.56	−3.0	1.5	−5.1	1,275	25.8	103.2	30.3
74 Mexico	64.2	3.3	−6.0	0.0	1.83	−2.0	0.9	−0.6	24,200	44.4	82.5	15.5
75 Brazil	27.6	3.6	−4.1	0.1	0.11	−0.2	0.3	−0.2	5,813	57.6	139.0	58.7
76 Georgia	102.3	5.9	−4.9	4.1	12.05	−5.8	4.4[k]	−5.8	5,392	48.9	124.9	93.8
77 Saint Kitts and Nevis	87.7	14.5	−14.0	3.9	6.73	..	10.5	−57.9	107	65.4	139.8	−4.2
78 Azerbaijan	75.6	3.6	−2.5	−0.1	2.36	0.0	3.4[l]	−6.6	2,130	61.0	110.9	28.5
79 Grenada	75.8	8.9	−10.6	1.2	3.55	−8.1	10.7	54.1	116	37.4	126.5	15.3
80 Jordan	113.8	5.3	−10.0	4.2	10.82	11.3	40.2	3.7	3,945	44.0	147.8	51.9
81 The former Yugoslav Republic of Macedonia	126.7	4.1	−1.6	2.5	3.69	−0.5	6.6	−5.2	400	68.1	109.1	17.9
81 Ukraine	102.2	2.5	−7.3	0.4	5.45	−0.2	11.4	0.5	24,700	43.4	144.1	21.3
83 Algeria	63.4	0.8	−1.0	0.1	0.10	−0.3	0.7	−1.1	2,634	18.1	93.3	3.7
84 Peru	48.4	4.6	−7.1	0.2	1.34	−2.0	0.3	..	3,164	40.2	102.9	20.6
85 Albania	87.9	9.7	−8.3	2.3	8.46	−3.2	3.1	−12.1	2,857	60.1	105.5	34.9
85 Armenia	75.0	3.5	−10.0	2.7	21.01	−3.4	10.6	−2.6	1,084	46.3	115.9	57.0
85 Bosnia and Herzegovina	85.0	1.8	−1.1	3.0	10.80	−0.3	0.6	−3.2	529	60.8	91.3	8.0
88 Ecuador	60.8	0.8	0.2	0.2	2.60	−0.4	2.3	−1.3	1,364	43.0	103.9	15.8
89 Saint Lucia	97.2	6.3	−13.3	1.9	2.26	0.0	6.7	−31.5	319	51.0	102.6	−5.3
90 China	50.3	3.8	−2.7	0.0	0.42	−0.2	0.1	−1.8	55,700	49.3	92.3	66.9
90 Fiji	136.4	4.1	−14.1	2.4	5.28	−6.6	2.6	20.1	658	41.8	98.8	31.6
90 Mongolia	112.2	18.7	−17.0	4.0	2.22	−1.1	0.6	−3.9	418	27.0	105.1	24.8
93 Thailand	143.8	3.3	0.4	0.0	1.47	0.3	5.6	−0.2	26,500	34.9	144.4	45.2
94 Dominica	81.0	3.5	−0.3	4.0	4.56	..	8.9	..	78	62.9	127.5	−8.1
94 Libya	94.8	0.9	3.5	0.1	0.03	−7.7	12.2	..	34	17.8	161.1	0.8
96 Tunisia	103.1	2.3	−2.4	1.6	4.87	−0.6	0.3	−3.2	6,269	46.2	128.5	37.9
97 Colombia	38.0	4.3	−4.1	0.2	1.09	−0.5	0.3	..	2,288	52.6	113.1	22.9
97 Saint Vincent and the Grenadines	87.2	17.9	−19.9	1.1	4.45	−9.1	9.4	..	72	56.5	105.2	−5.1
99 Jamaica	83.4	4.6	−5.5	0.5	15.05	−5.8	1.3	..	2,008	40.5	102.9	−5.0
100 Tonga	80.5	2.5	..	16.8	23.80	−15.4	5.2	..	45	40.0	64.3	25.6
101 Belize	127.2	5.5	−4.5	3.3	4.58	4.6	15.3	..	294	38.7	50.7	−5.6
101 Dominican Republic	56.8	2.6	−6.1	0.3	7.33	−2.7	3.9	2.7	4,690	49.6	78.9	−9.7
103 Suriname	75.6	2.6	−2.6	0.6	0.13	−1.9	7.7	..	249	40.1	170.6	16.2
104 Maldives	223.6	15.7	−13.4	1.2	0.14	0.0	24.4	−1,678.1	1,125	49.3	189.4	32.3
105 Samoa	80.9	3.0	−2.8	15.3	19.71	−13.4	3.0	..	116	21.2	55.5	..
MEDIUM HUMAN DEVELOPMENT												
106 Botswana	115.0	1.3	7.1	0.7	0.24	2.0	7.2	−5.4	2,145	18.5	167.3	74.2
107 Moldova (Republic of)	125.3	3.1	−2.7	4.2	24.91	−5.9	11.2[m]	−12.3	12	46.6	108.0	82.2
108 Egypt	42.3	2.0	−0.2	2.1	7.32	−0.5	0.4	1.0	9,174	31.7	114.3	58.6
109 Turkmenistan	117.7	7.3	..	0.1	..	−1.0	4.3	..	8	12.2	135.8	217.0
110 Gabon	93.6	4.4	−2.0	0.5	0.13	0.6	23.6	−38.3	269	9.8	210.4	120.4
110 Indonesia	49.5	2.7	−2.7	0.0	0.88	−0.6	0.1	−0.5	8,802	17.1	126.2	83.1
112 Paraguay	94.1	1.2	−3.0	0.5	2.04	−1.2	2.7	..	610	43.0	105.6	19.3
113 Palestine, State of	72.4	1.6	−1.2	19.1	18.29	−2.0	5.9[n]	−9.8	545	53.7	72.1	56.7
114 Uzbekistan	59.3	1.9	..	0.5	..	−1.4	4.4	−8.3	1,969	43.6	73.8	23.1
115 Philippines	59.9	1.3	−0.3	0.1	9.81	−1.4	0.2	−0.2	4,681	39.7	111.2	35.2
116 El Salvador	72.2	0.8	−0.7	0.7	16.37	−7.1	0.7	−1.3	1,283	29.7	144.0	17.7
116 South Africa	64.2	2.2	−2.5	0.4	0.27	−0.4	4.5	3.6	9,537	49.0	149.7	64.0

TABLE
15

TABLE 15 International integration | 263

TABLE 15   INTERNATIONAL INTEGRATION

	Trade	Financial flows				Human mobility				Communication		
	Exports and imports	Foreign direct investment, net inflows	Private capital flows	Net official development assistance received[a]	Remittances, inflows	Net migration rate	Stock of immigrants	International student mobility	International inbound tourists	Internet users	Mobile phone subscriptions	
	(% of GDP)	(% of GDP)	(% of GDP)	(% of GNI)	(% of GDP)	(per 1,000 people)	(% of population)	(% of total tertiary enrolment)	(thousands)	(% of population)	(per 100 people)	(% change)
HDI rank	2013[b]	2013[b]	2013[b]	2013[b]	2013[b]	2010/2015[c]	2013	2013[b]	2013[b]	2014	2014	2009–2014
116 Viet Nam	163.7	5.2	−4.9	2.5	6.35	−0.4	0.1	−2.2	7,572	48.3	147.1	32.1
119 Bolivia (Plurinational State of)	81.4	5.7	−4.3	2.4	3.93	−2.4	1.4	..	798	39.0	96.3	48.9
120 Kyrgyzstan	143.1	10.5	−10.6	7.7	31.52	−6.3	4.1	1.9	3,076	28.3	134.5	57.8
121 Iraq	66.7	1.2	1.3	0.7	0.13	2.7	0.3	−0.2	892	11.3	94.9	42.3
122 Cabo Verde	87.2	2.2	−1.4	13.4	9.34	−6.9	3.0	−34.5	503	40.3	121.8	103.6
123 Micronesia (Federated States of)	..	0.6	..	41.7	6.97	−15.7	2.5	..	42	..	..	..
124 Guyana	203.8	6.7	−7.0	3.4	10.98	−8.2	1.8	−16.9	177	37.4	70.5	12.8
125 Nicaragua	92.6	7.5	−7.9	4.5	9.61	−4.0	0.7	..	1,229	17.6	114.6	96.7
126 Morocco	80.5	3.2	−3.0	1.9	6.63	−2.7	0.2	−7.7	10,000	56.8	131.7	62.8
126 Namibia	104.1	6.9	−3.0	2.0	0.09	−0.3	2.2	−32.7	1,176	14.8	113.8	49.5
128 Guatemala	58.6	2.5	−3.2	0.9	9.98	−1.0	0.5	..	1,331	23.4	106.6	−13.8
129 Tajikistan	87.5	1.3	−1.0	4.5	47.50	−2.5	3.4	−4.2	208	17.5	95.1	44.6
130 India	53.3	1.5	−1.8	0.1	3.73	−0.4	0.4	−0.6	6,968	18.0	74.5	68.8
131 Honduras	117.5	5.8	−11.0	3.6	16.91	−1.2	0.3	−0.9	863	19.1	93.5	−16.8
132 Bhutan	103.7	2.8	..	8.1	0.66	2.7	6.7	−29.9	116	34.4	82.1	70.6
133 Timor-Leste	136.7	1.6	178.6	6.0	9.44	−13.3	1.0	..	58	1.1	58.7	78.1
134 Syrian Arab Republic	76.5	3.1	..	0.2	2.55	−13.7	6.4	..	5,070	28.1	71.0	48.9
134 Vanuatu	99.1	4.0	−2.0	11.4	2.86	0.0	1.2	..	110	18.8	60.4	5.9
136 Congo	142.6	14.5	−31.4	1.4	0.18	−2.1	9.7	−20.9	297	..	..	..
137 Kiribati	121.0	5.3	4.0	25.5	7.30	−2.0	2.6	..	6	12.3	17.4	69.2
138 Equatorial Guinea	156.8	12.3	..	0.1	..	5.3	1.3	..	0	18.9	66.4	124.7
139 Zambia	81.6	6.8	−6.4	4.4	0.20	−0.6	0.7	..	915	17.3	67.3	96.0
140 Ghana	89.4	6.7	−8.1	2.8	0.25	−0.8	1.4	0.5	931	18.9	114.8	80.0
141 Lao People's Democratic Republic	83.4	3.8	−3.9	4.0	0.53	−2.2	0.3	−3.4	2,510	14.3	67.0	29.8
142 Bangladesh	46.3	1.0	−0.9	1.6	9.24	−2.6	0.9	−1.1	148	9.6	75.9	121.0
143 Cambodia	139.5	8.8	−8.4	5.6	1.15	−2.3	0.5	−2.7	4,210	9.0	155.1	250.0
143 Sao Tome and Principe	58.3	3.4	−1.4	16.8	8.53	−1.6	3.3	..	12	24.4	64.9	39.2
LOW HUMAN DEVELOPMENT												
145 Kenya	50.9	0.9	−0.5	5.9	2.41	−0.2	2.2	..	1,434	43.4	73.8	51.9
145 Nepal	48.2	0.4	..	4.5	28.77	−2.9	3.5	−7.6	798	15.4	82.5	291.2
147 Pakistan	33.1	0.6	−0.5	0.9	6.30	−1.8	2.2	−5.3	966	13.8	73.3	32.2
148 Myanmar	..	..	..	..	..	−0.4	0.2	−1.0	2,044	2.1	49.5	..
149 Angola	96.5	−5.7	10.7	0.3	0.00	0.6	0.4	..	650	21.3	63.5	48.2
150 Swaziland	114.6	0.6	−0.3	3.4	0.79	−1.0	2.0	−32.8	968	27.1	72.3	27.8
151 Tanzania (United Republic of)	49.5	4.3	−5.7	7.9	0.14	−0.6	0.6	−9.3	1,063	4.9	62.8	56.8
152 Nigeria	31.0	1.1	−4.5	0.5	4.46	−0.4	0.7	..	600	42.7	77.8	62.3
153 Cameroon	49.6	1.1	−2.1	2.5	0.83	−0.5	1.3	−8.0	912	11.0	75.7	90.1
154 Madagascar	73.1	7.9	..	4.9	0.22	0.0	0.1	−2.6	196	3.7	38.2	24.7
155 Zimbabwe	86.4	3.0	..	6.5	..	5.7	2.6	−16.7	1,833	19.9	80.8	161.0
156 Mauritania	133.7	27.1	..	7.5	..	−1.0	2.3	−23.0	0	10.7	94.2	51.8
156 Solomon Islands	119.1	4.1	−4.0	30.0	1.51	−4.3	1.4	..	24	9.0	65.8	577.2
158 Papua New Guinea	..	0.1	−4.0	4.5	0.09	0.0	0.3	..	174	9.4	44.9	112.5
159 Comoros	78.2	2.3	−1.6	13.3	20.02	−2.8	1.7	−70.5	19	7.0	50.9	176.6
160 Yemen	82.1	−0.4	0.8	2.9	9.30	−1.1	1.3	−0.8	990	22.6	68.5	83.1
161 Lesotho	150.6	1.9	−1.1	11.2	19.81	−1.9	0.1	−11.7	433	11.0	101.9	206.8
162 Togo	95.8	1.9	−0.8	6.0	10.61	−0.3	3.0	−9.4	327	5.7	69.0	93.7
163 Haiti	71.1	2.2	..	13.7	21.05	−3.4	0.4	..	295	11.4	64.7	73.2
163 Rwanda	45.4	1.5	−1.5	14.6	2.26	−0.8	3.8	−5.7	864	10.6	64.0	177.5
163 Uganda	50.7	4.8	−5.5	7.0	3.77	−0.8	1.4	7.2	1,206	17.7	52.4	83.6
166 Benin	51.8	3.9	−0.9	7.9	2.75	−0.2	2.3	4.3	231	5.3	101.7	86.7
167 Sudan	25.7	3.3	−3.3	1.8	0.64	−4.3	1.2	..	591	24.6	72.2	99.9
168 Djibouti	134.2	19.6	..	9.6	2.45	−3.7	14.2	−39.3	63	10.7	32.4	106.7
169 South Sudan	61.0	..	..	13.4	..	15.7	5.6	..	0	15.9	24.5	..
170 Senegal	73.6	2.0	−8.1	6.7	11.18	−1.4	1.5	..	1,063	17.7	98.8	80.3
171 Afghanistan	55.4	0.3	0.0	25.7	2.65	−2.6	0.3	−5.8	..	6.4	74.9	97.6
172 Côte d'Ivoire	91.0	1.2	−3.1	4.2	1.50	0.5	12.0	0.2	289	14.6	106.3	49.9
173 Malawi	110.5	3.2	−1.9	31.5	0.67	0.0	1.3	−20.1	770	5.8	30.5	78.8
174 Ethiopia	41.5	2.0	..	8.1	1.44	−0.1	0.8	..	681	2.9	31.6	561.5
175 Gambia	87.8	2.8	..	12.7	15.45	−1.5	8.8	..	171	15.6	119.6	48.4
176 Congo (Democratic Republic of the)	74.7	5.2	−5.2	8.6	0.10	−0.2	0.7	−0.1	191	3.0	53.5	242.1

TABLE 15

HDI rank	Trade	Financial flows				Human mobility				Communication		
	Exports and imports	Foreign direct investment, net inflows	Private capital flows	Net official development assistance received[a]	Remittances, inflows	Net migration rate	Stock of immigrants	International student mobility	International inbound tourists	Internet users	Mobile phone subscriptions	
	(% of GDP)	(% of GDP)	(% of GDP)	(% of GNI)	(% of GDP)	(per 1,000 people)	(% of population)	(% of total tertiary enrolment)	(thousands)	(% of population)	(per 100 people)	(% change)
	2013[b]	2013[b]	2013[b]	2013[b]	2013[b]	2010/2015[c]	2013	2013[b]	2013[b]	2014	2014	2009–2014
177 Liberia	121.9	35.9	..	30.5	19.65	−0.9	5.3	..	0	5.4	73.4	158.3
178 Guinea-Bissau	..	1.5	−1.1	10.8	4.76	−1.2	1.1	..	30	3.3	63.5	75.7
179 Mali	69.0	3.7	−5.7	13.5	7.36	−4.0	1.3	−5.3	142	7.0	149.0	353.0
180 Mozambique	70.5	42.8	−44.3	14.9	1.39	−0.2	0.8	−1.1	1,886	5.9	69.7	172.6
181 Sierra Leone	107.5	3.5	−5.4	9.8	1.63	−0.7	1.6	..	81	2.1	76.7	272.8
182 Guinea	83.1	2.2	−10.6	8.8	1.51	−0.2	3.2	−5.1	56	1.7	72.1	118.9
183 Burkina Faso	57.1	2.9	1.8	8.1	1.34	−1.5	4.1	−2.0	218	9.4	71.7	183.2
184 Burundi	41.6	0.3	..	20.1	1.79	−0.4	2.5	−3.5	142	1.4	30.5	195.3
185 Chad	69.6	4.0	..	3.1	..	−1.9	3.4	−11.9	100	2.5	39.8	98.1
186 Eritrea	37.5	1.3	..	2.5	..	1.8	0.2	..	107	1.0	6.4	151.5
187 Central African Republic	33.4	0.1	..	12.3	..	0.4	2.9	6.2	71	4.0	31.4	54.9
188 Niger	64.7	8.5	−13.0	10.7	2.30	−0.3	0.7	−5.1	123	2.0	44.4	161.7
OTHER COUNTRIES OR TERRITORIES												
Korea (Democratic People's Rep. of)	..	..	..	..	..	0.0	0.2	..	0	0.0	11.2	..
Marshall Islands	..	11.9	..	41.4	11.93	..	3.2	−18.0	5	16.8	29.4	..
Monaco	..	..	..	..	..	..	64.2	..	328	92.4	88.5	39.7
Nauru	..	..	..	..	..	..	20.6	..	..	..	..	..
San Marino	..	..	..	..	..	..	15.4	..	70	..	118.9	21.8
Somalia	..	..	..	..	..	−2.9	0.2	..	0	1.6	50.9	644.9
Tuvalu	..	0.9	..	48.3	10.59	..	1.5	..	1	..	38.4	276.7
Human development groups												
Very high human development	62.5	1.9	−0.1	..	0.28	2.6	12.6	3.6	638,685	82.5	119.8	15.7
High human development	55.6	3.4	−2.8	0.1	0.76	−0.1	1.8	−1.1	317,832	49.8	104.6	26.2
Medium human development	62.3	2.3	−2.0	0.6	3.92	−0.9	0.7	−0.7	88,252	21.9	91.5	72.1
Low human development	48.4	1.5	−2.4	3.5	4.10	−0.7	1.6	−4.2	22,802	16.0	65.6	140.6
Developing countries	59.3	3.0	−2.6	0.4	1.50	−0.5	1.6	−1.2	428,877	31.9	91.2	65.5
Regions												
Arab States	92.6	1.7	2.0	0.9	2.06	0.4	8.3	−0.8	75,632	34.8	109.4	59.6
East Asia and the Pacific	58.5	3.6	−2.4	0.1	0.82	−0.3	0.4	−1.5	139,481	42.1	100.5	91.3
Europe and Central Asia	72.6	2.7	−4.0	0.6	2.26	−0.6	6.7	−1.3	87,474	47.4	113.0	45.4
Latin America and the Caribbean	44.9	3.5	−4.3	0.2	1.03	−1.0	1.3	−0.2	73,630	50.0	114.9	24.0
South Asia	51.4	1.3	−1.7	0.5	4.09	−0.9	0.9	−1.2	16,165	17.6	75.6	77.1
Sub-Saharan Africa	60.0	2.4	−3.4	3.0	2.18	−0.1	1.8	−2.1	33,865	19.3	71.1	111.5
Least developed countries	66.8	2.7	−1.1	5.6	4.42	−1.1	1.2	−2.3	23,829	8.6	63.1	157.5
Small island developing states	72.4	3.8	−6.7	1.9	6.11	−2.8	1.9	−7.8	17,532	28.2	64.8	44.0
Organisation for Economic Co-operation and Development	57.0	1.7	−0.5	..	0.31	1.9	9.6	3.4	602,443	78.1	110.4	9.7
World	60.4	2.3	−0.9	0.4	0.71	0.0	3.2	0.3	1,067,976	40.5	96.2	50.0

NOTES

a A negative value refers to net official development assistance disbursed by donor countries.

b Data refer to 2013 or the most recent year available.

c Data are average annual projected for 2010-2015.

d Includes Svalbard and Jan Mayen Islands.

e Includes Christmas Island, Cocos (Keeling) Islands and Norfolk Island.

f Includes Canary Islands, Ceuta and Melilla.

g Includes Northern Cyprus.

h Includes Sabah and Sarawak.

i Includes Agalega, Rodrigues and Saint Brandon

j Includes Kosovo.

k Excludes Abkhazia and South Ossetia, which have declared independence from Georgia.

l Includes Nagorno-Karabakh.

m Includes Transnistria.

n Includes East Jerusalem. Refugees are not part of the foreign-born migrant stock in the State of Palestine.

DEFINITIONS

Exports and imports: The sum of exports and imports of goods and services, expressed as a percentage of gross domestic product (GDP). It is a basic indicator of openness to foreign trade and economic integration and indicates the dependence of domestic producers on foreign demand (exports) and of domestic consumers and producers on foreign supply (imports), relative to the country's economic size (GDP).

Foreign direct investment, net inflows: Sum of equity capital, reinvestment of earnings, other long-term capital and short-term capital, expressed as a percentage of GDP.

Private capital flows: Net foreign direct investment and portfolio investment, expressed as a percentage of GDP.

Net official development assistance received: Disbursements of loans made on concessional terms (net of repayments of principal) and grants by official agencies to promote economic development and welfare in countries and territories on the Development Assistance Committee list of aid recipients, expressed as a percentage of the recipient country's GNI.

Remittances, inflows: Earnings and material resources transferred by international migrants or refugees to recipients in their country of origin or countries in which the migrant formerly resided.

Net migration rate: Ratio of the difference between the number of in-migrants and out-migrants from a country to the average population, expressed per 1,000 people.

Stock of immigrants: Ratio of the stock of immigrants into a country, expressed as a percentage of the country's population. The definition of immigrant varies across countries but generally includes the stock of foreign-born people, the stock of foreign people (according to citizenship) or a combination of the two.

International student mobility: Total number of tertiary students from abroad (inbound students) studying in a given country minus the number of students at the same level of education from that country studying abroad (outbound students), expressed as a percentage of total tertiary enrolment in that country.

International inbound tourists: Arrivals of nonresident visitors (overnight visitors, tourists, same-day visitors and excursionists) at national borders.

Internet users: People with access to the worldwide network.

Mobile phone subscriptions: Number of subscriptions for the mobile phone service expressed per 100 people.

MAIN DATA SOURCES

Columns 1, 2, 4, 5 and 9: World Bank (2015b).

Column 3: HDRO calculations based on data from World Bank (2015b).

Columns 6 and 7: UNDESA (2013b).

Column 8: UNESCO Institute of Statistics (2015).

Columns 10 and 11: ITU (2015)

Column 12: HDRO calculations based on data from ITU (2015).

TABLE 15

TABLE 15 International integration | 265

TABLE 16

Supplementary indicators: perceptions of well-being

		Perceptions of individual well-being							Perceptions of work and labour market				Perceptions of government		
		Education quality	Health care quality	Standard of living	Feeling safe	Freedom of choice		Overall life satisfaction, index	Ideal job	Feeling active and productive	Volunteered time	Local labour market	Trust in national government	Actions to preserve the environment	Confidence in judicial system
					(% answering yes)	(% satisfied)		(0, least satisfied, to 10, most satisfied)	(% answering yes)	(% answering agree or strongly agree)	(% answering yes)	(% answering good)	(% answering yes)		(% answering yes)
		(% satisfied)	(% satisfied)	(% satisfied)		Female	Male							(% satisfied)	
HDI rank		2014	2014	2014	2014	2014	2014	2014	2013	2013	2014	2014	2014	2014	2014
VERY HIGH HUMAN DEVELOPMENT															
1	Norway	82	82	95	86	95	96	7.4	85 [a]	..	32	52	70	56	83
2	Australia	67	82	83	62	91	93	7.3	70	60	40	25	46	58	60
3	Switzerland	81	94	94	81	96	93	7.5	84 [a]	..	27	47	75	86	81
4	Denmark	75	85	91	80	93	94	7.5	79	70	21	32	46	75	83
5	Netherlands	78	86	85	81	89	93	7.3	60	64	36	24	53	74	65
6	Germany	66	85	90	80	88	91	7.0	80	56	32	55	60	71	67
6	Ireland	83	67	77	75	93	91	7.0	68	63	41	40	46	75	67
8	United States	68	77	74	73	87	86	7.2	65 [a]	67	44	51	35	60	59
9	Canada	73	77	79	80	94	94	7.3	71	67	44	50	52	59	67
9	New Zealand	73	84	83	65	93	93	7.3	66	65	45	41	63	67	63
11	Singapore	87	89	89	91	83	76	7.1	71 [a]	..	27	48	84	84	85
12	Hong Kong, China (SAR)	51	62	75	91	84	83	5.5	60 [a]	..	15	46	46	51	76
13	Liechtenstein	..	..	..	..	..	..	..	..	..	..	..	..	..	..
14	Sweden	64	78	89	76	94	94	7.2	79	64	17	33	56	62	69
14	United Kingdom	65	77	79	79	86	81	6.8	71	59	32	43	42	65	60
16	Iceland	83 [b]	73 [b]	81 [b]	78 [b]	92 [b]	90 [b]	7.5 [b]	66	55	29 [b]	42 [b]	46 [b]	64 [b]	63 [b]
17	Korea (Republic of)	49	62	63	61	55	61	5.8	51	39	21	19	28	34	19
18	Israel	67	72	67	77	68	69	7.4	57	60	21	36	44	45	60
19	Luxembourg	74	88	86	68	94	91	6.9	58	37	34	20	66	81	76
20	Japan	60	71	61	68	79	75	5.9	69	39	26	30	38	51	64
21	Belgium	83	89	81	72	85	87	6.9	73	58	25	27	47	70	49
22	France	66	81	74	70	77	82	6.5	74	52	30	11	26	59	48
23	Austria	75	89	84	81	88	88	6.9	84	63	32	31	41	68	66
24	Finland	81	69	76	81	92	94	7.4	71	49	30	16	47	68	74
25	Slovenia	76	81	61	84	89	87	5.7	65	60	35	12	18	66	30
26	Spain	54	67	68	85	71	76	6.5	62	63	20	12	21	44	36
27	Italy	55	48	64	58	59	64	6.0	66	49	17	3	31	29	29
28	Czech Republic	63	75	68	61	76	76	6.5	70	40	13	19	34	64	39
29	Greece	45	35	36	62	33	39	4.8	59	45	7	10	19	25	44
30	Estonia	52	51	46	65	73	73	5.6	50	57	19	26	41	61	54
31	Brunei Darussalam	..	..	..	..	..	..	..	..	..	..	..	..	..	..
32	Cyprus	56	52	59	67	66	73	5.6	67	48	25	10	24	48	31
32	Qatar	72 [a]	90 [c]	86	92 [a]	89 [a]	91 [a]	6.4	73 [a]	..	19 [a]	66 [a]	..	91 [a]	..
34	Andorra	..	..	..	..	..	..	..	..	..	..	..	..	..	..
35	Slovakia	58	56	58	56	57	58	6.1	61	47	11	8	31	45	30
36	Poland	59	43	71	63	86	85	5.8	44	62	13	23	25	55	36
37	Lithuania	54	52	34	47	46	43	6.1	50	41	8	16	34	49	30
37	Malta	81	79	76	74	90	90	6.5	80	69	26	17 [b]	72	64	47
39	Saudi Arabia	66	72	78	..	76	74	6.3	61	49	15	59	..	65	..
40	Argentina	62	61	69	42	73	73	6.7	69	68	12	30	41	44	34
41	United Arab Emirates	70	84	81	90 [d]	94	91	6.6	70	53	22	59	..	93	..
42	Chile	41	34	81	55	69	74	6.8	74	69	16	41	40	40	23
43	Portugal	66	62	44	72	82	86	5.1	71	58	15	19	23	63	33
44	Hungary	53	56	48	47	48	46	5.2	71	54	11	19	31	41	46
45	Bahrain	74	83	74	..	80	84	6.2	69	54	30	42	..	69	..
46	Latvia	55	48	49	57	60	62	5.7	41	47	9	25	23	50	38
47	Croatia	57	49	48	60	56	42	5.4	42	41	17	14	16	42	45
48	Kuwait	53	75	81	..	82	78	6.2	69	49	15	47	..	66	..
49	Montenegro	52	45	41	75	50	47	5.3	50	41	7	13	40	26	39
HIGH HUMAN DEVELOPMENT															
50	Belarus	48	37	49	62	56	58	5.8	46	43	16	31	51	44	43
50	Russian Federation	48	38	55	51	65	66	6.0	48	56	19	31	64	28	36
52	Oman	70 [c]	78 [c]	87 [c]	..	92 [c]	..	6.9 [c]	..	..	22 [c]	69 [c]	..	..	..
52	Romania	55	55	48	55	67	72	5.7	41	47	7	16	24	26	36
52	Uruguay	59	69	74	49	90	88	6.6	63	70	18	40	60	64	45
55	Bahamas	..	..	..	..	..	..	..	..	..	..	..	..	..	..
56	Kazakhstan	46	39	66	53	70	70	6.0	48	47	25	40	60	34	43
57	Barbados	..	..	..	..	..	..	..	..	..	..	..	..	..	..

		Perceptions of individual well-being							Perceptions of work and labour market				Perceptions of government		
		Education quality	Health care quality	Standard of living	Feeling safe	Freedom of choice		Overall life satisfaction, index	Ideal job	Feeling active and productive	Volunteered time	Local labour market	Trust in national government	Actions to preserve the environment	Confidence in judicial system
					(% answering yes)	(% satisfied)		(0, least satisfied, to 10, most satisfied)	(% answering yes)	(% answering agree or strongly agree)	(% answering yes)	(% answering good)	(% answering yes)		(% answering yes)
		(% satisfied)	(% satisfied)	(% satisfied)		Female	Male							(% satisfied)	
HDI rank		2014	2014	2014	2014	2014	2014	2014	2013	2013	2014	2014	2014	2014	2014
58	Antigua and Barbuda	..	..	..	..	..	..	..	..	..	..	..	..	..	..
59	Bulgaria	42	38	37	54	53	53	4.4	51	49	4	13	14	22	19
60	Palau	..	..	..	..	..	..	..	..	..	..	..	..	..	..
60	Panama	71	62	77	40	89	88	6.6	76	85	31	60	44	56	38
62	Malaysia	76	85	62	48	77	79	6.0	76	58	37	55	63	67	57
63	Mauritius	81	78	71	64	81	80	5.6	74 c	..	34	31	56	75	61
64	Seychelles	..	..	..	..	..	..	..	..	..	..	..	..	..	..
64	Trinidad and Tobago	64 b	54 b	54 b	57 b	82 b	83 b	6.2 b	56	68	37 b	43 b	38 b	34 b	33 b
66	Serbia	50	37	36	70	52	48	5.1	45	49	6	7	45	22	28
67	Cuba	..	..	..	..	..	..	..	..	..	..	..	..	..	..
67	Lebanon	67	54	48	52	63	67	5.2	64	50	8	15	24	28	32
69	Costa Rica	79	66	82	42	93	91	7.2	80	74	27	24	40	55	42
69	Iran (Islamic Republic of)	55	44	55	..	..	..	4.7	72	38	24	24	..	58	..
71	Venezuela (Bolivarian Republic of)	61	40	54	22	56	57	6.1	78	70	11	22	20	27	22
72	Turkey	53	71	57	60	62	67	5.6	61	36	5 b	30	56	43	48
73	Sri Lanka	83	82	60	70	81	81	4.3	73	50	48	51	77	71	74
74	Mexico	66	55	70	52	78	73	6.7	72	60	18	41	33	56	39
75	Brazil	46	33	75	36	70	71	7.0	76	72	13	44	36	41	41
76	Georgia	59	51	27	79	64	67	4.3	33 c	41	21	11	53	45	45
77	Saint Kitts and Nevis	..	..	..	..	..	..	..	..	..	..	..	..	..	..
78	Azerbaijan	51	34	51	74	66	66	5.3	45	53	20	35	78	54	42
79	Grenada	..	..	..	..	..	..	..	..	..	..	..	..	..	..
80	Jordan	58	73	62	75	73	69	5.3	62	46	5	27	..	51	..
81	The former Yugoslav Republic of Macedonia	64	62	46	73	59	65	5.2	56	49	9	19	44	45	34
81	Ukraine	49	28	27	46	45	48	4.3	39	43	13	8	24	17	12
83	Algeria	70	47	72	..	57 a	56 a	6.4	51 a	..	8 a	43	..	48 a	..
84	Peru	48	37	62	45	63	77	5.9	67	62	16	42	24	39	18
85	Albania	52	47	46	56	68	77	4.8	32	34	9	15	50	40	24
85	Armenia	49	41	31	82	46	46	4.5	30	41	7	15	21	31	26
85	Bosnia and Herzegovina	60	55	49	66	41	32	5.2	47	48	8	5	10	18	25
88	Ecuador	77	56	77	56	70	73	5.9	77	73	10	43	65	61	47
89	Saint Lucia	..	..	..	..	..	..	..	..	..	..	..	..	..	..
90	China	64 b	65 b	77	75 b	76 b	77 b	5.2	51	45	4	38 b	..	63 b	..
90	Fiji	..	..	..	..	..	..	..	..	..	..	..	..	..	..
90	Mongolia	54	37	62	51	66	70	4.8	74	58	34	11	34	29	31
93	Thailand	88	83	76	72	88	91	7.0	80	65	14	58	72	80	81
94	Dominica	..	..	..	..	..	..	..	..	..	..	..	..	..	..
94	Libya	33 a	..	..	..	67 a	70 a	5.8 a	62 a	..	37 a	49 a	..	37 a	..
96	Tunisia	49	39	60	61	55	57	4.8	55	26	6	36	39	34	58
97	Colombia	62	45	79	44	79	79	6.4	69	78	20	40	30	40	26
97	Saint Vincent and the Grenadines	..	..	..	..	..	..	..	..	..	..	..	..	..	..
99	Jamaica	65	53	42	65	82	79	5.3	50	53	38	22	28	35	29
100	Tonga	..	..	..	..	..	..	..	..	..	..	..	..	..	..
101	Belize	62	50	66	50	88	84	6.0	..	..	26	40	38	62	37
101	Dominican Republic	84	57	70	38	89	91	5.4	56	57	35	31	56	61	31
103	Suriname	82 a	78 a	64 a	60 a	85 a	88 a	6.3 a	70 a	..	22 a	34 a	72 a	65 a	71 a
104	Maldives	..	..	..	..	..	..	..	..	..	..	..	..	..	..
105	Samoa	..	..	..	..	..	..	..	..	..	..	..	..	..	..
MEDIUM HUMAN DEVELOPMENT															
106	Botswana	56	52	32	35	78	79	4.0	48	42	26	27	71	71	72
107	Moldova (Republic of)	49	42	46	41	57	56	5.9	32	50	14	7	18	23	19
108	Egypt	36	33	70	74	55	59	4.9	58	30	7	30	70	36	68
109	Turkmenistan	77 c	..	92	71	74	76	5.8	76	61	21	73	..	83	..
110	Gabon	33	20	37	29	61	60	3.9	46	41	10	44	32	39	38
110	Indonesia	78	74	70	85	68	70	5.6	76	61	38	49	65	53	54
112	Paraguay	60	41	82	32	70	76	5.1	79	69	10	54	19	41	9
113	Palestine, State of	67	65	56	61	63	67	4.7	56	38	7	12	47	43	42
114	Uzbekistan	85	85	75	81	92	93	6.0	66	..	43	62	..	84	..
115	Philippines	83	80	70	62	89	91	5.3	87	64	42	66	69	88	63

TABLE 16   Supplementary indicators: perceptions of well-being | 267

TABLE **16**

TABLE 16 SUPPLEMENTARY INDICATORS: PERCEPTIONS OF WELL-BEING

	Perceptions of individual well-being							Perceptions of work and labour market				Perceptions of government		
	Education quality	Health care quality	Standard of living	Feeling safe	Freedom of choice		Overall life satisfaction, index	Ideal job	Feeling active and productive	Volunteered time	Local labour market	Trust in national government	Actions to preserve the environment	Confidence in judicial system
				(% answering yes)	(% satisfied)		(0, least satisfied, to 10, most satisfied)	(% answering yes)	(% answering agree or strongly agree)	(% answering yes)	(% answering good)	(% answering yes)		(% answering yes)
	(% satisfied)	(% satisfied)	(% satisfied)		Female	Male							(% satisfied)	
HDI rank	2014	2014	2014	2014	2014	2014	2014	2013	2013	2014	2014	2014	2014	2014
116 El Salvador	69	59	72	45	78	75	5.9	78	75	19	32	32	52	28
116 South Africa	73	57	44	31	80	77	4.8	51	33	28	25	49	51	44
116 Viet Nam	85	72	78	61	80 [b]	82 [b]	5.1	65	46	14	42	81 [b]	73	66 [b]
119 Bolivia (Plurinational State of)	66	47	72	47	88	87	5.9	76	68	22	51	47	62	23
120 Kyrgyzstan	58	57	76	49	70	63	5.3	55	49	36	41	37	33	28
121 Iraq	45	50	67	60	66	62	4.5	64	42	18	30	51	35	51
122 Cabo Verde	..	..	..	..	..	..	..	..	..	..	..	..	..	..
123 Micronesia (Federated States of)	..	..	..	..	..	..	..	..	..	..	..	..	..	..
124 Guyana	..	..	..	..	..	..	..	..	..	..	..	..	..	..
125 Nicaragua	80	54	69	56	81	80	6.3	76	67	20	38	58	68	46
126 Morocco	34	27	76	66	58	65	5.2	40	56	5	18	38 [b]	45	28
126 Namibia	71	58	43	44	85	85	4.6	..	..	21	46	78	64	68
128 Guatemala	70	49	72	50	82	84	6.5	77	72	41	36	37	55	41
129 Tajikistan	71	60	82	83	73	73	4.9	67	49	37	49	..	51	..
130 India	69	58	58	52	75	79	4.4	80	47	17	34	73	54	67
131 Honduras	70	50	63	49	66	71	5.1	78	62	33	21	33	46	25
132 Bhutan	94	90	92	63	84	80	4.9	88	42	38	54	96	95	97
133 Timor-Leste	..	..	..	..	..	..	..	..	..	..	..	..	..	..
134 Syrian Arab Republic	36 [b]	37 [b]	35 [b]	33 [b]	40 [b]	39 [b]	2.7 [b]	28	52	21 [b]	15 [b]	..	38 [b]	..
134 Vanuatu	..	..	..	..	..	..	..	..	..	..	..	..	..	..
136 Congo	36	23	41	48	65	61	4.1	55	55	14	45	48	32	47
137 Kiribati	..	..	..	..	..	..	..	..	..	..	..	..	..	..
138 Equatorial Guinea	..	..	..	..	..	..	..	..	..	..	..	..	..	..
139 Zambia	62	45	30	36	79	82	4.3	53	45	29	39	61	44	59
140 Ghana	44	41	24	71	67	64	3.9	59	63	27	14	34	30	50
141 Lao People's Democratic Republic	73 [a]	66 [a]	73 [a]	75 [a]	87 [c]	..	4.9 [a]	80 [a]	..	20 [a]	66 [a]	..	90 [c]	..
142 Bangladesh	87	59	80	81	63	69	4.6	85	49	10	39	72	50	72
143 Cambodia	87	79	78	42	94	92	3.9	80	49	9	65	..	89	..
143 Sao Tome and Principe	..	..	..	..	..	..	..	..	..	..	..	..	..	..
LOW HUMAN DEVELOPMENT														
145 Kenya	68	53	45	52	82	81	4.9	63	46	43	44	64	60	51
145 Nepal	83	60	73	59	62	71	5.0	87	37	27	50	59	60	63
147 Pakistan	53	39	57	50	52	53	5.4	74	65	12	30	43	31	57
148 Myanmar	78	76	70	81	73 [b]	74 [b]	4.8	52	47	50	42 [b]	..	68 [b]	..
149 Angola	46	29	35	46	30	37	3.8	60	39	17	43	57	37	44
150 Swaziland	77 [c]	..	..	42 [c]	62 [c]	..	..	56 [c]	..	..	25 [c]	35 [c]	56 [c]	56 [c]
151 Tanzania (United Republic of)	40	23	27	52	64	65	3.5	60	37	13	37	65	40	50
152 Nigeria	51	46	40	52	60	66	4.8 [b]	48	55	32	35	29	37	45
153 Cameroon	52	32	55	51	76	81	4.2	56	48	12	43	61	55	51
154 Madagascar	49	30	23	42	55	50	3.7	56	49	34	47	51	42	43
155 Zimbabwe	64	57	43	55	67	57	4.2	52	45	21	25	57	46	52
156 Mauritania	40	30	57	52	45	46	4.5	52	63	22	43	35	39	28
156 Solomon Islands	..	..	..	..	..	..	..	..	..	..	..	..	..	..
158 Papua New Guinea	..	..	..	..	..	..	..	..	..	..	..	..	..	..
159 Comoros	49 [a]	24 [a]	38 [a]	72 [a]	50 [a]	57 [a]	4.0 [a]	64 [a]	..	18 [a]	30 [a]	46 [a]	39 [a]	34 [a]
160 Yemen	44	21	51	57	60	63	4.0	47	31	3	14	34	27	29
161 Lesotho	40 [c]	21 [c]	27 [c]	38 [c]	61 [c]	..	..	41 [c]	..	16 [c]	21 [c]	40 [c]	23 [c]	64 [c]
162 Togo	37	24	23	52	65	66	2.8	43 [c]	..	9	28	48	49	44
163 Haiti	37	21	18	41	46	53	3.9	31	27	24	17	33	34	24
163 Rwanda	84	80	44	85	90	89	3.6	63	61	11	50	..	92	..
163 Uganda	47	38	35	46	83	81	3.8	53	49	24	31	58	52	36
166 Benin	42	32	27	51	79	76	3.3	51	49	11	37	57	47	59
167 Sudan	28	22	52	71	25	29	4.1	51 [a]	..	23	18	..	11	65 [c]
168 Djibouti	67 [c]	49 [c]	63 [c]	72 [c]	76 [c]	..	..	59 [c]	..	8 [c]	55 [c]	68 [c]	58 [c]	..
169 South Sudan	33	21	25	44	51	55	3.8	..	..	24	23	45	30	43
170 Senegal	40	26	39	64	68	70	4.4	43	57	14	32	66	39	67
171 Afghanistan	52	32	32	34	45	51	3.1	87	58	9	19	41	41	27
172 Côte d'Ivoire	47	32	29	51	79	77	3.6	50	45	8	54	63	44	50
173 Malawi	54	44	41	43	78	79	4.6	48	46	33	47	62	60	57

TABLE 16

	Perceptions of individual well-being							Perceptions of work and labour market				Perceptions of government		
	Education quality	Health care quality	Standard of living	Feeling safe	Freedom of choice		Overall life satisfaction, index	Ideal job	Feeling active and productive	Volunteered time	Local labour market	Trust in national government	Actions to preserve the environment	Confidence in judicial system
					(% satisfied)		(0, least satisfied, to 10, most satisfied)		(% answering agree or strongly agree)					
	(% satisfied)	(% satisfied)	(% satisfied)	(% answering yes)	Female	Male		(% answering yes)		(% answering yes)	(% answering good)	(% answering yes)	(% satisfied)	(% answering yes)
HDI rank	2014	2014	2014	2014	2014	2014	2014	2013	2013	2014	2014	2014	2014	2014
174 Ethiopia	75	58	53	68	65	64	4.5	65	61	13	42	68	79	56
175 Gambia	..	..	..	..	..	..	..	..	..	..	..	..	..	..
176 Congo (Democratic Republic of the)	37	22	35	30	49	59	4.4	49	47	14	25	31	40	29
177 Liberia	39	29	29	35	56	56	4.6	31	45	46	30	35	28	27
178 Guinea-Bissau	..	..	..	..	..	..	..	..	..	..	..	..	..	..
179 Mali	34	30	35	64	62	68	4.0	62	49	5	58	62	32	45
180 Mozambique	65 c	47 c	38 c	42 c	63 c	..	5.0 c	59 c	..	17 c	45 c	63 c	55 c	62 c
181 Sierra Leone	33 b	35	33 b	56	66	67	4.5	49	46	25	27	59	41	38
182 Guinea	41	26	37	51	64	70	3.4	53	54	20	41	57	47	42
183 Burkina Faso	67	45	41	67	68	72	3.5	50	53	11	47	54	55	52
184 Burundi	54	37	26	43	47	39	2.9	..	..	10	10	..	41	..
185 Chad	49	26	43	51	57	55	3.5	70	42	9	38	37	53	30
186 Eritrea	..	..	..	..	..	..	..	..	..	..	..	..	..	..
187 Central African Republic	39 c	..	34 c	60 c	75 c	80 c	3.7 c	62 c	..	15 c	36 c	78 c	69 c	..
188 Niger	49	29	50	69	62	70	4.2	62	54	9	56	58	50	59
OTHER COUNTRIES OR TERRITORIES														
Korea (Democratic People's Rep. of)	..	..	..	..	..	..	..	..	..	..	..	..	..	..
Marshall Islands	..	..	..	..	..	..	..	..	..	..	..	..	..	..
Monaco	..	..	..	..	..	..	..	..	..	..	..	..	..	..
Nauru	..	..	..	..	..	..	..	..	..	..	..	..	..	..
San Marino	..	..	..	..	..	..	..	..	..	..	..	..	..	..
Somalia	49	34	67	71	82	85	..	..	..	16	38	63	75	49
Tuvalu	..	..	..	..	..	..	..	..	..	..	..	..	..	..
Human development groups														
Very high human development	64	72	73	71	..	..	6.6	68	57	30	36	38	56	53
High human development	61	58	71	65	..	..	5.6	69	50	10	37	45	55	41
Medium human development	70	60	63	59	..	..	4.7	76	49	20	38	69	55	63
Low human development	53	41	45	53	..	..	4.4	56	52	20	35	48	45	49
Developing countries	63	56	63	61	..	..	5.0	72	50	16	37	58	54	..
Regions														
Arab States	46	41	65	66	..	..	5.0	53	41	12	31	..	41	..
East Asia and the Pacific	..	..	..	..	..	..	..	..	..	..	..	..	..	..
Europe and Central Asia	57	55	53	62	..	..	5.3	56	42	17	30	46	42	35
Latin America and the Caribbean	57	44	71	43	..	..	6.5	74	..	17	39	35	46	35
South Asia	68	55	60	55	..	..	4.5	81	49	17	34	69	52	66
Sub-Saharan Africa	54	41	39	51	..	..	4.3	52	50	22	37	50	48	47
Least developed countries	60	43	50	59	..	..	4.2	66	49	17	37	58	51	52
Small island developing states	..	..	..	..	..	..	..	..	..	..	..	..	..	..
Organisation for Economic Co-operation and Development	63	70	72	69	..	..	6.6	68	56	28	35	38	55	52
World	**63**	**58**	**64**	**62**	..	..	**5.3**	**71**	**52**	**18**	**37**	**54**	**54**	**54**

NOTES

a Refers to 2012.

b Refers to 2013.

c Refers to 2011.

d Refers to 2010.

DEFINITIONS

Satisfaction with education quality: Percentage of respondents answering "satisfied" to the Gallup World Poll question, "Are you satisfied or dissatisfied with the education system?"

Satisfaction with health care quality: Percentage of respondents answering "satisfied" to the Gallup World Poll question, "Are you satisfied or dissatisfied with the availability of quality health care?"

Satisfaction with standard of living: Percentage of respondents answering "satisfied" to the Gallup World Poll question, "Are you satisfied or dissatisfied with your standard of living, all the things you can buy and do?"

Perception of safety: Percentage of respondents answering "yes" to the Gallup World Poll question, "Do you feel safe walking alone at night in the city or area where you live?"

Satisfaction with freedom of choice: Percentage of respondents answering "satisfied" to the Gallup World Poll question, "In this country, are you satisfied or dissatisfied with your freedom to choose what you do with your life?"

Overall life satisfaction index: Average response to the Gallup World Poll question, "Please imagine a ladder, with steps numbered from zero at the bottom to ten at the top. Suppose we say that the top of the ladder represents the best possible life for you, and the bottom of the ladder represents the worst possible life for you. On which step of the ladder would you say you personally feel you stand at this time, assuming that the higher the step the better you feel about your life, and the lower the step the worse you feel about it? Which step comes closest to the way you feel?"

Ideal job: Percentage of employed respondents answering "yes" to the Gallup World Poll question, "Would you say that your job is the ideal job for you, or not?"

Feeling active and productive: Percentage of respondents answering that they agree or strongly agree to the Gallup World Poll question, "In the last seven days have you felt active and productive every day?"

Volunteered time: Percentage of respondents answering "yes" to the Gallup World Poll question, "In the past month have you volunteered your time to an organization?"

Satisfaction with local labour market: Percentage of respondents answering "good" to the Gallup World Poll question, "Thinking about the job situation in the city or area where you live today, would you say that it is now a good time or a bad time to find a job?"

Trust in national government: Percentage of respondents answering "yes" to the Gallup World Poll question, "In this country, do you have confidence in the national government?"

Satisfaction with actions to preserve the environment: Percentage of respondents answering "satisfied" to Gallup World Poll question, "In this country, are you satisfied or dissatisfied with the efforts to preserve the environment?"

Confidence in judicial system: Percentage of respondents answering "yes" to the Gallup World Poll question, "In this country, do you have confidence in the judicial system and courts?"

MAIN DATA SOURCES

Columns 1–14: Gallup (2015).

TABLE 16

TABLE 16 Supplementary indicators: perceptions of well-being | 269

Regions

Arab States (20 countries or territories)
Algeria, Bahrain, Djibouti, Egypt, Iraq, Jordan, Kuwait, Lebanon, Libya, Morocco, State of Palestine, Oman, Qatar, Saudi Arabia, Somalia, Sudan, Syrian Arab Republic, Tunisia, United Arab Emirates, Yemen

East Asia and the Pacific (24 countries)
Cambodia, China, Fiji, Indonesia, Kiribati, Democratic People's Republic of Korea, Lao People's Democratic Republic, Malaysia, Marshall Islands, Federated States of Micronesia, Mongolia, Myanmar, Nauru, Palau, Papua New Guinea, Philippines, Samoa, Solomon Islands, Thailand, Timor-Leste, Tonga, Tuvalu, Vanuatu, Viet Nam

Europe and Central Asia (17 countries)
Albania, Armenia, Azerbaijan, Belarus, Bosnia and Herzegovina, Georgia, Kazakhstan, Kyrgyzstan, Republic of Moldova, Montenegro, Serbia, Tajikistan, The former Yugoslav Republic of Macedonia, Turkey, Turkmenistan, Ukraine, Uzbekistan

Latin America and the Caribbean (33 countries)
Antigua and Barbuda, Argentina, Bahamas, Barbados, Belize, Plurinational State of Bolivia, Brazil, Chile, Colombia, Costa Rica, Cuba, Dominica, Dominican Republic, Ecuador, El Salvador, Grenada, Guatemala, Guyana, Haiti, Honduras, Jamaica, Mexico, Nicaragua, Panama, Paraguay, Peru, Saint Kitts and Nevis, Saint Lucia, Saint Vincent and the Grenadines, Suriname, Trinidad and Tobago, Uruguay, Bolivarian Republic of Venezuela

South Asia (9 countries)
Afghanistan, Bangladesh, Bhutan, India, Islamic Republic of Iran, Maldives, Nepal, Pakistan, Sri Lanka

Sub-Saharan Africa (46 countries)
Angola, Benin, Botswana, Burkina Faso, Burundi, Cabo Verde, Cameroon, Central African Republic, Chad, Comoros, Congo, Democratic Republic of the Congo, Côte d'Ivoire, Equatorial Guinea, Eritrea, Ethiopia, Gabon, Gambia, Ghana, Guinea, Guinea-Bissau, Kenya, Lesotho, Liberia, Madagascar, Malawi, Mali, Mauritania, Mauritius, Mozambique, Namibia, Niger, Nigeria, Rwanda, São Tomé and Príncipe, Senegal, Seychelles, Sierra Leone, South Africa, South Sudan, Swaziland, United Republic of Tanzania, Togo, Uganda, Zambia, Zimbabwe

Note: Countries included in aggregates for Least Developed Countries and Small Island Developing States follow UN classifications, which are available at www.unohrlls.org.

Statistical references

Aguna, C., and M. Kovacevic. 2011. "Uncertainty and Sensitivity Analysis of the Human Development Index." Human Development Research Paper 2010/11. UNDP–HDRO, New York. http://hdr.undp.org/en/content/uncertainty-and-sensitivity-analysis-human-development-index. Accessed 15 April 2015.

Alkire, S., and G. Robles. 2015. "Multidimensional Poverty Index 2015: Brief Methodological Note and Results." Oxford Poverty and Human Development Initiative, Oxford University. http://ophi.qeh.ox.ac.uk. Accessed 10 September 2015.

Alkire, S., and M. Santos. 2010. "Acute Multidimensional Poverty: A New Index for Developing Countries." Human Development Research Papers 2010/11. UNDP–HDRO, New York. http://hdr.undp.org/en/content/acute-multidimensional-poverty. Accessed 15 April 2015.

Barro, R.J., and J.-W. Lee. 2013a. Dataset of educational attainment, April 2013 revision. www.barrolee.com. Accessed 9 April 2013.

———. **2013b.** "A New Data Set of Educational Attainment in the World, 1950–2010." *Journal of Development Economics* 104: 184–198.

———. **2014.** Dataset of educational attainment, June 2014 revision. www.barrolee.com. Accessed 15 December 2014.

Charmes, J. 2015. "Time Use across the World: Findings of a World Compilation of Time-Use Surveys." Working paper. UNDP–HDRO, New York.

CRED EM-DAT (Centre for Research on the Epidemiology of Disasters). 2015. The International Disaster Database. www.emdat.be. Accessed 31 March 2015.

ECLAC (United Nations Economic Commission for Latin America and the Caribbean). 2014. *Preliminary Overview of the Economies of Latin America and the Caribbean, 2014.* Santiago. http://repositorio.cepal.org/bitstream/handle/11362/37345/S1420977_en.pdf?sequence=31. Accessed 15 March 2015.

Eurostat. 2015. European Union Statistics on Income and Living Conditions (EUSILC). Brussels. http://ec.europa.eu/eurostat/web/microdata/european-union-statistics-on-income-and-living-conditions. Accessed 15 January 2015.

FAO (Food and Agriculture Organization). 2011. Unpublished table prepared for the *State of the World's Land and Water Resources for Food and Agriculture* Background Thematic Report 3 "Land degradation." Rome.

———. **2015a.** FAOSTAT database. http://faostat3.fao.org. Accessed 2 June 2015.

———. **2015b.** AQUASTAT database. www.fao.org/nr/water/aquastat/main/. Accessed 4 June 2015.

Gallup. 2015. Gallup World Poll Analytics database. www.gallup.com/products/170987/gallup-analytics.aspx. Accessed 15 May 2015.

Høyland, B., K. Moene, and F. Willumsen. 2011. "The Tyranny of International Index Rankings." *Journal of Development Economics* 97(1): 1–14.

ICF Macro. Various years. Demographic and Health Surveys. www.measuredhs.com. Accessed 15 March 2015.

ICPS (International Centre for Prison Studies). 2014. "World Prison Population List (10th edition)." London. www.prisonstudies.org/highest-to-lowest/prison_population_rate. Accessed 21 November 2014.

IDMC (Internal Displacement Monitoring Centre). 2015. IDPs worldwide. www.internal-displacement.org. Accessed 9 June 2015.

ILO (International Labour Organization). 2012. Global Wage Report data. www.ilo.org/ilostat/faces/help_home/global_wage?_adf.ctrl-state=18djc8t28q_78&_afrLoop=305541918060446. Accessed 15 May 2015.

———. **2013.** *Domestic Workers across the World: Global and Regional Statistics and the Extent of Legal Protection.* Geneva.

———. **2015a.** *Key Indicators on the Labour Market: 8th edition.* Geneva. www.ilo.org/empelm/what/WCMS_114240/. Accessed 18 May 2015.

———. **2015b.** ILOSTAT database. www.ilo.org/ilostat. Accessed 30 March 2015.

———. **2015c.** ILO Social Protection Sector databases. www.ilo.org/protection/information-resources/databases/. Accessed 15 May 2015.

———. **2015d.** NORMLEX database. www.ilo.org/dyn/normlex/en/f?p=1000:12001:0::NO. Accessed 15 June 2015.

IMF (International Monetary Fund). 2015. World Economic Outlook database. Washington, DC. https://www.imf.org/external/pubs/ft/weo/2015/01/weodata/index.aspx. Accessed 15 April 2015.

IPU (Inter-Parliamentary Union). 2015. Women in national parliaments. www.ipu.org/wmn-e/classif-arc.htm. Accessed 12 March 2015.

ITU (International telecommunication union). 2015. *ICT Facts and Figures: The World in 2015.* www.itu.int/en/ITU-D/Statistics/Pages/stat/. Accessed 1 July 2015.

LIS (Luxembourg Income Study). 2014. Luxembourg Income Study Project. www.lisdatacenter.org/data-access/. Accessed 15 December 2014.

National Institute for Educational Studies of Brazil. 2013. Correspondence on school life expectancy. Brasilia.

OECD (Organisation for Economic Co-operation and Development). 2014. PISA 2012 results. www.oecd.org/pisa/keyfindings/pisa-2012-results.htm. Accessed 20 November 2014.

Palma, J.G. 2011. "Homogeneous Middles vs. Heterogeneous Tails, and the End of the 'Inverted-U': The Share of the Rich Is What It's All About." Cambridge Working Paper in Economics 1111. Cambridge University, UK. www.econ.cam.ac.uk/dae/repec/cam/pdf/cwpe1111.pdf. Accessed 15 September 2013.

Samoa Bureau of Statistics. 2013. Census table. sbs.gov.ws. Accessed 15 November 2014.

Timor-Leste Ministry of Finance. 2015. National account 2000–2013. Dili. www.statistics.gov.tl/timor-leste-nacional-account-2000-2013. Accessed 25 June 2015.

UNDESA (United Nations Department of Economic and Social Affairs). 2011. *World Population Prospects: The 2010 Revision.* New York. www.un.org/en/development/desa/population/publications/trends/population-prospects_2010_revision.shtml. Accessed 15 October 2013.

———. **2013a.** *World Population Prospects: The 2012 Revision.* New York. http://esa.un.org/unpd/wpp/. Accessed 15 April 2013.

———. **2013b.** *Trends in International Migrant Stock: The 2013 Revision.* New York. http://esa.un.org/unmigration/migrantstocks2013.htm?msax. Accessed 15 April 2015.

———. **2014.** *World Urbanization Prospects: The 2014 Revision.* CD-ROM edition. http://esa.un.org/unpd/wup/CD-ROM. Accessed 15 May 2015.

———. **2015.** World Population Prospects database. Extracted 9 July 2015.

UNESCO (United Nations Educational, Scientific and Cultural Organization) Institute for Statistics. 2011. *Global Education Digest 2011: Comparing Education Statistics across the World.* Paris. www.uis.unesco.org/Library/Documents/global_education_digest_2011_en.pdf. Accessed 15 March 2015.

———. **2012.** *Global Education Digest 2012: Opportunities Lost: The Impact of Grade Repetition and Early School Leaving.* Paris. www.uis.unesco.org/Education/Documents/ged-2012-en.pdf. Accessed 15 March 2015.

———. **2013.** International Literacy Data 2013. www.uis.unesco.org/literacy/Pages/data-release-map-2013.aspx. Accessed 15 March 2015.

———. **2015.** Data Centre. http://data.uis.unesco.org. Accessed 26 March 2015.

UNESCWA (United Nations Economic and Social Commission for Western Asia). 2014. *Survey of Economic and Social Developments in Western Asia, 2013-2014.* Beirut. www.escwa.un.org/information/publications/edit/upload/E_ESCWA_EDGD_14_3_E.pdf. Accessed 15 March 2015.

UNHCR (Office of the United Nations High Commissioner for Refugees). 2015. *UNHCR Mid-Year Trends 2014.* www.unhcr.org/54aa91d89.html. Accessed 7 April 2015.

UNICEF (United Nations Children's Fund). 2015. *The State of the World's Children 2015.* New York. http://sowc2015.unicef.org. Accessed 2 April 2015.

———. **Various years.** Multiple Indicator Cluster Surveys. New York. www.unicef.org/statistics/index_24302.html. Accessed 15 April 2015.

UN Maternal Mortality Estimation Group (World Health Organization, United Nations Children's Fund, United Nations Population Fund and World Bank). 2014. Maternal mortality data. www.childinfo.org/maternal_mortality_ratio.php. Accessed 15 April 2015.

UNODC (United Nations Office on Drugs and Crime). 2014. Global Study on Homicide: UNODC Homicide Statistics 2013. www.unodc.org/gsh/en/data.html. Accessed 2? November 2014.

UNSD (United Nations Statistics Division). 2015. National Accounts Main Aggregates Database. http://unstats.un.org/unsd/snaama. Accessed 1 July 2015.

UN Women. 2014. "Violence against Women Prevalence Surveys by Country." New York. www.endvawnow.org/en/browser/files/vawprevalence_matrix_june2013.pdf. Accessed 19 November 2014.

WHO (World Health Organization). 2015. Global Health Observatory. www.who.int/gho/. Accessed 31 March 2015.

World Bank. 2014. World Development Indicators database. Washington, DC. http://data.worldbank.org/data-catalog/world-development-indicators/wdi-2014. Accessed 7 May 2015.

———. World Development Indicators database. Washington, DC. http://data.worldbank.org. Accessed

———. World Development Indicators database. Washington, DC. http://data.worldbank.org. Accessed

———. "Getting a Job." http://wbl.worldbank.org/data/exploretopics/getting-a-job#Parental. Accessed